W9-AGO-198

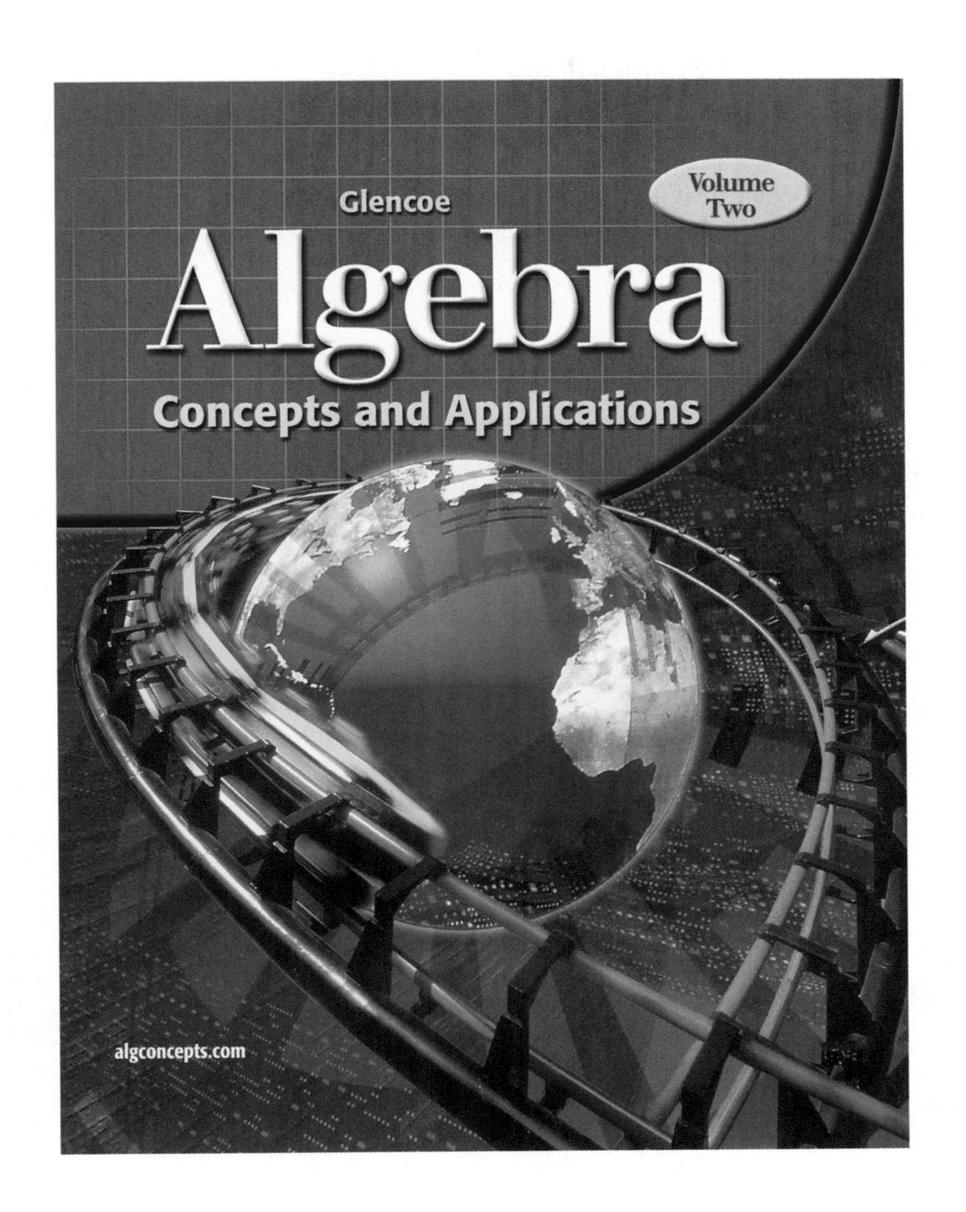

New York, New York   Columbus, Ohio   Chicago, Illinois   Peoria, Illinois   Woodland Hills, California

Visit the Glencoe Mathematics Internet Site for
Algebra: Concepts and Applications at

www.algconcepts.com

You'll find:

- Self-Check Quizzes
- Research Helps
- Data Updates
- Career Data
- Investigations
- Review Activities
- Test Practice

links to Web sites relevant to Problem-Solving Workshops, Investigations, Math In the Workplace features, exercises, and much more!

The McGraw·Hill Companies

Send all inquiries to:
Glencoe/McGraw-Hill
8787 Orion Place
Columbus, OH 43240

ISBN 0-07-860775-2 *Algebra: Concepts and Applications,* Volume Two
Printed in the United States of America.
1 2 3 4 5 6 7 8 9 10 027/055 12 11 10 09 08 07 06 05 04 03

Dear Students, Teachers, and Parents,

Algebra: Concepts and Applications is designed to help you learn algebra and apply it to the real world. Throughout the text, you will be given opportunities to make connections from concrete models to abstract concepts. The real-world photographs and realistic data will help you see algebra in your world. You will also have plenty of opportunities to review and use arithmetic and geometry concepts as you study algebra. And for those of you who love a good debate, you will find plenty of opportunities to communicate your understanding of algebra.

We know that most of you haven't yet decided which careers you would like to pursue, so we've also included a little career guidance. This text offers real examples of how mathematics is used in many types of careers.

You may have to take an end-of-course exam for algebra, a proficiency test for graduation, the SAT, and/or the ACT. When you enter the workforce, you may also have to take job placement tests that include a section on mathematics. All of these tests include algebra problems. Because all Algebra 1 concepts are covered in this text, this program will prepare you for all of those tests.

Each day, as you use ***Algebra: Concepts and Applications,*** you will see the practical value of algebra. You will grow to appreciate how often algebra is used in ways that relate directly to your life. You will have meaningful experiences that will prepare you for the future. If you don't already see the importance of algebra in your life, you soon will!

Sincerely,
The Authors

Contents in Brief

Authors

Jerry Cummins
Staff Development Specialist
Bureau of Education and Research
State of Illinois
President, National Council of Supervisors of Mathematics
Western Springs, IL

Carol Malloy
Assistant Professor of Mathematics Education
University of North Carolina at Chapel Hill
Chapel Hill, NC

Kay McClain
Lecturer
George Peabody College
Vanderbilt University
Nashville, TN

Yvonne Mojica
Mathematics Teacher and Mathematics Department Chairperson
Verdugo Hills High School
Tujunga, CA

Jack Price
Professor, Mathematics Education
California State Polytechnic University
Pomona, CA

Academic Consultants and Teacher Reviewers

Each of the Academic Consultants read all 15 chapters, while each Teacher Reviewer read two chapters. The Consultants and Reviewers gave suggestions for improving the Student Editions and the Teacher's Wraparound Editions.

Academic Consultants

Richie Berman, Ph.D.
Mathematics Lecturer & Supervisor
University of California at Santa Barbara
Santa Barbara, California

Judith Cubillo
Mathematics Teacher & Department Chairperson
Northgate High School
Walnut Creek, California

Mary C. Enderson
Faculty
Middle Tennessee State University
Murfreesboro, Tennessee

Alan G. Foster
Former Mathematics Teacher & Department Chairperson
Addison Trail High School
Addison, Illinois

Deborah A. Haver, Ed.D.
Assistant Principal
Great Bridge Middle School
Chesapeake, Virginia

Nicki Hudson
Mathematics Teacher
West Linn High School
West Linn, Oregon

Daniel Marks, Ph.D.
Associate Professor of Mathematics
Auburn University at Montgomery
Montgomery, Alabama

Donald McGurrin
Senior Administrator for Secondary Mathematics
Wake County Public Schools
Raleigh, North Carolina

C. Vincent Pané, Ed.D.
Associate Professor of Education
Molloy College
Rockville Centre, New York

Gary Shannon, Ph.D.
Professor of Mathematics
California State University
Sacramento, California

Marianne Weber
National Mathematics Consultant
St. Louis, Missouri

Teacher Reviewers

Breta J. Brown
Mathematics Teacher
Hillsboro High School
Hillsboro, North Dakota

Kimberly A. Brown
Mathematics Teacher
McGehee High School
McGehee, Arkansas

Helen Carpini
Mathematics Department Chairperson
Middletown High School
Middletown, Connecticut

Terry Cepaitis
Mathematics Resource Teacher
Anne Arundel County Public Schools
Annapolis, Maryland

Kindra R. Cerfoglio
Mathematics Teacher
Reed High School
Sparks, Nevada

Tom Cook
Mathematics Department Chairperson
Carlisle High School
Carlisle, Pennsylvania

Donna L. Cooper
Mathematics Department Chairperson
Walter E. Stebbins High School
Riverside, Ohio

James D. Crawford
Instructional Coordinator—Mathematics
Manchester Memorial High School
Manchester, New Hampshire

David A. Crine
Mathematics Department Chairperson
Basic High School
Henderson, Nevada

Carol Damiano
Mathematics Teacher
Morton High School
Hammond, Indiana

Douglas D. Dolezal, Ph.D.
Mathematics Educator/ Methods Instructor
Crete High School/Doane College
Crete, Nebraska

Richard F. Dube
Mathematics Supervisor
Taunton High School
Taunton, Massachusetts

Dianne Foerster
Mathematics Department Chairperson
Riverview High School
Riverview, Florida

Victoria G. Fortenberry
Mathematics Teacher
Lincoln High School
Tallahassee, Florida

Candace Frewin
Mathematics Teacher
East Lake High School
Tarpon Springs, Florida

Linda Glover
Mathematics Department Chairperson
Conway High School East
Conway, Arkansas

Karin Sorensen Grandone
Mathematics Department Supervisor
Bremen District 228
Midlothian, Illinois

R. Emilie Greenwald
Mathematics Teacher
Worthington Kilbourne High School
Worthington, Ohio

John Scott Griffith
Mathematics Department Chairperson
New Castle Middle School
New Castle, Indiana

Rebecca M. Gummerson
Mathematics Department Chairperson
Burley High School
Burley, Idaho

T. L. Watanabe Hall
Mathematics Teacher
Northwood Junior High School
Kent, Washington

T. B. Harris
Mathematics Teacher
Luray High School
Luray, Virginia

Jerome D. Hayden
Mathematics Department Chairperson
McLean County Unit District 5
Normal, Illinois

Karlene M. Hubbard
Mathematics Teacher
Rich East High School
Park Forest, Illinois

Joseph Kavanaugh
Academic Head of Mathematics
Scotia-Glenville Central School District
Scotia, New York

Ruth C. Keefe
Mathematics Department Chairperson
Litchfield High School
Litchfield, Connecticut

Roger M. Marchegiano
Supervisor of Mathematics
Bloomfield School District
Bloomfield, New Jersey

Marilyn Martau
Mathematics Teacher (Retired)
Lakewood High School
Lakewood, Ohio

Jane E. Morey
Mathematics Department Chairperson
Washington High School
Sioux Falls, South Dakota

Grace Clover Mullen
Talent Search Tutor
Dabney Lancaster Community College
Lexington, Virginia

Laurie D. Newton
Mathematics Teacher
Crossler Middle School
Salem, Oregon

Rinda Olson
Mathematics Department Chairperson
Skyline High School
Idaho Falls, Idaho

Catherine E. Oppio
Mathematics Teacher
Truckee Meadows Community College High School
Reno, Nevada

Peter Pace
Mathematics Teacher
Life Center Academy
Burlington, New Jersey

LaVonne Peterson
Instructional Leader
Minico High School
Rupert, Idaho

Donna H. Preston
Mathematics Teacher
Thomas Worthington High School
Worthington, Ohio

Matthew D. Quibell
Mathematics Teacher
Horseheads Middle School
Horseheads, New York

Linda Rohleder
Mathematics Teacher
Jasper Middle School
Jasper, Indiana

Steve Sachs
Mathematics Department Chairperson
Lawrence North High School
Indianapolis, Indiana

Elizabeth Schopp
Mathematics Department Chairperson
Moberly Senior High School
Moberly, Missouri

Terry Anagnos Serbanos
Mathematics Department Chairperson
Countryside High School
Clearwater, Florida

Eli Shaheen
Mathematics Department Chairperson
Plum Borough School District
Pittsburgh, Pennsylvania

Sue A. Shattuck
Mathematics Teacher
Otsego Middle School
Otsego, Michigan

William Shutters
Mathematics Teacher
Urbandale High School
Urbandale, Iowa

Richard R. Soendlin
Mathematics Department Chairperson
Arsenal Technical High School
Indianapolis, Indiana

Frank J. Sottile, Ed.D.
Supervisor of Mathematics and Computer Science
Scranton School District
Scranton, Pennsylvania

Lana Stahl
K–12 Mathematics Instructional Specialist
Pasadena ISD
Pasadena, Texas

Randall Starewicz
Mathematics Department Chairperson
Lake Central High School
St. John, Indiana

Mary M. Stewart
Secondary Mathematics Curriculum Specialist
Kansas City Kansas Public Schools
Kansas City, Kansas

Dora M. Swart
Mathematics Department Chairperson
W. F. West High School
Chehalis, Washington

Elizabeth Sweeney
Mathematics Teacher
Clayton Midde School
Reno, Nevada

Ruby M. Sylvester
Assistant Principal
High School of Transit Technology
Brooklyn, New York

Lisa Thomson
Mathematics Teacher
Reed High School
Reno/Sparks, Nevada

John Mark Tollett
Mathematics Teacher
Smackover High School
Smackover, Arkansas

Ronald R. Vervaecke
Mathematics Coordinator
Warren Consolidated Schools
Warren, Michigan

Ronald C. Voss
Mathematics Teacher
Sparks High School
Sparks, Nevada

Karen S. Whiteside
Mathematics Teacher
Dunedin High School
Dunedin, Florida

Dale I. Winters
Mathematics Teacher
Worthington Kilbourne High School
Worthington, Ohio

Michael J. Zelch
Mathematics Teacher
Worthington Kilbourne High School
Worthington, Ohio

Table Of Contents

Chapters 1–8 are contained in Volume One. Chapters 8–15 are contained in Volume Two.

Lesson 1–1, page 7

Chapter 2 Integers

Lesson 2–2, page 61

Lesson 3–4, page 115

Chapter 3 Addition and Subtraction Equations

Lesson 4–2, page 150

Chapter 4 Multiplication and Division Equations

Chapter 5 Proportional Reasoning and Probability 186

Lesson 5–5, page 217

Lesson 6–3, page 254

Chapter 6 Functions and Graphs 236

Chapter 7 Linear Equations

Lesson 7–4, page 302

Chapter 8 Powers and Roots

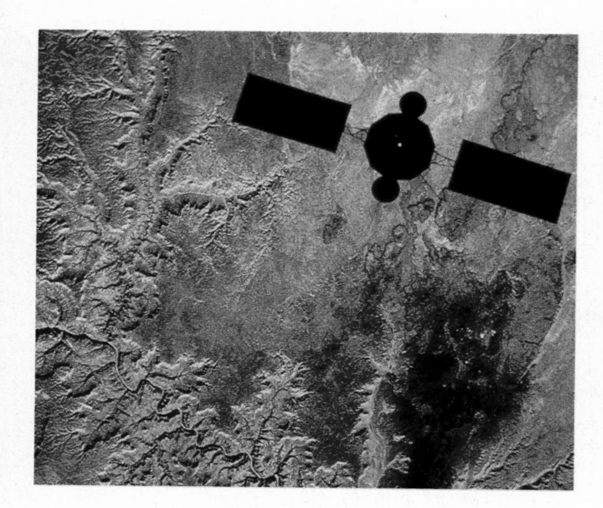

Lesson 8–2, page 345

Investigation, page 411

Chapter 9 Polynomials

Lesson 10–3, page 434

Chapter 10 Factoring

Lesson 11–2, page 467

Lesson 12–3, page 517

Chapter 12 Inequalities

Chapter 13 Systems of Equations and Inequalities 548

Lesson 13–4, page 567

Chapter 14 Radical Expressions

Lesson 14–2, page 608

Lesson 15–5, page 666

Preparing for Standardized Test Success

The **Preparing for Standardized Tests** pages at the end of each chapter have been created in partnership with **The Princeton Review**, the nation's leader in test preparation materials, to help you get ready for the mathematics portions of your standardized tests. On these pages, you will find strategies for solving problems and test-taking advice to help you maximize your score.

It is important to remember that there are many different standardized tests given by schools and states across the country. Find out as much as you can about your test. Start by asking your teacher and counselor for any information, including practice materials, that may be available to help you prepare.

To help you get ready for these tests, do the **Standardized Test Practice** question in each lesson. Also review the concepts and techniques contained in the **Preparing for Standardized Tests** pages at the end of each chapter listed below. This will help you become familiar with the types of math questions that are asked on various standardized tests.

The **Preparing for Standardized Tests** pages are part of a complete test preparation course offered in this text. The test items on these pages were written in the same style as those in state proficiency tests and standardized tests like ACT and SAT. The 15 topics are closely aligned with those tests, the algebra curriculum, and this text. These topics cover all of the types of problems you will see on these tests.

With some practice, studying, and review, you will be ready for standardized test success. Good luck from Glencoe/McGraw-Hill and The Princeton Review! The Princeton Review is not affiliated with Princeton University nor Educational Testing Service.

To The Student

Chapter B contains four sections: pretest, review lessons, chapter tests, and posttest. The pretest is a review of the concepts that you will need to succeed in the second half of *Algebra: Concepts and Applications.* You should take the pretest to determine which concepts you need to review. The review lessons allow you to develop, and eventually master, the individual skills the pretest identified as needing to be reinforced. You should take the posttest to make sure you understand all of the concepts and to measure your progress.

Pretest

Circle the letter of the correct answer.

1. Write the expression *two less than three times c* algebraically.

A $2 - 3c$ **B** $\frac{3c}{2}$ **C** $3c - 2$ **D** $2c - 3$

2. Find the value of x if $-17 + 8 = x$.

F -11 **G** -9 **H** 9 **I** 25

3. Simplify $-0.47 + 2.31 + (-8.07)$.

A -6.23 **B** -5.29 **C** 5.29 **D** 9.91

4. The line graph shows the number of annual visitors to Yosemite National Park. Which is a reasonable conclusion from the information on the graph?

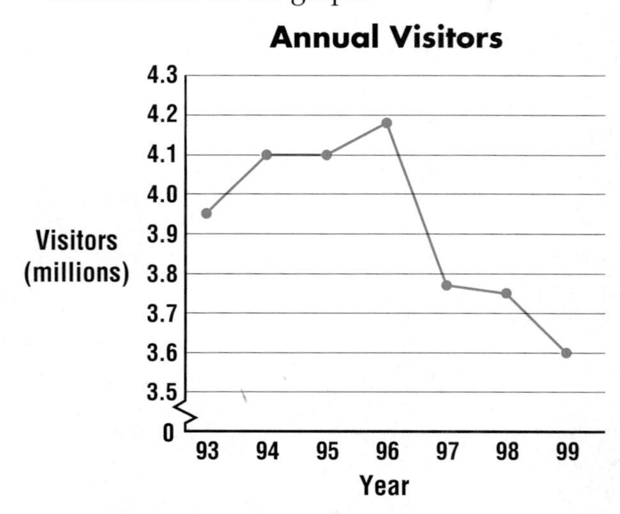

F The average number of visitors each year is about 4 million.

G The number of visitors increases each year.

H More people visited the park in 1995 than in any other year.

I Since 1997, the number of visitors has decreased by one million each year.

5. Name the property shown by $(m \cdot n) \cdot 7 = m \cdot (n \cdot 7)$.

A Associative Property of Addition

B Associative Property of Multiplication

C Commutative Property of Multiplication

D Distributive Property

6. What is the ordered pair for point M?

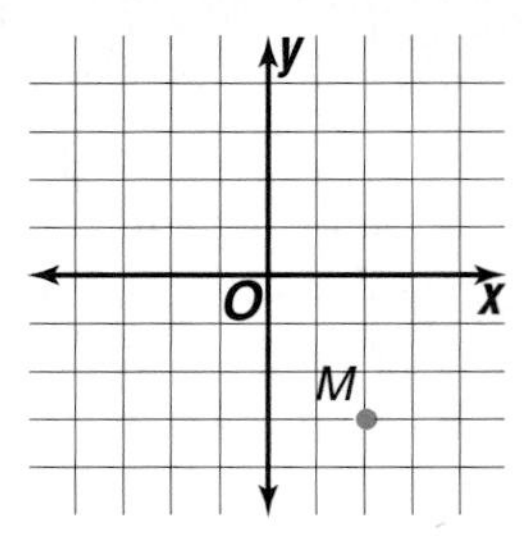

F $(-3, 2)$ **G** $(-2, -3)$

H $(2, -3)$ **I** $(2, 3)$

7. Find the value of $[a + 2(3 + b)] \div 4$ if $a = 8$ and $b = 5$.

A 6 **B** 8 **C** 12 **D** 20

8. Evaluate $\frac{-5y}{x}$ if $x = -3$ and $y = -9$.

F -18 **G** -15 **H** 15 **I** 18

9. What is the solution of $58 = 29 + z$?

A 29 **B** 31 **C** 39 **D** 87

10. Solve $|x - 2| + 7 = 13$.

F $\{-18, 22\}$ **G** $\{-4, 8\}$

H $\{-8, 4\}$ **I** $\varnothing$

11. The table shows the number of acres for the six largest city-owned parks in the U.S. Find the mean size of the parks.

Park	Size
Cullen (Houston)	10,534
Fairmount (Philadelphia)	8700
Griffith (Los Angeles)	4218
Eagle Creek (Indianapolis)	3800
Pelham Bay (Bronx, N.Y.)	2764
Mission Bay (San Diego)	2300

Source: Urban Land Institute/Trust for Public Land

A 4009 acres **B** 4218 acres

C 5386 acres **D** 8234 acres

12. What is the solution of $-\frac{1}{5}w = -45$?

F -225 **G** -9 **H** 9 **I** 225

Pretest

13. The weight of an object on Mars is $\frac{3}{8}$ of its weight on Earth. Suppose an eagle weighs $12\frac{2}{3}$ pounds on Earth. How much would it weigh on Mars?

A $4\frac{2}{3}$ lb **B** $4\frac{5}{8}$ lb
C $4\frac{3}{4}$ lb **D** $33\frac{7}{9}$ lb

14. Solve $4(a - 3) - 5(3a + 6) = 13$.
F 5 **G** −1 **H** −5 **I** −7

15. According to the scale on a map, 1 centimeter represents 20.5 kilometers. Suppose two towns are 150 kilometers apart. Which proportion could be used to find d, the distance between the two cities on the map?

A $\frac{1}{20.5} = \frac{d}{150}$ **B** $\frac{20.5}{1} = \frac{d}{150}$
C $\frac{1}{d} = \frac{150}{20.5}$ **D** $\frac{150}{1} = \frac{20.5}{d}$

16. Suppose the spinner shown is spun once. Find the odds of landing on a number greater than or equal to 4.

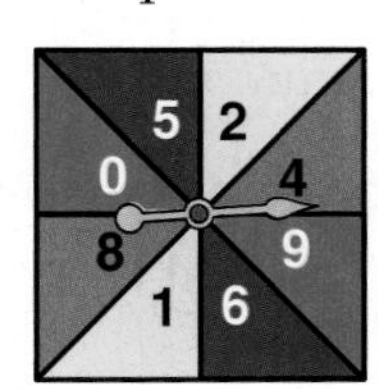

F 1:2 **G** 3:5 **H** 4:4 **I** 5:3

17. What is the range of the relation {(3, 4), (5, 5), (−1, 7), (6, 8), (0, −5)}?
A {3, 5, −1, 6, −2, 0} **B** {−5, −4, 7, 5, 4}
C {4, 5, 7, 8} **D** {4, 5, 7, 8, −5}

18. What number is 36% of 165?
F 5.94 **G** 58.4 **H** 59.4 **I** 63

19. In May 1997, farmers in the U.S. received an average of \$8.40 per bushel of soybeans. In December 1999, they received \$4.29 per bushel. What was the percent of decrease? Round to the nearest percent.
A 47% **B** 49% **C** 51% **D** 53%

20. Which relation is a function?
F {(5, −3), (5, 5), (0, 2), (6, 4)}
G

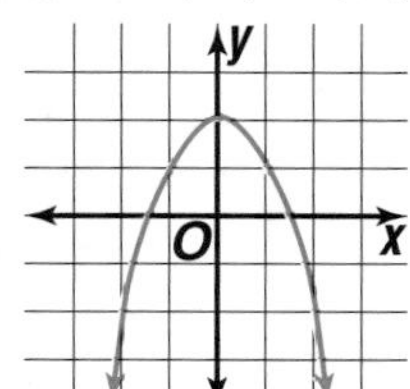

H

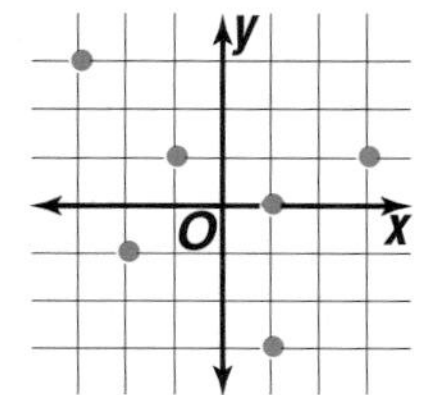

I

x	y
0	2
1	3
−5	−2
2	−3
1	4

21. Suppose y varies directly as x and $y = 18$ when $x = 9$. Find x when $y = 32$.
A 16 **B** 12 **C** 8 **D** 6

22. What is the y-intercept of $2x - 3y = 6$?
F (3, 0) **G** (0, 2)
H (−3, 0) **I** (0, −2)

23. Find the slope of the line that passes through (8, 3) and (4, 5).

A $-\frac{1}{2}$ **B** $\frac{1}{2}$ **C** $\frac{2}{3}$ **D** 2

24. Find an equation of the line through (−8, 3) with slope $-\frac{1}{2}$.

F $y = -\frac{1}{2}x + 7$ **G** $y = -\frac{1}{2}x + 3$
H $y = -\frac{1}{2}x - 1$ **I** $y = -\frac{1}{2}x + 1$

25. A line that is parallel to the graph of $4x - 5y = 8$ has which of the following slopes?

A $-\frac{5}{4}$ **B** $-\frac{4}{5}$ **C** $\frac{4}{5}$ **D** $\frac{5}{4}$

REVIEW 1-1

Writing Expressions and Equations

Any letter used to represent an unknown number is called a **variable.** You can use variables to translate verbal expressions into **algebraic expressions.**

Words	Symbols
4 more than a number	$y + 4$
a number decreased by 12	$b - 12$
the product of 3 and a number	$3t$
a number divided by 8	$h \div 8$ or $\frac{h}{8}$

An **equation** is a sentence that contains an equals sign (=). Phrases such as *is equal to, is the same as,* and *is equivalent to* indicate the equals sign.

Example

Write an equation for *Five subtracted from n is 13.*

Five subtracted from n is 13. → $n - 5 = 13$

Exercises **Write an algebraic expression for each verbal expression.**

1. A number increased by 14
2. six times a number
3. 12 more than a number
4. a number divided by 4
5. the sum of a number and 17
6. 25 less than a number
7. three plus the product of a number and 13
8. 74 decreased by twice a number

Write a verbal expression for each algebraic expression.

9. $z + 1$
10. $5b$
11. $57 - 3q$

Write an equation for each sentence.

12. The product of 4 and x is equivalent to 12.
13. Twenty divided by y is equal to 10.
14. Eleven is the same as the sum of a, b, and 7.

15. **Environment** According to the American Water Works Association, it takes an average of 20 gallons of water to wash dishes by hand and 25 gallons of water to run a dishwasher.
    a. Write an expression that represents the amount of water used to wash dishes by hand for x washes.
    b. Write an expression that represents the amount of water needed to run a dishwasher for x washes.
    c. Write an expression that represents the difference between the amount of water needed to run a dishwasher and the amount of water used to wash dishes by hand for x washes.

REVIEW 1-2

Order of Operations

Ian said Shelley won first prize and I won second prize. Without punctuation, this sentence has three possible meanings.

> Ian said, "Shelley won first prize and I won second prize."
>
> "Ian," said Shelley, "won first prize and I won second prize."
>
> Ian said Shelley won first prize and I [that is, the speaker] won second prize.

In mathematics, in order to avoid confusion about meaning, an agreed-upon **order of operations** tells us whether a mathematical expression such as $45 - 15 \div 5$ means $(45 - 15) \div 5$ or $45 - (15 \div 5)$.

Order of Operations	
	1. Find the values of expressions inside grouping symbols, such as parentheses ( ), brackets [ ], and as indicated by fraction bars. 2. Do all multiplications and/or divisions from left to right. 3. Do all additions and/or subtractions from left to right.

You can **evaluate** an algebraic expression when the value of each variable is known. Replace each variable with its known value and then use the order of operations. Remember to do all operations within grouping symbols first.

Examples

1 **Simplify $45 - 15 \div 5$.**

$$45 - 15 \div 5 = 45 - 3$$
$$= 42$$

2 **Evaluate $a - (b + 7)$ if $a = 10$ and $b = 2$.**

$$a - (b + 7) = 10 - (2 + 7)$$
$$= 10 - 9 \text{ or } 1$$

Exercises Find the value of each expression.

1. $18 - 2 \cdot 3$
2. $16 \div 4 - 3 + 4 \cdot 6$
3. $5(36 - 31) + 2 \cdot 7$
4. $\dfrac{70 - 25}{7 + 8}$

Evaluate each algebraic expression if $a = 4$, $b = 2$, $x = 1$, and $y = 6$.

5. $8a - 2b$
6. $10x + 9y$
7. $48x - (3a - 5b)$
8. $\dfrac{a + y}{5b}$

9. **Accounting** Keisha and Mark are selling tickets for a dinner theater performance. Combination tickets for dinner and the performance cost $15. Performance only tickets cost $9.
   a. If Keisha and Mark sell both types of tickets, write an expression for the total amount of money they collect.
   b. Keisha sells 20 combination tickets and 15 performance tickets. Mark sells 10 combination tickets and 25 performance tickets. How much money have they collected?

REVIEW **1-3**

Commutative and Associative Properties

The Commutative and Associative Properties can be used to simplify algebraic expressions. To **simplify** an expression, eliminate all parentheses first and then add, subtract, multiply, or divide.

Commulative Properties of Addition and Multiplication	For any numbers a and b, $a + b = b + a$ and $a \cdot b = b \cdot a$.
Associative Properties of Addition and Multiplication	For any numbers a, b, and c, $(a + b) + c = a + (b + c)$ and $(a \cdot b) \cdot c = a \cdot (b \cdot c)$

Example

Simplify $(3 + 5n) + 12$. Identify the properties used in each step.

$(3 + 5n) + 12 = (5n + 3) + 12$ *Commutative Property of Addition*

$= 5n + (3 + 12)$ *Associative Property of Addition*

$= 5n + 15$ *Substitution Property*

Exercises **Name the property shown by each statement.**

1. $13a + 4b = 4b + 13a$
2. $1 \cdot z = z \cdot 1$
3. $2(x \cdot 7) = (2 \cdot x)7$
4. $(5w + 7) + 6z = 5w + (7 + 6z)$

Simplify each expression.

5. $5 + 6b + 12$
6. $(3m + 4) + 9$
7. $(3q + 2) + (1 + 5)$
8. $8 + (9y + 3)$
9. $(6st + 9) + 7$
10. $(3x + 5) + 10$
11. $6 \times (4 \times s)$
12. $2(w) \cdot 4$
13. $8 \cdot v \cdot 3$
14. $(4r) \cdot 3$
15. $(2q \cdot 4) \cdot 4$
16. $(2 \cdot 12n) \cdot 3$

17. **Automobiles** Consider the steps necessary to fill a car's gasoline tank. One step is to remove the gasoline tank cap and the other step is to pump the gasoline. Would you say that these steps are commutative? Explain.

REVIEW 1-4 Distributive Property

To find the product of two integers, you can find the sum of two partial products. For example, write

$$\begin{array}{r} 63 \\ \times\ 7 \\ \hline 441 \end{array} \quad \text{as} \quad \begin{array}{r} 60 + 3 \\ \times \quad\ 7 \\ \hline 420 + 21 \end{array} \leftarrow (60 \times 7) + (3 \times 7).$$

The statement $(60 + 3) \times 7 = (60 \times 7) + (3 \times 7)$ illustrates the **Distributive Property.** The multiplier 7 is distributed over the addition of 60 and 3.

Distributive Property	For any numbers a, b, and c, $a(b + c)$ 5 $ab + ac$ and $a(b - c) = ab - ac$.

You can use the Distributive Property to simplify algebraic expressions.

Example

Simplify $6(x + z) + 9z$.

$6(x + z) + 9z = 6x + 6z + 9z$	*Distributive Property*
$= 6x + (6 + 9)z$	*Distributive Property*
$= 6x + 15z$	*Substitution Property*

Name the coefficient of each term. Then name the like terms in each list of terms.

1. $12r, 7r, 3r, 4rs$
2. $3xy, 11x, 3y, 5x$

Use the Distributive Property to rewrite each expression.

3. $7(5w + 3)$
4. $5a - 5b$

Simplify each expression, if possible. If not possible, write in simplest form.

5. $12c - 5c$
6. $4b + 9b$
7. $5p + 3 - p$
8. $6(4x + 9)$
9. $2t + 15t + 16y - 9y$
10. $26rs + 17st$
11. $3(1 - x) - 2$
12. $2w - 3(4v + 3v)$
13. $9(3z - 2) - 7z$
14. $9(a + b) + 6b$
15. $12x - y + 4$
16. $5(2g + 1) + 6(4g - h)$
17. **Business** Each month, a business owner pays \$23 for telephone service and \$9.95 for an on-line service.
    a. Write an expression representing the owner's total costs for telephone and on-line services for an entire year.
    b. What amount does the business owner pay for these services for a year?

REVIEW 1-5

A Plan for Problem Solving

A **formula** is an equation that states a rule for the relationship between quantities. Using a formula is one way to solve problems. However, any problem can be solved using a problem-solving plan.

1. **Explore** Read the problem carefully. Identify all information given and determine what you are asked to find.

2. **Plan** Select a strategy for solving the problem. Some strategies are shown in the box below.

Look for a pattern.	Use an equation or formula.
Draw a diagram.	Make a graph.
Make a table.	Guess and check.
Work backward.	

3. **Solve** Use your strategy to solve the problem. You may have to choose a variable for the unknown, and then write an expression. Be sure to answer the question asked.

4. **Examine** Check your answer. Does it make sense?

Exercises Solve.

1. **Geometry** The perimeter P of a square is four times the length of a side s.
   a. Write a formula for the perimeter of a square.

   5 in.

   b. What is the perimeter of the square shown?

2. **Savings** An Internet service offers two different payment plans. Plan A offers unlimited use for $19.95 a month. Plan B offers 20 free hours a month, but then $2.50 for each additional hour.
   a. If you use the Internet, on average, 30 hours a month, which plan is the better deal? Explain.

   b. What is the maximum number of monthly hours that you can use the Internet where the better deal is with Plan B?

REVIEW 1-6

Collecting Data

Sampling is a method used to gather data in which a small group, or **sample,** is used to represent a much larger **population.** Three important characteristics of a good sample are listed below.

Sampling Criteria

A good sample is:
- representative of the larger population,
- selected at random, and
- large enough to provide accurate data.

A survey can be biased and give false results if these criteria are not followed.

Example

Twenty random people leaving a school board meeting were asked whether or not they would vote for the upcoming school levy. Eighteen or 90% of those sampled said they would vote for the levy. Is this a good sample? Explain.

No, this is not a good sample since the people leaving a school board meeting are more likely than the average person to be active in school issues and would vote for a school levy. It does not represent the larger population.

Determine whether each is a good sample. Describe what caused the bias in each poor sample. Explain.

1. The first eight people leaving a movie are asked to rate the movie on a scale from one to ten, with ten being the highest rating.

2. All people leaving a shopping center are asked to name their favorite brand of clothing.

3. Every other student on school files is surveyed to determine how many students can be expected to attend the next school dance.

4. Two hundred randomly chosen cars of a certain make are used to determine the average number of miles per gallon of gas that make of car gets on the freeway.

Displaying and Interpreting Data

Mrs. Cortez interviewed some students who were willing to help her with some typing. The number of words typed per minute by each student is listed below.

16 21 18 17 18 20 15 16

These data can be organized in a **stem-and-leaf plot.**

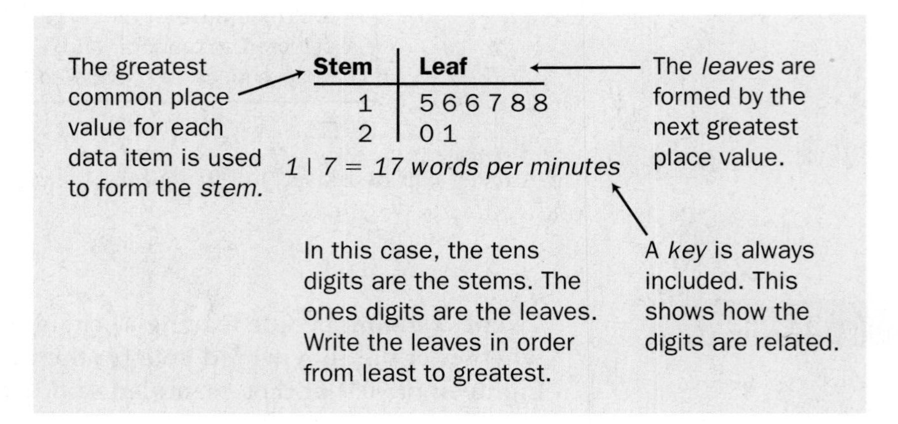

In the stem-and-leaf plot at the right, the data are represented by three-digit numbers. In this case, use the digits in the first two place values to form the stems. For example, the values for 102, 114, 115, 108, 139, 127, 125, and 131 are shown in the stem-and-leaf plot at the right.

Stem	Leaf
10	2 8
11	4 5
12	5 7
13	1 9

11 | 5 = 115

Exercises Solve.

1. The stem-and-leaf plot at the right shows the height, in feet, of buildings in Boston that are at least 500 feet tall.

Stem	Leaf
5	0 0 1 2 3 5
6	0 0 0 0 1
7	5
8	0

5 | 2 = 520 feet

   a. How tall is the tallest building?
   b. What is the height of the shortest building represented in the plot?
   c. What building height occurs most frequently?

2. Danielle is a sales clerk at a bagel shop. Her sales for the first week were $58, $64, $26, $79, and $55. Her sales for the second week were $39, $72, $64, $52, and $78.
   a. Make a stem-and-leaf plot of Danielle's two weeks' sales.
   b. What is the greatest value of sales that Danielle made in one day during the two-week period?

CHAPTER 1 Test

1. Write an equation for the sentence *a number x subtracted from 16 is equivalent to 2.*
2. Write an algebraic expression for *the sum of 3 times a number and 7.*

Find the value of each expression.

3. $15 - 30 \div 3 + 6$
4. $4[(20 - 4) \div 8]$

Simplify each expression.

5. $6c + 4g - 2c$
6. $3(7q + 1)$
7. $2 \cdot b \cdot 5$
8. $9 + (r + 4)$
9. $10h + 6(2 + h)$
10. $7y + (5y - 8)$

Name the property shown by each statement.

11. $4 + (1 + 6) = (4 + 1) + 6$
12. $9t - 3t = t(9 - 3)$
13. $x \cdot y = y \cdot x$

Evaluate each expression if $p = 3$, $r = 5$, and $s = 2$.

14. $s + 4r - 5$
15. $r(p - s)$

16. Write the members of the data set used to make the stem-and-leaf plot.
17. Find the difference of the highest and lowest values of the stem-and-leaf plot.
18. Which number occurs most often in the stem-and-leaf plot?

Stem	Leaf
7	0 4 6 7
8	1 2 9
9	3 3 5 8

8 | 1 = 81

19. Every person in a British literature class is asked to name the average number of books read in a month to determine how literate the students in a school are. Is this a good sample? Explain.

20. **Savings** You can use the formula $I = prt$ to find the interest earned I after depositing an amount p into an account with an interest rate r after so many years t. If you deposit \$100 into an account with an interest rate of 5%, after how many years will you have earned \$20?

REVIEW

2-1 Graphing Integers on a Number Line

The figure at the right is part of a number line. On a number line, the set of numbers that include 0 and numbers to the right of 0 are called the set of **whole numbers.**

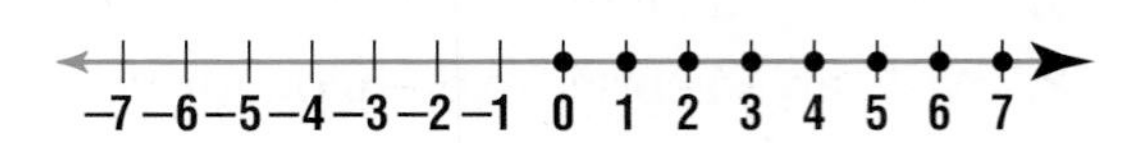

The set of numbers used to name the points marked on the number line at the right is called the set of **integers.**

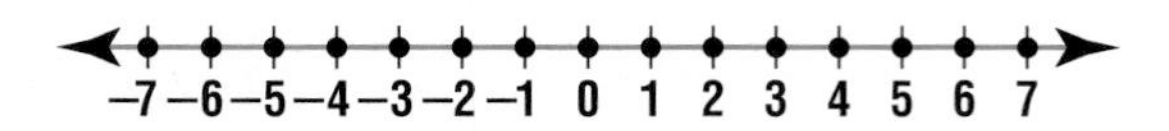

To graph a set of numbers means to locate the points named by those numbers on the number line. The number that corresponds to a point on the number line is called the **coordinate** of the point.

The coordinate of K is 3.

The numbers on a number line increase as you move to the right and decrease as you move to the left. When two integers are graphed on a number line, the number to the right is always greater. The number to the left is always less than the other.

Name the coordinate of each point.

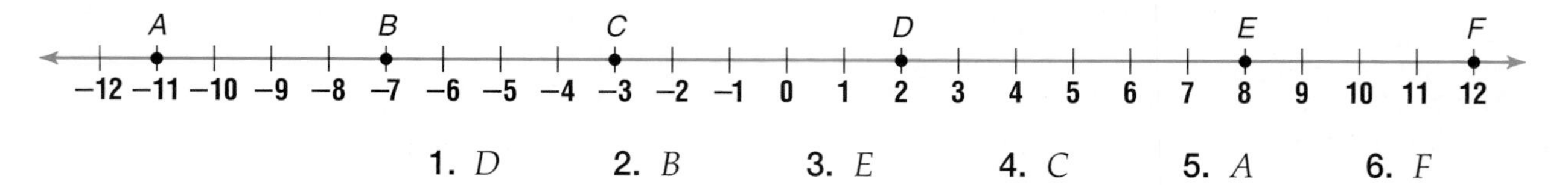

1. D **2.** B **3.** E **4.** C **5.** A **6.** F

Graph each set of numbers on a number line.

7. $\{-4, -2, 0, 3\}$ **8.** $\{-5, -1, 2, 5\}$ **9.** $\{-2, -1, 0, \ldots\}$

10. {integers that are also whole numbers}

11. {integers that are a multiple of 2}

Replace each ● with < or > to make a true sentence.

12. -3 ● -9 **13.** -5 ● 7 **14.** 12 ● 14

15. -3 ● 3 **16.** -8 ● -4 **17.** -1 ● -2

18. **Temperature** With the wind chill factor considered, a temperature of 8°F felt like −5°F. The next day, a temperature of 10°F felt like −11°F. On which day, the first or second, did it feel colder?

REVIEW 2-2

The Coordinate Plane

In the diagram at the right, the two perpendicular lines, called the **x-axis** and the **y-axis,** divide the coordinate plane into four **quadrants.** The point where the two axes intersect is called the **origin.** The origin is represented by the **ordered pair** (0, 0).

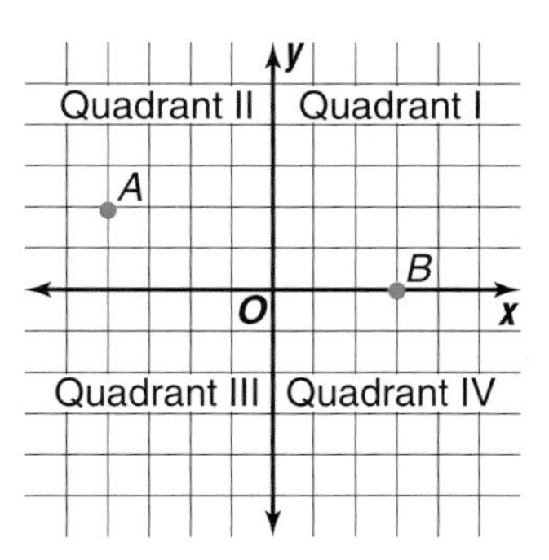

Every other point in the coordinate plane is also represented by an ordered pair of numbers. The ordered pair for point A is $(-4, 2)$. We say that -4 is the **x-coordinate** of A and 2 is the **y-coordinate** of A.

Example

Refer to the diagram above. What ordered pair names point B?

The x-coordinate is 3 and the y-coordinate is 0. Thus, the ordered pair is (3, 0).

To graph any ordered pair (x, y), begin at the origin. Move left or right x units, and then up or down y units. Draw a dot at that point.

Graph each point on a coordinate plane.

1. $C(-4, 0)$
2. $D(2, 2)$
3. $E(3, -5)$
4. $F(-2, 4)$
5. $G(-1, -3)$
6. $H(0, -3)$

Write the ordered pair that names each point shown at the right. Name the quadrant in which the point is located.

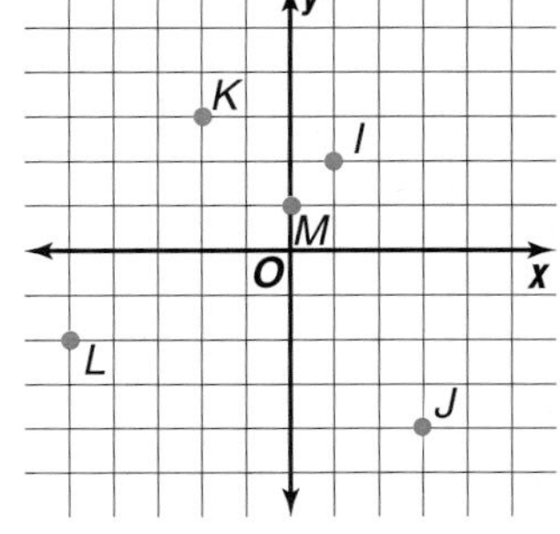

7. I
8. J
9. K
10. L
11. M

12. **Computers** Juan is a computer programmer. He is developing multimedia math software. Each screen in the program must be mapped out pixel by pixel. In the screen shown at the right, find the coordinates of each pixel making up the variable X.

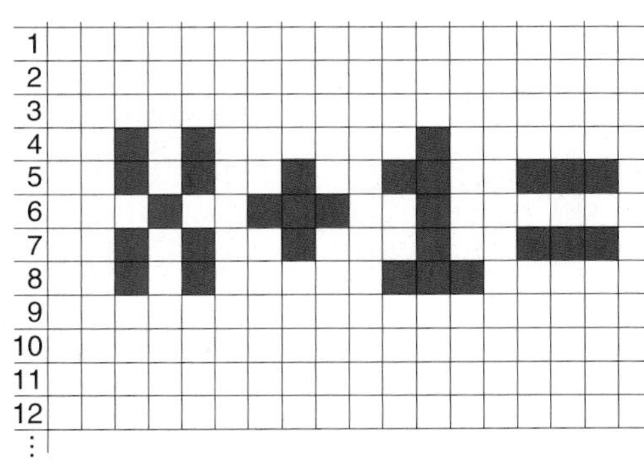

REVIEW 2-3

Adding Integers

Use the following definitions, rules, and properties when adding integers.

Definition, Rule, or Property		Numbers
Definition of Absolute Value	The absolute value of a number is the distance it is from 0 on the number line.	$\lvert 6 \rvert = 6$ $\lvert -6 \rvert = 6$
Adding Integers with the Same Sign	To add integers with the same sign, add their absolute values. Give the result the same sign as the integers.	$3 + 2 = 5$ $-3 + (-2) = -5$
Additive Inverse Property	The sum of any number and its additive inverse is 0.	$3 + (-3) = 0$ $-7 + 7 = 0$
Adding Integers with Different Signs	To add integers with different signs, find the difference of their absolute values. Give the result the same sign as the integer with the greater absolute value.	$3 + (-2) = 1$ $-3 + 2 = -1$

You can use the Distributive Property and the rules for adding integers to simplify expressions with like terms.

Example

Simplify $-3w + (-2w)$.

$$\begin{aligned} -3w + (-2w) &= [-3 + (-2)]w \\ &= (-5)w \\ &= -5w \end{aligned}$$

Find each sum.

1. $-13 + (-7)$
2. $96 + (-54)$
3. $39 + 62$
4. $18 + (-3)$
5. $26 + (-40)$
6. $-27 + (-1)$
7. $12a + (-17a)$
8. $-2t + 20t$
9. $-6x + 15x$

Evaluate each expression if $a = 2$, $b = -3$, and $c = -1$.

10. $212 + |b|$
11. $(-23) + |c|$
12. $-34 + a$
13. $16 + (-a)$

14. **Elevators** The Sears Tower in Chicago has 110 stories. If an employee rode the elevator to her office on the 83rd floor, then went up 15 floors to deliver a report, which floor is she on now?

REVIEW 2-4

Subtracting Integers

Use the following rule to subtract integers.

Rule	Words	Numbers
Subtracting Integers	To subtract an integer, add its additive inverse.	$13 - (-8) = 13 + 8$ or 21 $4 - 3 = 4 + (-3)$ or 1

Example

Evaluate $a - b$ if $a = -9$ and $b = 5$.

$a - b = -9 - 5$	*Replace a with −9 and b with 5.*
$= -9 + (-5)$	*Write 9 − 5 as 9 + (−5).*
$= -14$	*−9 + (−5) = −14*

Find each difference.

1. $11 - 2$
2. $7 - 4$
3. $-5 - 5$
4. $3 - (-1)$
5. $12 - (-8)$
6. $-14 - 7$
7. $-2 - (-6)$
8. $17 - 9$
9. $0 - 3$
10. $12 - (-15)$
11. $25 - 10$
12. $-4 - 13$

Evaluate each expression if $g = 6$, $h = -2$, and $k = 2$.

13. $g - h$
14. $h - k$
15. $g - k$
16. $h - h$
17. $k - h + (-3)$
18. $5 - g + h$

Simplify each expression.

19. $8y - 7y$
20. $6b - (-8b)$
21. $-3v - v$
22. $4x - 5x$
23. $-4r - 7r - 6r$
24. $9f - 5f - (-2f)$

25. **Budgeting** Terri spent $22 of the $35 she had saved to buy a video game. How much more money would she have to save to buy a video game that costs the same price as the first one she bought?

REVIEW 2-5

Multiplying Integers

You can use the following rules when multiplying integers.

Rule	Words	Numbers
Multiplying Two Integers with Different Signs	The product of two integers with different signs is negative.	$(-3)9 = -27$ $9(-3) = -27$
Multiplying Two Integers with the Same Sign	The product of two integers with the same sign is positive.	$-11(-4) = 44$ $11(4) = 44$

To find the product of three or more numbers, multiply the first two numbers. Then multiply the result by the next number, until you come to the end.

Example

Find $-3(7)(-1)(-5)$.

$$-3(7)(-1)(-5) = -21(-1)(-5) \qquad -3(7) = -21$$
$$= 21(-5) \qquad -21(-1) = 21$$
$$= -105 \qquad 21(-5) = -105$$

Find each product.

1. $(-8)(-3)$
2. $(10)(-4)$
3. $(-16)(2)$
4. $-5(0)$
5. $3(3) \quad 9$
6. $-6(-4)$
7. $(-1)(-5)(-3)$
8. $(-4)(-5)(-3)$
9. $(5)(-2)(-1)(-3)$

Evaluate each expression if $x = 4$, $y = -1$, and $z = -2$.

10. $-8y$
11. $5yz$
12. $3x + 2y$
13. $xyz + x$

14. **Nutrition** The recommended daily allowance (RDA) of calcium for a teenaged female is 1200 mg. An eight-ounce serving of skim milk contains 302 mg of calcium. A one-ounce serving of cheddar cheese contains 204 mg of calcium. A medium orange contains 50 mg of calcium.
    a. How much calcium is in 3 eight-ounce servings of skim milk?
    b. Do 3 eight-ounce servings of skim milk satisfy the RDA of calcium for a teenaged female? Explain your reasoning.
    c. Would a serving of cheddar cheese and 2 oranges, in addition to 3 servings of skim milk, satisfy the RDA of calcium for a teenaged female? Explain your reasoning.

REVIEW 2-6

Dividing Integers

Use the following rules when dividing integers.

Rule	Words	Numbers
Dividing Two Integers with the Same Sign	The quotient of two integers with the same sign is positive. $-40 \div (-8) = 5$	$40 \div 8 = 5$
Dividing Two Integers with Different Signs	The quotient of two integers with different signs is negative. $63 \div (-9) = -7$	$-63 \div 9 = -7$

Recall that fractions are another way of showing division. So, you can use the division rules above to evaluate rational expressions.

Example

Evaluate $\frac{-b}{4a}$ if $b = -16$ and $a = -2$.

$\frac{-b}{4a} = \frac{-(-16)}{4(-2)}$ *Replace b with −16 and a with −2.*

$= \frac{16}{-8}$ *$-(-16) = 16$ and $4(-2) = -8$*

$= -2$ *$\frac{16}{-8}$ means $16 \div -8$.*

Find each quotient.

1. $49 \div (-7)$
2. $-36 \div 9$
3. $18 \div 2$
4. $-24 \div (-3)$
5. $30 \div (-6)$
6. $-15 \div (-5)$
7. $32 \div 8$
8. $-64 \div 8$
9. $-42 \div (-6)$

Evaluate each expression if $x = 3$, $y = -1$, and $z = -12$.

10. $\frac{z}{x}$
11. $\frac{8 - x}{y}$
12. $\frac{yz}{x}$

13. **Parties** Caitlyn brought in a bag of 45 chocolates to share with the class for her birthday. She would like to distribute the chocolates evenly among her classmates.
    a. If there are 15 students in the class, including herself, how many chocolates does each person get?
    b. Suppose Caitlyn bought the jumbo bag of 80 chocolates. This time, she also shares the chocolates with her teacher. How many chocolates does each person get?

CHAPTER 2 Test

1. Graph $\{-3, -1, 0, 2, 5\}$ on a number line.
2. Name the coordinate of point P.

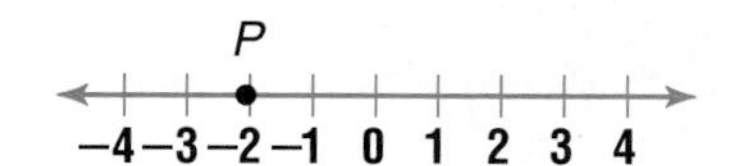

Use the coordinate plane shown below for Exercises 3–5.

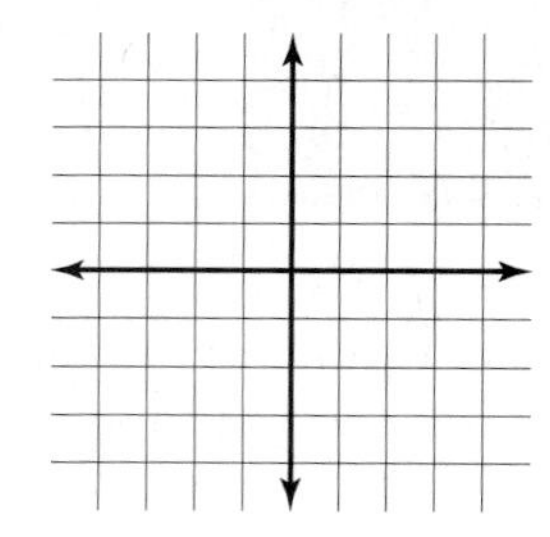

3. Copy the grid at the right. Label the x-axis, the y-axis, and the origin.
4. Plot point $A(-2, -1)$ and name the quadrant in which it is located.
5. Plot point $B(4, -3)$ and name the quadrant in which it is located.

Find each sum or difference.

6. $-4 + 36$
7. $13 + (-5) + (-6)$
8. $8 - (-3)$
9. $4 - 13$

10. Find the value of $|y - x|$ if $x = 2$ and $y = -3$.

11. Write $-3, 1, -1, 0$ in order from least to greatest.

Simplify.

12. $(-8)(-7)$
13. $3(-6)$
14. $(-9)(4) + (2)(1)$
15. $3(-3) - 4(2)$
16. $21 \div 7$
17. $-16 \div 4$
18. $\frac{-63}{9}$
19. $\frac{-32}{-4}$

20. **Nutrition** The table shows the amount of sodium a can of regular soda pop contains compared with the amount of sodium a can of diet soda pop contains for three different brands.
    a. Find the differences in sodium for each brand.
    b. Which brand has the greatest difference in sodium?

Amount of Sodium in a Can of Soda Pop (mg)		
Brand	Regular	Diet
A	50	25
B	35	30
C	25	10

Rational Numbers

Rational Number	A rational number is any number that can be expressed as a fraction where the numerator and denominator are integers and the denominator is not zero

Comparison Property	For any two numbers a and b, exactly one of the following sentences is true: $a < b$, $a = b$, or $a > b$

To compare two fractions, you can use **cross products**. Cross products are the products of the diagonal terms.

Comparison Property for Rational Numbers	For any rational numbers $\frac{a}{b}$ and $\frac{c}{d}$, with $b > 0$ and $d > 0$: **1.** if $\frac{a}{b} < \frac{c}{d}$, then $ad < cd$, and **2.** if $ad < bc$, then $\frac{a}{b} < \frac{c}{d}$.

This property also holds if < is replaced by >, ≤, ≥, or =.

Example

Replace ● with <, >, or = to make a true sentence.

$\frac{8}{10}$ ● $\frac{3}{4}$

8(4) ● 10(3)

$32 > 30$

If $32 > 30$, then $\frac{8}{10} > \frac{3}{4}$.

Replace each ● with <, >, or = to make a true sentence.

1. -6 ● 3
2. $\frac{4}{7}$ ● $\frac{5}{8}$
3. -3.35 ● -3.40

Write the numbers in each set from least to greatest.

4. $-\frac{1}{5}$, 0.3, -2.5
5. $\frac{1}{8}$, -1, $-\frac{9}{7}$
6. Find a number between $\frac{1}{3}$ and $\frac{7}{9}$.
7. Which is the better buy: a 15-ounce box of cereal for \$1.99 or a 20-ounce box of cereal for \$2.49?

REVIEW 3-2

Adding and Subtracting Rational Numbers

The rules for adding and subtracting integers also apply to adding and subtracting rational numbers.

Rational Number	Form $\frac{a}{b}$
5	$\frac{5}{1}$
$-1\frac{2}{3}$	$-\frac{5}{3}$
0.75	$\frac{3}{4}$

Examples

1 **Find $\left(-3\frac{1}{4}\right) + 6\frac{1}{2}$.**

$$\left(-3\frac{1}{4}\right) + 6\frac{1}{2} = +\left(\left|6\frac{2}{4}\right| - \left|-3\frac{1}{4}\right|\right)$$
$$= +\left(6\frac{2}{4} - 3\frac{1}{4}\right) \text{ or } 3\frac{1}{4}$$

2 **Find $-6.1 - 9.8$.**

$$-6.1 - 9.8 = -6.1 + (-9.8)$$
$$= -15.9$$

3 **Find $-\frac{1}{4} + \frac{5}{6} + \left(-\frac{7}{4}\right)$.**

$$-\frac{1}{4} + \frac{5}{6} + \left(-\frac{7}{4}\right) = \left[-\frac{1}{4} + \left(-\frac{7}{4}\right)\right] + \frac{5}{6}$$ *Comm. & Assoc. Properties of Addition*

$$= -\frac{8}{4} + \frac{5}{6}$$

$$= -\frac{24}{12} + \frac{10}{12}$$ *The LCD is 12.*

$$= -\frac{14}{12} \text{ or } -1\frac{1}{6}$$

Find each sum or difference.

1. $-\frac{4}{13} + \left(-\frac{7}{13}\right)$
2. $\frac{3}{4} + \left(-\frac{7}{8}\right)$
3. $-0.01 + 0.06$
4. $\frac{4}{7} - \left(-\frac{2}{7}\right)$
5. $3.97 - 1.55$
6. $-\frac{2}{3} - \frac{4}{9}$

Evaluate each expression if $a = -0.25$ and $b = \frac{1}{3}$.

7. $a + 0.56$
8. $2 - b$
9. $-3 - a$
10. $b + \frac{2}{7}$

Find each sum.

11. $-40.1 + 62.7 + (-16.8)$
12. $-13q + (-4q) + 18q$
13. $\frac{1}{4} + \left(-\frac{3}{8}\right) + \frac{1}{2}$
14. **Fishing** The saltwater fish record for redeye bass is 8.75 pounds. The record for black sea bass is 0.75 pounds more than the redeye bass record. The record for barred sand bass is 3.75 pounds more than the black sea bass record. What is the record for barred sand bass?

REVIEW 3-3

Mean, Median, Mode, and Range

In working with statistical data, it is often useful to have one value represent the complete set of data. For example, **measures of central tendency** represent the center or middle of the data. Three measures of central tendency are the **mean, median**, and **mode. Measures of variation** describe the distribution of data. **Range** is a measure of variation.

	Definitions	Numbers
Mean	The mean, or *average*, of a set of data is the sum of the data divided by the number of pieces of data.	Data: 14, 16, 17, 14, 22, 22 $\frac{14 + 16 + 17 + 14 + 22 + 22}{6} = 17.5$
Median	The median of a set of data is the middle number when the data are arranged numerically. For an even number of data items, the median is the mean of the two middle values.	Data: 14, 14, 16, 17, 22, 22 $\frac{16 + 17}{2} = 16.5$
Mode	The mode of a set of data is the number that occurs most often.	Data: 14, 16, 17, 14, 22, 22 There are two modes, 14 and 22.
Range	The range of a set of data is the difference between the greatest and the least values of the set.	Data: 14, 16, 17, 14, 22, 22 $22 - 14$ or 8

Find the mean, median, mode(s), and range for each set of data.

1. 9, 8, 9, 7, 5
2. 2, 1, 0.5, 3, 4.5, 1, 2, 3, 5, 3
3. 10, 18, 18, 18, 10, 10
4. 12, 7, 14, 30
5. 5, 7, 7, 13, 11, 5
6. 12.4, 6.8, 19.1, 30.4, 7.3
7.

Stem	Leaf
5	3 8 8 9
6	1 2 2 2 4 6
7	7 7 7 8 9 9

6 | 2 = 62

8.

Stem	Leaf
11	1 3 4 5
12	2 2 4 4
13	1 3 3 3 4 5

13 | 3 = 133

9. **Football** The table at the right shows the teams that have played in the Super Bowl most often.
   a. Find the mean, median, mode(s), and range of the data.
   b. Would you choose any measure of central tendency over another to represent these data? Why?

Team	Times in Super Bowl
Dallas Cowboys	8
Denver Broncos	6
Miami Dolphins	5
Pittsburgh Steelers	5
San Francisco 49ers	5
Washington Redskins	5
Buffalo Bills	4
Minnesota Vikings	4
Green Bay Packers	4

Source: *The World Almanac*

REVIEW 3-4

Equations

Mathematical statements with one or more variables are called **open sentences**. Open sentences are **solved** by finding a replacement for the variable that results in a true sentence. The replacement is called a **solution**.

Example 1 **Replace y in $5y - 9 = 21$ with the value 6.**

$5y - 9 = 21$
$5(6) - 9 = 21$
$30 - 9 = 21$
$21 = 21$ true

Since $y = 6$ makes the sentence $5y - 9 = 21$ true, 6 is a solution.

A set of numbers from which replacements for a variable may be chosen is called a **replacement set**. The set of all replacements for the variable in an open sentence that results in a true sentence is called the **solution set** for the sentence.

A sentence that contains an equals sign, =, is called an **equation** and sometimes may be solved by applying the order of operations.

Example 2 **Solve $\frac{3(4 + 5)}{2 \cdot 4 - 3} = w$.**

$\frac{3(4 + 5)}{2 \cdot 4 - 3} = w$

$\frac{3(9)}{8 - 3} = w$ *$4 + 5 = 9$; $2 \cdot 4 = 8$*

$\frac{27}{5} = w$ *$3 \cdot 9 = 27$; $8 - 3 = 5$*

Find the solution of each equation if the replacement sets are $x = \{-4, 0, 2, 3\}$ and $y = \{-5, -2, 1, 4, 8\}$.

1. $y - 6 = 2$
2. $6y = -30$
3. $1 - x = -1$
4. $2x + (-5) = -5$
5. $2x - 3x = 4$
6. $\frac{4y + 5}{y} = 9$

Solve each equation.

7. $d = 2(4 \cdot 5 - 3)$
8. $m = 5\frac{2}{5} - 1\frac{3}{10}$
9. $\frac{40 - 7}{5 + 6} = z$
10. **Business** Saundra must sell 4 T-shirts to make a $10 profit.
    a. Write an equation that represents the number of T-shirts she must sell to make a $60 profit.
    b. How many T-shirts would she have to sell?

REVIEW 3-5 Solving Equations by Using Models

You can use algebra tiles to solve equations.

the variable

x

x

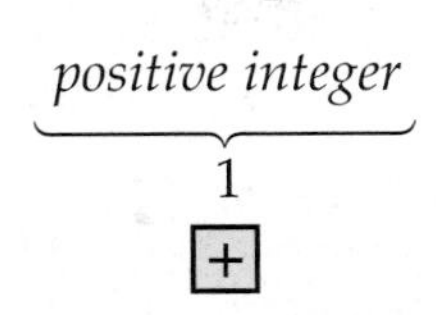

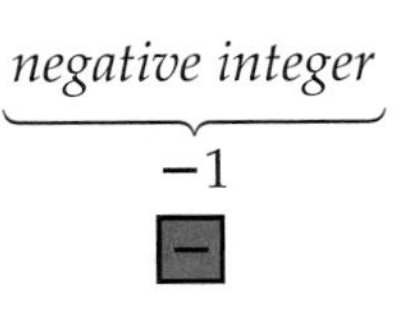

Example

Use algebra tiles to solve $x + 3 = -2$.

Step 1 Model $x + 3 = -2$ by placing 1 green tile and 3 yellow tiles on one side of the mat to represent $x + 3$. Place 2 red tiles on the other side to represent -2.

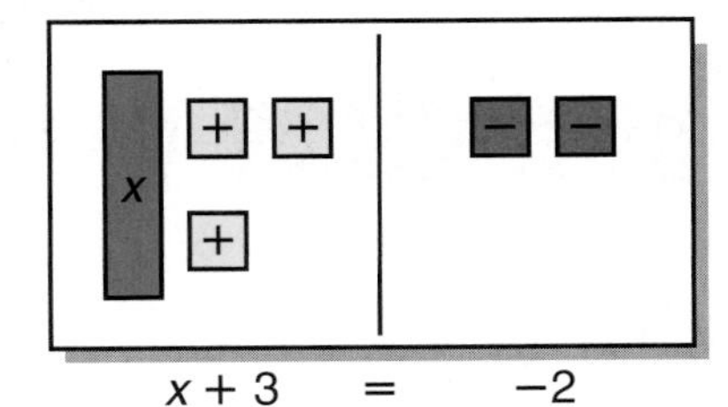

Steps 2 & 3 Since you cannot simply remove 3 yellow tiles from each side of the mat to solve for x, you must add 3 red tiles to each side to make 3 zero pairs on the left side of the mat. Remove the zero pairs.

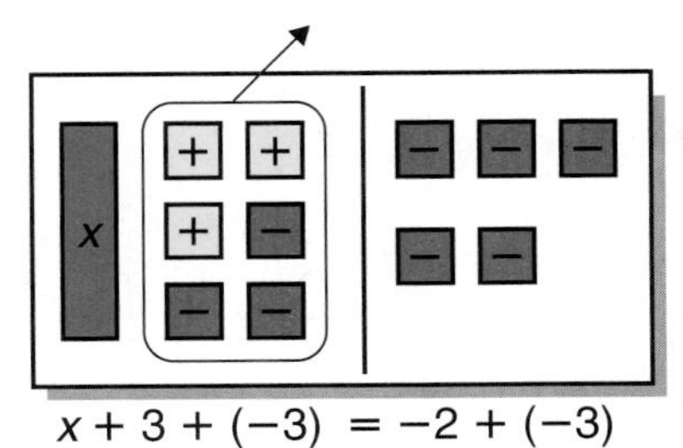

Step 4 The green tile is matched with 5 red tiles. Therefore, $x = -5$.

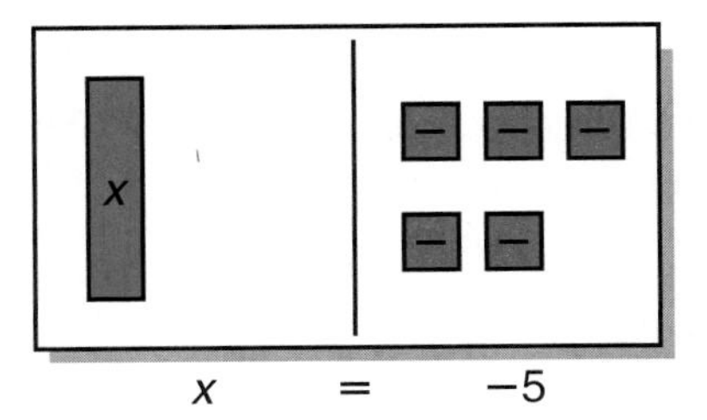

Solve each equation. Use algebra tiles if necessary.

1. $x + (-3) = 1$
2. $x - 2 = 5$
3. $x - 4 = 4$
4. $6 + x = 3$
5. $-2 + x = -1$
6. $x - 1 = -5$
7. $x + 7 = 10$
8. $4 + x = -2$
9. $-3 + x = 0$

10. **Golf** In the 2000 U.S. Open, runners-up Ernie Els and Miguel Angel Jimenez scored 3 above par. This number is 15 more than what the winner of the tournament, Tiger Woods, scored to set a record low for the tournament.
    a. Write an equation that represents this situation.
    b. What was Tiger Woods' score?

REVIEW 3-6

Solving Addition and Subtraction Equations

You can use the Addition and Subtraction Properties of Equality to solve equations. To check, substitute the solution for the variable in the original equation. If the resulting sentence is true, your solution is correct.

Addition Property of Equality	For any numbers a, b, and c, if $a = b$, then $a + c = b + c$.
Subtraction Property of Equality	For any numbers a, b, and c, if $a = b$, then $a - c = b - c$.

Examples

1 Solve $z - 15 = -3$.

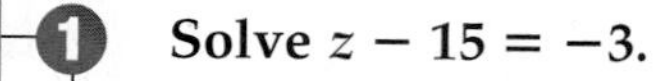

$$z - 15 = -3$$
$$z - 15 + 15 = -3 + 15$$
$$z = 12$$

Check:
$$z - 15 = -3$$
$$12 - 15 \stackrel{?}{=} -3$$
$$-3 = -3 \quad \checkmark$$

2 Solve $q + 5 = -2$.

$$q + 5 = -2$$
$$q + 5 - 5 = -2 - 5$$
$$q = -7$$

Check:
$$q + 5 = -2$$
$$-7 + 5 \stackrel{?}{=} -2$$
$$-2 = -2 \quad \checkmark$$

Sometimes an equation can be solved more easily if it is rewritten first. Recall that subtracting a number is the same as adding its inverse. For example, the equation $b - (-3) = 11$ may be rewritten as $b + 3 = 11$.

Solve each equation. Check your solution.

1. $w - 7 = -3$
2. $y + 8 = 3$
3. $q + 5 = -20$
4. $-6 = m + 3$
5. $d + (-12) = 4$
6. $t - (-2) = 14$
7. $16 + a = -11$
8. $62 = 41 + s$
9. $-7.4 = c + (-1.2)$
10. $-\frac{3}{8} + x = 2\frac{5}{8}$
11. $-\frac{1}{6} + b = \frac{2}{3}$
12. $\frac{4}{9} = n + \frac{1}{3}$
13. **Clothing Design** A clothing designer is designing a blazer. The blazer will have three buttons on the front. The length of each buttonhole is to be $\frac{1}{8}$-inch longer than the sum of the diameter and the thickness of the button.
    a. Translate the description of the buttonhole length into a formula.
    b. Suppose the clothing designer decides that each buttonhole is to be $\frac{15}{16}$-inch long and each button is to be $\frac{5}{8}$-inch in diameter. What thickness should the button be?

REVIEW

3-7 Solving Equations Involving Absolute Value

Consider the graph of $|x| = 3$. The distance from 0 to x is 3 units. Thus, $x = -3$ or $x = 3$.

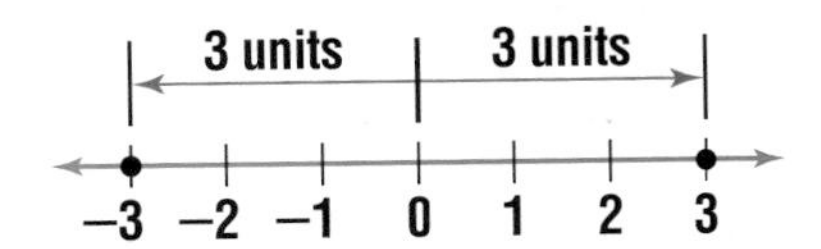

Equations that involve absolute value can be solved by writing them as a compound sentence and solving. Two cases must be considered: when the number inside the absolute value symbol is negative and when it is positive. You can check your solutions by substituting them for the variable in the original equation.

Example

Solve $|r + 9| + 2 = 3$.

First, simplify the expression.

$|r + 9| + 2 = 3$

$|r + 9| + 2 - 2 = 3 - 2$ *Subtract 2 from each side.*

$|r + 9| = 1$

Then write a compound sentence and solve it.

$r + 9 = 1$ or $r + 9 = -1$

$r + 9 - 9 = 1 - 9$ *Subtract 9 from each side.* $r + 9 - 9 = -1 - 9$

$r = -8$ $r = -10$

The solution set is $\{-10, -8\}$.

Remember that the absolute value of any integer is always positive or zero. So, for example, $|a| = -8$ is never true. Its solution set $\varnothing$ is called the **empty set**.

Solve each equation. Check your solution.

1. $|c| = 13$
2. $|f| = 100$
3. $|g| = 0$
4. $|x + 7| = 9$
5. $|-3 + y| = -6$
6. $|b - 5| = 8$
7. $|h| - 10 = -8$
8. $|-4 + u| = 12$
9. $|k| + 1 = 0$
10. $|7 + p| + 4 = 4$
11. $-6 + |v - 3| = -2$

12. What is the absolute value of the sum of 9 and -5?
13. The absolute value of 2 and x is 15. What is x?

14. **Games** On a game show, contestants guess the price of a prize. If they come within \$100 of the actual price, they win two prizes. If the actual prize is \$24,560.25, write and solve an equation to find the greatest and least amounts a person could bid to win two prizes.

CHAPTER 3 Test

Replace each ● with <, >, or = to make a true sentence.

1. $-6 \bullet -8$

2. $-3.65 \bullet -3.60$

3. $\frac{4}{9} \bullet \frac{4}{7}$

4. Write $\frac{1}{4}, \frac{2}{5}, \frac{3}{4}$, and $\frac{3}{8}$ in order from least to greatest.

Find each sum or difference.

5. $2.5 + (-0.3)$

6. $-8.2 - 6.0$

7. $3.4 - (-0.64)$

8. $-\frac{1}{6} + \left(-\frac{5}{6}\right)$

9. $\frac{2}{3} - \left(-\frac{5}{12}\right)$

10. $\frac{3}{5} + 1\frac{1}{10}$

Solve each equation.

11. $y = (5 - 4) \div 3 \cdot 6$

12. $q = 3(2.1) - 5.4$

13. $\frac{2 \cdot 1 + 3}{20 \div 2} = r$

Solve each equation. Use algebra tiles if necessary.

14. $x + 7 = 2$

15. $3 + x = -1$

Solve each equation. Check your solution.

16. $r - 4 = -48$

17. $h + (-9.5) = 1.3$

18. $-\frac{1}{3} + a = \frac{4}{5}$

19. $|b| + 2 = 8$

20. $|n - 4| = -10$

21. $|s - (-9)| = 5$

The list below gives the prices of the same pair of sneakers at different stores. Use these data to complete Exercises 22–25.

72, 34, 65, 72, 78, 70, 68, 72, 68

22. Find the mean of the sneaker prices.

23. Find the median of the sneaker prices.

24. Find the mode of the sneaker prices.

25. Which measure of central tendency, mean, median, or mode, least represents the sneaker data? Explain.

Multiplying Rational Numbers

You can use the same rules for multiplying rational numbers as you did for multiplying integers.

Multiplying Two Rational Numbers	Words	Numbers
Different Signs	The product of two rational numbers having different signs is negative.	$0.2(-5.4) = -1.08$ $-0.2(5.4) = -1.08$
Same Sign	The product of two rational numbers having the same sign is positive.	$8(0.6) = 4.8$ $-8(-0.6) = 4.8$

The rule for multiplying fractions is given below.

Multiplying Fractions	To multiply fractions, multiply the numerators and multiply the denominators.

Example

Find $-\frac{1}{2} \cdot \frac{3}{8}(-4)$.

$$-\frac{1}{2} \cdot \frac{3}{8}(-4) = \left(\frac{-1 \cdot 3}{2 \cdot 8}\right)\left(-\frac{4}{1}\right)$$
$$= -\frac{3}{16}\left(-\frac{4}{1}\right)$$
$$= \frac{3}{4}$$

Find each product.

1. $3 \cdot 2.5$
2. $4.1(6)$
3. $-0.8 \cdot 5$
4. $\frac{2}{3} \cdot \frac{1}{4}$
5. $\frac{5}{6}(-2)$
6. $\left(-\frac{3}{5}\right)(-55)$

Simplify each expression.

7. $(-9m)(4.4)$
8. $-0.2n(-0.7)$
9. $-6g \cdot -\frac{3}{2}$
10. $\left(\frac{2}{7}x\right)\left(\frac{1}{6}y\right)$

Counting Outcomes

A result of an experiment or situation is called an **outcome**. You can make a **tree diagram** to find the list of all possible outcomes, commonly called the **sample space**.

Example

A car dealership offers the car options shown in the table. Make a tree diagram to find the sample space.

Option 1	Option 2
radio tape cd	air no air

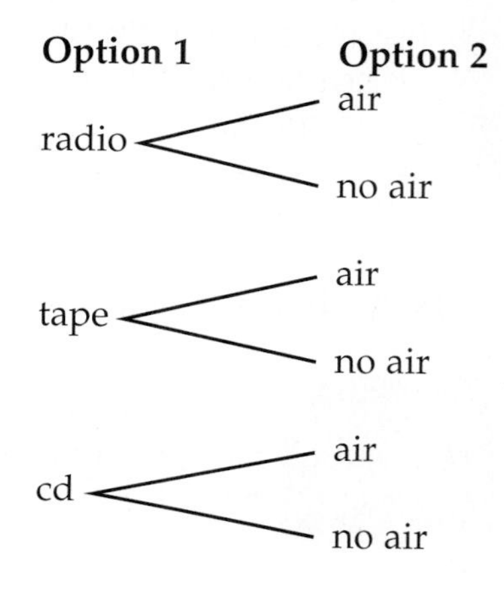

Outcome
radio, air
radio, no air
tape, air
tape, no air
cd, air
cd, no air

Six possible outcomes make up the sample space.

Find the number of possible outcomes by drawing a tree diagram and listing the sample space.

1. two tosses of a coin

2. Three people have been nominated for the positions of secretary or treasurer of the student council. How many different ways can the positions be filled?

Person	Position
Amy Ben Pete	secretary treasurer

3. Four people are swimming on a relay team in which each person swims one of four legs. How many different ways can the team swim the relay?

Person	Leg
Ed Kazu Mario Jay	first second third fourth

REVIEW 4-3

Dividing Rational Numbers

Use the following rules to divide rational numbers.

Rule or Property	Words	Numbers
Dividing Two Rational Numbers	The quotient of two numbers having the same sign is positive. The quotient of two numbers having different signs is negative.	$-08 \div -4 = 0.2$ $-6.3 \div 7 = 0.9$
Multiplicative Inverse Property	The product of a number and its multiplicative inverse is 1.	$\frac{3}{8} \cdot \frac{8}{3} = 1$ $-\frac{2}{5} \cdot -\frac{5}{2} = 1$
Dividing Fractions	To divide a fraction by any nonzero number, multiply by its reciprocal.	$\frac{3}{7} \div \frac{3}{4} = \frac{3}{7} \cdot \frac{4}{3} = \frac{4}{7}$

Since the fraction bar indicates division, you can use the division rules and the Distributive Property to simplify rational expressions.

Example

Find the quotient of $\frac{2}{5}$ and $1\frac{1}{3}$.

$$\frac{2}{5} \div 1\frac{1}{3} = \frac{2}{5} \div \frac{4}{3}$$

$$= \frac{2}{5} \cdot \frac{3}{4}$$

$$= \frac{6}{20} \text{ or } \frac{3}{10}$$

Find each quotient.

1. $5.4 \div 9$
2. $48 \div -1.6$
3. $-1.2 \div -6$
4. $-\frac{4}{5} \div \frac{7}{10}$
5. $-\frac{7}{3} \div \frac{1}{2}$
6. $\frac{-96}{-37} \div \frac{96}{37}$
7. $\frac{2}{3} \div \frac{3}{5}$
8. $-\frac{1}{5} \div 5$
9. $-\frac{2}{4} \div \frac{3}{6}$
10. $\frac{18}{2} \div \frac{1}{3}$
11. $4 \div \frac{1}{8}$
12. $\frac{-\frac{3}{7}}{3}$
13. **Cooking** Theo has a recipe for chocolate chip cookies that makes 6 dozen cookies. He only has enough flour to make 3 dozen cookies.
    a. If the original recipe calls for $2\frac{1}{4}$ cups of margarine, how much margarine will he need to make 3 dozen cookies?
    b. If the original recipe calls for $3\frac{1}{2}$ teaspoons of baking powder, how much baking powder will he need to make 3 dozen cookies?

REVIEW 4-4 Solving Multiplication and Division Equations

You can solve equations in which a variable has a coefficient by using the Multiplication and Division Properties of Equality.

Division Property of Equality	If you divide each side of an equation by the same nonzero number, the two sides remain equal.
Multiplication Property of Equality	If you multiply each side of an equation by the same number, the two sides remain equal.

Examples

1 **Solve $4x = 28$.**

$$4x = 28$$
$$\frac{4x}{4} = \frac{28}{4}$$
$$x = 7$$

Check:
$$4x = 28$$
$$4(7) \stackrel{?}{=} 28$$
$$28 = 28 \quad \checkmark$$

2 **Solve $\frac{1}{7}p = 3$.**

$$\frac{1}{7}p = 3$$
$$7\left(\frac{1}{7}p\right) = 7(3)$$
$$p = 21$$

Check:
$$\frac{1}{7}p = 3$$
$$\frac{1}{7}(21) \stackrel{?}{=} 3$$
$$3 = 3 \quad \checkmark$$

Solve each equation. Check your solution.

1. $-6z = -72$
2. $-4q = 48$
3. $5r = -35$
4. $\frac{1}{7}a = -3$
5. $8s = \frac{4}{7}$
6. $\frac{4}{3}t = -\frac{8}{3}$

Write an equation and solve.

7. Seven times a number x is 63. What is the number?
8. One third of a number n is twelve. What is the number?
9. Negative three times a number y is -33. What is the number?
10. Tyrone paid $82.50 for 3 ballet tickets. What is the cost of each ticket t?

Complete.

11. If $5y = 10$, then $7y =$ __?__.
12. If $-3r = 27$, then $4r =$ __?__.
13. If $8b = -40$, then $-2b =$ __?__.
14. If $3m = 30$, then $6m =$ __?__.

REVIEW 4-5

Solving Multi-Step Equations

To solve some equations you must perform more than one operation. First, determine what operations have been done to the variable. Then undo these operations in the reverse order. In other words, work backward.

Examples

1 How would you solve $\frac{x}{4} + 5 = 13$?

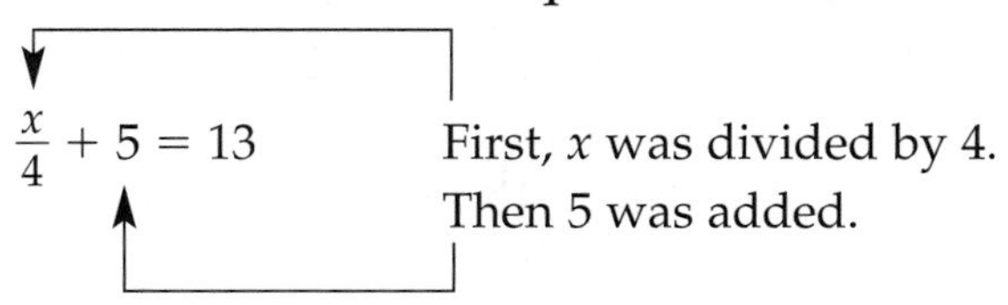

Step 1 Subtract 5 from each side.

$$\frac{x}{4} + 5 - 5 = 13 - 5$$
$$\frac{x}{4} = 8$$

Step 2 Multiply each side by 4.

$$4\left(\frac{x}{4}\right) = 4(8)$$
$$x = 32$$

2 **$3w + 7 = 34$** *Addition of 7 is indicated.*

$3w + 7 - 7 = 34 - 7$ *Therefore, subtract 7 from each side.*

$3w = 27$ *Multiplication by 3 is also indicated.*

$\frac{3w}{3} = \frac{27}{3}$ *Therefore, divide each side by 3.*

$w = 9$

Check:

$3w + 7 = 34$

$3(9) + 7 \stackrel{?}{=} 34$

$27 + 7 \stackrel{?}{=} 34$

$34 = 34$ ✓

Solve each equation. Check your solution.

1. $5r - 21 = 44$
2. $6y + 2 = 50$
3. $9a - 6 = 3$
4. $\frac{3m}{2} = 9$
5. $2 = \frac{q + 5}{8}$
6. $1.4 + 4z = 6.6$
7. $\frac{3}{5}c - 6 = 9$
8. $\frac{t}{-2} + 7 = -9$
9. $-6 = \frac{5p - (-2)}{-7}$

Define a variable, write an equation, and solve each problem. Check your solution.

10. Find three consecutive integers whose sum is 12.
11. Find two consecutive odd integers whose sum is 76.
12. Deborah, Travis, and Tito were each born in one of three consecutive years. The sum of their ages is 45. What are the three ages?

REVIEW 4-6

Solving Equations with Variables on Both Sides

If an equation has variables on both sides, use the Addition or Subtraction Property of Equality to write an equivalent equation that has all the variables on one side. Then solve the equation.

Example

Solve $3z - 4 = -5z - 20$.

$$3z - 4 = -5z - 20$$

$$3z + 5z - 4 = -5z + 5z - 20 \quad \text{Add } 5z \text{ to each side.}$$

$$8z - 4 = -20$$

$$8z - 4 + 4 = -20 + 4 \quad \text{Add 4 to each side.}$$

$$8z = -16$$

$$\frac{8z}{8} = \frac{-16}{8} \quad \text{Divide each side by 8.}$$

$$z = -2$$

Check:

$$3z - 4 = -5z - 20$$

$$3(-2) - 4 \stackrel{?}{=} 5(-2) - 20$$

$$-6 - 4 \stackrel{?}{=} 10 - 20$$

$$-10 = -10 \quad \checkmark$$

Some equations may have *no solution*, and some equations may have *every number* in their solution set. An equation that is true for every value of the variable is called an **identity**.

Solve each equation. Check your solution.

1. $b + 16 = b - 4$
2. $8 - z = 3z$
3. $12c - 5c = 4c + 15$
4. $40 - 5s = 3s$
5. $2.8w + 5.3 = 3.3w - 0.7$
6. $m + 9 = 8 - m$
7. $k + 32 = \frac{1}{2}k - 1$
8. $\frac{3}{4}x = 2x + 10$
9. $7t + 1 = 5t - 5$
10. $3d + 1.1 = 2.3 - d$
11. $-7a + 1 = 1 - 7a$
12. $3n - 1 = 9n + 4$
13. **Sales** Rundell owns a chain of coffee shops. His shops sell both doughnuts and bagels. Over the past few years he has found that the sales of doughnuts have been decreasing by 0.01 million dollars per year, and sales of bagels have been increasing by 0.06 million dollars per year. Last year, Rundell's shops sold 1.6 million dollars in doughnuts and 0.9 million dollars in bagels. If the sales trends continue, after how many years will sales of bagels and doughnuts be equal?

REVIEW 4-7

Solving Equations with Grouping Symbols

When an equation contains parentheses or other grouping symbols, first use the Distributive Property to remove the grouping symbols. Then solve the equation as normal.

Example

Solve $4(m + 9) = 3(8 + m)$.

$4(m + 9) = 3(8 + m)$	
$4m + 36 = 24 + 3m$	*Distributive Property*
$4m - 3m + 36 = 24 + 3m - 3m$	*Subtract 3m from each side.*
$m + 36 = 24$	
$m + 36 - 36 = 24 - 36$	*Subtract 36 from each side.*
$m = -12$	

Check: $4(m + 9) = 3(8 + m)$

$4(-12 + 9) \stackrel{?}{=} 3(8 + (-12))$

$4(-3) \stackrel{?}{=} 3(-4)$

$-12 = -12$ ✓

Solve each equation. Check your solution.

1. $5(x - 4) = 20$
2. $-4 = 2(3t - 8)$
3. $(h + 1)(-2) = 6$
4. $5 = -(b - 9)$
5. $2z(5 - (-4)) = 18$
6. $4(2g + 8) = 32$ 0
7. $10 = (y + 7) - 1$
8. $6(x + 1) - 5 = 13$
9. $40 - 5s = -2(-1 + 3s)$
10. $3(2m - 6) = 2(3m - 9)$
11. $k + 11 = \frac{1}{2}(-14k - 10)$
12. $\frac{3v + 15}{3} = 6$
13. The length of a base of the trapezoid shown at the right is twice the length of the other base.
    a. Express the lengths of both bases in terms of x.
    b. Find the length of each base of the trapezoid given that the area of the trapezoid is 60 square feet. (*Hint:* The formula for the area of a trapezoid is $A = \frac{1}{2}h(b_1 + b_2)$.)

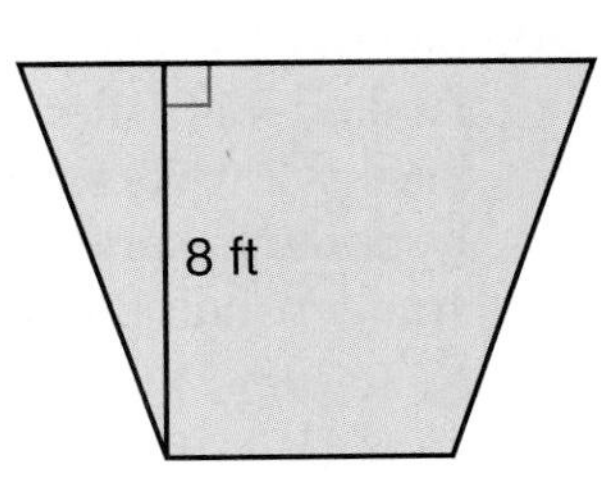

CHAPTER 4 Test

Find each product or quotient.

1. $-4.7 \cdot -3$

2. $2.1(-0.6)$

3. $-4.5 \div 9$

4. $\frac{7}{12} \cdot \frac{3}{4}$

5. $-8 \div -\frac{2}{9}$

6. $-\frac{1}{2} \div \frac{2}{5}$

Simplify each expression.

7. $-0.48x(10)$

8. $3t\left(\frac{5}{6}\right)$

9. $\frac{-s}{\frac{5}{3}}$

10. Draw a tree diagram to show the different styles of glasses that are possible from the choices given in the table. How many outcomes are in the sample space?

Frame	Lens Shape
wire plastic	oval circular rectangular

Solve each equation. Check your solution.

11. $7w = -42$

12. $-3z = -3.9$

13. $\frac{1}{8}x = -5$

14. $11y + 6 = -38$

15. $32 = 7q + 4$

16. $\frac{3}{7}c + 3 = 9$

17. $w + 5 = 9w - 3$

18. $-3(b + 8) = 3(b - 1)$

19. $4x = -(2x + 48)$

20. **Geometry** The side of the triangle labeled h is called the *hypotenuse.* The base is 2 units smaller than the hypotenuse and the height is 4 units smaller than the hypotenuse.

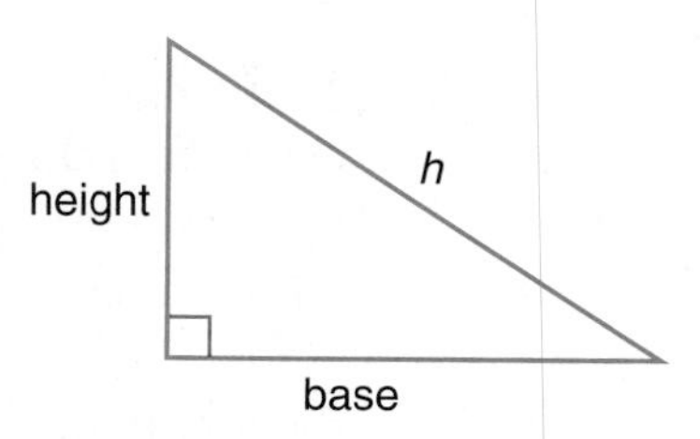

   a. Express the lengths of the base and height of the triangle in terms of the hypotenuse h.

   b. If the perimeter is 24 inches, find the dimensions of the triangle.

REVIEW 5-1

Solving Proportions

In mathematics, a **ratio** compares two numbers by division. A ratio that compares a number x to a number y can be written in the following ways.

$$x \text{ to } y \qquad x{:}y \qquad \frac{x}{y}$$

When a ratio compares two quantities with different units of measure, that ratio is called a **rate.** For example, a 7°F rise in temperature per hour is a rate and can be expressed as $\frac{7 \text{ degrees}}{1 \text{ hour}}$, or 7 degrees per hour.

Proportions are often used to solve problems involving ratios. You can use the cross products of proportions to solve equations that have the form of a proportion.

Proportion	An equation of the form $\frac{a}{b} = \frac{c}{d}$ stating that two ratios are equal.
Property of Proportions	The cross products of a proportion are equal. If $\frac{a}{b} = \frac{c}{d}$, then $ad = bc$.

Example

Solve $\frac{r}{6} = \frac{14}{3}$.

$\frac{r}{6} = \frac{14}{3}$

$3r = 84$ *Find the cross products.*

$\frac{3r}{3} = \frac{84}{3}$ *Divide each side by 3.*

$r = 28$

The solution is 28.

Solve each proportion.

1. $\frac{3}{11} = \frac{9}{z}$
2. $\frac{a}{7} = \frac{5}{35}$
3. $\frac{0.25}{4} = \frac{0.5}{y}$
4. $\frac{w-1}{9} = \frac{7}{9}$
5. $\frac{4}{x+2} = \frac{16}{28}$
6. $\frac{9-2t}{7} = \frac{15}{7}$
7. $\frac{2b}{3} = \frac{8}{20}$
8. $\frac{m+3}{-5} = \frac{-m}{-4}$

Use a proportion to solve each problem.

9. To make a model of the Rio Grande River, Angelica used 1 inch of clay for 50 miles of the actual river's length. Her model river was 38 inches long. How long is the Rio Grande?
10. Jorge finished 28 math problems in one hour. At that rate, how many hours will it take him to complete 70 math problems?

Scale Drawings and Models

A **scale drawing** or **scale model** is used to represent an object that is too large or too small to be drawn or built at actual size. The **scale** is the ratio of a length on the drawing or model to the corresponding length of the real object.

Example

In a movie about dinosaurs, the dinosaurs were scale models and so was the sport utility vehicle that the T-Rex overturned. The vehicle was made to the scale of 1 inch to 8 inches. The actual vehicle was about 168 inches long. What was the length of the model vehicle?

$$\frac{1 \text{ inch}}{8 \text{ inches}} = \frac{x \text{ inches}}{168 \text{ inches}}$$ *← model length* *← actual length*

$168(1) = 8x$ *Find the cross products.*

$168 = 8x$

$\frac{168}{8} = \frac{8x}{8}$

$21 = x$

The length of the model vehicle was 21 inches.

On a map, the scale is 1 inch = 40 miles. Find the actual distance for each map distance.

	From	To	Map Distance
1.	Buffalo, NY	Rochester, NY	2 inches
2.	San Diego, CA	Fresno, CA	8.5 inches
3.	Dallas, TX	Amarillo, TX	9 inches
4.	Columbus, OH	Springfield, OH	1 inch
5.	St. Paul, MN	Duluth, MN	$3\frac{3}{4}$ inches
6.	Waterloo, IA	Ames, IA	2.5 inches

7. **Photos** In a photograph of a building, one story measures one-fourth inch. If the building is 6 inches tall in the photograph, how many stories high is it?

8. **Architecture** On a blueprint of a house, one bathroom is 2.5 inches long. When the house is built, the room will be 10 feet long. Find the scale of the blueprint.

REVIEW 5-3

The Percent Proportion

Percent is a ratio that compares a number to 100. A percent problem may be easier to solve if a proportion is used.

Percent Proportion	If P is the percentage, B is the base, and r is the percent, the percent proportion is $\frac{P}{B} = \frac{r}{100}$.

Examples

1 48 is what percent of 60?

$$\frac{\text{percentage}}{\text{base}} \begin{matrix}\rightarrow \\ \rightarrow\end{matrix} \frac{48}{60} = \frac{r}{100} \leftarrow \text{rate}$$

$$4800 = 60r$$

$$80 = r$$

48 is 80% of 60.

2 What number is 30% of 180?

$$\frac{\text{percentage}}{\text{base}} \begin{matrix}\rightarrow \\ \rightarrow\end{matrix} \frac{n}{180} = \frac{30}{100} \leftarrow \text{rate}$$

$$n = \frac{30}{100}(180)$$

$$n = 54$$

54 is 30% of 180.

Use the percent proportion to find each number.

1. Sixty is what percent of 80?
2. Twenty is what percent of 60?
3. What is 40% of 90?
4. Find 36% of 240.
5. Fifty-five is 44% of what number?
6. 8.4 is 12% of what number?
7. On Wednesday, Eric's Machine Shop received a shipment of 24 drill bits. Eric had ordered 40 drill bits. What percent of his order arrived on Wednesday?
8. Mackenzie received a commission of 6% on the sale of a house. If the amount of her commission was $8400, what was the selling price of the house?
9. According to the book *Are You Normal?*, 33% of Americans are afraid of snakes, 8% are afraid of thunder and lightning, 4% each are afraid of crowds and dogs, 2% of cats, 3% of driving a car or leaving the house, and $\frac{1}{3}$ are afraid of flying.
   a. In a group of 1025 people, about how many are afraid of snakes?
   b. About how many are afraid of crowds?
   c. About how many are afraid of flying?
10. In 1995, 351 million pairs of athletic shoes were purchased. Women accounted for 42% of those sales, and men for about 33% of those sales. About how many pairs of athletic shoes were purchased by women and by men?

REVIEW 5-4

The Percent Equation

The equation $P = RB$ is called the **percent equation.** In this equation, R is the **rate.** The rate is the decimal form of the percent. The percent equation can be used to solve any percent problem.

Examples

1 What is 20% of 30?

$P = RB$

$= 0.2(30)$

$= 6$

6 is 20% of 30.

2 2 is what percent of 16?

$P = RB$

$2 = R(16)$

$R = \frac{2}{16} = 0.125$ or 12.5%

2 is 12.5% of 16.

Use the percent equation to find each number.

1. What number is 50% of 120?
2. Find 25% of 48.
3. Find 15% of 12.
4. 8 is what percent of 24?
5. 40 is 40% of what number?
6. Find 150% of 76.
7. 200 is 80% of what number?
8. 45 is what percent of 50?
9. What number is 75% of 180?
10. Find 0.3% of 500.
11. **Jobs** Alissa is a busser at a restaurant. She makes 15% of all the tips her waiter earns. If her waiter earns $85 in tips, how much money will Alissa receive?
12. You can use the formula $I = prt$ to find the interest earned I after depositing an amount p into an account with an interest rate r after so many years t. Suppose you open an account with $1000 that pays an annual interest rate of 4%. After 2 years, how much interest will your account have earned?
13. **Sales** A sale is running where you can take 30% off the regular price of an item. For a particular shirt regularly priced at $28, there is a tag that says take an additional 10% off the sale price. How much money will you save if you buy the shirt?

Percent of Change

Some percent problems involve finding a percent of increase or decrease.

Percent of Increase	Percent of Decrease
A basketball that cost \$20 last year costs \$22 this year. The price increased by \$2 since last year. $\frac{\text{amount of increase}}{\text{original price}} \rightarrow \frac{2}{20} = \frac{r}{100}$ $200 = 20r$ $10 = r$ or $r = 10$ The percent of increase is 10%.	A jacket that originally cost \$80 is now on sale for \$60. $\frac{\text{amount of increase}}{\text{original price}} \rightarrow \frac{20}{80} = \frac{r}{100}$ $2000 = 80r$ $25 = r$ or $r = 25$ The percent of decrease is 25%.

The **sales tax** on a purchase is a percent of the regular price. To find the total price, you must calculate the amount of sales tax and add it to the regular price. **Discount** is the amount by which the regular price of a purchase is reduced. To find the reduced price, calculate the discount and subtract it from the regular price.

Find the final price of each item to the nearest cent. When there is a discount and sales tax, compute the discount first.

1. compact disc: \$12.00
   discount: 20%
2. two concert tickets: \$35.00
   student discount: 15%
3. airline ticket: \$348
   early booking discount: 30%
4. photo calendar: \$12.95
   sales tax: 6%
5. class ring: \$110
   group discount: 12%
   sales tax: 7%
6. multimedia software: \$49.95
   discount: 25%
   sales tax: 5%

Solve each problem. Round to the nearest tenth of a percent.

7. **Consumerism** According to the U.S. Bureau of Economic Analysis, Americans spent \$402.5 billion on recreation in 1995. In 1996, spending on recreation in the United States had grown to \$431.1 billion. What was the percent of increase in spending on recreation from 1995 to 1996?

8. **Agriculture** According to the U.S. Department of Agriculture, the number of farms in the United States in 1992 was 1,925,000. By 1997, this number had fallen to 1,912,000 farms. Find the percent of decrease in the number of farms from 1992 to 1997.

REVIEW 5-6

Probability and Odds

The **probability** of an event is a ratio that tells how likely it is that the event will take place.

Definition of Probability

$P(\text{event}) = \dfrac{\text{number of favorable outcomes}}{\text{number of possible outcomes}}$

Example 1

Ms. Michalski picks 8 of the 24 students in her class at random for a special project. What is the probability of being picked?

$$P(\text{being picked}) = \frac{\text{number of students picked}}{\text{total number of students}}$$

The probability of being picked is $\frac{8}{24}$ or $\frac{1}{3}$.

The probability of any event has a value from 0 to 1. If the probability of an event is 0, it is impossible for the event to occur. An event that is certain to occur has a probability of 1. This can be expressed as $0 \leq P(\text{event}) \leq 1$.

The odds of an event occurring is the ratio of the number of ways an event can occur (successes) to the number of ways the event cannot occur (failures).

Definition of Odds

$\text{Odds} = \dfrac{\text{number of favorable outcomes}}{\text{number of unfavorable outcomes}}$

Example 2

Find the odds that a member of Ms. Michalski's class will be picked for the special project.

Number of successes: 8 Number of failures: 16

Odds of being picked = number of successes : number of failures

= 8:16 or 1:2

Solve.

1. There are 3 brown puppies, 4 white puppies, and 5 spotted puppies in a pen. What is the probability of pulling out a white puppy at random?
2. It will rain 5 times in March and snow 7 times. The other days it will be sunny. What is the probability of sun? What are the odds of sun?

There are 300 freshmen, 250 sophomores, 200 juniors, and 250 seniors at a high school. Solve each problem.

3. If one student is chosen at random, what is the probability that a freshman will be chosen?
4. What would be the probability of choosing a freshman if all of the seniors were eliminated?
5. What are the odds of choosing a senior at random?
6. What are the odds that a junior will not be chosen?

Compound Events

A **compound event** consists of two or more simple events connected by the words *and* or *or.* Compound events connected by *and* are **independent events.** The outcome of one event does not effect the outcome of the other event.

Probability of Independent Events	$P(A \text{ and } B) = P(A) \cdot P(B)$

Example 1 **Two coins are tossed. Find the probability that they both land heads up.**

$P(\text{a head and a head}) = P(\text{a head}) \cdot P(\text{a head})$

$= \frac{1}{2} \cdot \frac{1}{2} \text{ or } \frac{1}{4}$

Compound events can also be connected by *or.* In this case, you may have **mutually exclusive** or **inclusive** events. Mutually exclusive events are two events that cannot occur at the same time. Yet, inclusive events can occur at the same time.

Probability of Mutually Exclusive Events	$P(A \text{ or } B) = P(A) + P(B)$
Probability of Inclusive Events	$P(A \text{ or } B) = P(A) + P(B) - P(A \text{ and } B)$

Example 2 **A six-sided die is tossed. What is the probability that the face shows an even number or a 2?**

Since an even number can also be a two, find the probability of two inclusive events.

$P(\text{even number or } 2)$

$= P(\text{even number}) + P(2) - P(\text{even number and } 2)$

$= \frac{3}{6} + \frac{1}{6} - \frac{1}{6} \text{ or } \frac{1}{2}$

A bag contains 3 white marbles, 3 red marbles, and 2 blue marbles. Find the probability for each situation.

1. P(drawing a white or red marble)
2. P(drawing a white or blue marble)
3. P(drawing 2 white marbles, with replacement)
4. P(drawing 1 marble of each color, with replacement)

CHAPTER 5

Test

Solve each proportion.

1. $\frac{4}{8} = \frac{6}{z}$

2. $\frac{t-1}{7} = \frac{6}{7}$

3. $\frac{5}{x+2} = \frac{25}{30}$

Find each number.

4. What number is 20% of 40?

5. Find 3% of 500.

6. 1 is what percent of 20?

7. 15 is 75% of what number?

8. 150% of 14 is what number?

9. 6 is what percent of 40?

10. 54.6 is what percent of 130?

11. What number is 72% of 600?

12. Maps The scale on a map is 1 inch = 15 miles. If the distance between two points on the map is 6 inches, what is the actual distance between the two points?

13. Banking Ms. Morrison learned that $28 was withdrawn from her checking account when she purchased a scarf in Paris for 126 francs. About how much money would have been withdrawn from her account if she spent 172 francs at a restaurant?

14. Surveys In a survey, 29% of respondents reported that they have returned rented videotapes to the store late. If 3200 people responded to the survey, how many returned a videotape late?

15. Shopping The regular cost of a paperback book is $6.95. If all paperbacks are on sale at a 25% discount, what is the sale cost of a paperback book?

16. Auto Maintenance The price of an oil change increased from $19.95 to $22.95. What was the percent of increase?

17. Odds If the probability in favor of an event is $\frac{8}{15}$, what are the odds of that event occurring?

18. Probability A box of sports cards contains 9 baseball cards, 7 football cards, and 4 basketball cards.

a. What is the probability that a baseball card is chosen at random?

b. What is the probability of drawing a football or basketball card?

c. What is the probability of drawing a football card, replacing it, and then drawing a basketball card?

REVIEW 6-1 Relations

A **relation** is a set of ordered pairs. The **domain** is the set of all first coordinates of the ordered pairs, and the **range** is the set of all second coordinates.

Examples

State the domain and range of each relation.

1. $\{(7, 1), (7, 3), (7, 5)\}$ Domain $= \{7\}$; Range $= \{1, 3, 5\}$
2. $\{(2, 4), (3, 6), (4, 4)\}$ Domain $= \{2, 3, 4\}$; Range $= \{4, 6\}$

Relations can be expressed as ordered pairs, tables, and graphs. The relation $\{(-3, 4), (0, 6), (2, -1)\}$ can be expressed in each of the following ways.

Ordered Pairs

$(-3, 4)$
$(0, 6)$
$(2, -1)$

Table

x	y
−3	4
0	6
2	−1

Graph

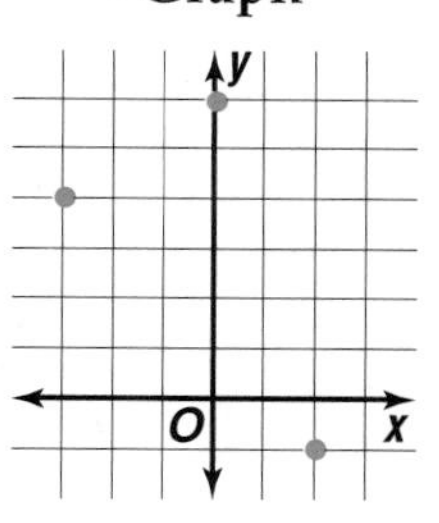

State the domain and range of each relation.

1. $\{(5, 4), (-2, 5), (-3, 0), (4, 5), (5, 0)\}$
2. $\{(1.25, -0.3), (-14, 12), (6, 1.25)\}$
3. $\left\{\left(\frac{1}{3}, \frac{3}{8}\right), \left(-\frac{7}{9}, 4\right), \left(2\frac{1}{5}, -\frac{1}{4}\right)\right\}$

Express each relation as a set of ordered pairs. Then state the domain and the range.

4.

x	y
3	−1
5	4
7	2

5. 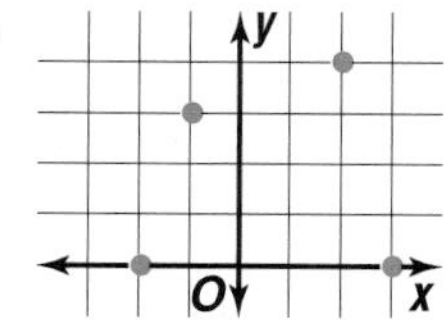

6. **Computers** The table gives the percentage of public schools in the United States that have computers capable of using CD-ROM software.

Percentage of Public Schools with CD-ROMs

Year	1995	1996	1997	1998
Percentage	41	54	54	74

a. Determine the domain and range of the relation.

b. Graph the data.

c. What conclusions might you make from the graph of the data?

REVIEW 6-2

Equations as Relations

An **equation in two variables** describes a relation since the solutions are ordered pairs.

Solution of an Equation in Two Variables	If a true statement results when the numbers in an ordered pair are substituted into an equation in two variables, then the ordered pair is a solution of an equation.

Example

Solve $y = -x + 4$ if the domain is $\{-4, -2, 6\}$.

Make a table. Substitute each value for x into the equation to determine the corresponding values of y.

x	$-x + 4$	y	(x, y)
-4	$-(-4) + 4$	8	$(-4, 8)$
-2	$-(-2) + 4$	6	$(-2, 6)$
6	$-(6) + 4$	-2	$(6, -2)$

The solution set is $\{(-4, 8), (-2, 6), (6, -2)\}$.

Which ordered pairs are solutions of each equation?

1. $a + 3b = 10$ a. $(3, 1)$ b. $(1, 3)$ c. $(10, 0)$ d. $(-2, 4)$
2. $5x + 2y = 18$ a. $(4, -1)$ b. $(3, 1)$ c. $(2, 4)$ d. $(1, 7)$

Solve each equation if the domain is $\{-2, -1, 0, 1, 2\}$.

3. $y = x - 3$
4. $y = 2x + 6$
5. $y = -2x + 7$
6. $3x + y = 1$
7. $2x - 9 = y$
8. $y = -\frac{x}{4} + 3$
9. **Cooking** For mashed potatoes, some cooks recommend using one more potato than the number of people to be served.
   a. Write an equation describing the number of potatoes to use when making mashed potatoes. Let x be the number of people to be served. Let y be the number of potatoes to use.
   b. Make a graph of the solution set if the domain is $\{3, 4, 5, 6\}$.

REVIEW 6-3

Graphing Linear Relations

An equation whose graph is a straight line is called a **linear equation**.

Linear Equation in Standard Form	A **linear equation** is an equation that can be written in the form $Ax + By = C$, where A, B, and C are any real numbers and A and B are not both zero. $Ax + By = C$ is called the standard form if A, B, and C are integers.

Graphing a Linear Equation

1. Solve the equation for one variable.
2. Make a table of values for the variables.
3. Graph the ordered pairs and connect them with a line.

Example

Graph $y + 2x = 2$.

$y + 2x = 2$

$y = 2 - 2x$

x	$2 - 2x$	y	(x, y)
−1	2 − 2(−1)	4	(−1, 4)
0	2 − 2(0)	2	(0, 2)
1	2 − 2(1)	0	(1, 0)

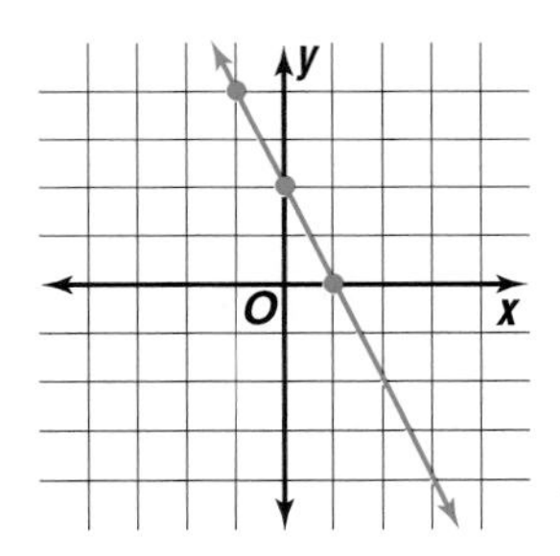

Determine whether each equation is a linear equation. If an equation is linear, identify *A*, *B*, and *C*.

1. $5x - 3y = 6$
2. $x^2 + 14 = 2y$
3. $9y = 3$

Graph each equation.

4. $3x + 3y = 9$
5. $x - y = 4$
6. $x + y = -3$
7. $2x - y = 4$
8. $x - y = -2$
9. $\frac{3}{4}x - \frac{1}{4}y = \frac{1}{8}$

10. **Employment** Maria Tomasso works as a sales representative. She receives a salary of \$2200 per month plus a 4% commission on monthly sales. She estimates that her sales in October, November, and December will be \$2000, \$3500, and \$2800.
    a. Graph ordered pairs that represent her incomes y for the three monthly sales figures x.
    b. Will her total monthly income be more than \$2300 in any of the three months? Explain.

REVIEW 6-4

Functions

A special type of relation is called a **function**.

Functions	A function is a relation in which each member of the domain is paired with *exactly* one element of the range.

Example 1 **Is {(4, 2), (7, 1), (5, −1), (3, 2)} a function?**

Since each element of the domain is paired with exactly one element of the range, the relation is a function.

The equation $y = 3x - 1$ can be written in **functional notation** as $f(x) = 3x - 1$. If $x = 4$, then $f(4) = 3(4) - 1$, or 11. Thus, $f(4)$, which is read "f of 4" is a way of referring to the value of y that corresponds to $x = 4$.

Example 2 **If $f(x) = 5x + 4$, find $f(2)$ and $f(-1)$.**

$$f(2) = 5(2) + 4 = 10 + 4 = 14$$

$$f(-1) = 5(-1) + 4 = -5 + 4 = -1$$

Determine whether each relation is a function.

1. 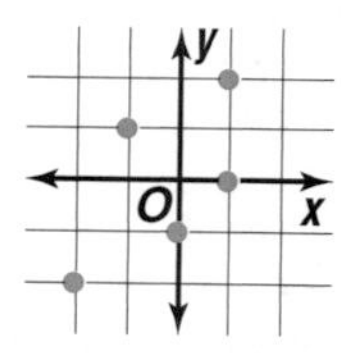

2. 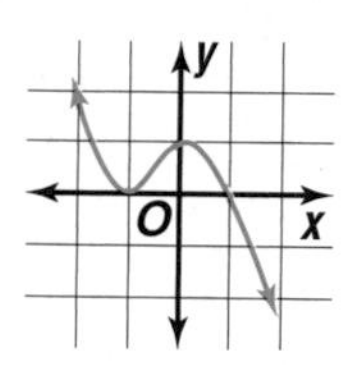

3. 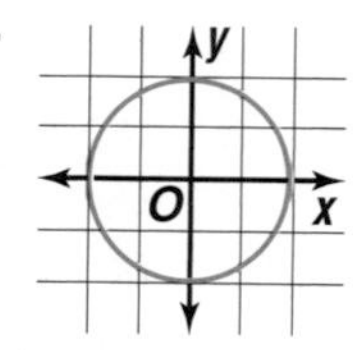

4. {(2, 1), (4, 4), (3, 1)}
5. {(7, 3), (6, 4), (7, 5)}
6. {(2, 1), (−2, 1), (3, 1)}
7. {(1, 3), (−1, 0), (3, 6)}
8. {(2, 6), (4, 4), (6, 2)}
9. {(3, 2), (0, 7), (−3, 2)}

If $f(x) = 6x + 4$ and $g(x) = 2x - 1$, find each value.

10. $f(2)$
11. $f(-3)$
12. $g(0)$
13. $g(2)$
14. $f(1)$
15. $g(-1)$

16. **Business** Viktor Alessandrovich owns a small ice cream parlor. He has noticed that his daily sales are dependent on the high temperature for the day. The formula for the relationship is $s = 50t + 1000$, where t represents the daily high temperature in degrees Fahrenheit and s is the amount of daily sales in dollars.
    a. Suppose the domain for the function includes the set {60, 70, 80, 90}. Make a table using these values for t and graph the function.
    b. Describe the graph of the function. What trends do you see?

REVIEW 6-5

Direct Variation

If two variables x and y are related by the equation $y = kx$, where k is a nonzero constant, then the equation is called a **direct variation**, and k is called the **constant of variation**. As values for x increase, values for y increase. Similarly, as values for x decrease, values for y decrease.

Example

If y varies directly as x, and $y = 24$ when $x = 6$, find y when $x = 4$.

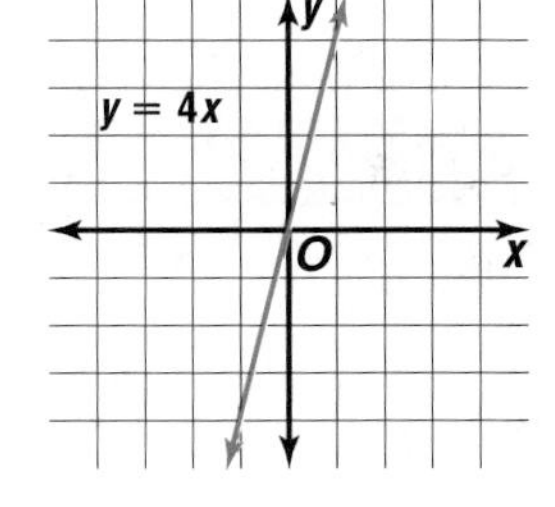

$y = kx$ *Definition of direct variation*

$24 = 6k$ *Replace y with 24 and k with 6.*

$\frac{24}{6} = \frac{6k}{6}$ *Divide each side by 6.*

$4 = k$

$y = kx$ *Definition of direct variation*

$= 4(4)$ *Replace k with 4 and x with 4.*

$= 16$

Notice that the graph of direct variation passes through the origin.

Determine whether each equation is a direct variation. Verify the answer with a graph.

1. $y = x$
2. $y = 2x$
3. $y = -4x + 1$
4. $y = -0.25x$
5. $y = -1$
6. $\frac{x}{y} = \frac{1}{3}$

Solve. Assume that y varies directly as x.

7. If $y = 15$ when $x = 3$, find y when $x = 8$.
8. If $y = -49$ when $x = 7$, find x when $y = 91$.
9. If $y = 0.3$ when $x = 2$, find y when $x = 2.5$.
10. If $y = \frac{4}{5}$ when $x = -\frac{3}{4}$, find x when $y = -\frac{7}{10}$.

11. **Space** The weight of an object on Mars varies directly as its weight on Earth. An unmanned probe that weighs 500 pounds on Earth weighs 190 pounds on Mars. In the future, NASA hopes to send a manned mission to Mars. How much will an astronaut with gear weighing 220 pounds on Earth weigh on Mars?

REVIEW 6-6

Inverse Variation

If two variables x and y are related by the equation $xy = k$, where k is a nonzero constant, then the equation is called an **inverse variation**, and k is called the **constant of variation**. You may see the equation written as $y = \frac{k}{x}$. As values for x increase, values for y decrease. Similarly, as values for x decrease, values for y increase.

Example

If y varies inversely as x, and $y = 8$ when $x = 9$, find x when $y = 24$.

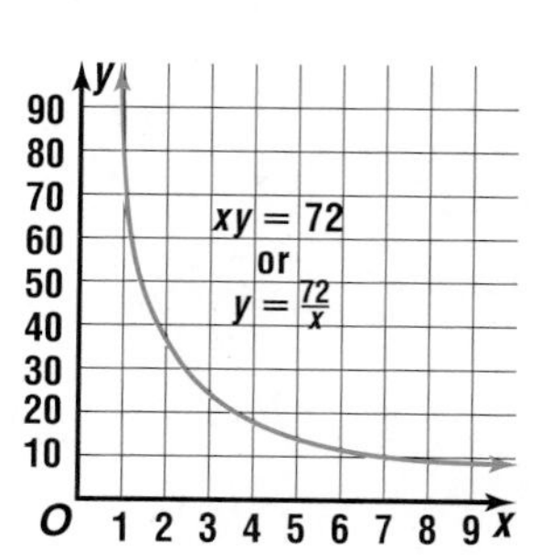

$xy = k$	*Definition of inverse variation*
$9(8) = k$	*Replace x with 9 and y with 8.*
$72 = k$	
$xy = k$	*Definition of inverse variation*
$24x = 72$	*Replace y with 24 and k with 72.*
$\frac{24x}{24} = \frac{72}{24}$	*Divide each side by 24.*
$x = 3$	

Find the constant of variation. Then write an equation for each statement.

1. y varies inversely as x, and $y = 12$ when $x = 3$.
2. y varies inversely as x, and $y = -5$ when $x = 7$.
3. y varies inversely as x, and $y = 0.2$ when $x = 10$.

Solve. Assume that y varies inversely as x.

4. If $y = 10$ when $x = 9$, find y when $x = 3$.
5. If $y = -6$ when $x = 14$, find y when $x = -4$.
6. If $y = 18.1$ when $x = 12.4$, find y when $x = 20$.
7. If $y = \frac{3}{8}$ and $x = \frac{1}{9}$, find y when $x = \frac{1}{6}$.

8. **Electricity** In the formula $V = IR$, where V is voltage from a power source, the current I varies inversely as the resistance R. For a particular light bulb, the resistance is 150 ohms while the current flowing through it is equal to 0.8 amperes. Suppose you would like to dim the light. If you increase the resistance to 200 ohms, find the measure of current.

CHAPTER 6 Test

Use the table to complete Exercises 1 and 2.

1. Write the relation as a set of ordered pairs.
2. State the domain and range of the relation.

x	y
3	12
0	5
−4	−8

3. Which ordered pairs are solutions of the equation $2x + 5y = -18$?
   a. $(-3, -2)$ b. $(10, -1)$ c. $(-9, 0)$ d. $(12, -2)$

Solve each equation if the domain is {−2, 0, 2, 4}.

4. $y = x - 6$
5. $y = 3x + 10$
6. $x - 2y = 8$

Graph each equation.

7. $-2x + y = 1$
8. $x + y = -6$
9. $4x - 2y = 0$

Determine whether each relation is a function.

10. 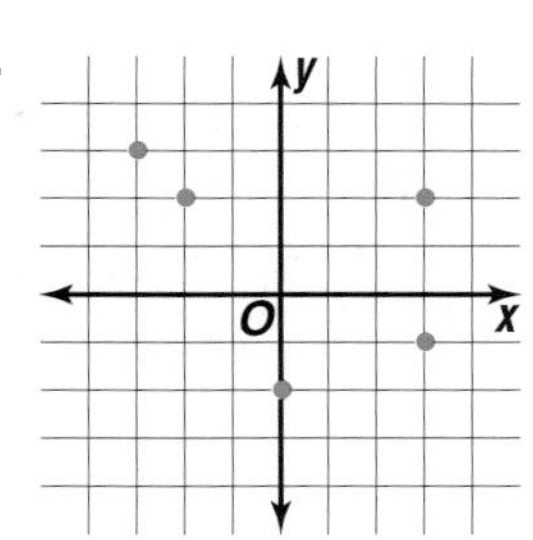

11. 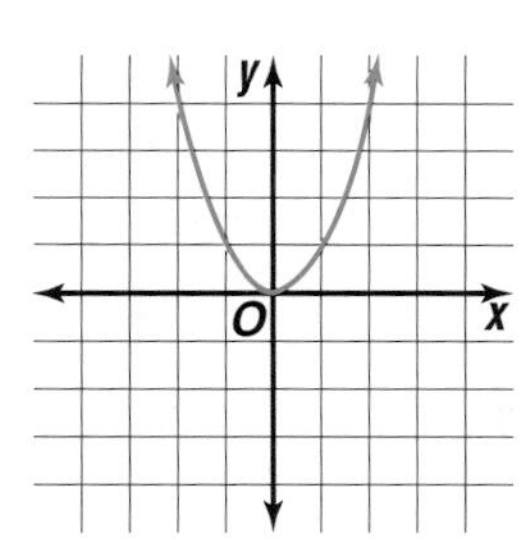

12. $\{(3, 4), (5, 5), (6, 4)\}$
13. $\{(1, 0), (2, 0), (5, 0)\}$
14. Given $h(x) = -6x$, find $h(-2)$.
15. Given $f(x) = x - 4$, find $f(-5)$.
16. Given $g(x) = 4x + 5$, find $g(8)$.
17. Given $f(x) = 7x + 12$, find $f(3)$.
18. If y varies directly as x and $y = 12$ when $x = 14$, find y when $x = 49$.
19. If y varies inversely as x and $y = 3$ when $x = 60$, find y when $x = 100$.
20. **Variation** The diameter of a circle of light y varies directly as the distance from the light source x. If you stand 2 meters from a wall and shine a flashlight against it, the circle of light on the wall measures 4 meters in diameter.
    a. What does the circle's diameter measure if you stand 3 meters from the wall? four meters? five meters?
    b. Write an equation that represents this direct variation.
    c. Draw a graph of the equation and describe the pattern that you see.

REVIEW 7-1

Slope

The ratio of *rise* to *run* is called **slope**. The slope of a line describes its steepness, or rate of change.

On a coordinate plane, a line extending from lower left to upper right has a positive slope. A line extending from upper left to lower right has a negative slope.

The slope of a horizontal line is zero. A vertical line has *no slope.* The slope of a nonvertical line can be determined from the coordinates of any two points on the line.

Slope

The slope of a line is the ratio of the change in the y-coordinates to the corresponding change in the x-coordinates.

$$\text{slope} = \frac{\text{change in } y}{\text{change in } x}$$

Determining Slope Given Two Points

Given the coordinates of two points (x_1, y_1) and (x_2, y_2) on a line, the slope m can be found as follows:

$$m = \frac{y_2 - y_1}{x_2 - x_1}, \text{ where } x_1 \neq x_2.$$

Example

Determine the slope of the line passing through points at (−3, 2) and (6, −4).

$$m = \frac{y_2 - y_1}{x_2 - x_1}$$
$$= \frac{-4 - 2}{6 - (-3)}$$
$$= \frac{-6}{9} = -\frac{2}{3}$$

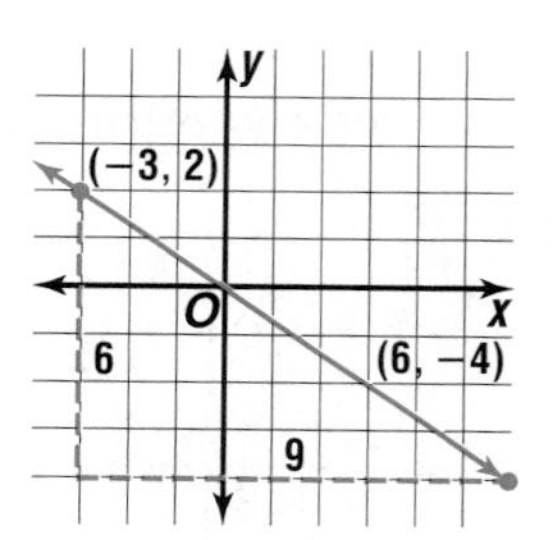

Determine the slope of the line passing through each pair of points whose coordinates are given.

1.
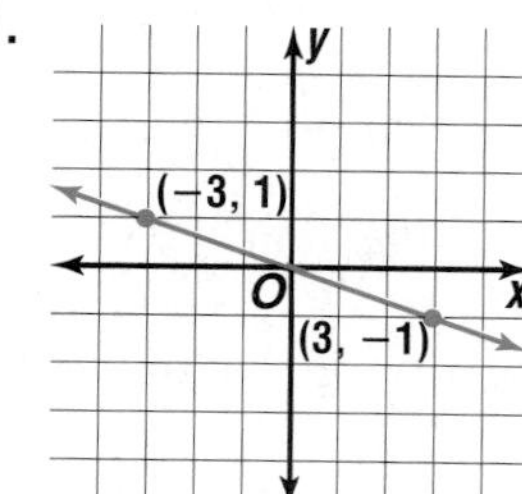

2.
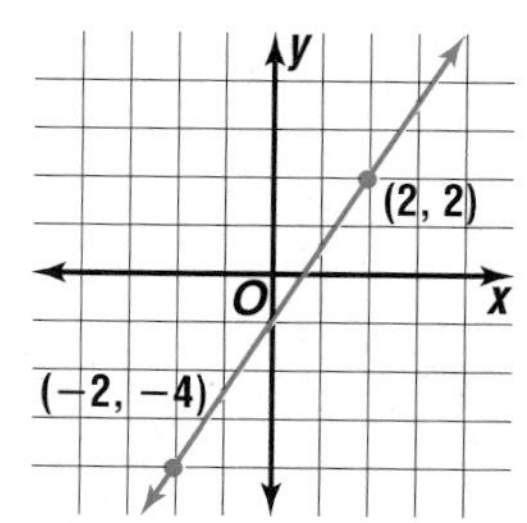

3.
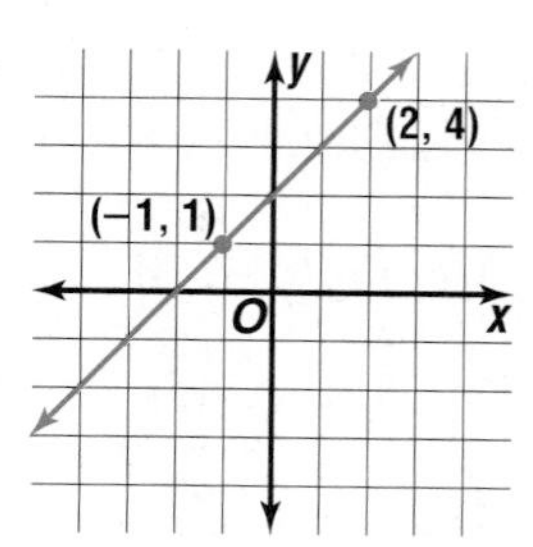

4. (3, 7), (4, 5)
5. (10, −2), (6, 3)
6. (3, 0), (9, 1)
7. (0, 5), (8, −9)
8. (−2, −2), (−1, 7)
9. (4, − 3), (−9, 0)

10. **Road Construction** A portion of the John Scott Memorial Highway in Steubenville, Ohio, has a grade (slope) of 12%. The length of this portion is approximately 1.1 miles. What is the change in elevation from the top of the grade to the bottom of the grade in feet?

Writing Equations in Point-Slope Form

If you know the slope of a line and the coordinates of one point on the line, you can write an equation of the line by using the **point-slope form**.

Point-Slope Form	Example
For a nonvertical line through the point at (x_1, y_1) with slope m, the point-slope form of a linear equation is $y - y_1 = m(x - x_1)$.	Given: point at $(-1, 3)$ and slope $m = 2$. $(y - 3) = 2(x - (-1))$ $(y - 3) = 2(x + 1)$

Example

Write the point-slope form of an equation of the line that passes through (−3, 8) and has a slope of $-\frac{2}{7}$.

$y - y_1 = m(x - x_1)$

$y - 8 = -\frac{2}{7}(x - (-3))$ or $y - 8 = -\frac{2}{7}(x + 3)$

You can also find an equation of a line if you know the coordinates of two points on the line. First, find the slope of the line. Then write an equation of the line by using the point-slope form.

Write the point-slopze form of an equation for each line passing through the given point and having the given slope.

1. $(5, 6), m = \frac{7}{10}$
2. $(-1, 0), m = 3$
3. $(7, -8), m = -3$
4. $(1, 4), m = -\frac{8}{9}$
5. $(6, -5), m = \frac{1}{6}$
6. $(9, 6), m = 0$

Write the point-slope form of an equation of the line passing through each pair of points.

7. $(7, 5), (-3, 2)$
8. $(-6, 8), (12, 4)$
9. $(4, 4), (2, -8)$
10. $(-9, -4), (4, -2)$
11. $(6, 7), (7, 6)$
12. $(-4, -9), (-6, -3)$

13. **Painting** Vanessa is painting a wall mural. To paint the perspective correctly, she must connect two points on the wall with a line. One point is two feet from the bottom of the painting, three feet from the left side of the painting, and has coordinates (3, 2). The other point is seven feet from the bottom of the painting and ten feet from the left of the painting.
    a. Write the point-slope form of an equation of the line that connects these two points.
    b. Vanessa has painted the line on her mural but is worried that the line isn't straight. The line she has painted goes through the point (6, 5). Has she painted the line correctly? Explain.

REVIEW 7-3

Writing Equations in Slope-Intercept Form

The x-coordinate of a point where a line crosses the x-axis is called the **x-intercept**. The y-coordinate of the point where the line crosses the y-axis is called the **y-intercept**.

You can always write an equation of a line if you know the slope and y-intercept.

Slope-Intercept Form	Model
Given the slope m and y-intercept b of a line, the slope-intercept form of an equation of the line is $y = mx + b$.	y-intercept; y; (0, b); O; x

Examples

Write an equation of the line in slope-intercept form for each situation.

1 $m = -2, b = 0$

$y = mx + b$

$y = -2(x) + 0$

$y = -2x$

2 $m = \frac{3}{2}, (-2, 5)$

$y - y_1 = m(x - x_1)$

$y - (5) = \frac{3}{2}(x - (-2))$

$y - 5 = \frac{3}{2}x + 3$

$y = \frac{3}{2}x + 8$

Write an equation of the line in slope-intercept form for each situation.

1. $m = 0, b = 5$
2. $m = 3, b = 1$
3. $m = -9, b = 8$
4. $m = 2, b = -7$
5. $m = \frac{1}{3}, b = -7$
6. $m = -1, (2, 4)$
7. $m = -\frac{5}{3}, (6, 0)$
8. $m = 0, (8, -5)$
9. $m = -2, \left(4, -\frac{7}{2}\right)$
10. $(-3, 0)$ and $(1, 0)$
11. $(-1, 4)$ and $(3, 2)$
12. **Temperature** Water freezes at 32°F and boils at 212°F. Likewise, water freezes at 0°C and boils at 100°C. Assume that temperature in Fahrenheit y is a linear function of temperature in degrees Celsius x.
    a. Write an equation in slope-intercept form for this function.
    b. What would the temperature be in degrees Fahrenheit if it is 25°C?

Scatter Plots

A **scatter plot** is a graph that shows the relationship between paired data. The scatter plot may reveal a pattern, or association, between the paired data.

- Data points that appear to go uphill show a *positive* relationship.
- Data points that appear to go downhill show a *negative* relationship.

The scatter plot at the right represents the relationship between the amount of time Anita spends on her Spanish homework each week and her score on her weekly Spanish quiz. Since the points appear to go uphill, there seems to be a positive relationship between the paired data. In general, the scatter plot seems to show that the more Anita studies, the better her quiz score.

Quiz Score vs. Study Time

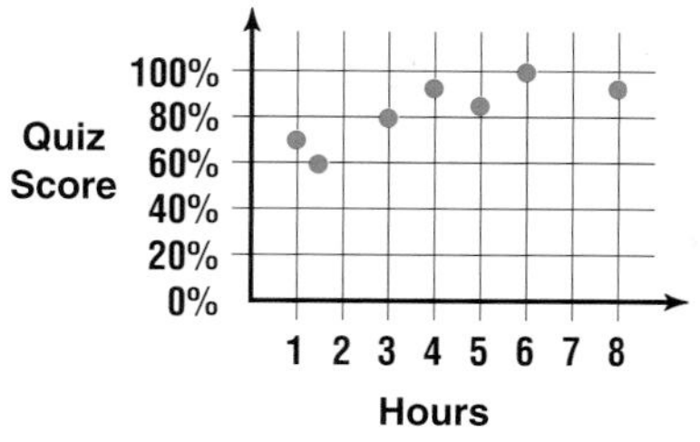

Solve.

1. The table at the right shows the gasoline/mileage record for a certain car. At each gasoline fill-up, the car's owner recorded the amount of gasoline used since the previous fill-up and the distance traveled on that amount of gasoline.
   a. Draw a scatter plot from the data in the table.
   b. What are the paired data?
   c. Is there a relationship between the gasoline used and the distance driven?

Gasoline/Mileage Record	
Gallons of Gasoline	Miles Traveled
9.8	250
7.9	200
9.2	240
8.0	210
7.5	185
9.3	255
9.5	250

2. Alex's biking speed after 10 minutes was 25 mi/h; at 30 minutes, 22 mi/h; at 45 minutes, 20 mi/h; and at 60 minutes, 19 mi/h.
   a. Make a scatter plot pairing time biked with biking speed.
   b. How is the data related, positively, negatively, or not at all?

3. The table shows the percent of children who are overweight ranked by the hours per day they spend watching television.
   a. Draw a scatter plot for these data.
   b. How is the data related, positively, negatively, or not at all?
   c. What conclusion might you draw from these data?

Hours of TV Per Day	Percent Overweight
1	12
2	23
3	28
4	30
5 or more	33

REVIEW 7-5

Graphing Linear Equations

There are three methods you can use for graphing equations. You can find two ordered pairs that satisfy the equation, the x- and y-intercepts, or the slope and y-intercept.

Examples

1 Graph $2x - 3y = 12$ by using the x- and y-intercepts.

To find the x-intercept, let $y = 0$.

$$2x - 3(0) = 12$$
$$2x = 12$$
$$x = 6$$

To find the y-intercept, let $x = 0$.

$$2(0) - 3y = 12$$
$$-3y = 12$$
$$y = -4$$

Thus, the graph contains the points (6, 0) and (0, −4).

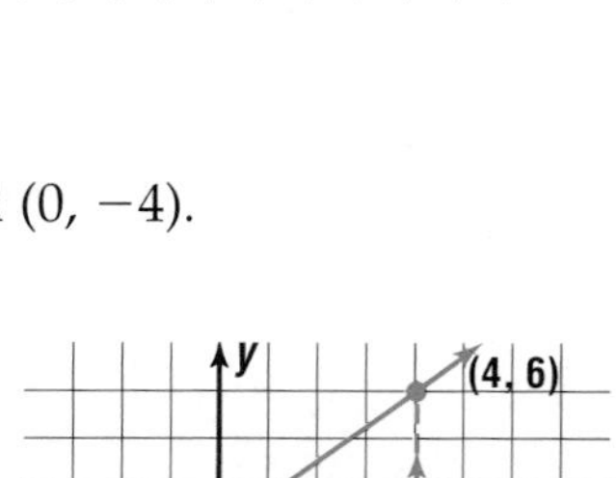

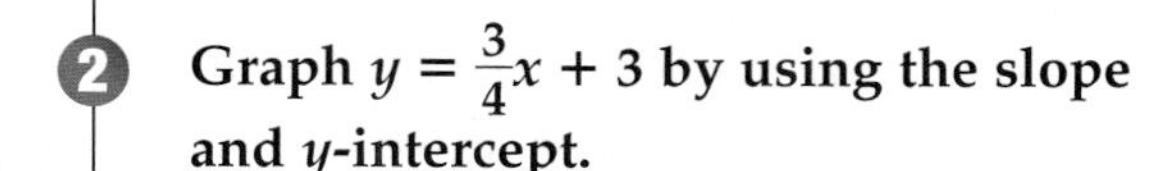

2 Graph $y = \frac{3}{4}x + 3$ by using the slope and y-intercept.

The y-intercept is 3, and the slope is $\frac{3}{4}$. Graph a point at (0, 3). From that point, move over 4 units and up 3 units to graph a second point at (4, 6).

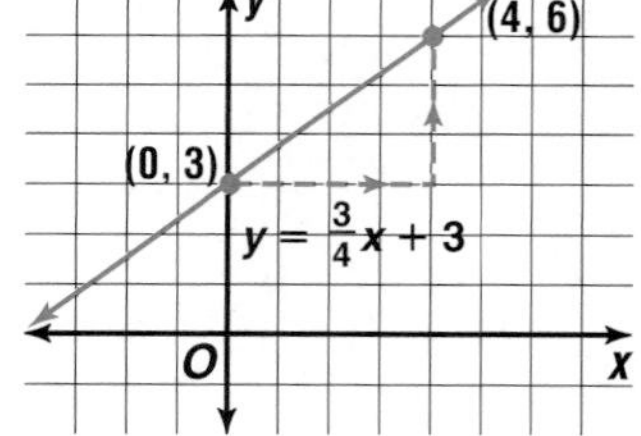

Determine the x-intercept and y-intercept of the graph of each equation. Then graph the equation.

1. $3x - 5y = 15$
2. $x + 7y = 14$

Determine the slope and y-intercept of the graph of each equation. Then graph the equation.

3. $y = \frac{1}{5}x + 2$
4. $y = \frac{3}{4}x - 3$

5. **Biology** As the temperature increases, the number of times a cricket chirps per minute also increases. The temperature can be approximated by the equation $T = \frac{1}{4}c + 40$, where c represents the number of cricket chirps per minute and T is the temperature in degrees Fahrenheit.
   a. Graph the equation.
   b. On a summer evening, you hear a cricket chirping outside your window. Suppose the cricket chirps 132 times per minute. What is the temperature outside?

Families of Linear Graphs

Families of graphs are graphs that have at least one characteristic in common. Families of linear graphs are those with the same slope or those with the same x- or y-intercept. These families of graphs are illustrated below.

same slope

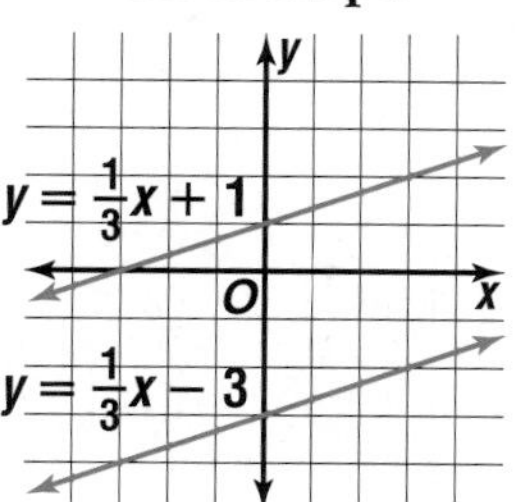

same y-intercept

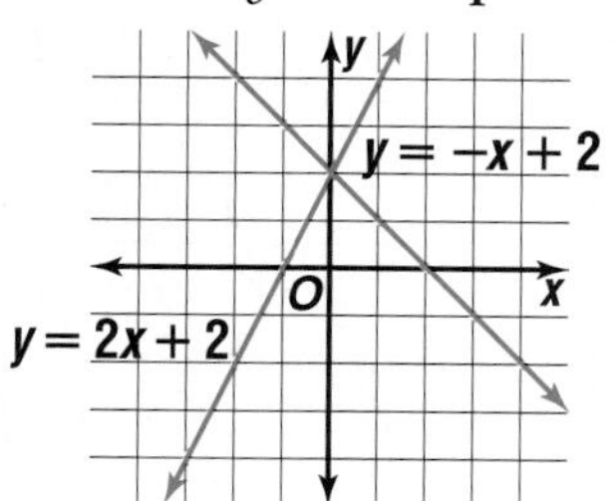

same x-intercept

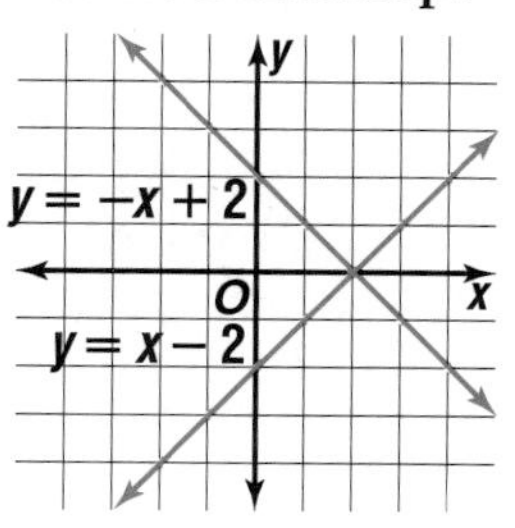

Example

Graph the pair of equations. Describe any similarities or differences. Explain why they are a family of graphs.

$y = \frac{5}{3}x + 3$

$y = -\frac{5}{3}x + 3$

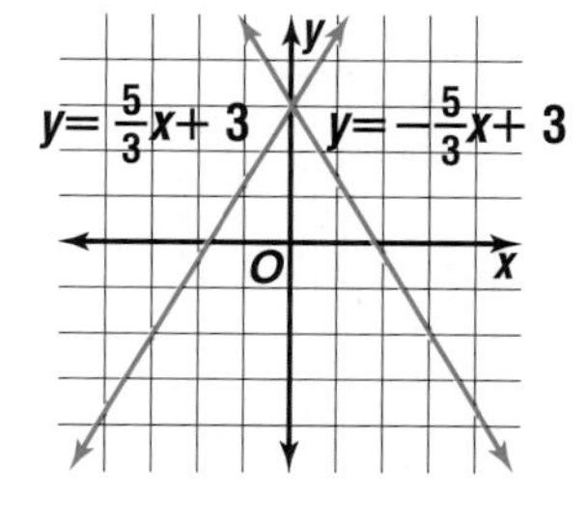

The graphs have slopes equal to $\frac{5}{3}$ and $-\frac{5}{3}$, respectively.

They are a family of graphs because the y-intercept of each line is 3.

Graph each pair of equations. Describe any similarities or differences. Explain why they are a family of graphs.

1. $y = 4x - 5$
   $y = 4x$
2. $y = 3x + 1$
   $y = 3x - 1$
3. $y = 2x - 3$
   $y = 3 + 2x$
4. $y = 3x + 2$
   $y = x + 2$
5. $y = -\frac{1}{3}x + 7$
   $y = -\frac{2}{3}x + 7$
6. $y = \frac{4}{5}x - 4$
   $y = -x + 5$

7. **Manufacturing** Companies often order supplies in bulk from vendors. The graph represents the cost of large sheets of plastic for two different vendors.
   a. Are these graphs a family of graphs? Explain.
   b. What does the slope represent?
   c. Which vendor's plastic sheets cost more? Explain.

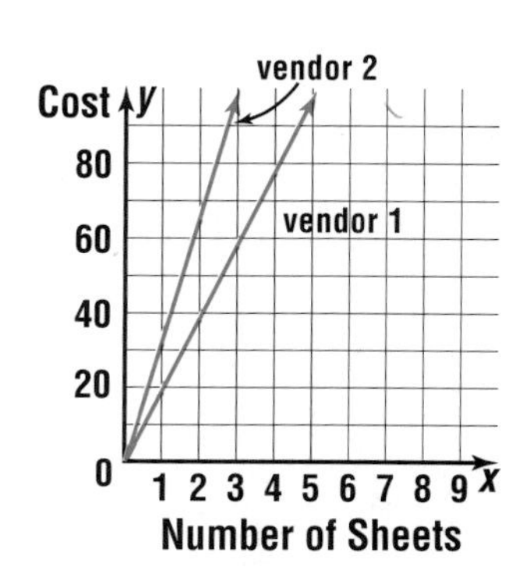

REVIEW 7-7

Parallel and Perpendicular Lines

Two special types of graphs are described below.

Parallel Lines	Perpendicular Lines
If two lines have the same slope, then they are parallel. $y = 2x + 3$; $y = 2x - 4$	If the products of the slopes of two lines is -1, then the lines are perpendicular. $y = 2x + 1$; $y = -\frac{1}{2}x - 3$

Example

Write an equation in slope-intercept form of the line that is parallel to the graph of $-2y = 6x + 1$ and passes through (3, 0).

Write $-2y = 6x + 1$ in slope-intercept form as $y = -3x - 0.5$.
The slope is -3. The slope of the equation of a parallel line will also be -3 and will go through the point at (3, 0).

$$y - y_1 = m(x - x_1)$$
$$y - (0) = -3(x - 3)$$
$$y = -3x + 9$$

Determine whether the graphs of each pair of equations are *parallel*, *perpendicular*, or *neither*.

1. $y = 2x + 3$
   $y = -2x + 3$
2. $y = x$
   $y = x - 5$
3. $y = 3x - 8$
   $3y = -x + 6$

Write an equation in slope-intercept form of the line that is parallel to the graph of each equation and passes through the given point. Repeat for a line that is perpendicular to the graph of each equation.

4. $y = 4x - 5$; (1, 6)
5. $y = 5x + 12$; (−6, 2)
6. $y = -\frac{1}{3}x + 7$; (8, 0)
7. $x + y = 0$; (3, −2)

8. **Transportation** In an aerial photograph, one rail in a set of railroad tracks has the equation $y = 0.75x + 3.5$. Find the equation of the other rail.

CHAPTER 7

Test

Determine the slope of the line that passes through each pair of points.

1. (6, 9), (1, 10)
2. (8, −3), (2, 5)
3. 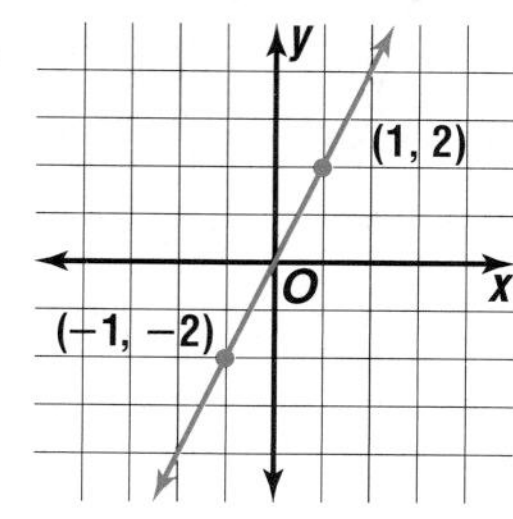

4. 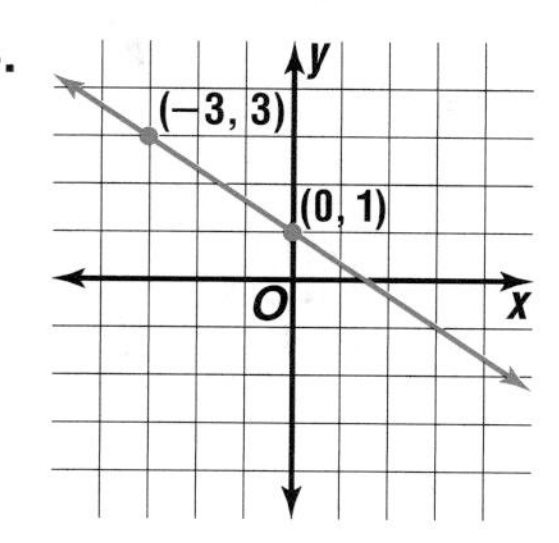

Write the point-slope form of an equation of the line satisfying the given conditions.

5. passes through (1, 3) and has a slope of $\frac{5}{8}$
6. passes through (12, 3) and has a slope of 0
7. passes through (4, 7) and (−3, 2)

The value of U.S. exports and imports (in billions of dollars) for selected years from 1970 to 1994 are listed in the table at the right. Use the data for Exercises 8 and 9.

8. Make a scatter plot of the data with the value of the exports on the horizontal axis and the value of the imports on the vertical axis.
9. Does the data have a *positive, negative,* or *no* correlation?

Year	Exports	Imports
1970	43	40
1975	108	99
1980	221	245
1985	213	345
1990	394	495
1995	585	743
1996	625	795
1997	689	871
1998	682	912

Source: U.S. International Trade Administration

10. Write an equation in slope-intercept form of a line that has a slope of −4 and y-intercept of 7.
11. Find the slope and y-intercept of the graph of $2x + 0.5y = 1$.
12. Find the x- and y-intercepts of the graph of $9x - 3y = 5$.
13. Graph $4x - y = 8$ by using the x- and y-intercepts.
14. Graph $y = \frac{3}{5}x - 4$ by using the slope and y-intercept.
15. Write an equation in slope-intercept form of the line that is parallel to the graph of $y = -\frac{3}{4}x + 6$ and passes through (−8, −5).
16. Write an equation in slope-intercept form of the line that is perpendicular to the graph of $y = 3x + 17$ and passes through (−7, 4).

Posttest

Circle the letter of the correct answer.

1. Write the expression *four less than the quotient of m and n.*

A $\frac{m}{n} - 4$

B $mn - 4$

C $4 - \frac{m}{n}$

D $4 - mn$

2. If $-3 - (-8) = y$, what is the value of y?

F -11

G -5

H 5

I 11

3. Find $-4\frac{3}{8} - 2\frac{3}{4}$.

A $-1\frac{3}{4}$

B $-6\frac{1}{2}$

C $-7\frac{1}{8}$

D $-8\frac{3}{8}$

4. A store's monthly CD sales are shown in the graph. If the average cost of a CD is $13, including tax, what would be a reasonable sales total for the month of August?

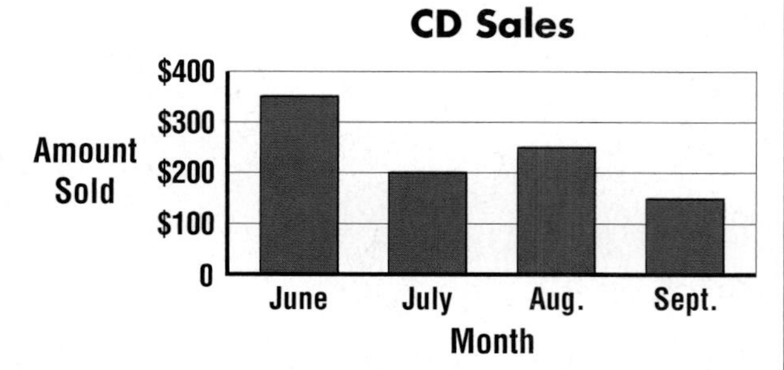

F $2000

G $2600

H $3200

I $4000

5. Which property is shown by $3(a + 6) = (3 \cdot a) + (3 \cdot 6)$?

A Associative Property of Addition

B Associative Property of Multiplication

C Commutative Property of Addition

D Distributive Property

6. Which ordered pair names point G?

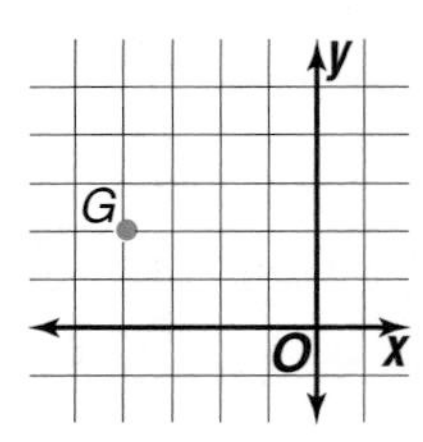

F $(-4, 2)$

G $(4, 2)$

H $(2, -4)$

I $(2, 4)$

7. Evaluate $3[-(12 \div 4)] + 7$.

A -5

B -2

C 2

D 16

8. Simplify $(-5a)(-3b)$.

F $-15ab$

G $-8ab$

H $8ab$

I $15ab$

9. Solve $-2.83 + z = 0.47$.

A 3.3

B 3.2

C 2.3

D -2.36

10. Solve $|x| + 6 = 3$.

F $\{3, -3\}$

G $\{3\}$

H $\{-3\}$

I $\varnothing$

Posttest

11. The table lists the states having the most Alpine ski areas in operation. What is the median of the data?

State	Amount
New York	54
Pennsylvania	34
California	32
Michigan	40
Wisconsin	38

A 38
B 39.6
C 40
D 54

12. Find the value of n in the equation $-28 = \left(-3\frac{1}{2}\right)n$.

F -8
G 2
H 6
I 8

13. Julie and her three friends purchased $2\frac{2}{3}$ pounds of nuts. How much will each person get if they split the nuts equally?

A $\frac{8}{9}$ lb
B $\frac{3}{4}$ lb
C $\frac{2}{3}$ lb
D $\frac{1}{3}$ lb

14. What is the solution of $4(2.5 - n) + 6 = 5n - n$?

F -2
G 2
H 4
I no solution

15. In 10 seconds, light travels about 1.86 million miles. The distance from the sun to Mercury is about 36 million miles. Which proportion could be used to find t, the time it takes for light to travel from the sun to Mercury?

A $\frac{1.86}{10} = \frac{36}{t}$
B $\frac{1.86}{36} = \frac{t}{10}$
C $\frac{1}{t} = \frac{1.86}{36}$
D $\frac{t}{36} = \frac{1.86}{10}$

16. Suppose one disk is chosen at random from the disks shown. What is the probability of selecting an even number greater than 5?

1 2 3 4 5 6
7 8 9 10 11 12

F $\frac{1}{4}$
G $\frac{1}{3}$
H $\frac{1}{2}$
I $\frac{7}{12}$

17. What is the domain of the relation shown in the table?

x	y
-1	5
2	-2
8	1
-3	-1
0	3

A $\{-1, 2, 1, -1, 3\}$
B $\{5, -2, 1, -1, 3\}$
C $\{-1, 2, 8, -3, 0\}$
D $\{-1, 2, 8, -3\}$

Posttest

18. 72 is 18% of what number?

F 250

G 400

H 1296

I 4000

19. Between 1987 and 1997, the number of students taking a college-prep course increased from 38% to 61%. Find the percent of increase to the nearest percent.

A 58%

B 61%

C 62%

D 165%

20. Which relation is *not* a function?

F

x	6	−1	0	6	1
y	4	0	5	4	3

G

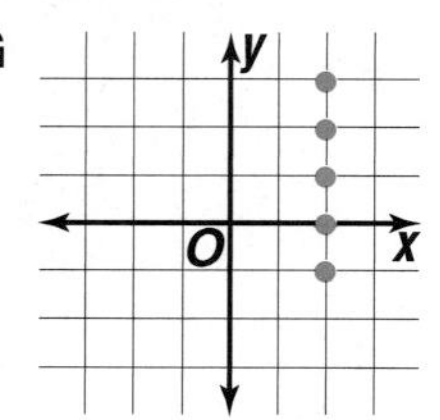

H {(1, 1), (2, 2), (3, 3), (4, 4)}

I

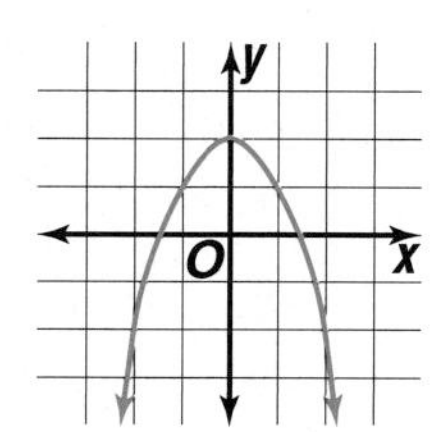

21. Suppose y varies inversely as x. If $y = 16$ when $x = -6$, what is x when $y = 24$?

A −14

B −9

C −4

D 4

22. Find the x-intercept for $3x - 2y = 8$.

F (0, 0)

G $\left(2\frac{2}{3}, 0\right)$

H (0, −4)

I $\left(-2\frac{1}{3}, 0\right)$

23. Find the slope of the line that passes through (4, 2) and (−4, 4).

A −4

B $-\frac{1}{4}$

C $\frac{1}{4}$

D 4

24. Find the equation of the line through (−6, 3) with slope $\frac{2}{3}$.

F $y = \frac{2}{3}x - 8$

G $y = \frac{2}{3}x + 4$

H $y = \frac{2}{3}x + 7$

I $y = \frac{2}{3}x - 1$

25. What is the slope of a line that is perpendicular to the graph of $3x + 7y = 4$?

A $-\frac{7}{3}$

B $\frac{1}{3}$

C $\frac{3}{7}$

D $\frac{7}{3}$

Graphing Calculator Quick Reference Guide

Throughout this text, **Graphing Calculator Explorations** have been included so you can use technology to solve problems. These activities use the TI–83 Plus graphing calculator. Graphing calculators have a wide variety of applications and features. If you are just beginning to use a graphing calculator, you will not need to use all of its features. This page is designed to be a quick reference for the features you will need to use as you study from this text.

To darken or lighten the screen:	2nd ▲ or 2nd ▼
To clear an entry:	CLEAR
To get to the home screen:	2nd [QUIT]
To recall an entry:	2nd [ENTRY]
To recall an answer:	2nd [ANS]
To turn the calculator off:	2nd [OFF]

Task	Keystrokes
Using tables	2nd [TABLE]
Use lists	STAT ENTER
Find the mean and median of listed data	STAT ▶ ENTER ENTER
Solve equations	MATH 0
Set the viewing window	WINDOW
Plot points	2nd [DRAW] ▶ ENTER
Enter an equation	Y=
Enter an inequality symbol	2nd [TEST]
Graph an equation	GRAPH
Zoom in	ZOOM 2
Zoom out	ZOOM 3
Graph in the viewing window x: [−10, 10] and y: [−10, 10]	ZOOM 6
Graph an equation with integer coordinates	ZOOM 8
Place a statistical graph in a good viewing window	ZOOM 9
Trace a graph	TRACE
Find the intersection of two graphs	2nd [CALC] 5
Shade an inequality	2nd [DRAW] 7
Enter a program	PRGM ▶ ▶ ENTER

CHAPTER 8 Powers and Roots

Pages 2–333 are contained in Volume One.

Make this Foldable to help you organize information about the material in this chapter. Begin with four sheets of grid paper.

1. **Fold** each sheet of grid paper in half along the width. Then cut along the crease.

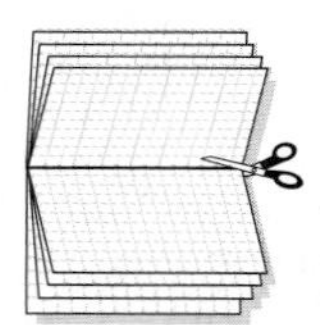

2. **Staple** the eight half-sheets together to form a booklet.

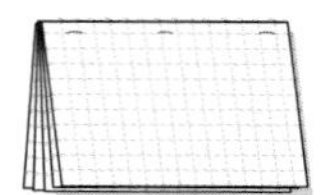

3. **Cut** seven lines from the bottom of the top sheet, six lines from the second sheet, and so on.

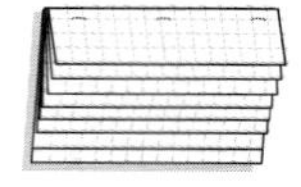

4. **Label** the tabs with lesson topics as shown.

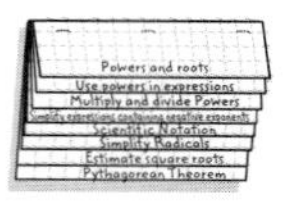

Reading and Writing As you read and study the chapter, use each page to write notes and to graph examples for each lesson.

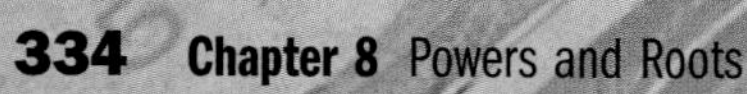

Problem-Solving Workshop

Project

Winners of the *Super Sweepstakes Contest*, a fictitious contest, can choose between two prizes. The first choice is $1 million each year for 25 years. The second choice is $1 the first year, $2 the second year, $4 the third year, $8 the fourth year, and so on, for 25 years. At the end of 25 years, which prize awards more money?

Working on the Project

Work with a partner and choose a strategy to help analyze and solve the problem. Here are some questions to help you get started.

- For each prize, how much money is awarded each year for the first five years?
- For each prize, what is the total amount of money awarded at the end of five years?

Strategies

Look for a pattern.
Draw a diagram.
Make a table.
Work backward.
Use an equation.
Make a graph.
Guess and check.

Technology Tools

- Use a **spreadsheet** to perform the calculations.
- Use a **graphing calculator** or **graphing software** to show the data in a graph.
- Use **word processing software** to write a report about your solution.

*inter*NET CONNECTION **Research** For more information about linear and exponential functions, visit: www.algconcepts.com

Presenting the Project

Write a report that includes any spreadsheets or graphs that you have made. As part of your project, answer the following questions.

- At the end of 25 years, what is the difference in the total amount of money awarded?
- Suppose you are the prize winner. Which prize would you choose? Explain your reasoning.
- Suppose you are the sponsor of the *Super Sweepstakes Contest*. Which prize would you prefer the winner to choose? Explain your reasoning.

8–1 Powers and Exponents

What You'll Learn
You'll learn to use powers in expressions.

Why It's Important
Landscaping Landscape architects use the formula $A = \pi r^2$ to find the area of a circular region. The 2 in the formula is an exponent. *See Exercise 43.*

Perfect squares like 1, 4, 9, and 16 can be represented by a square array of dots.

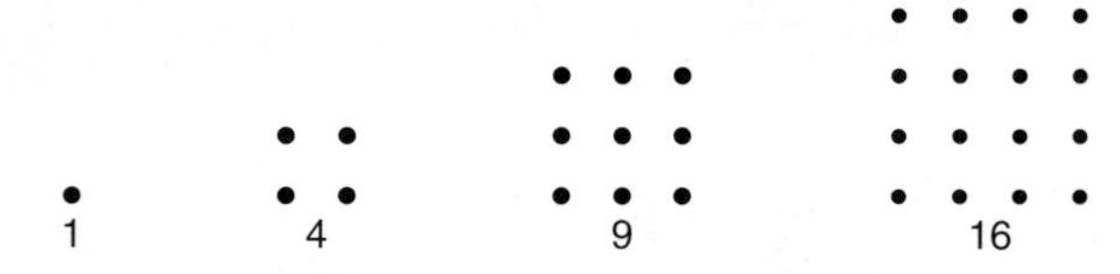

A perfect square is the product of a number and itself. For example, 16 is a perfect square because $16 = 4 \times 4$. The expression 4×4 can be written using exponents. An **exponent** tells how many times a number, called the **base**, is used as a factor. Numbers that are expressed using exponents are called **powers**. The expression 4×4 can be written as 4^2.

base $\rightarrow 4^2 \leftarrow$ *exponent*

Symbols	Words	Meaning
4^1	4 to the first power	4
4^2	4 to the second power or 4 squared	$4 \cdot 4$
4^3	4 to the third power or 4 cubed	$4 \cdot 4 \cdot 4$
4^4	4 to the fourth power	$4 \cdot 4 \cdot 4 \cdot 4$
4^n	4 to the *n*th power	$\underbrace{4 \cdot 4 \cdot 4 \cdot \ldots \cdot 4}_{n \text{ factors}}$

Examples

Write each expression using exponents.

1 $2 \cdot 2 \cdot 2 \cdot 2 \cdot 2$

The base is 2. It is a factor 5 times.
$2 \cdot 2 \cdot 2 \cdot 2 \cdot 2 = 2^5$

2 $m \cdot m \cdot m \cdot m$

The base is m. It is a factor 4 times.
$m \cdot m \cdot m \cdot m = m^4$

When no exponent is shown, it is understood to be 1. For example, $10 = 10^1$.

3 7

The base is 7. It is a factor 1 time.
$7 = 7^1$

Your Turn

a. $4 \cdot 4 \cdot 4 \cdot 4$ b. $x \cdot x \cdot x$ c. 10

Example 4 Write $(2)(2)(2)(-5)(-5)$ using exponents.

Use the Associative Property to group the factors with like bases.

$$(2)(2)(2)(-5)(-5) = [(2)(2)(2)][(-5)(-5)]$$
$$= (2)^3(-5)^2$$

Your Turn

d. Write $(-1)(-1)(-1)(-1)(3)(3)$ using exponents.

You can use the definition of exponent to write a power as a multiplication expression.

Examples Write each power as a multiplication expression.

5 10^2

The base is 10. The exponent 2 means that 10 is a factor 2 times.
$10^2 = 10 \cdot 10$

6 b^3

The base is b. The exponent 3 means that b is a factor 3 times.
$b^3 = b \cdot b \cdot b$

7 $10x^2y^4$

10 is used as a factor once, x is used twice, and y is used 4 times.
$10x^2y^4 = 10 \cdot x \cdot x \cdot y \cdot y \cdot y \cdot y$

Your Turn

e. 2^3 f. y^4 g. $5a^3b^2$

You can also use the definition of exponent to evaluate expressions.

Real World **Example 8** Science Link

The distance between the Earth and the Sun is about 10^8 kilometers. Write this number as a multiplication expression and then evaluate the expression.

$$10^8 = 10 \cdot 10 \cdot 10 \cdot 10 \cdot 10 \cdot 10 \cdot 10 \cdot 10$$
$$= 100{,}000{,}000$$

The distance between Earth and the sun is about 100 million kilometers.

When an expression contains an exponent, simplify the expression using the rules for order of operations.

Order of Operations: Lesson 1–2

Order of Operations

1. Do all operations within grouping symbols first; start with the innermost grouping symbols.
2. Evaluate all powers in order from left to right.
3. Do all multiplications and divisions from left to right.
4. Do all additions and subtractions from left to right.

In any expression, an exponent goes with the number or the quantity in parentheses that immediately precedes it.

$2 \cdot 5^3$ means $2 \cdot 5 \cdot 5 \cdot 5$ *The exponent 3 goes with the 5.*

$(2 \cdot 5)^3$ means $(2 \cdot 5)\,(2 \cdot 5)\,(2 \cdot 5)$ *The exponent goes with $(2 \cdot 5)$.*

Examples

Evaluate each expression.

9 $4m^3$ **if** $m = 2$

$4m^3 = 4(2)^3$ *Replace m with 2.*

$= 4(8)$ *Evaluate the power: $2 \cdot 2 \cdot 2 = 8$.*

$= 32$ *Multiply.*

10 $3x + y^2$ **if** $x = -2$ **and** $y = -3$

$3x + y^2 = 3(-2) + (-3)^2$ *Replace x with −2 and y with −3.*

$= 3(-2) + (9)$ *$(-3)^2 = (-3)(-3)$ or 9*

$= (-6) + (9)$ *Multiply.*

$= 3$ *Add.*

Your Turn

h. $3a^3$ if $a = -2$

i. $-5(m + n)^2$ if $m = 4$ and $n = 2$

In the following activity, you will use exponents to find perimeters and areas of squares.

Graphing Calculator Exploration

Graphing Calculator Tutorial
See pp. 724–727.

The pattern at the right contains perfect squares.

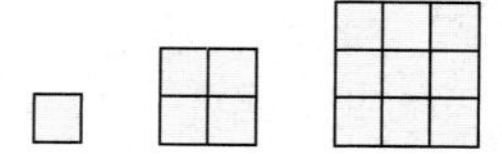

Step 1 Draw the next three squares in the pattern.

Step 2 Find the perimeter and area of each square. Organize your data in a table.

Length of Side	Perimeter	Area

Try These

1. Write an equation that shows the relationship between the length of a side of a square x and its perimeter y.
2. Write an equation that shows the relationship between the length of a side of a square x and its area y.
3. Graph the two equations. Compare and contrast the graphs.
4. When is the value of the perimeter greater than the value of the area? When is the value of the perimeter equal to the area?
5. If the length of each side of a square is doubled, how does its perimeter change? How does its area change?

Check for Understanding

Communicating Mathematics

1. **Write** a definition of *perfect square.*
2. **Explain** what the 2 represents in 10^2.
3. **You Decide?** Jan thinks that $(6n)^3$ is equal to $6n^3$. Becky thinks they are not equal. Who is correct? Explain your reasoning.

Vocabulary

perfect squares
exponent
base
powers

Guided Practice

Write each expression using exponents. *(Examples 1–3)*

4. $9 \cdot 9 \cdot 9 \cdot 9$
5. $a \cdot a \cdot a \cdot a \cdot a$
6. 3

Write each power as a multiplication expression. *(Examples 5–7)*

7. 12^4
8. x^5
9. m^4n^3

Evaluate each expression if $a = 3$, $b = -2$, and $c = 4$. *(Examples 8–10)*

10. c^3
11. $2a^4$
12. $3a^2b$

13. **Number Theory** The prime factorization of 360 is $2 \cdot 2 \cdot 2 \cdot 3 \cdot 3 \cdot 5$. Write the prime factorization using exponents. *(Example 4)*

Exercises

Practice

Write each expression using exponents.

14. $10 \cdot 10 \cdot 10$
15. $(-2)(-2)(-2)(-2)$
16. 6
17. 7 cubed
18. $4 \cdot 4 \cdot 4 \cdot 6 \cdot 6$
19. $2 \cdot 3 \cdot 5 \cdot 2 \cdot 3 \cdot 3$
20. $a \cdot a \cdot a \cdot a \cdot b \cdot b$
21. $3 \cdot x \cdot x \cdot y \cdot y$
22. $(-5)(m)(m)(m)(n)$

Homework Help	
For Exercises	See Examples
14–22	1–4
23–30	5–7
31–41, 43	9, 10
Extra Practice	
See page 707.	

Write each power as a multiplication expression.

23. 3^5
24. $(-2)^2$
25. $2^4 \cdot 3^2$
26. $2 \cdot 3^5$
27. y^3
28. x^2y^2
29. $6ab^4$
30. $-2y^4$

Evaluate each expression if $x = -2$, $y = 3$, $z = -1$, and $w = 0.5$.

31. x^5
32. $4y^2$
33. $-2x^6$
34. $x^2 - y^2$
35. $3(y^2 + z)$
36. $-2(x^3 + 1)$
37. $2w^2$
38. wx^3y

39. Find the value of $x^2 + 2x + 1$ if $x = -3$.
40. Which is greater, 2^5 or 5^2?

Applications and Problem Solving

41. **Number Theory** The prime factorization of a number is $2 \cdot 3^5$. Find the number.
42. **Geometry** Use exponents to write an expression that represents the total number of unit cubes in the large cube. Then evaluate the expression.

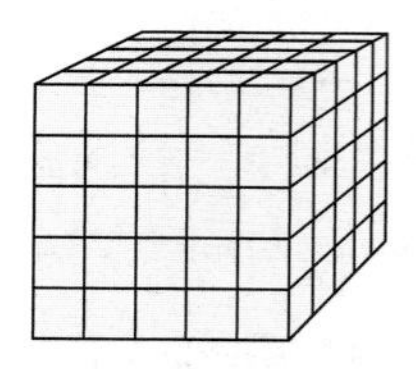

Exercise 42

43. **Landscaping** Landscape architects use the formula $A = \pi r^2$ to find the area of circular flower beds. In the formula, $\pi \approx 3.14$, and r is the radius of the circle.
    a. Estimate the area of a circular flower bed with a radius of 8 feet.
    b. About how many bags of mulch will the landscape architect need to cover the bed if each bag covers about 10 square feet?
44. **Critical Thinking** Suppose you raise a negative integer to a positive power. When is the result negative? When is the result positive?

Mixed Review

45. Write an equation of the line that is parallel to the graph of $y = -2x + 3$ and passes through (1, 4). *(Lesson 7–7)*

Graph each equation using the slope and y-intercept. *(Lesson 7–6)*

46. $y = 3x - 2$
47. $y = x + 2$
48. $-x + 2y = 8$

Solve each problem. *(Lesson 5–4)*

49. Find 26% of 120.
50. 17 is 40% of what number?
51. 9 is what percent of 18?
52. 98% of 40 is what number?

Standardized Test Practice

A B C D

53. **Multiple Choice** A mechanic charges an initial fee of \$40 plus \$30 for each hour she works. Which equation represents the cost c of a repair job that lasts h hours? *(Lesson 4–5)*

    A $c = 30 + 40h$
    B $c = 40 + 30h$
    C $c = 40 - 30h$
    D $c = 30 - 40h$

www.algconcepts.com/self_check_quiz

8–2 Multiplying and Dividing Powers

What You'll Learn
You'll learn to multiply and divide powers.

Why It's Important
Movie Industry
The intensity of sound is measured using *decibels*, a unit that is based on powers of ten. *See page 346.*

Powers can be multiplied and divided. In the example below, we will use powers of 2 to form a rule about multiplying exponents. The table shows several powers of 2 and their values.

Power of 2	2^1	2^2	2^3	2^4	2^5	2^6
Value	2	4	8	16	32	64

You can use the table to substitute exponents for the factors of multiplication equations. What do you notice about the exponents in the following products?

Numerical Products	$4 \cdot 2 = 8$	$4 \cdot 8 = 32$	$8 \cdot 8 = 64$
Products of Powers	$2^2 \cdot 2^1 = 2^3$	$2^2 \cdot 2^3 = 2^5$	$2^3 \cdot 2^3 = 2^6$

These examples suggest that you can multiply powers with the same base by adding the exponents. Think about $a^2 \cdot a^3$.

$a^2 \cdot a^3 = (a \cdot a)(a \cdot a \cdot a)$ *a^2 has two factors; a^3 has three factors.*

$= a \cdot a \cdot a \cdot a \cdot a$ *Substitution Property*

$= a^5$ *The product has 2 + 3 or 5 factors.*

Product of Powers

Words:	You can multiply powers with the same base by adding the exponents.
Numbers:	$3^3 \cdot 3^2 = 3^{3+2}$ or 3^5
Symbols:	$a^m \cdot a^n = a^{m+n}$

Examples

An expression is *simplified* when each base appears only once and all of the fractions are in simplest form.

Simplify each expression.

1 $4^3 \cdot 4^5$

$4^3 \cdot 4^5 = 4^{3+5}$ *To multiply powers that have the same base, write the common base, then add the exponents.*
$= 4^8$

2 $x^3 \cdot x^4$

$x^3 \cdot x^4 = x^{3+4}$ *To multiply powers that have the same base, write the common base, then add the exponents.*
$= x^7$

3 $(4y^2)(3y)$

$$
\begin{aligned}
(4y^2)(3y) &= (4 \cdot 3)(y^2 \cdot y) && \textit{Use the Commutative and Associative Properties.}\\
&= 12y^{2+1} && y = y^1\\
&= 12y^3
\end{aligned}
$$

4 $(a^3b^2)(a^2b^4)$

$$
\begin{aligned}
(a^3b^2)(a^2b^4) &= (a^3 \cdot a^2)(b^2 \cdot b^4) && \textit{Use the Commutative and Associative Properties.}\\
&= a^{3+2} \cdot b^{2+4}\\
&= a^5b^6
\end{aligned}
$$

Your Turn

a. $10^4 \cdot 10^2$ **b.** $y^4 \cdot y^2$ **c.** $(-3x^2)(5x)$ **d.** $(x^5y^2)(x^4y^6)$

You can use powers of 2 to help find a rule for dividing powers. Study each quotient in the table. What do you notice about the exponents?

Numerical Quotients	$16 \div 8 = 2$	$32 \div 4 = 8$	$64 \div 2 = 32$
Quotient of Powers	$2^4 \div 2^3 = 2^1$	$2^5 \div 2^2 = 2^3$	$2^6 \div 2^1 = 2^5$

These examples suggest that you can divide powers with the same base by subtracting the exponents. Think about $a^5 \div a^2$. Remember that you can write a division expression as a fraction.

$$
\begin{aligned}
\frac{a^5}{a^2} &= \frac{a \cdot a \cdot a \cdot a \cdot a}{a \cdot a} && \textit{a^5 has five factors; a^2 has two factors.}\\
&= \frac{\overset{1}{\cancel{a}} \cdot \overset{1}{\cancel{a}} \cdot a \cdot a \cdot a}{\underset{1}{\cancel{a}} \cdot \underset{1}{\cancel{a}}} && \textit{Notice that } \frac{a \cdot a}{a \cdot a} = 1.\\
&= a \cdot a \cdot a && \textit{The quotient has 5 − 2 or 3 factors.}\\
&= a^3
\end{aligned}
$$

Quotient of Powers		
	Words:	You can divide powers with the same base by subtracting the exponents.
	Numbers:	$\frac{5^7}{5^4} = 5^{7-4}$ or 5^3
	Symbols:	$\frac{a^m}{a^n} = a^{m-n}$ *The value of a cannot be zero.*

www.algconcepts.com/extra_examples

Examples

Simplify each expression.

5 $\frac{4^3}{4^2}$

$\frac{4^3}{4^2} = 4^{3-2}$ *To divide powers that have the same base, write the common base. Then subtract the exponents.*

$= 4^1$ or 4

6 $\frac{x^6}{x^4}$

$\frac{x^6}{x^4} = x^{6-4}$ *Write the common base. Then subtract the exponents.*

$= x^2$

7 $\frac{8m^4n^5}{2m^3n^2}$

$\frac{8m^4n^5}{2m^3n^2} = \left(\frac{8}{2}\right)\left(\frac{m^4}{m^3}\right)\left(\frac{n^5}{n^2}\right)$ *Group the powers that have the same base.*

$= 4m^{4-3}n^{5-2}$

$= 4m^1n^3$

$= 4mn^3$

Your Turn

e. $\frac{10^5}{10^2}$ f. $\frac{y^5}{y^4}$ g. $\frac{a^4b^3}{ab^2}$ h. $\frac{-30m^5n^2}{10m^3n}$

A special case results when you divide a power by itself. Consider the following two ways to simplify $\frac{b^4}{b^4}$, where $b \neq 0$.

Method 1 Definition of Power

$$\frac{b^4}{b^4} = \frac{\overset{1}{\cancel{b}} \cdot \overset{1}{\cancel{b}} \cdot \overset{1}{\cancel{b}} \cdot \overset{1}{\cancel{b}}}{\underset{1}{\cancel{b}} \cdot \underset{1}{\cancel{b}} \cdot \underset{1}{\cancel{b}} \cdot \underset{1}{\cancel{b}}}$$

$$= 1$$

Method 2 Quotient of Powers

$$\frac{b^4}{b^4} = b^{4-4}$$

$$= b^0$$

Since $\frac{b^4}{b^4}$ cannot have two different values, you can conclude that $b^0 = 1$. Therefore, any nonzero number raised to the zero power is equal to 1.

Example 8

Simplify $\frac{x^4y^2}{xy^2}$.

$$\frac{x^4y^2}{xy^2} = \left(\frac{x^4}{x^1}\right)\left(\frac{y^2}{y^2}\right)$$
$$= x^{4-1}y^{2-2}$$
$$= x^3y^0 \qquad y^0 = 1$$
$$= x^3 \cdot 1 \text{ or } x^3$$

Your Turn

i. $\frac{a^3b^4}{a^3b}$ j. $\frac{10x^4y^3}{5x^4y^2}$ k. $\frac{m^{10}n^5}{m^{10}n^5}$

Check for Understanding

Communicating Mathematics

1. **Explain** why $a^4 \cdot a^7$ can be simplified but $a^4 \cdot b^7$ cannot.
2. **You Decide** Tonia says $10^3 \times 10^2 = 100^5$, but Emilio says $10^3 \times 10^2 = 10^5$. Who is correct? Explain your reasoning.

Guided Practice

Simplify each expression.

3. $y^2 \cdot y^5$ 4. $m^5(m)$ 5. $(t^2)(t^2)(t)$ *(Examples 1 & 2)*

6. $(x^3y)(xy^3)$ 7. $(3a^2)(4a^3)$ 8. $(-5x^3)(4x^4)$ *(Examples 3 & 4)*

9. $\frac{n^8}{n^5}$ 10. $\frac{b^6c^5}{b^3c^2}$ 11. $\frac{xy^7}{y^4}$ *(Examples 5 & 6)*

12. $\frac{12x^5}{4x^4}$ 13. $\frac{ab^5c}{ac}$ 14. $\frac{22a^4b^5c^7}{-11abc^2}$ *(Examples 7 & 8)*

15. **Measurement** There are 10^1 millimeters in 1 centimeter and 10^2 centimeters in 1 meter. How many millimeters are in 1 meter? Write your answer as a power and then evaluate the expression. *(Example 1)*

Exercises

Practice

Homework Help	
For Exercises	See Examples
16–31, 45, 46, 48	1–4
32–43, 47	5–7
45	8
Extra Practice	
See page 708.	

Simplify each expression.

16. $2^6 \cdot 2^8$ 17. $5^3 \cdot 5$ 18. $y^7 \cdot y^7$ 19. $d \cdot d^5$

20. $(b^4)(b^2)$ 21. $(a^2b)(ab^4)$ 22. $(m^3n)(mn^2)$ 23. $(r^3t^4)(r^4t^4)$

24. $(2xy)(3x^2y^2)$ 25. $(-5a)(-3a)$ 26. $(-10x^3y)(2x^2)$ 27. $(4x^2y^3)(2xy^2)$

28. $(-8x^3y)(2x^4)$ 29. $m^4(m^3b^2)$ 30. $(ab)(ac)(bc)$ 31. $(m^2n)(am)(an^2)$

32. $\frac{10^9}{10^3}$ 33. $\frac{9^6}{9^5}$ 34. $\frac{w^7}{w^3}$ 35. $\frac{k^9}{k^4}$

36. $\frac{a^4b^8}{ab^2}$ 37. $\frac{24a^3b^6}{-2a^2b^2}$ 38. $\frac{24x^2y^7z^3}{-6x^2y^3z}$ 39. $\frac{-40mn^2}{-10mn^2}$

40. $\frac{3}{4}a(12b^2)$ 41. $\frac{5}{6}c(18a^3)$ 42. $\left(\frac{1}{2}a^2\right)\left(6ab^2\right)$ 43. $x^0(2x^3)$

44. Evaluate 5^0.

45. Find the product of $2x$ and $-8x$.

Applications and Problem Solving

46. **Geometry** The measure of the length of a rectangle is $5x$ and the measure of the width is $3x$. Find the measure of the area.

47. **Earth Science** At a distance of 10^7 meters from Earth, a satellite can see almost all of our planet. At a distance of 10^{13} meters, a satellite can see all of our solar system. How many times as great is a distance of 10^{13} meters as 10^7 meters?

48. **Manufacturing** The Pizza Parlor uses square boxes to package their pizzas. The drawing shows that a pizza with radius r just fits inside the box. Write an expression for the area of the bottom of the box.

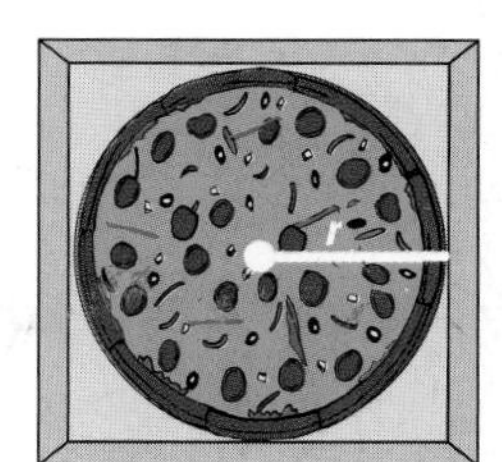

49. **Critical Thinking** Study the following pattern.

$(5^2)^3 = (5^2)(5^2)(5^2)$ or 5^6

Simplify each expression.

a. $(10^3)^4$ b. $(4^5)^3$ c. $(x^2)^4$

d. Write a rule for finding the *power of a power*.

Mixed Review

Evaluate each expression if $x = -1$, $y = 2$, and $z = -3$. *(Lesson 8–1)*

50. z^3 51. $3x^4$ 52. $5xy^3z$ 53. $3(x^2 + y^2)$

54. Write an equation of the line that is perpendicular to the graph of $y = 3x + 5$ and passes through (0, 0). *(Lesson 7–7)*

Solve each proportion. *(Lesson 5–1)*

55. $\frac{96}{6} = \frac{152}{x}$ 56. $\frac{9}{m} = \frac{15}{10}$ 57. $\frac{8.6}{25.8} = \frac{1}{n}$ 58. $\frac{3}{7} = \frac{2.1}{d}$

Standardized Test Practice

59. **Multiple Choice** Choose the expression that has a value of 28. *(Lesson 1–2)*

A $4 + 3 \cdot 4$ B $75 \div 3 - 2$ C $8 \cdot 4 - 8$ D $(5 + 3) \cdot 7 \div 2$

Broadcast Technician

Do you dream of being in the movies? If you don't make it onto the big screen, you may find a career behind the scenes as a sound mixer. Sound mixers are broadcast technicians who develop movie sound tracks. Using a process called *dubbing*, they sit at sound consoles and fade in and fade out each sound by regulating its volume. All of the sounds for each scene are blended on a master sound track.

Sound intensity is measured in *decibels*. The decibel scale is based on powers of ten. The softest audible sound is represented by 10^0. The chart lists several common sounds and their intensity as compared to the softest audible sound.

Sound	Decibels	Intensity
jet airplane	140	10^{14}
rock band	120	10^{12}
motorcycle	110	10^{11}
circular saw	100	10^{10}
busy traffic	80	10^8
vacuum cleaner	70	10^7
noisy office	60	10^6
talking	40	10^4
whispering	20	10^2
breathing	10	10^1
softest sound	0	10^0

Find how many times as intense the first sound is as the second.

1. vacuum cleaner, noisy office
2. motorcycle, busy traffic
3. rock band, talking
4. jet airplane, whispering
5. Find two sounds, one of which is 10^2 times as intense as the other.

FAST FACTS About Broadcast Technicians

Working Conditions

- usually work indoors in pleasant conditions
- work a 40-hour week, but some overtime required to meet deadlines
- evening, weekend, and holiday work

Education

- high school math, physics, and electronics
- postsecondary training in engineering or electronics at a technical school or community college

Job Outlook

Expected Growth in Employment

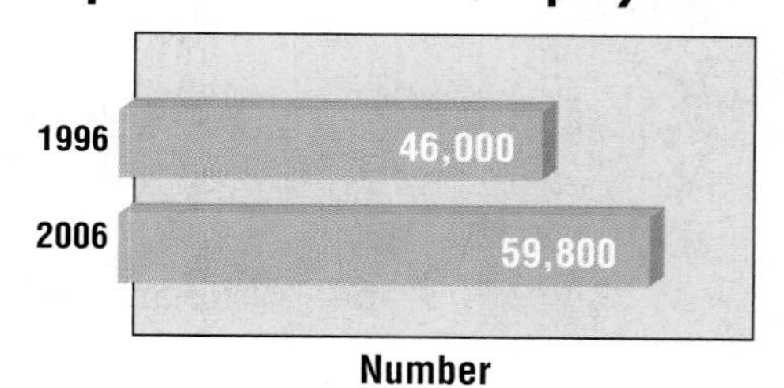

Number

Source: *Occupational Outlook Handbook*

*inter*NET CONNECTION **Career Data** For up-to-date information about a career as a broadcast technician, visit: www.algconcepts.com

8–3 Negative Exponents

What You'll Learn
You'll learn to simplify expressions containing negative exponents.

Why It's Important
Electronics Electric current is measured using the prefixes *micro*, which is 10^{-6}, and *milli*, which is 10^{-3}.
See Exercise 14.

Not all exponents are positive integers. Some exponents are negative integers. Study the pattern at the right to find the value of 10^{-1} and 10^{-2}. Extending the pattern suggests that $10^{-1} = \frac{1}{10}$ and $10^{-2} = \frac{1}{10^2}$ or $\frac{1}{100}$.

$$\begin{aligned} 10^3 &= 1000 \\ 10^2 &= 100 \\ 10^1 &= 10 \\ 10^0 &= 1 \\ 10^{-1} &= ? \\ 10^{-2} &= ? \end{aligned}$$

÷ 10 ÷ 10 ÷ 10 ÷ 10 ÷ 10

When you multiply by the base, the exponent in the result increases by one. For example, $10^3 \times 10 = 10^4$. When you divide by the base, the exponent in the result decreases by one. For example, $10^{-2} \div 10 = 10^{-3}$.

You can use the Quotient of Powers rule and the definition of power to simplify the expression $\frac{x^3}{x^5}$ and write a definition of negative exponents.

Method 1
Quotient of Powers

$$\frac{x^3}{x^5} = x^{3-5} = x^{-2}$$

Method 2
Definition of Power

$$\frac{x^3}{x^5} = \frac{\overset{1}{\cancel{x}} \cdot \overset{1}{\cancel{x}} \cdot \overset{1}{\cancel{x}}}{\underset{1}{\cancel{x}} \cdot \underset{1}{\cancel{x}} \cdot \underset{1}{\cancel{x}} \cdot x \cdot x} = \frac{1}{x \cdot x} \text{ or } \frac{1}{x^2}$$

You can conclude that x^{-2} and $\frac{1}{x^2}$ are equal because $\frac{x^3}{x^5}$ cannot have two different values. This and other examples suggest the following definition.

Negative Exponents	
	Numbers: $5^{-2} = \frac{1}{5^2}; \frac{1}{4^{-3}} = 4^3$
	Symbols: $a^{-n} = \frac{1}{a^n}; \frac{1}{a^{-n}} = a^n$

The value of a cannot be zero.

Example 1 Write 10^{-3} using positive exponents. Then evaluate the expression.

$10^{-3} = \frac{1}{10^3}$ *Definition of negative exponent*

$= \frac{1}{1000}$ or 0.001 $10 \cdot 10 \cdot 10 = 1000$

Your Turn

a. 2^{-4}

b. 10^{-2}

c. 5^{-1}

To simplify an expression with a negative exponent, write an equivalent expression that has positive exponents. Each base should appear only once and all fractions should be in simplest form.

Examples

Simplify each expression.

2 xy^{-2}

$xy^{-2} = x \cdot y^{-2}$

$= x \cdot \frac{1}{y^2}$ *Definition of negative exponent*

$= \frac{x}{y^2}$

3 $\frac{a^5b}{a^3b^4}$

$\frac{a^5b}{a^3b^4} = \frac{a^5}{a^3} \cdot \frac{b^1}{b^4}$

$= a^{5-3} \cdot b^{1-4}$ *Quotient of powers*

$= a^2 \cdot b^{-3}$

$= a^2 \cdot \frac{1}{b^3}$ *Definition of negative exponent*

$= \frac{a^2}{b^3}$

Your Turn

d. mn^{-3}

e. $\frac{x^4y}{x^2y^5}$

4 Simplify $\frac{-6r^3s^5}{18r^{-7}s^5t^{-2}}$.

$\frac{-6r^3s^5}{18r^{-7}s^5t^{-2}} = \left(\frac{-6}{18}\right)\left(\frac{r^3}{r^{-7}}\right)\left(\frac{s^5}{s^5}\right)\left(\frac{1}{t^{-2}}\right)$

$= \left(\frac{-1}{3}\right)\left(\frac{r^3}{r^{-7}}\right)\left(\frac{s^5}{s^5}\right)\left(\frac{t^0}{t^{-2}}\right)$ $\frac{-6}{18} = -\frac{1}{3}$, $1 = t^0$

$= -\frac{1}{3}r^{3-(-7)}s^{5-5}t^{0-(-2)}$ *Quotient of powers*

$= -\frac{1}{3}r^{10}s^0t^2$ $3 - (-7) = 10, 0 - (-2) = 2$

$= -\frac{r^{10}t^2}{3}$ $s^0 = 1$

Your Turn

Simplify each expression.

f. $\dfrac{5a^4b^6}{-25a^{-2}b^6c^{-3}}$

g. $\dfrac{8x^3y^5}{10x^{-3}y^5z}$

Biology Link

5 **The *E. coli* bacteria has a width of 10^{-3} millimeter. The head of a pin has a diameter of 1 millimeter. How many *E. coli* bacteria can fit across the head of a pin?**

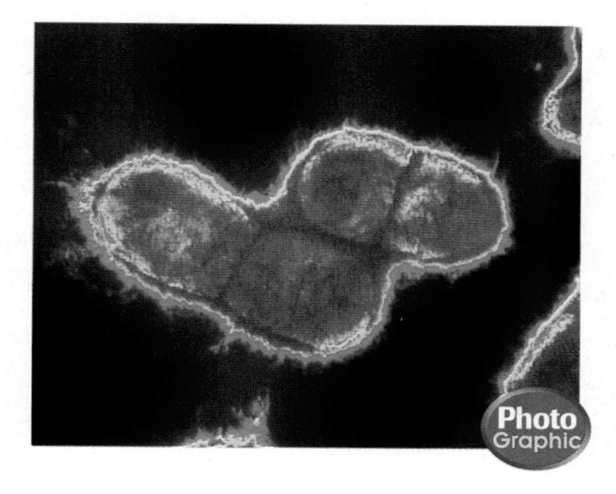

To find the number of bacteria, divide 1 by 10^{-3}.

$$\frac{1}{10^{-3}} = \frac{10^0}{10^{-3}} \qquad 1 = 10^0$$
$$= 10^{0-(-3)}$$
$$= 10^3$$

Since $10^3 = 1000$, about 1000 bacteria could fit across the head of a pin.

Your Turn

h. The figure below shows the electromagnetic wave spectrum. An ultraviolet wave has a length of 10^{-5} centimeter. An FM radio wave has a length of 10^2 centimeters. How many times as long is the FM wave as the ultraviolet wave?

Wavelength (centimeters)

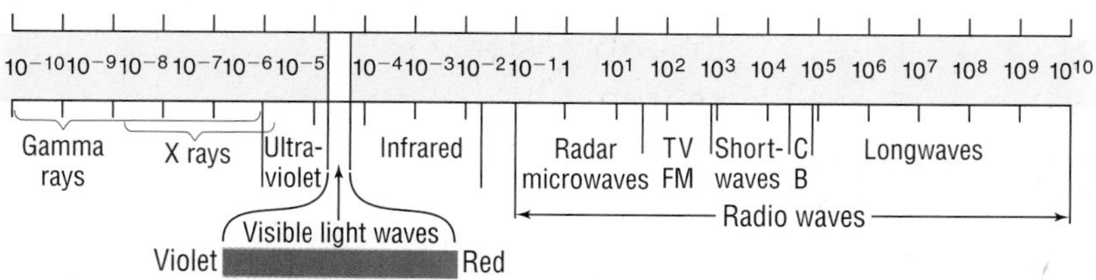

Check for Understanding

Communicating Mathematics

1. **Express** 5^{-2} using positive integers. Then evaluate the expression.
2. **Evaluate** $6t^{-2}$ if $t = 3$.

3. **You Decide?** Booker and Antonio both correctly simplified $a^{-2} \cdot a^3$.

Booker's Method	Antonio's Method
$a^{-2} \cdot a^3 = a^{-2+3}$	$a^{-2} \cdot a^3 = \frac{1}{a^2} \cdot a^3$
$= a^1$ or a	$= \frac{a^3}{a^2}$ or a

a. Which student used the Product of Powers rule?
b. Which student used the definition of negative exponents?
c. Whose method do you prefer? Explain your reasoning.

Guided Practice

Write each expression using positive exponents. Then evaluate the expression. *(Example 1)*

4. 10^{-4}
5. 3^{-3}

Simplify each expression.

6. z^{-1}
7. n^{-3}
8. $s^{-2}t^3$
9. $p^{-1}q^{-2}r^2$ *(Example 2)*
10. $\frac{x^2}{x^3}$
11. $\frac{a^3}{a^{-4}}$
12. $\frac{m^4n^{-2}}{m^6n^2}$
13. $\frac{3a^3bc^5}{27a^4bc^2}$ *(Examples 3–5)*

14. **Electronics** Electric current can be measured in amperes, milliamperes, or microamperes. The prefixes *milli* and *micro* mean 10^{-3} and 10^{-6}, respectively. Express 10^{-3} and 10^{-6} using positive exponents. *(Example 1)*

Exercises

Practice

Write each expression using positive exponents. Then evaluate the expression.

15. 2^{-5}
16. 10^{-5}
17. 4^{-1}
18. 6^{-2}

Homework Help

For Exercises	See Examples
15–18, 45–47	1
19–26	2
27–42	3

Extra Practice

See page 708.

Simplify each expression.

19. r^{-10}
20. p^{-6}
21. $a^4(a^{-2})$
22. $x^{-7}(x^4)$
23. $s^{-2}t^4$
24. $a^0b^{-1}c^{-2}$
25. $15rs^{-2}$
26. $10x^{-4}y^{-5}z$
27. $\frac{m^2}{m^{-4}}$
28. $\frac{x^2}{x^3}$
29. $\frac{k^{-2}}{k^6}$
30. $\frac{1}{r^{-3}}$
31. $\frac{an^3}{n^5}$
32. $\frac{bm^2}{m^6}$
33. $\frac{x^3y^{-3}}{x^3y^6}$
34. $\frac{a^5b^{-3}}{a^7b^3}$
35. $\frac{12b^5}{4b^{-4}}$
36. $\frac{24c^6}{4c^{-2}}$
37. $\frac{7x^4}{28x}$
38. $\frac{20y^5}{40y^{-2}}$
39. $\frac{4x^3}{28x}$
40. $\frac{-15r^5s^8}{5r^5s^2}$
41. $\frac{5ac}{8ab^5c^2}$
42. $\frac{12c^3d^4f^6}{60cd^6f^3}$

43. Evaluate $4x^{-3}y^2$ if $x = 2$ and $y = 6$.
44. Find the value of $(2b)^{-3}$ if $b = -2$.
45. Which is greater, 2^{-4} or 2^{-6}?

Applications and Problem Solving

46. **Physical Science** Visible light waves have wavelengths between 10^{-5} centimeter and 10^{-4} centimeter. Express 10^{-5} and 10^{-4} using positive exponents. Then evaluate each expression.

47. **Physical Science** Refer to the figure on page 349 in Your Turn, part h. Some infrared waves have lengths of 10^{-3} centimeter. Which kind of wave has a length that is 1000 times as long as an infrared wave?

48. **Critical Thinking** Which point on the number line could be the graph of n^{-2} if n is a positive integer?

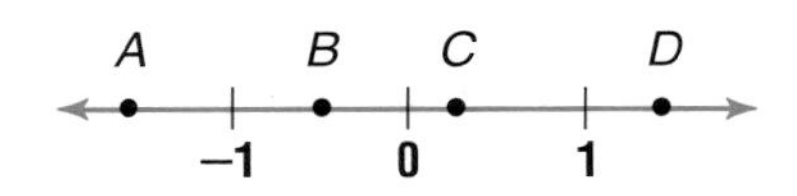

Mixed Review

Simplify each expression. *(Lesson 8–2)*

49. $(5a^3)(-2a^4)$
50. $x^4 \cdot x$
51. $\dfrac{n^{10}}{n^4}$
52. $\dfrac{15a^3b^6c^2}{-5ab^2c^2}$

Write each expression using exponents. *(Lesson 8–1)*

53. $z \cdot z \cdot z \cdot z$
54. $(3)(3)(-2)(-2)(-2)$
55. 9

Standardized Test Practice

56. **Grid In** Determine the slope of the line passing through (1, −2) and (6, 2). *(Lesson 7–1)*

57. **Multiple Choice** The Jaguars softball team played 8 games and scored a total of 96 runs. What was the mean number of runs scored per game? *(Lesson 3–3)*

A 8 B 12 C 88 D 104

Quiz 1 Lessons 8–1 through 8–3

Write each expression using exponents. *(Lesson 8–1)*

1. $6 \cdot 6 \cdot 6$
2. $x \cdot x \cdot x \cdot x$
3. 10
4. $(-3)(-3)(-3)$
5. **Number Theory** The prime factorization of 96 is $2 \cdot 2 \cdot 2 \cdot 2 \cdot 2 \cdot 3$. Write the prime factorization using exponents. *(Lesson 8–1)*

Simplify each expression. *(Lessons 8–2 & 8–3)*

6. $(m^4)(m^6)$
7. $\dfrac{x^7}{x^2}$
8. $(-3x^2y^3)(-2x^4y^{-1})$
9. $\dfrac{3a^2b}{-9a^2b^4}$
10. **Biology** The length of a *Euglena* protist is about 10^{-2} centimeter. Express 10^{-2} using positive exponents. Then evaluate the expression.

8–4 Scientific Notation

What You'll Learn

You'll learn to express numbers in scientific notation.

Why It's Important

Fiber Optics
Laser technicians use scientific notation when they deal with the speed of light, 3×10^8 meters per second.
See Example 7.

The prefixes *mega*, *giga*, and *kilo* are metric prefixes. They are used with very large measures. Other prefixes are used with very small measures. The chart shows some metric prefixes.

Metric Prefixes

Prefix	Power of 10	Meaning	Prefix	Power of 10	Meaning
tera	10^{12}	1,000,000,000,000	pico	10^{-12}	0.000000000001
giga	10^{9}	1,000,000,000	nano	10^{-9}	0.000000001
mega	10^{6}	1,000,000	micro	10^{-6}	0.000001
kilo	10^{3}	1000	milli	10^{-3}	0.001

Scientific fields and industry use metric units because calculations are easier with powers of ten. This is also why metric units, based on powers of ten, are widely used in science.

When you multiply a number by a power of ten, the nonzero digits in the original number and the product are the same. Only the position of the decimal point is different.

$$5 \times 10^2 = 5 \times 100$$
$$= 500 \quad \textit{2 places right}$$

$$8.23 \times 10^4 = 8.23 \times 10{,}000$$
$$= 82{,}300 \quad \textit{4 places right}$$

$$4 \times 10^{-1} = 4 \times 0.1$$
$$= 0.4 \quad \textit{1 place left}$$

$$1.23 \times 10^{-3} = 1.23 \times 0.001$$
$$= 0.00123 \quad \textit{3 places left}$$

Prerequisite Skills Review
Operations with Decimals, p. 684

These and similar examples suggest the following rules for multiplying a number by a power of ten.

Multiplying by Powers of 10

- If the exponent is *positive*, move the decimal point to the *right*.
- If the exponent is *negative*, move the decimal point to the *left*.

Examples

Express each measurement in standard form.

1 **2 megabytes**

2 megabytes $= 2 \times 10^6$ bytes — *The prefix mega- means 10^6.*

$= 2{,}000{,}000$ bytes — *Move the decimal point 6 places right.*

Reading Algebra

The absolute value of the exponent tells the number of places to move the decimal point.

2 **3.6 nanoseconds**

3.6 nanoseconds $= 3.6 \times 10^{-9}$ seconds — *The prefix nano- means 10^{-9}.*

$= 0.0000000036$ seconds — *Move the decimal point 9 places left.*

Your Turn

a. 2 gigabytes

b. 3.4 milliseconds

When you use very large numbers like 5,800,000 or very small numbers like 0.000076, it is difficult to keep track of the place value. Numbers such as these can be written in **scientific notation**.

Scientific Notation	A number is expressed in scientific notation when it is in the form $a \times 10^n$, where $1 \leq a < 10$ and n is an integer.

Follow these steps to write a number in scientific notation.

- First, move the decimal point after the first nonzero digit.
- Then, find the power of ten by counting the decimal places.
- When the number is greater than one, the exponent of 10 is *positive*.
- When the number is between zero and one, the exponent of 10 is *negative*.

Examples

Express each number in scientific notation.

3 **5,800,000**

$5{,}800{,}000 = 5.8 \times 10^?$ — *The decimal point moves 6 places.*

$= 5.8 \times 10^6$ — *Since 5,800,000 is greater than one, the exponent is positive.*

4 **0.000076 in scientific notation.**

$0.000076 = 7.6 \times 10^?$ — *The decimal point moves 5 places.*

$= 7.6 \times 10^{-5}$ — *Since 0.000076 is between zero and one, the exponent is negative.*

Your Turn

c. 3,900,000,000

d. 0.0000035

You can use scientific notation to simplify computation.

Examples

Evaluate each expression.

5 **$400 \times 2{,}000{,}000{,}000$**

First express each number in scientific notation. Then use the Associative and Commutative Properties to regroup terms.

$$
\begin{aligned}
400 \times 2{,}000{,}000{,}000 &= (4 \times 10^2)(2 \times 10^9) \\
&= (4 \times 2)(10^2 \times 10^9) \quad \textit{Associative and Commutative Properties} \\
&= 8 \times 10^{11} \\
&= 800{,}000{,}000{,}000
\end{aligned}
$$

Technology Tip

On a graphing calculator, 8×10^{11} is shown as 8E11.

6 $\dfrac{4.8 \times 10^3}{1.6 \times 10^1}$

$$
\begin{aligned}
\frac{4.8 \times 10^3}{1.6 \times 10^1} &= \left(\frac{4.8}{1.6}\right)\left(\frac{10^3}{10^1}\right) \quad \frac{4.8}{1.6} = 3 \\
&= 3 \times 10^2 \text{ or } 300
\end{aligned}
$$

Your Turn

e. $2000 \times 3{,}000{,}000{,}000$

f. $\dfrac{7.5 \times 10^7}{1.5 \times 10^4}$

Physics Link

7 **The light from a laser beam travels at a speed of 300,000,000 meters per second. How far does the light travel in 2 nanoseconds? Use the formula $d = rt$, where d is the distance in meters, r is the speed of light, and t is the time in seconds.**

Express 300,000,000 in scientific notation.
Express 2 nanoseconds in seconds.

$300{,}000{,}000 = 3 \times 10^8$ and 2 nanoseconds $= 2 \times 10^{-9}$ seconds

Method 1 Paper and Pencil

$d = rt$

$d = (3 \times 10^8)(2 \times 10^{-9})$

$d = (3 \times 2)(10^8 \times 10^{-9})$ *Associative Property*

$d = 6 \times 10^{-1}$

Method 2 Calculator

3 [2nd] [EE] 8 [×] 2 [2nd] [EE] [(–)] 9 [ENTER] *0.6*

The light travels 6×10^{-1} meter, or 0.6 meter, in 2 nanoseconds.

Check for Understanding

Communicating Mathematics

Vocabulary
scientific notation

1. **Tell** whether 23.5×10^3 is expressed in scientific notation. Explain your answer.
2. **Explain** an advantage of expressing very large or very small numbers in scientific notation.

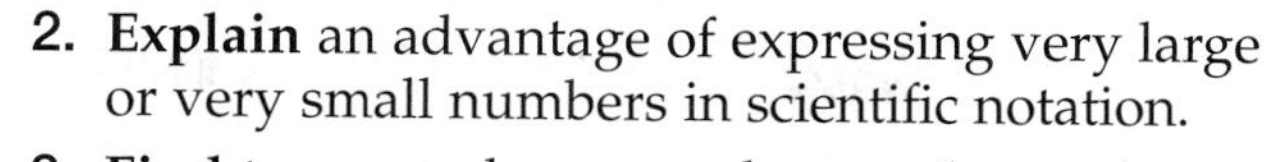

3. **Find** two very large numbers and two very small numbers in a newspaper. Write each number in standard form and in scientific notation.

Guided Practice

Getting Ready **Find each product.**

Sample 1: 2×10^3
Solution: 2000 *three places right*

Sample 2: 3.4×10^{-1}
Solution: 0.34 *one place left*

4. 2.45×10^2
5. 6.8×10^4
6. 2×10^6
7. 6.4×10^{-1}
8. 9.23×10^{-2}
9. 3×10^{-4}

Express each measure in standard form. *(Examples 1 & 2)*

10. 5 megaohms
11. 6.5 milliamperes

Express each number in scientific notation. *(Examples 3 & 4)*

12. 9500
13. 56.9
14. 0.0087
15. 0.000023

Evaluate each expression. Express each result in scientific notation and standard form. *(Examples 5–7)*

16. $(2 \times 10^5)(3 \times 10^{-8})$
17. $\frac{5.2 \times 10^5}{2 \times 10^2}$

18. **Health** The length of the virus that causes AIDS is 0.00011 millimeter. Express 0.00011 in scientific notation. *(Example 4)*

Exercises

Practice

Homework Help

For Exercises	See Examples
19–24, 45–47	1, 2
25–36, 48	3, 4
37–40	5
41–43	6
44, 49	7

Extra Practice
See page 708.

Express each measure in standard form.

19. 5.8 billion dollars
20. 4 megahertz
21. 3.9 nanoseconds
22. 82 kilobytes
23. 9 milliamperes
24. 2.3 micrograms

Express each number in scientific notation.

25. 5280
26. 240,000
27. 268.3
28. 25,000,000
29. 0.00032
30. 0.08
31. 0.004296
32. 15.9
33. 0.012
34. 1,000,000
35. 0.000000022
36. 0.0000946

Evaluate each expression. Express each result in scientific notation and standard form.

37. $(3 \times 10^2)(2 \times 10^4)$
38. $(4 \times 10^2)(1.5 \times 10^6)$
39. $(3 \times 10^{-2})(2.5 \times 10^4)$
40. $(7.8 \times 10^{-6})(1 \times 10^{-2})$

Evaluate each expression. Express each result in scientific notation and standard form.

41. $\dfrac{6.4 \times 10^5}{3.2 \times 10^2}$

42. $\dfrac{8 \times 10^2}{2 \times 10^{-3}}$

43. $\dfrac{13.2 \times 10^{-6}}{2.4 \times 10^{-1}}$

44. Find the product of (1.2×10^5) and (5×10^{-4}) mentally.

45. Express 5×10^6 in standard form.

Applications and Problem Solving

46. **Biology** The mass of an orchid seed is 3.5×10^{-6} grams. Express 3.5×10^{-6} in standard form.

47. **Astronomy** The diameter of Venus is 1.218×10^4 km, the diameter of Earth is 1.276×10^4 km, and the diameter of Mars is 6.76×10^3 km. List the planets in order from greatest to least diameter.

48. **Electronics** Engineering notation is similar to scientific notation. However, in engineering notation, the powers of ten are always multiples of 3, such as 10^3, 10^6, 10^{-9}, and 10^{-12}. For example, 240,000 is expressed as 240×10^3. Express 15,000 ohms in scientific and engineering notation.

49. **Biology** Laboratory technicians look at bacteria through microscopes. A microscope set on 1000× makes an organism appear to be 1000 times larger than its actual size. Most bacteria are between 3×10^{-4} and 2×10^{-3} millimeter in diameter. How large would the bacteria appear under a microscope set on 1000×?

50. **Critical Thinking** Express each number in scientific notation.
   a. 32×10^5 b. 284×10^3 c. 0.76×10^{-2} d. 0.09×10^{-3}

Orchid

Mixed Review

Simplify each expression. *(Lessons 8–2, 8–3)*

51. $y^5(y^{-2})$

52. $15a^{-1}b^3c^{-2}$

53. $\dfrac{30c^4}{-5c^{-2}}$

54. $\dfrac{5r^2s^7}{25r^2s^{10}}$

55. $x^2 \cdot x^3 \cdot x$

56. $\dfrac{2}{3}r(15s^2)$

57. $\dfrac{a^4b^5}{ab^2}$

58. $(-15y^2z)(-2y^2z^0)$

59. $(3x^5y^2)(-2x^{-3}y^4)$

Standardized Test Practice

60. **Grid In** Julia's wages vary directly as the number of hours she works. If her wages for 5 hours are \$34.75, how much will they be for 30 hours in dollars? *(Lesson 6–5)*

61. **Multiple Choice** Choose the graph that represents a function. *(Lesson 6–4)*

A
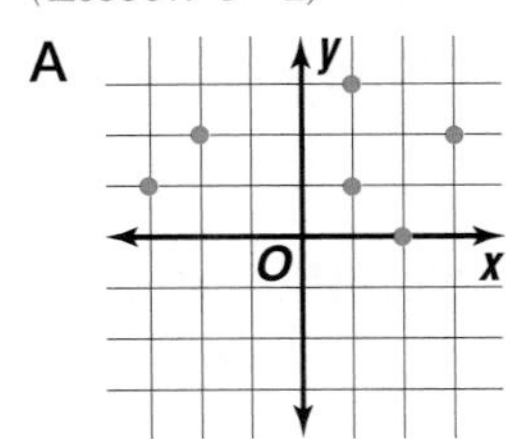

B
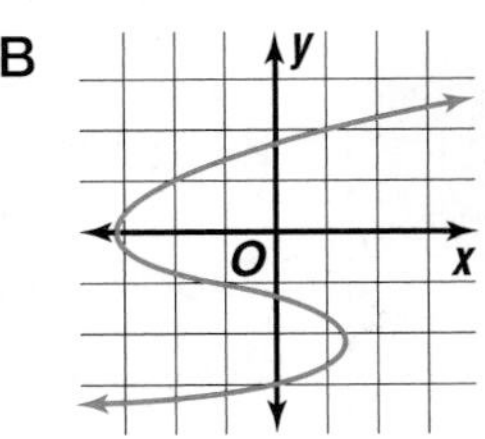

C
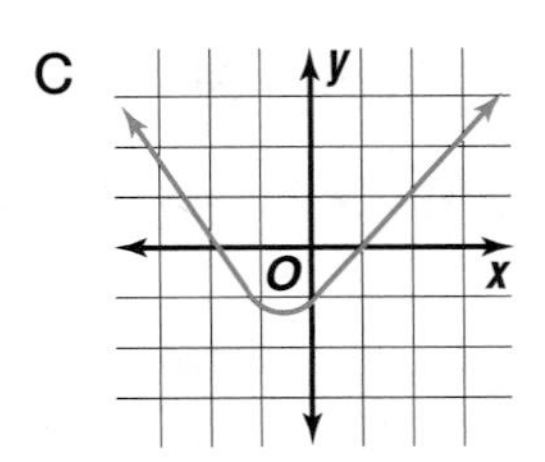

D
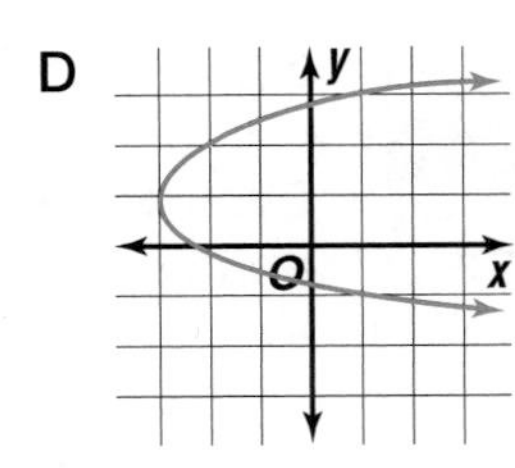

8–5 Square Roots

What You'll Learn

You'll learn to simplify radicals by using the Product and Quotient Properties of Square Roots.

Why It's Important

Aviation Pilots can use the formula $d = 1.5\sqrt{h}$ to determine the distance to the horizon. The formula contains a square root symbol. *See Example 6.*

Can you help this character from *Shoe* take his math test?

SHOE

You will find the square root of 225 in Example 3.

In Lesson 8–1, you learned that *squaring* a number means using that number as a factor twice. The opposite of squaring is finding a **square root**. To find a square root of 36, you must find *two equal factors* whose product is 36.

$6 \times 6 = 36$ → The square root of 36 is 6.

Square Root	**Words:** A square root of a number is one of its two equal factors. **Symbols:** $\sqrt{a} = b$, where $a = b \cdot b$.

The symbol $\sqrt{\ }$, called a **radical sign**, is used to indicate the square root.

Read $\sqrt{a}$ as *the square root of a.*

$\sqrt{36} = 6$ *$\sqrt{36}$ indicates the positive square root of 36.*

$-\sqrt{36} = -6$ *$-\sqrt{36}$ indicates the negative square root of 36.*

Exponents can also be used to indicate the square root. $9^{\frac{1}{2}}$ means the same thing as $\sqrt{9}$. $9^{\frac{1}{2}}$ is read *nine to the one half power.* $9^{\frac{1}{2}} = 3$.

Examples

Simplify each expression.

1 $\sqrt{49}$

Since $7^2 = 49$, $\sqrt{49} = 7$.

2 $-\sqrt{64}$

Since $8^2 = 64$, $-\sqrt{64} = -8$.

Your Turn

a. $\sqrt{25}$ b. $\sqrt{121}$ c. $-\sqrt{25}$ d. $-\sqrt{9}$

A **radical expression** is an expression that contains a square root. You can simplify a radical expression like $\sqrt{225}$ by using prime numbers.

A **prime number** is a whole number that has exactly two factors, the number itself and 1. A **composite number** is a whole number that has more than two factors. Every composite number can be written as the product of prime numbers. The tree diagram shows one way to find the prime factors of 225.

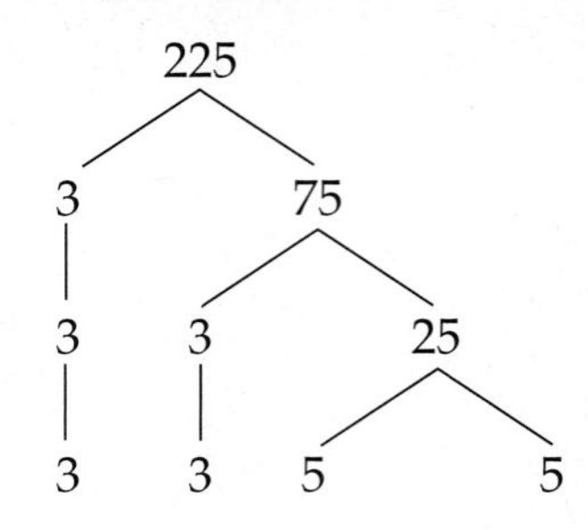

When a number is expressed as a product of prime factors, the expression is called the **prime factorization** of the number. Since 3 and 5 are prime numbers, the prime factorization of 225 is $3 \times 3 \times 5 \times 5$.

To simplify $\sqrt{225}$, use the following property.

Product Property of Square Roots		
	Words:	The square root of a product is equal to the product of each square root.
	Numbers:	$\sqrt{4 \cdot 9} = \sqrt{4} \cdot \sqrt{9}$
	Symbols:	$\sqrt{ab} = \sqrt{a} \cdot \sqrt{b} \quad a \geq 0, b \geq 0$

Examples

Simplify each expression.

3 $\sqrt{225}$

$\sqrt{225} = \sqrt{3 \cdot 3 \cdot 5 \cdot 5}$ *Find the prime factorization of 225.*

$= \sqrt{9 \cdot 25}$ *$3 \times 3 = 9, 5 \times 5 = 25$*

$= \sqrt{9} \cdot \sqrt{25}$ *Use the Product Property of Square Roots.*

$= 3 \cdot 5$ or 15 *Simplify each radical.*

4 $\sqrt{576}$

$\sqrt{576} = \sqrt{2 \cdot 2 \cdot 2 \cdot 2 \cdot 2 \cdot 2 \cdot 3 \cdot 3}$ *Find the prime factorization of 576.*

$= \sqrt{64 \cdot 9}$ *$2 \cdot 2 \cdot 2 \cdot 2 \cdot 2 \cdot 2 = 64, 3 \cdot 3 = 9$*

$= \sqrt{64} \cdot \sqrt{9}$ *Use the Product Property of Square Roots.*

$= 8 \cdot 3$ or 24 *Simplify each radical.*

Your Turn

e. $\sqrt{144}$

f. $\sqrt{324}$

A similar property for quotients can be used to simplify radicals.

Quotient Property of Square Roots		
	Words:	The square root of a quotient is equal to the quotient of each square root.
	Numbers:	$\sqrt{\frac{4}{9}} = \frac{\sqrt{4}}{\sqrt{9}}$
	Symbols:	$\sqrt{\frac{a}{b}} = \frac{\sqrt{a}}{\sqrt{b}}$ $\quad a \geq 0, b > 0$

Example 5 Simplify $\sqrt{\frac{81}{121}}$.

$\sqrt{\frac{81}{121}} = \frac{\sqrt{81}}{\sqrt{121}}$ *Use the Quotient Property of Square Roots.*

$= \frac{9}{11}$

Your Turn

g. $\sqrt{\frac{64}{81}}$

h. $\sqrt{\frac{36}{4}}$

You can use a graphing calculator to evaluate square roots. Press 2nd [$\sqrt{\ }$] and then the number to find its positive square root.

Example 6

Aviation Link

Pilots use the formula $d = 1.5\sqrt{h}$ to determine the distance in miles that an observer can see under ideal conditions. In the formula, d is the distance in miles and h is the height in feet of the plane. If an observer is in a plane that is flying at a height of 3600 feet, how far can he or she see?

$d = 1.5\sqrt{h}$

$d = 1.5 \times \sqrt{3600}$ *Replace h with 3600.*

1.5 2nd [$\sqrt{\ }$] 3600 ENTER 90

The observer can see a distance of 90 miles.

Technology Tip

On a graphing calculator, it is not necessary to use the 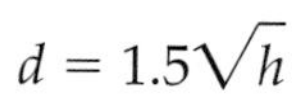 key when you multiply square roots.

Check for Understanding

Communicating Mathematics

1. **Find** the first ten perfect squares.
2. **Write** the symbol for the negative square root of 9.
3. **Explain** why finding a square root and squaring are inverse operations.

Vocabulary

square root
radical sign
radical expression
prime number
composite number
prime factorization

Guided Practice

Getting Ready **Find the prime factorization of each number.**

Sample: 81 **Solution:** $81 = 3 \cdot 3 \cdot 3 \cdot 3$ or 3^4

4. 100 5. 121 6. 169 7. 196 8. 256

Simplify.

9. $\sqrt{4}$ 10. $-\sqrt{49}$ 11. $-\sqrt{121}$ *(Examples 1 & 2)*

12. $\sqrt{256}$ 13. $\sqrt{\frac{1}{4}}$ 14. $-\sqrt{\frac{49}{121}}$ *(Examples 2 & 3)*

15. **Buildings** A famous 1933 movie about a gorilla helped make the Empire State Building in New York City a popular tourist attraction. If the gorilla's eyes were at a height of 1225 feet, how far could he see in the distance on a clear day? Use the formula $d = 1.5\sqrt{h}$, where d is the visible distance in miles and h is the height in feet. *(Example 5)*

Exercises

Practice

Simplify.

16. $-\sqrt{81}$ 17. $\sqrt{100}$ 18. $\sqrt{144}$ 19. $-\sqrt{196}$

20. $-\sqrt{169}$ 21. $\sqrt{529}$ 22. $\sqrt{25}$ 23. $-\sqrt{676}$

24. $\sqrt{441}$ 25. $-\sqrt{484}$ 26. $-\sqrt{1024}$ 27. $\sqrt{289}$

28. $\sqrt{\frac{81}{64}}$ 29. $-\sqrt{\frac{9}{100}}$ 30. $\sqrt{\frac{36}{196}}$ 31. $-\sqrt{\frac{25}{400}}$

32. $\sqrt{\frac{225}{25}}$ 33. $\sqrt{\frac{144}{196}}$ 34. $\sqrt{\frac{196}{289}}$ 35. $\sqrt{\frac{0.09}{0.16}}$

36. $\sqrt{0.16}$ 37. $-\sqrt{0.0025}$ 38. $\sqrt{0.0036}$ 39. $\sqrt{0.0009}$

40. Find the negative square root of 49.
41. If $x = \sqrt{36}$, what is the value of x?

Homework Help

For Exercises	See Examples
16–27, 36–41	1, 2
28–35	5

Extra Practice
See page 709.

Applications and Problem Solving

42. **Geometry** The area of a square is 25 square inches. Find the length of one of its sides.

25 in^2

43. **Geometry** The formula for the perimeter P of a square is $P = 4s$, where s is the length of a side. The area of a square is 169 square meters. Find its perimeter.

44. **Firefighting** The velocity of water sprayed from a nozzle is given by the formula $V = 12.14\sqrt{P}$, where V is the velocity in feet per second and P is the pressure at the nozzle in pounds per square inch. Find the velocity of water if the nozzle pressure is 64 pounds per square inch.

45. **Critical Thinking** *True* or *false*: $\sqrt{-36} = -6$. Explain.

Mixed Review

Express each number in scientific notation. *(Lesson 8–4)*

46. 350
47. 63,000
48. 0.023
49. 0.00076

Write each expression using positive exponents. *(Lesson 8–3)*

50. 7^{-2}
51. x^{-4}
52. $ab^{-2}c$
53. $x^{-2}y^{-3}$

***inter*NET CONNECTION**

Data Update For the latest information about high school enrollment, visit: www.algconcepts.com

54. **Business** Among high school students ages 15–18, 41% say they are employed. The graph shows the number of hours per week these students work. If you survey 50 high school students who have jobs, predict how many work between 11 and 20 hours per week. *(Lesson 5–3)*

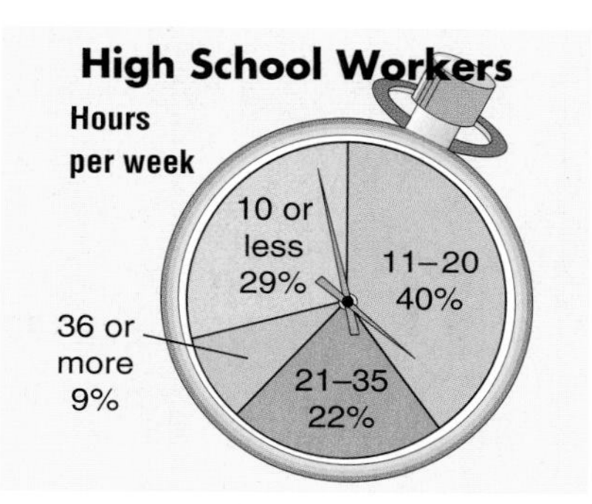

Source: Michaels Opinion Research

Standardized Test Practice

55. **Short Response** Write an equation with variables on both sides in which the solution is -5. *(Lesson 4–6)*

Quiz 2 Lessons 8–4 and 8–5

Express each number in scientific notation. *(Lesson 8–4)*

1. 700,000
2. 0.000053

Simplify each expression. *(Lesson 8–5)*

3. $\sqrt{441}$
4. $-\sqrt{\frac{9}{36}}$

5. **Architecture** A square house is the most energy-efficient because it has the least outside wall space for its area. What is the length and width of the most energy-efficient house you could build with an area of 900 square feet? *(Lesson 8–5)*

8-6 Estimating Square Roots

What You'll Learn

You'll learn to estimate square roots.

Why It's Important

Law Enforcement Police officers use the formula $s = \sqrt{30df}$ when they investigate accidents. *See Exercise 38.*

Numbers like 25 and 81 are perfect squares because their square roots are whole numbers ($\sqrt{25} = 5$ and $\sqrt{81} = 9$). But there are many other numbers that are *not* perfect squares.

Notice what happens when you find $\sqrt{2}$ and $\sqrt{15}$ with a calculator.

2nd [$\sqrt{\ }$] 2 ENTER 1.414213562...

2nd [$\sqrt{\ }$] 15 ENTER 3.872983346...

Numbers like $\sqrt{2}$ and $\sqrt{15}$ are not integers or rational numbers because their decimal values do not terminate or repeat. They are **irrational numbers**. You can estimate irrational square roots by using perfect squares.

Hands-On Algebra

Materials: base-ten tiles

You can use base-ten tiles to estimate the square root of 60.

Step 1 Arrange 60 tiles into the largest square possible. The square has 49 tiles, with 11 left over.

Step 2 Add tiles until you have the next larger square. You need to add 4 tiles. This square has 64 tiles.

7

7

Step 1

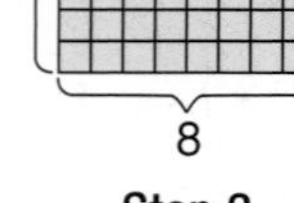

Step 2

Step 3 Now use these models to estimate $\sqrt{60}$.

- 60 is between 49 and 64.
- $\sqrt{60}$ is between 7 and 8.
- Since 60 is closer to 64 than to 49, $\sqrt{60}$ is closer to 8 than to 7.
- To the nearest whole number, $\sqrt{60} \approx 8$.

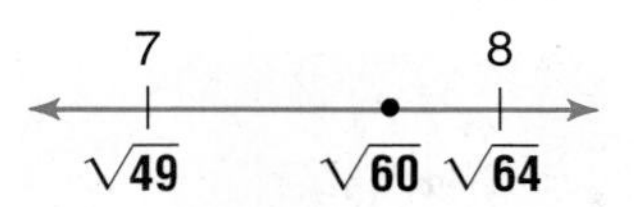

Try These

For each number, arrange base-ten tiles into the largest square possible. Then add tiles until you have the next larger square. To the nearest whole number, estimate the square root of each number.

1. 20 **2.** 76 **3.** 150 **4.** 3

Examples

Estimate each square root to the nearest whole number.

1 $\sqrt{22}$

List some perfect squares to find the two perfect squares closest to 22.

1, 4, 9, 16, 25, 36, . . .

22 is between 16 and 25.

The expression 16 < 22 < 25 means that *16 is less than 22 and 22 is less than 25.* It also means that 22 is *between* 16 and 25.

$$16 < 22 < 25$$
$$\sqrt{16} < \sqrt{22} < \sqrt{25}$$
$$4 < \sqrt{22} < 5$$

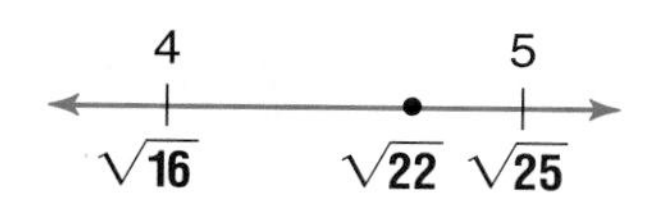

Since 22 is closer to 25 than to 16, the best whole number estimate for $\sqrt{22}$ is 5.

2 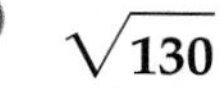

$$121 < 130 < 144$$
$$\sqrt{121} < \sqrt{130} < \sqrt{144}$$
$$11 < \sqrt{130} < 12$$

. . . 64, 81, 100, 121, 144, . . .

11　12
√121　√130　√144

Since 130 is closer to 121 than to 144, the best whole number estimate for $\sqrt{130}$ is 11.

Your Turn

a. $\sqrt{45}$

b. $\sqrt{190}$

Gardening Link

3 **A box of fertilizer covers 250 square feet of garden. Find the length in whole feet of the largest square garden that can be fertilized with one box of fertilizer.**

area ≤ 250 ft^2

Find the largest perfect square that is less than 250. Then find its square root.

$225 < 250$　　*$15^2 = 225$*
　　　　　　　$16^2 = 256$

The square root of 225 is 15. Therefore, the largest square garden that can be fertilized with one box of fertilizer has a length of 15 feet.

Check for Understanding

Communicating Mathematics

1. **Explain** why $\sqrt{10}$ is an irrational number.
2. **Graph** $\sqrt{75}$ on a number line.
3. **Write a problem** that can be solved by estimating $\sqrt{200}$.

Vocabulary
irrational number

Guided Practice

Getting Ready **Find two consecutive perfect squares between which each number lies.**

Sample: 60 **Solution:** 60 is between 49 and 64.

4. 56
5. 85
6. 175
7. 500

Estimate each square root to the nearest whole number.
(Examples 1 & 2)

8. $\sqrt{85}$
9. $\sqrt{71}$
10. $\sqrt{149}$
11. $\sqrt{255}$

12. **Geometry** The area of a square is 200 square inches. Find the length of each side. Round to the nearest whole number. *(Example 3)*

Exercises

Practice

Homework Help	
For Exercises	See Examples
13–34	1, 2
35–38	3

Extra Practice
See page 709.

Estimate each square root to the nearest whole number.

13. $\sqrt{3}$
14. $\sqrt{7}$
15. $\sqrt{13}$
16. $\sqrt{19}$
17. $\sqrt{33}$
18. $\sqrt{56}$
19. $\sqrt{113}$
20. $\sqrt{175}$
21. $\sqrt{410}$
22. $\sqrt{500}$
23. $\sqrt{575}$
24. $\sqrt{1000}$
25. $\sqrt{60.3}$
26. $\sqrt{94.5}$
27. $\sqrt{131.4}$
28. $\sqrt{2.314}$
29. $\sqrt{152.75}$
30. $\sqrt{189.2}$
31. $\sqrt{0.08}$
32. $\sqrt{0.76}$

33. Tell whether 6 is closer to $\sqrt{34}$ or $\sqrt{44}$.
34. Which is closer to $\sqrt{43}$, 6 or 7?

Applications and Problem Solving

35. **Weather** Meteorologists use the formula $t = \sqrt{\frac{D^3}{216}}$ to describe violent storms such as hurricanes. In the formula, D is the diameter of the storm in miles, and t is the number of hours it will last. A typical hurricane has a diameter of about 40 miles. Estimate how long a typical hurricane lasts.

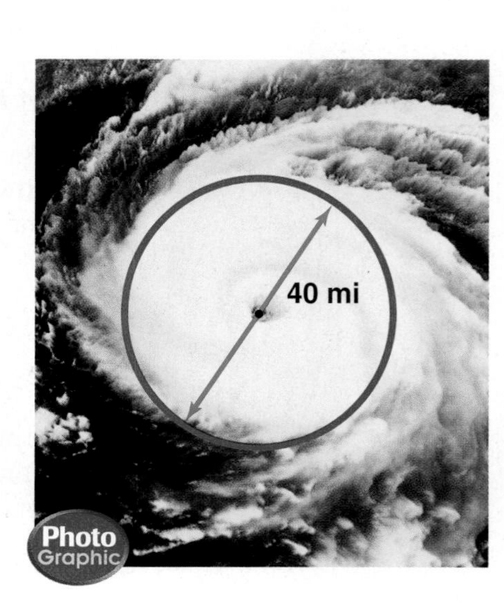

36. **Aviation** The British Airways Concorde flies at a height of 60,000 feet. To the nearest mile, about how far can the pilot see when he or she looks out the window on a clear day? Use the formula $d = 1.5\sqrt{h}$, where d is the visible distance in miles and h in the height in feet.

37. **Geometry** You can use the formula $A = \sqrt{s(s - a)(s - b)(s - c)}$ to find the area of a triangle given the measures of its sides. In the formula, a, b, and c, are the measures of the sides, A is the area, and s is one-half the perimeter. The figure shows a triangular plot of land that contains a large lake. Estimate the area of the triangle to the nearest square mile.

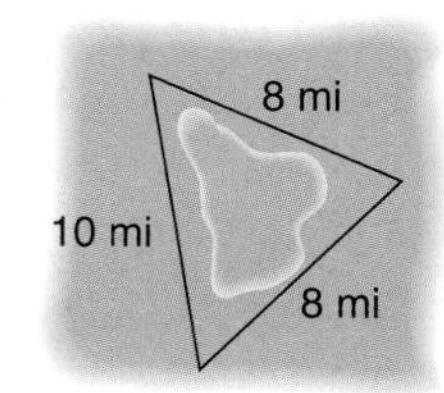

38. **Law Enforcement** When police officers investigate traffic accidents, they need to determine the speed of the vehicles involved in the accident. They can measure the skid marks and then use the formula $s = \sqrt{30df}$. In the formula, s is the speed in miles per hour, d is the length of the skid marks in feet, and f is the friction factor that depends on road conditions. The table gives some values for f.

	Friction Factor (f)	
	concrete	asphalt
wet	0.4	0.5
dry	0.8	1.0

Use the formula to determine the speed in miles per hour for each skid length and road condition. Round to the nearest tenth.

a. 40-foot skid, wet concrete
b. 150-foot skid, dry asphalt
c. 350-foot skid, wet asphalt
d. 45-foot skid, dry concrete

39. **Critical Thinking** Find three numbers that have square roots between 3 and 4.

Mixed Review

Find each square root. *(Lesson 8–5)*

40. $\sqrt{0.36}$ 41. $\sqrt{256}$ 42. $\sqrt{729}$ 43. $\sqrt{\frac{169}{121}}$

Express each measure in standard form. *(Lesson 8–4)*

44. 2 gigabytes 45. 4 nanoseconds 46. 6.5 megahertz

Write an equation in slope-intercept form of the line having the given slope that passes through the given point. *(Lesson 7–3)*

47. 5; (3, −2) 48. $\frac{1}{4}$; (0, 8) 49. −5; (5, 4)

Standardized Test Practice

A B C D

50. **Grid In** Tim drove 4 hours at an average speed of 60 miles per hour. How many hours would it take Tim to drive the same distance at an average speed of 50 miles per hour? *(Lesson 6–6)*

51. **Multiple Choice** Which statement is *not* correct? *(Lesson 3–1)*

A $\frac{1}{2} > \frac{3}{8}$ B $\frac{4}{5} < \frac{5}{6}$ C $\frac{8}{11} < \frac{7}{13}$ D $\frac{1}{3} < \frac{10}{13}$

8–7 The Pythagorean Theorem

What You'll Learn
You'll learn to use the Pythagorean Theorem to solve problems.

Why It's Important
Carpentry Carpenters use the Pythagorean Theorem to determine whether the corners of a deck are right angles.
See Example 4.

The sides of the right triangle below have lengths of 3, 4, and 5 units. The relationship among these lengths forms the basis for one of the most famous theorems in mathematics.

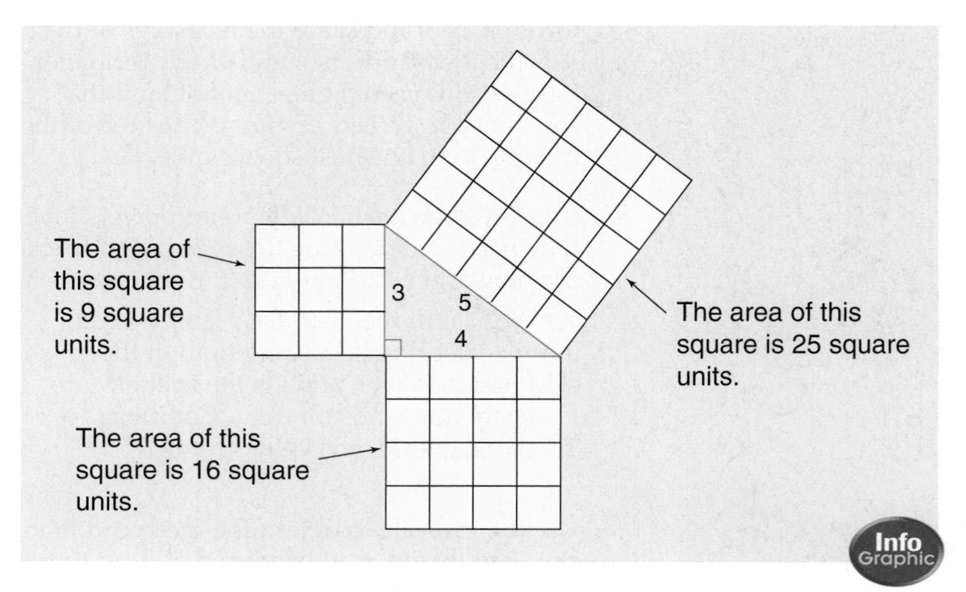

The two sides that form the right angle are called the **legs**. In the triangle above, the lengths of the legs are 3 units and 4 units. The side opposite the right angle is called the **hypotenuse**. The hypotenuse of this triangle has a length of 5 units.

The squares drawn along each side of the triangle illustrate the Pythagorean Theorem geometrically. Study the areas of the squares. Do you notice a relationship between them? The area of the larger square is equal to the total area of the two smaller squares.

$$25 = 9 + 16$$

$$5^2 = 3^2 + 4^2$$

This relationship is true for *any* right triangle and is called the **Pythagorean Theorem**.

Pythagorean Theorem

Words: In a right triangle, the square of the length of the hypotenuse, c, is equal to the sum of the squares of the lengths of the legs, a and b.

Model:

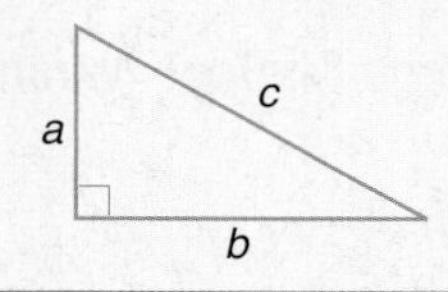

Symbols: $c^2 = a^2 + b^2$

Example 1

Find the length of the hypotenuse of the right triangle.

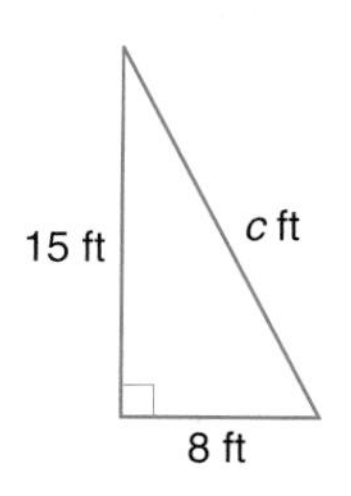

$c^2 = a^2 + b^2$ *Pythagorean Theorem*

$c^2 = 15^2 + 8^2$ *Replace a with 15 and b with 8.*

$c^2 = 225 + 64$

$c^2 = 289$

$c = \sqrt{289}$ *Find the square root of each side.*

$c = 17$

The length of the hypotenuse is 17 feet.

Your Turn

a. Find the length of the hypotenuse of a right triangle if the lengths of the legs are 6 meters and 8 meters.

You can also use the Pythagorean Theorem to find the length of a leg of a right triangle.

Example 2

Find the length of one leg of a right triangle if the length of the hypotenuse is 14 meters and the length of the other leg is 6 meters. Round to the nearest tenth.

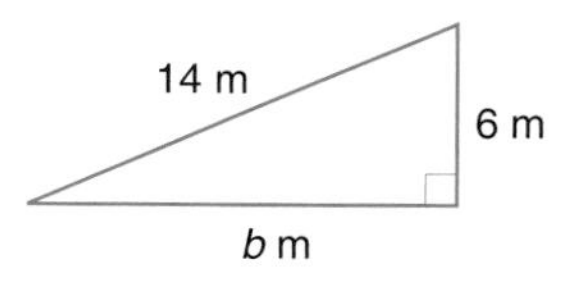

Reading Algebra

The hypotenuse is *always* the longest side of a right triangle.

$c^2 = a^2 + b^2$ *Pythagorean Theorem*

$14^2 = 6^2 + b^2$ *Replace c with 14 and a with 6.*

$196 = 36 + b^2$

$196 - 36 = 36 - 36 + b^2$ *Subtract 36 from each side.*

$160 = b^2$ ***Estimate:*** *Since $10^2 = 100$ and $15^2 = 225$, $\sqrt{160}$ is between 10 and 15.*

$\sqrt{160} = b$

160 [2nd] [$\sqrt{\ }$] [ENTER] 12.64911064

To the nearest tenth, the length of the leg is 12.6 meters.

Your Turn

Find each missing measure. Round to the nearest tenth.

b.

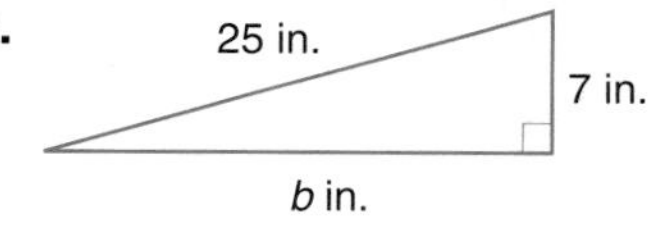

c.

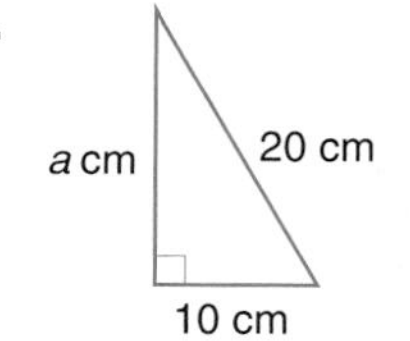

A **converse** of a theorem is the reverse, or opposite, of the theorem. You can use the converse of the Pythagorean Theorem to test whether a triangle is a right triangle.

Converse of the Pythagorean Theorem	If c is the measure of the longest side of a triangle and $c^2 = a^2 + b^2$, then the triangle is a right triangle.

Examples

3 **The measures of the three sides of a triangle are 5, 7, and 9. Determine whether this triangle is a right triangle.**

$c^2 = a^2 + b^2$ *Pythagorean Theorem*

$9^2 \stackrel{?}{=} 5^2 + 7^2$ *Replace c with 9, a with 5, and b with 7.*

$81 \stackrel{?}{=} 25 + 49$

$81 \neq 74$

Since $c^2 \neq a^2 + b^2$, the triangle is *not* a right triangle.

Your Turn

The measures of three sides of a triangle are given. Determine whether each triangle is a right triangle.

d. 20, 21, 28

e. 37, 12, 34

Carpentry Link

4 **A carpenter checks whether the corners of a deck are square by using a 3-4-5 system. He or she measures along one side in 3-foot units and along the adjacent side in the same number of 4-foot units. If the measure of the hypotenuse is the same number of 5-foot units, the corner of the deck is square.** *If a corner is square, it is formed by a right angle. So, the triangle is a right triangle.*

Suppose a carpenter measures along one side of a deck, a distance of 9 feet, and along the adjacent side, a distance of 12 feet. The measure of the hypotenuse is 15 feet. Is the corner of the deck square?

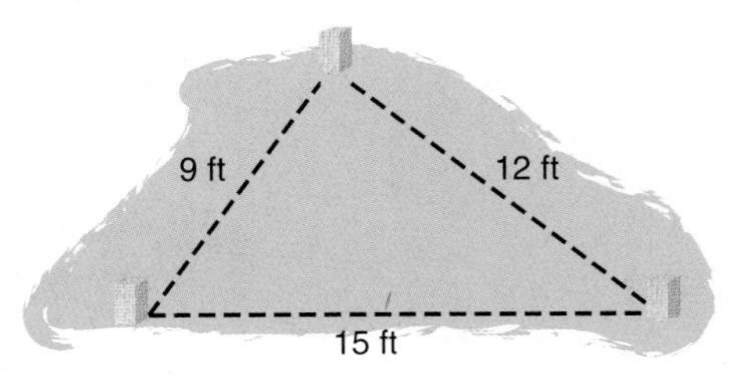

Explore You know that the measures of the sides are 9, 12, and 15. You want to know if the sides form a right triangle.

Plan The measures of the legs are 9 and 12, and the measure of the hypotenuse is 15. Use these numbers in the Pythagorean Theorem.

Solve	$c^2 = a^2 + b^2$	*Pythagorean Theorem*
	$15^2 \stackrel{?}{=} 9^2 + 12^2$	*Replace c with 15, a with 9, and b with 12.*
	$225 \stackrel{?}{=} 81 + 144$	
	$225 = 225$	

Since $c^2 = a^2 + b^2$, the triangle is a right triangle and the corner of the deck is square.

Examine The measures of the sides are 9, 12, and 15. They are multiples of a 3-4-5 triangle. The answer is reasonable.

Check for Understanding

Communicating Mathematics

1. **Draw** a right triangle and label the right angle, the hypotenuse, and the legs.
2. **Explain** how you know whether a triangle is a right triangle if you know the lengths of the three sides.

Vocabulary
hypotenuse
leg
Pythagorean Theorem
converse

Guided Practice

Getting Ready **Determine whether each sentence is *true* or *false*.**

Sample: $4^2 + 5^2 = 6^2$ **Solution:** $16 + 25 \neq 36$; false

3. $9^2 + 10^2 = 11^2$
4. $7^2 + 24^2 = 25^2$
5. $3^2 + 12^2 = 20^2$

If *c* is the measure of the hypotenuse and *a* and *b* are the measures of the legs, find each missing measure. Round to the nearest tenth if necessary.

6. 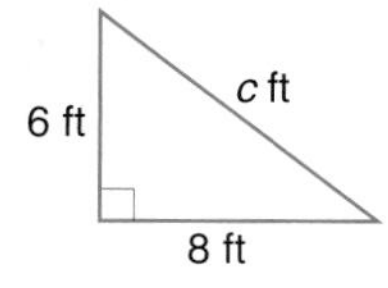

7. 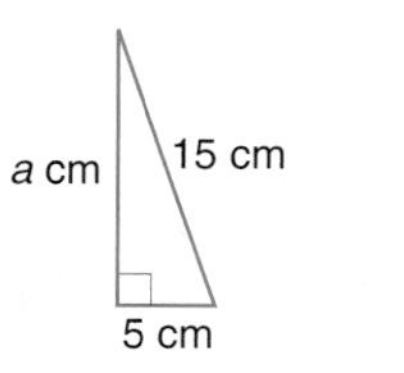

(Example 1)

8. $a = 30, c = 34, b = ?$
9. $a = 7, b = 4, c = ?$ *(Example 2)*

The lengths of three sides of a triangle are given. Determine whether each triangle is a right triangle. *(Example 3)*

10. 9 ft, 16 ft, 20 ft
11. 9 mm, 40 mm, 41 mm

12. **Construction** A builder is laying out the foundation for a house. The measure along one side is 12 feet, the adjacent side is 16 feet, and the hypotenuse is 21 feet. Determine whether the corner is a right angle. *(Example 4)*

Exercises

Practice

Homework Help	
For Exercises	See Examples
13, 15, 17–19, 23, 32, 34, 36	1
14, 16, 20–22, 24, 35	2
25–31	3, 4

Extra Practice

See page 709.

If c is the measure of the hypotenuse and a and b are the measures of the legs, find each missing measure. Round to the nearest tenth if necessary.

13. 5 cm, 12 cm, c cm

14.

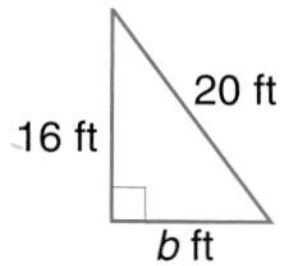

15.

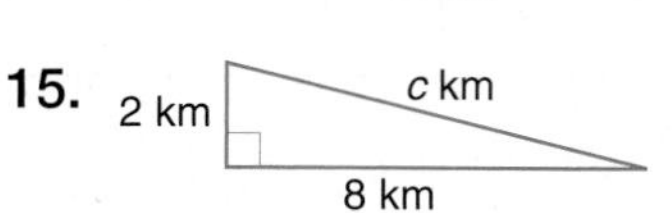

16. 25 in., a in., 24 in.

17.

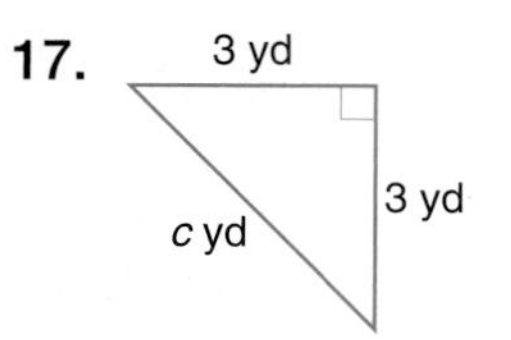

18. 5 m, 10 m, c m

19. $a = 6, b = 3, c = ?$

20. $b = 10, c = 11, a = ?$

21. $c = 29, a = 20, b = ?$

22. $a = 5, c = 30, b = ?$

23. $a = 7, b = 9, c = ?$

24. $a = 11, c = 20, b = ?$

The lengths of three sides of a triangle are given. Determine whether each triangle is a right triangle.

25. 11 in., 12 in., 16 in.

26. 11 cm, 60 cm, 61 cm

27. 6 ft, 8 ft, 9 ft

28. 6 mi, 7 mi, 12 mi

29. 45 m, 60 m, 75 m

30. 1 mm, 1 mm, 2 mm

31. Is a triangle with measures 30, 40, and 50 a right triangle?

32. Find the length of the hypotenuse of a right triangle if the lengths of the legs are 6 miles and 11 miles. Round to the nearest tenth.

Applications and Problem Solving

33. Geometry Find the length of the diagonal of a rectangle whose length is 8 meters and whose width is 5 meters. Round to the nearest tenth.

34. Carpentry The rise, the run, and the rafter of a pitched roof form a right triangle. Find the length of the rafter that is needed if the rise of the roof is 6 feet and the run is 12 feet. Round your answer to the nearest tenth.

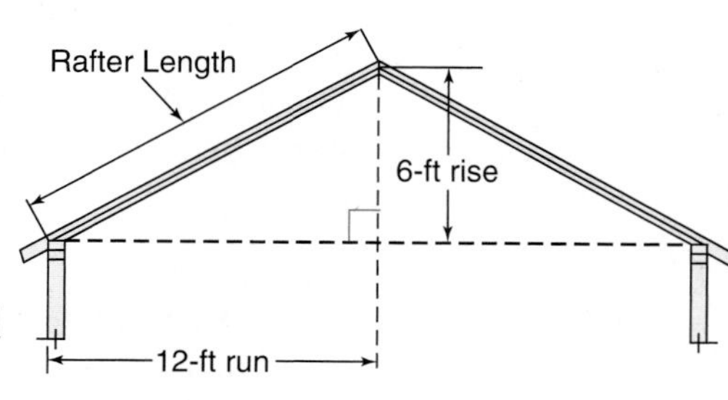

35. **Architecture** Architects often use arches when they design windows and doors. The Early English, or pointed, arch is based on an *equilateral triangle*. An equilateral triangle has three sides of equal measure.

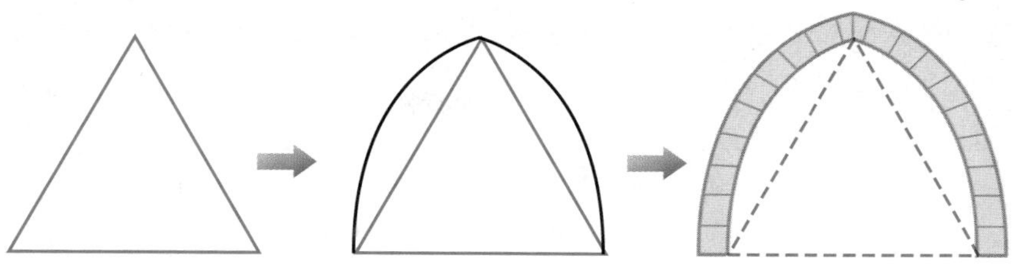

In the figure at the right, the height h separates the equilateral triangle into two right triangles of equal size. Find the height of the arch. Round to the nearest tenth.

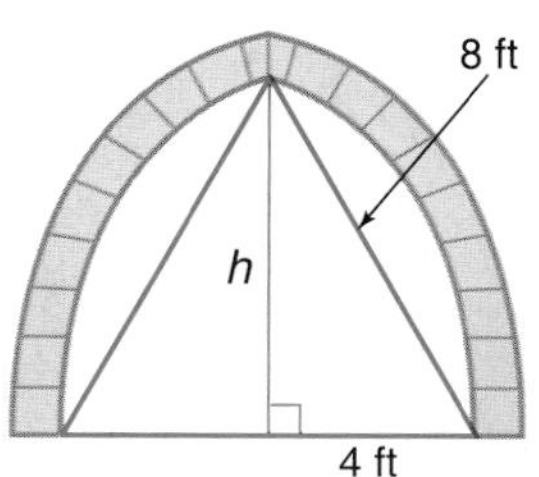

36. **Critical Thinking** The figure shows three squares. The two smaller squares each have an area of 1 square unit.

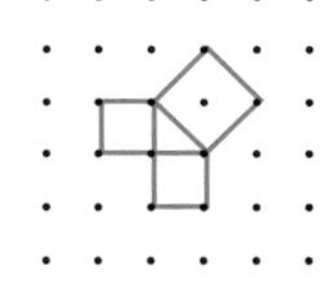

a. Find the area of the third square.

b. Draw a square on rectangular dot paper with an area of 5 square units.

Mixed Review

Estimate each square root to the nearest whole number. *(Lesson 8–6)*

37. 65 38. 95 39. 200 40. 500

Simplify. *(Lesson 8–5)*

41. $\sqrt{225}$ 42. $-\sqrt{49}$ 43. $\sqrt{100}$ 44. $-\sqrt{\frac{9}{25}}$

45. **Finance** To save for a new bicycle, Katie begins a savings plan. Her plan can be described by the equation $s = 5w + 100$, where s represents the total savings in dollars and w represents the number of weeks since the start of the savings plan. *(Lesson 7–3)*

a. How much money had Katie already saved when the plan started?

b. How much does Katie save each week?

Standardized Test Practice

46. **Short Response** Write the point-slope form of an equation for the line that passes through the point at (2, −4) and has a slope of −1. *(Lesson 7–2)*

47. **Multiple Choice** Which set is the domain of the relation shown on the graph? *(Lesson 6–1)*

A {−3, 0, 2}

B {−4, −2, 3, 4}

C neither A nor B

D both A and B

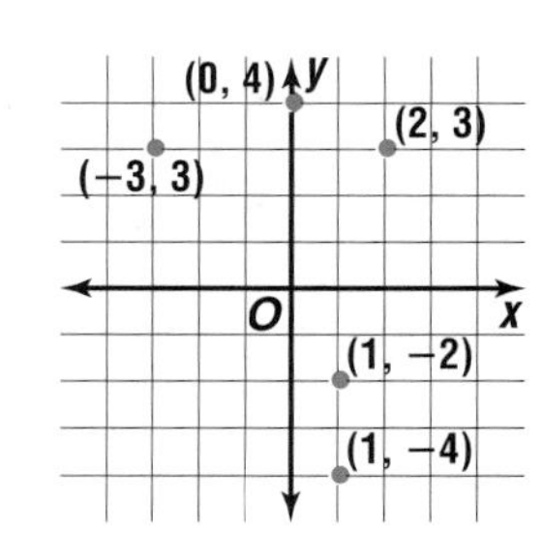

Chapter 8 Investigation

Must Be TV

Materials

uncooked spaghetti

ruler

grid paper

scissors

calculator

Using the Pythagorean Theorem

Look at an advertisement for an electronics store and you will find television sets that have 10-inch, 32-inch, and 36-inch screens. Did you ever wonder what these numbers represent? They represent the hypotenuse of a right triangle.

A television set is shaped like a rectangle with length ℓ and width w. The diagonal is a line segment from one corner to the opposite corner, which separates the screen into two right triangles of equal size. The diagonal is the hypotenuse of the right triangles. So, a 10-inch television set has a screen in which the diagonal is 10 inches.

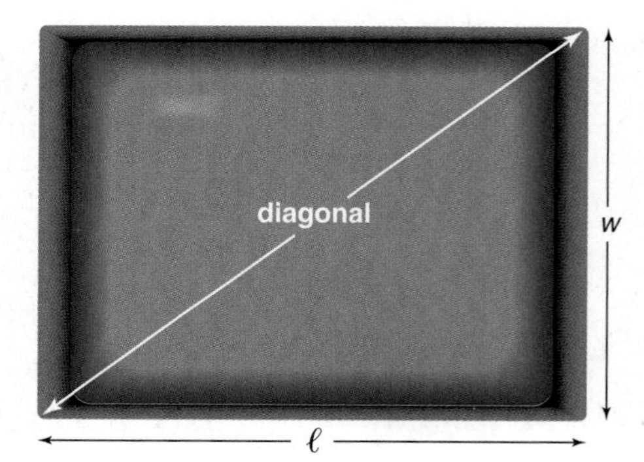

Investigate

1. Use a piece of uncooked spaghetti and grid paper to find several rectangles that have a 10-inch diagonal.
   a. Break the spaghetti so that one piece measures 10 inches.
   b. Place the spaghetti diagonally on the grid paper as shown below.
   c. Using the ends of the spaghetti as endpoints of the diagonal, draw horizontal and vertical lines from the endpoints to complete the rectangle.

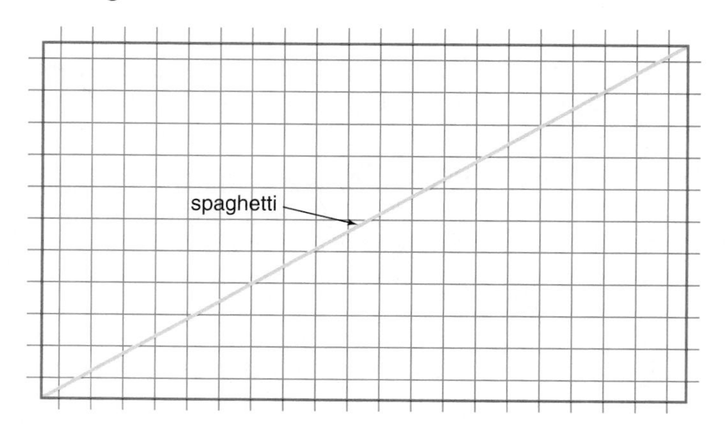

d. Repeat Steps 1b and 1c four more times using the same piece of spaghetti and different pieces of grid paper. For each rectangle, change the angle at which you place the spaghetti on the grid paper.

e. Cut out the rectangles. Choose the rectangle that looks most like a television screen. Describe how this rectangle is different from the others.

2. Television manufacturers have set a standard ratio for the size of a television screen. The ratio of the height to the width is 3:4. For which of your rectangles is the ratio of the height to the width closest to 3:4?

Extending the Investigation

In this extension, you will predict the dimensions of a 19-inch television screen.

- Use a calculator and the Pythagorean Theorem to find the measures of five different rectangles that have a 19-inch diagonal.
- Find the areas of each rectangle and the ratios of each height to width.
- Analyze the data and make a prediction about the dimensions of a 19-inch television screen.

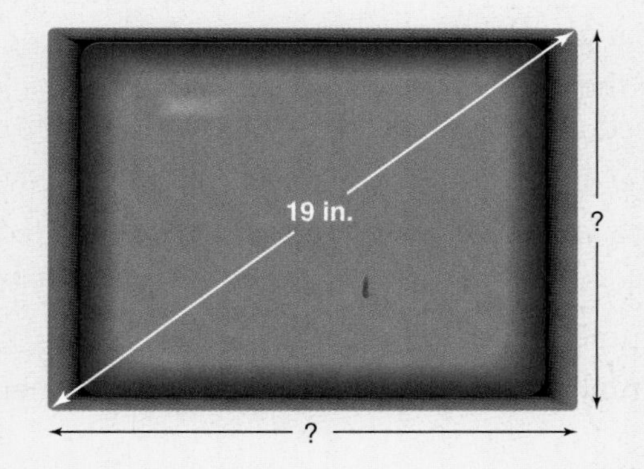

Presenting Your Investigation

Here are some ideas to help you present your conclusions to the class.

- Make a poster that shows scale drawings of your rectangles.
- Write a report in which you explain how you made your prediction. Be sure to include the terms *ratio* and *Pythagorean Theorem* in your report. Then compare your prediction to the dimensions of an actual 19-inch television.

Investigation For more information on ratios, visit: www.algconcepts.com

CHAPTER 8

Study Guide and Assessment

Understanding and Using the Vocabulary

After completing this chapter, you should be able to define each term, property, or phrase and give an example or two of each.

interNET CONNECTION **Review Activities**
For more review activities, visit: www.algconcepts.com

Algebra

base *(p. 336)*
composite number *(p. 358)*
exponent *(p. 336)*
irrational numbers *(p. 362)*
negative exponent *(p. 348)*
perfect square *(p. 336)*
power *(p. 336)*
prime factorization *(p. 358)*
prime number *(p. 358)*
radical expression *(p. 358)*
radical sign *(p. 357)*
scientific notation *(p. 353)*
square root *(p. 357)*

Geometry

hypotenuse *(p. 366)*
leg *(p. 366)*
Pythagorean Theorem *(p. 366)*

Logic

converse *(p. 268)*

Complete each sentence using a term from the vocabulary list.

1. When a number is expressed as a product of factors that are all prime, the expression is called the ___?___ of the number.
2. The symbol $\sqrt{\ }$ is called a ___?___.
3. The square roots of numbers such as 2, 3, and 7 are ___?___.
4. The ___?___ tells the relationship among the measures of the legs of a right triangle and the hypotenuse.
5. In the expression 2^3, the 3 is called the ___?___. It tells how many times 2 is used as a factor.
6. The ___?___ of a number is one of its two equal factors.
7. The longest side of a right triangle is called the ___?___.
8. When you express 5,000,000 as 5×10^6, you are using ___?___.
9. Numbers like 25, 36, and 100 are called ___?___.
10. Numbers that are expressed using exponents are called ___?___.

Skills and Concepts

Objectives and Examples

• **Lesson 8–1** Use powers in expressions.

Write $n \cdot n \cdot n \cdot n$ using exponents.
$n \cdot n \cdot n \cdot n = n^4$

Write $2^3 \cdot 3^2$ as a multiplication expression.
$(2^3)(3^2) = (2)(2)(2)(3)(3)$

Review Exercises

Write each expression using exponents.

11. $9 \cdot 9 \cdot 9 \cdot 9 \cdot 9$
12. 5 squared
13. $2 \cdot 2 \cdot 3 \cdot 3 \cdot 3$
14. $(-2)(x)(x)(y)(y)$

Write each power as a multiplication expression.

15. 12^3
16. $(-2)^5$
17. y^2
18. a^3b^2
19. $5x^2y^3z^2$
20. $-5m^2$

www.algconcepts.com/vocabulary_review

Chapter 8 Study Guide and Assessment

Objectives and Examples

• **Lesson 8–2** Multiply and divide powers.

$$(2y^4)(5y^5) = (2 \cdot 5)(y^4 \cdot y^5) = (10)(y^{4+5}) = 10y^9$$

$$\frac{x^7y^4}{x^3y^3} = \left(\frac{x^7}{x^3}\right)\left(\frac{y^4}{y^3}\right) = (x^{7-3})(y^{4-3}) = x^4y$$

Review Exercises

Simplify each expression.

21. $b^2 \cdot b^5$ **22.** $y^3 \cdot y^3 \cdot y$

23. $(a^2b)(a^2b^2)$ **24.** $(3ab)(-4a^2b^3)$

25. $\frac{y^{10}}{y^6}$ **26.** $\frac{a^2b^4}{a^3b^2}$

27. $\frac{42x^7}{14x^4}$ **28.** $\frac{-15x^3y^5z^4}{5x^2y^2z}$

• **Lesson 8–3** Simplify expressions containing negative exponents.

$$a^2b^{-3} = a^2 \cdot \frac{1}{b^3} = \frac{a^2}{b^3}$$

Write each expression using positive exponents. Then evaluate the expression.

29. 2^{-3} **30.** 10^{-2}

Simplify each expression.

31. x^{-4} **32.** $m^{-2}n^5$

33. $\frac{y^3}{y^{-4}}$ **34.** $\frac{rs^2}{s^3}$

35. $\frac{-25a^5b^6}{5a^5b^3}$ **36.** $\frac{27b^{-2}}{14b^{-3}}$

• **Lesson 8–4** Express numbers in scientific notation.

$3600 = 3.6 \times 10^3$

$0.023 = 2.3 \times 10^{-2}$

Express each measure in standard form.

37. 1.5 nanoseconds **38.** 5 kilobytes

Express each number in scientific notation.

39. 240,000 **40.** 0.000314

41. 4,880,000,000 **42.** 0.000015

Evaluate each expression.

43. $(2 \times 10^5)(3 \times 10^6)$ **44.** $\frac{8 \times 10^{-3}}{2 \times 10^4}$

• **Lesson 8–5** Simplify radicals.

$$\sqrt{400} = \sqrt{2 \cdot 2 \cdot 2 \cdot 2 \cdot 5 \cdot 5} = \sqrt{16 \cdot 25} = \sqrt{16} \cdot \sqrt{25} = 4 \cdot 5 \text{ or } 20$$

Simplify.

45. $\sqrt{121}$ **46.** $-\sqrt{324}$

47. $\sqrt{\frac{4}{81}}$ **48.** $-\sqrt{\frac{100}{225}}$

• Extra Practice
See pages 707–709.

Objectives and Examples

• **Lesson 8–6** Estimate square roots.

Estimate $\sqrt{150}$.

$144 < 150 < 169$ *. . . 100, 121, 144, 169, . . .*

$\sqrt{144} < \sqrt{150} < \sqrt{169}$

$12 < \sqrt{150} < 13$ *150 is closer to 144 than to 169.*

The best whole number estimate for $\sqrt{150}$ is 12.

Review Exercises

Estimate each square root to the nearest whole number.

49. $\sqrt{19}$

50. $\sqrt{108}$

51. $\sqrt{200}$

52. $\sqrt{125.52}$

53. Which is closer to $\sqrt{50}$, 7 or 8?

• **Lesson 8–7** Use the Pythagorean Theorem to solve problems.

Find the length of the missing side.

$$c^2 = a^2 + b^2$$
$$25^2 = 15^2 + b^2$$
$$625 = 225 + b^2$$
$$625 - 225 = 225 - 225 + b^2$$
$$400 = b^2$$
$$20 = b$$

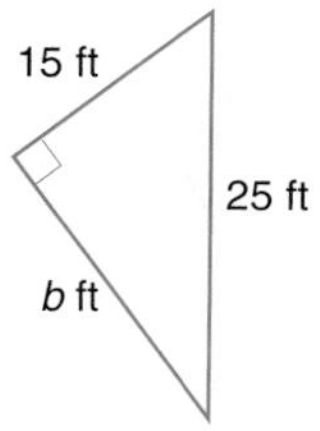

(Figure: right triangle with legs 15 ft and b ft, hypotenuse 25 ft)

If c is the measure of the hypotenuse and a and b are the measures of the legs, find each missing measure. Round to the nearest tenth if necessary.

54.

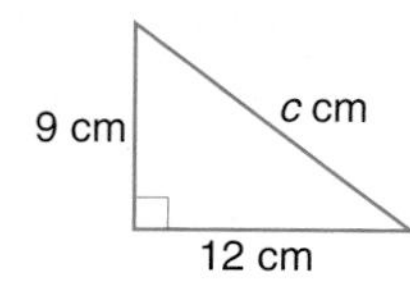

55.

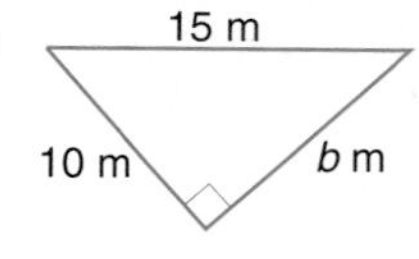

56. $a = 6$ mi, $b = 10$ mi, $c = ?$

57. $b = 6$, $c = 12$, $a = ?$

Applications and Problem Solving

58. Biology *E. coli* is a fast-growing bacteria that splits into two identical cells every 15 minutes. If you start with one *E. coli* bacterium, how many bacteria will there be at the end of 2 hours? *(Lesson 8–1)*

59. Geometry The area of a square is 90 square meters. Estimate the length of its side to the nearest whole number. *(Lesson 8–6)*

60. Construction Park managers want to construct a road from the park entrance directly to a campsite. Use the figure below to find the length of the proposed road. *(Lesson 8–7)*

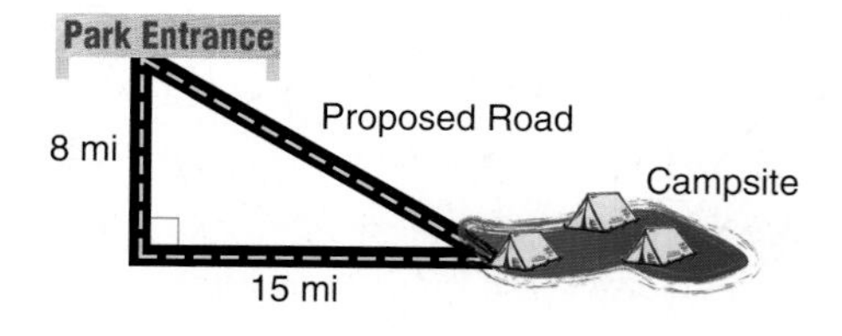

CHAPTER 8 Test

1. **Write** a multiplication problem whose product is $12x^5$.
2. **Explain** why 3 is the square root of 9.
3. **Draw** a figure that can be used to estimate $\sqrt{96}$.

Evaluate each expression if $x = 2$, $y = 3$, $z = -1$, and $w = -2$.

4. $3x^5$
5. $5w^3y$
6. $12y^{-2}$
7. z^0

Simplify each expression.

8. $(a^3b^2)(a^5b)$
9. $\dfrac{xy^3}{y^5}$
10. m^{-10}
11. $\dfrac{-3r^2s^8}{12r^2s^5}$

Express each number in scientific notation.

12. 0.00000125
13. 36,000,000,000

Evaluate each expression. Express each result in scientific notation and standard form.

14. $(8.2 \times 10^{-5})(1 \times 10^{-3})$
15. $\dfrac{6 \times 10^3}{2 \times 10^{-2}}$

Estimate each square root to the nearest whole number.

16. $\sqrt{5}$
17. $\sqrt{20}$
18. $\sqrt{63}$
19. $\sqrt{395}$

Find each missing measure. Round to the nearest tenth if necessary.

20.

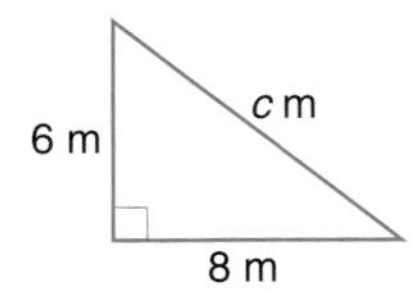

21.

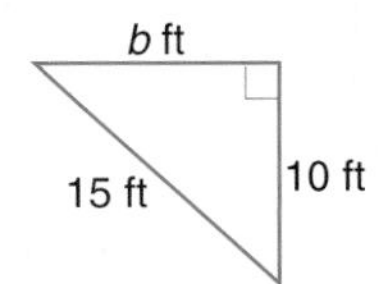

22.

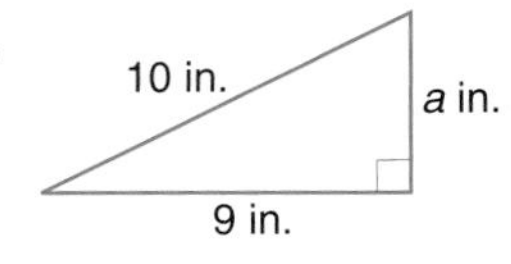

23. **Chemistry** The diameter of an atom is about 10^{-8} centimeter and the diameter of its nucleus is about 10^{-13} centimeter. How many times as large is the diameter of the atom as the diameter of its nucleus?

24. **Sports** Gymnastic routines are performed on a 40-foot by 40-foot square mat. Gymnasts usually use the diagonal of the mat because it gives them more distance to complete their routine. To the nearest whole number, how much longer is the diagonal of the mat than one of its sides?

25. **Sequences** The first term of the sequence $1, \frac{1}{2}, \frac{1}{4}, \frac{1}{8}, \ldots$ can be written as 2^0. Write the next three terms using negative exponents.

CHAPTER 8

Preparing for Standardized Tests

Pythagorean Theorem Problems

All standardized tests contain several problems that you can solve using the Pythagorean Theorem. The **Pythagorean Theorem** states that in a right triangle, the sum of the squares of the measures of the legs equals the square of the measure of the hypotenuse.

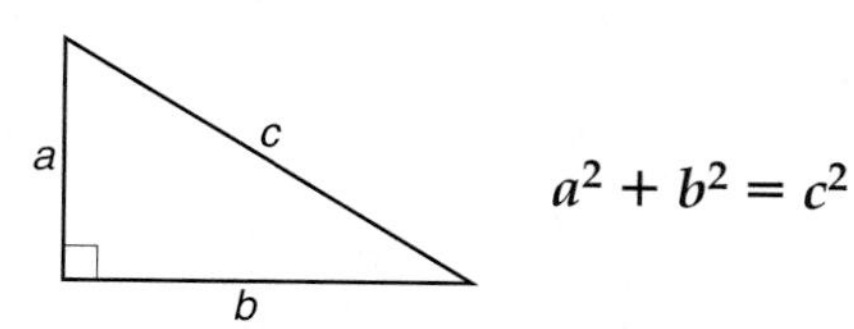

$$a^2 + b^2 = c^2$$

Test-Taking Tip

The 3-4-5 right triangle and its multiples, like 6-8-10 and 9-12-15, occur frequently on standardized tests. Other commonly used Pythagorean triples include 5-12-13 and 7-24-25.

State Test Example

Use the information in the figure below. Find *BC* to the nearest centimeter.

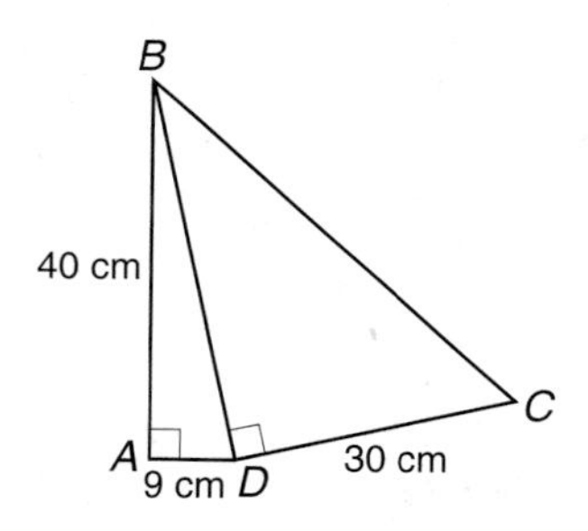

A 31 cm **B** 41 cm

C 51 cm **D** 80 cm

Hint Use the Pythagorean Theorem twice.

Solution The figure is made up of two right triangles, $\triangle ABD$ and $\triangle BDC$. $\overline{BD}$ is the hypotenuse of $\triangle ABD$. $\overline{BD}$ is also a leg of $\triangle BDC$.

First, find BD. Then use BD to find BC.

$(BD)^2 = 9^2 + 40^2$
$(BD)^2 = 81 + 1600$
$(BD)^2 = 1681$
$BD = 41$

$(BC)^2 = 30^2 + 41^2$
$(BC)^2 = 900 + 1681$
$(BC)^2 = 2581$
$BC \approx 50.80$

The answer is C.

SAT Example

A 25-foot ladder is placed against a vertical wall of a building with the bottom of the ladder standing on concrete 7 feet from the base of the building. If the top of the ladder slips down 4 feet, then the bottom of the ladder will slide out how many feet?

A 5 ft **B** 6 ft

C 7 ft **D** 8 ft

Hint Start by drawing a diagram.

Solution The ladder placed against the wall forms a 7-24-25 right triangle. After the ladder slips down 4 feet, the new right triangle has sides that are multiples of a 3-4-5 right triangle, 15-20-25.

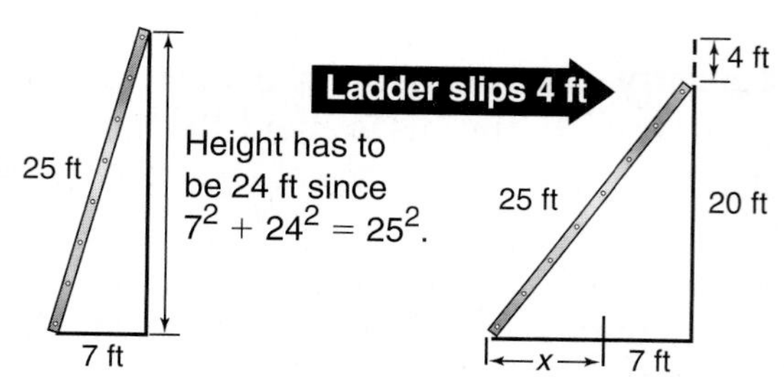

The bottom of the ladder will be 15 feet from the wall. This means the bottom of the ladder will slide out 15 − 7 or 8 feet.

The answer is D.

After you work each problem, record your answer on the answer sheet provided or on a sheet of paper.

Multiple Choice

1. Diego bought 3 fish. Every month, the number of fish doubles. The formula $f = 3 \cdot 2^m$ represents the number of fish he will have after m months. How many fish will he have after 4 months if he keeps all of them and none of them dies?

A 12 **B** 24
C 48 **D** 1296

2. Which number is greater than 2.8?

A 175% **B** $\sqrt{12}$
C -3.5 **D** 5.6×10^{-1}

3. A wire from the top of a 25-foot flagpole is attached to a point 20 feet from the base of the flagpole. To the nearest whole number, find the length of the wire.

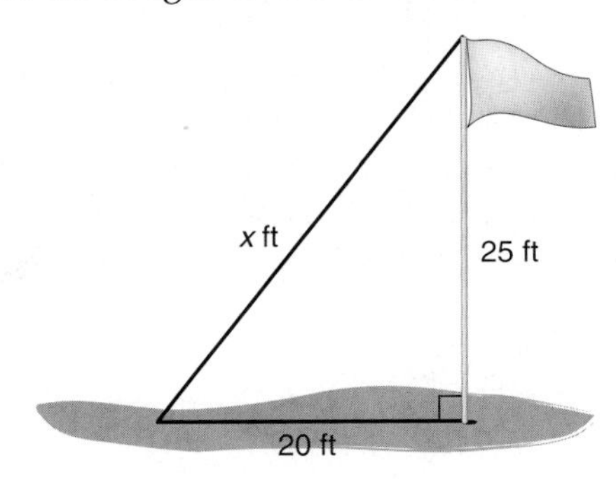

A 45 ft **B** 32 ft
C 15 ft **D** 35 ft

4. Which of the following are measures of three sides of a right triangle?

A 4, 7, 8 **B** 10, 15, 20
C 3, 7, 9 **D** 9, 12, 15

5. Which equation does *not* have 6 as a solution?

A $2(x - 6) = 0$ **B** $4x + 3 = 27$
C $2x + 1 = 14$ **D** $2x + x = 18$

6. In the figure, $\triangle ABC$ and $\triangle ACD$ are right triangles. If $AB = 20$, $BC = 15$, and $AD = 7$, then $CD =$ —

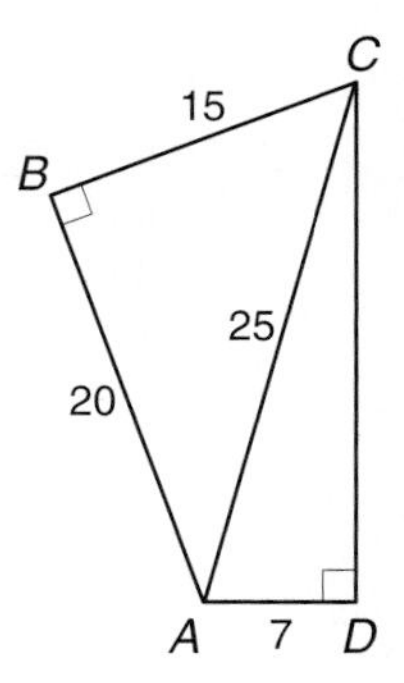

A 22.
B 23.
C 24.
D 25.

7. A farmer wants to put a fence around a square garden that has a total area of 1600 square feet. If fencing costs $1.50 per foot, how much will it cost to fence the garden? To solve this problem, begin by—

A dividing 1600 by $1.50.
B multiplying 1600 by $1.50 and taking the square root of the product.
C dividing 1600 by 4 and multiplying by $1.50.
D finding the square root of 1600 and multiplying by 4.

8. If $x = 3$, $y = -2$, $z = -1$, and $w = 5$, then the value of $9xy^2z^{10}w^0$ is—

A -590. **B** -108.
C 1. **D** 108.

Grid In

9. On the first test of the semester, Jennifer scored a 60. On the last test of the semester, she scored a 75. By what percent did Jennifer's score improve?

Short Response

10. The table shows the squares of six whole numbers. Describe one way that the digits in the squares are related to the digits in the original number.

$15^2 = 225$	$55^2 = 3025$	$135^2 = 18{,}225$
$25^2 = 625$	$95^2 = 9025$	$175^2 = 30{,}625$

CHAPTER 9

Polynomials

FOLDABLES™ Study Organizer

Make this Foldable to help you organize information about polynomials. Begin with a sheet of notebook paper.

1. **Fold** lengthwise to the holes.

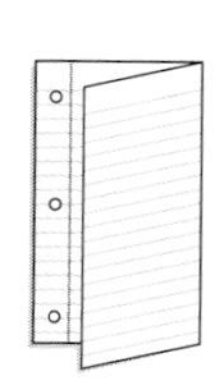

2. **Cut** along the top line and then cut three tabs.

3. **Label** the tabs using the lesson concepts as shown.

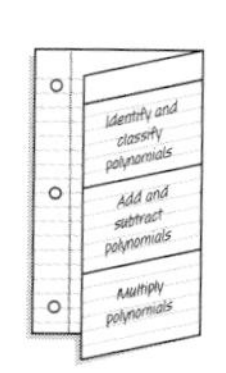

Reading and Writing Store the foldable in a 3-ring binder. As you read and study the chapter, write notes and examples under the tabs.

Problem-Solving Workshop

Project

Congratulations! You have been selected to design the box for a new caramel corn snack. The manufacturers would like you to present three possible box designs. Each box must be a rectangular prism with a surface area of 150 square inches. Make a pattern for each of the three box designs. In a one-page paper, explain which box design you think will be best for the new product snack.

Working on the Project

Work with a partner and choose a strategy. Here are some suggestions to help you get started.

- Cut apart an empty cereal box along its edges to make a pattern for the surface area of a rectangular prism.
- Find the area of each of the six faces and the surface area of the prism.
- Suppose you don't know the measurements of the prism. Let ℓ represent the length, w represent the width, and h represent the height of the prism. Write a polynomial expression for the surface area of the prism.

Strategies

Look for a pattern.
Draw a diagram.
Make a table.
Work backward.
Use an equation.
Make a graph.
Guess and check.

Technology Tools

- Use a **calculator** to find the surface area of your patterns.
- Use **drawing software** to make patterns for your boxes.

*inter***NET** CONNECTION **Research** For more information about packaging, visit: www.algconcepts.com

Presenting the Project

Prepare a portfolio of your box designs. Make sure your portfolio contains the following information:

- your calculations of the surface area of each box design,
- a one-page paper promoting the box design you think will be best, and
- a name for the new product and a logo or design for the box.

9–1 Polynomials

What You'll Learn

You'll learn to identify and classify polynomials and find their degree.

Why It's Important

Medicine Doctors can use polynomials to study the heart. See Exercise 57.

A **monomial** is a number, a variable, or a product of numbers and variables that have only positive exponents. A monomial cannot have a variable as an exponent. The tables below show examples of expressions that are and are not monomials.

Monomials	
-4	a number
y	a variable
a^2	the product of variables
$\frac{1}{2}x^2y$	the product of numbers and variables

Not Monomials	
2^x	has a variable as an exponent
$x^2 + 3$	a sum
$5a^{-2}$	includes a negative exponent
$\frac{3}{x}$	a quotient

Examples

Determine whether each expression is a monomial. Explain why or why not.

1 $-6ab$

$-6ab$ is a monomial because it is the product of a number and variables.

2 $m^2 - 4$

$m^2 - 4$ is *not* a monomial because it includes subtraction.

Your Turn

a. 10 b. $5z^{-3}$ c. $\frac{6}{x}$ d. x^2

A monomial or the sum of one or more monomials is called a **polynomial**. For example, $x^3 + x^2 + 3x + 2$ is a polynomial. Each monomial is a term. The terms of the polynomial are x^3, x^2, $3x$, and 2.

Recall that to subtract, you add the opposite.

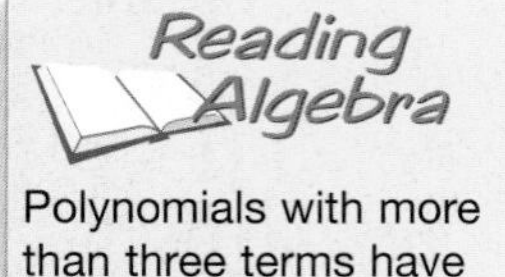

Polynomials with more than three terms have no special name.

Special names are given to polynomials with two or three terms. A polynomial with two terms is a **binomial**. A polynomial with three terms is a **trinomial**. Here are some examples.

Binomial	Trinomial
$x + 2$	$a + b + c$
$5c - 4$	$x^2 + 5x - 7$
$4w^2 - w$	$3a^2 + 5ab + 2b^2$

Examples

State whether each expression is a polynomial. If it is a polynomial, identify it as a *monomial, binomial,* or *trinomial.*

3 $2m - 7$

The expression $2m - 7$ can be written as $2m + (-7)$. So, it is a polynomial. Since it can be written as the sum of two monomials, $2m$ and -7, it is a binomial.

4 $x^2 + 3x - 4 - 5$

The expression $x^2 + 3x - 4 - 5$ can be written as $x^2 + 3x + (-9)$. So, it is a polynomial. Since it can be written as the sum of three monomials, it is a trinomial.

5 $\frac{5}{2x} - 3$

The expression $\frac{5}{2x} - 3$ is not a polynomial since $\frac{5}{2x}$ is not a monomial.

Your Turn

e. $5a - 9 + 3$ **f.** $4m^{-2} + 2$ **g.** $3y^2 - 6 + 7y$

The terms of a polynomial are usually arranged so that the powers of one variable are in descending or ascending order.

Polynomial	Descending Order	Ascending Order	
$2x + x^2 + 1$	$x^2 + 2x + 1$	$1 + 2x + x^2$	
$3y^2 + 5y^3 + y$	$5y^3 + 3y^2 + y$	$y + 3y^2 + 5y^3$	
$x^2 + y^2 + 3xy$	$x^2 + 3xy + y^2$	$y^2 + 3xy + x^2$	*in terms of x*
$2xy + y^2 + x^2$	$y^2 + 2xy + x^2$	$x^2 + 2xy + y^2$	*in terms of y*

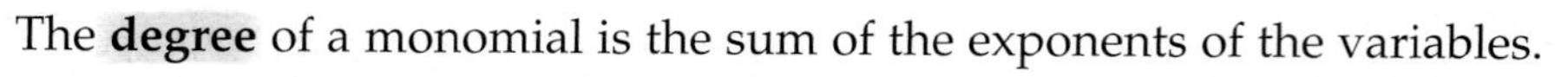
The **degree** of a monomial is the sum of the exponents of the variables.

Look Back
Zero Power: Lesson 8–2

Monomial	Degree
$-3x^2$	2
$5pq^2$	$1 + 2 = 3$
2	0

$p = p^1$
$2 = 2x^0$

To find the degree of a polynomial, find the degree of each term. The greatest of the degrees of its terms is the degree of the polynomial.

Polynomial	Terms	Degree of the Terms	Degree of the Polynomial
$2n + 7$	$2n$, 7	1, 0	1
$3x^2 + 5x$	$3x^2$, $5x$	2, 1	2
$a^6 + 2a^3 + 1$	a^6, $2a^3$, 1	6, 3, 0	6
$5x^4 - 4a^2b^6 + 3x$	$5x^4$, $-4a^2b^6$, $3x$	4, 8, 1	8

Examples

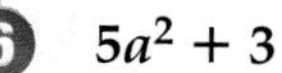

Find the degree of each polynomial.

6 $5a^2 + 3$

Term	Degree
$5a^2$	2
3	0

$3 = 3x^0$

The degree of $5a^2 + 3$ is 2.

7 $6x^2 - 4x^2y - 3xy$

Term	Degree
$6x^2$	2
$-4x^2y$	2 + 1 or 3
$-3xy$	1 + 1 or 2

$y = y^1$
$x = x^1, y = y^1$

The degree of $6x^2 - 4x^2y - 3xy$ is 3.

Your Turn

h. $3x^2 - 7x$

i. $8m^3 - 2m^2n^2 + 5$

Polynomials can be used to represent many real-world situations.

Example
Science Link

8 **The expression $14x^3 - 17x^2 - 16x + 34$ can be used to estimate the number of eggs that a certain type of female moth can produce. In the expression, x represents the width of the abdomen in millimeters. About how many eggs would you expect this type of moth to produce if her abdomen measures 3 millimeters?**

Explore You know the width of the abdomen. You need to determine the number of eggs that the moth can produce.

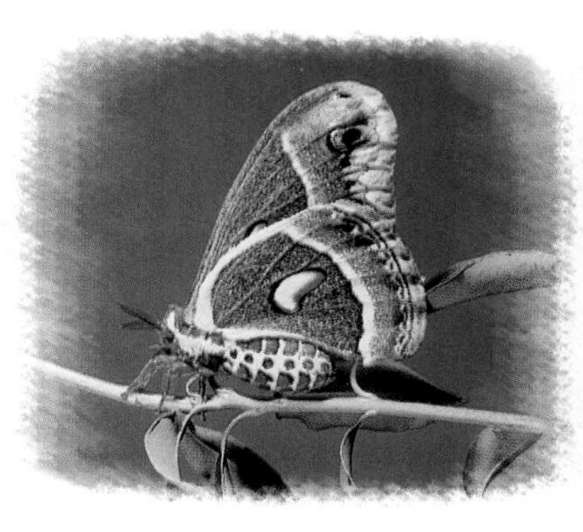
Female Cecropia Moth

Plan In the expression $14x^3 - 17x^2 - 16x + 34$, replace x with 3 to determine the number of eggs that the moth can produce.

Solve

$$\begin{aligned} 14x^3 - 17x^2 - 16x + 34 &= 14(3)^3 - 17(3)^2 - 16(3) + 34 \\ &= 14(27) - 17(9) - 16(3) + 34 \\ &= 378 - 153 - 48 + 34 \\ &= 211 \end{aligned}$$

Examine Check your computations by adding in a different order.

$$\begin{aligned} 378 - 153 - 48 + 34 &\stackrel{?}{=} 378 + 34 + (-153) + (-48) \\ &\stackrel{?}{=} 412 + (-201) \\ &= 211 \quad \checkmark \end{aligned}$$

The moth can produce about 211 eggs.

Check for Understanding

Communicating Mathematics

1. **Write** an expression that is *not* a monomial. Explain why it is *not* a monomial.
2. **Explain** how to find the degree of a polynomial.
3. **Arrange** the terms of the polynomial $3x^2 + 5xy + 2y^2$ so that the powers of x are in ascending order.

Vocabulary
monomial
polynomial
binomial
trinomial
degree

4. **Write** a definition for *monomial, polynomial, binomial,* and *trinomial.* Give three examples of each.

Guided Practice

Determine whether each expression is a monomial. Explain why or why not. *(Examples 1 & 2)*

5. $7xy$
6. $a^2 + 5$

State whether each expression is a polynomial. If it is a polynomial, identify it as a *monomial, binomial,* or *trinomial*. *(Examples 3–5)*

7. $2mn + 3$
8. 0
9. $\frac{4}{d} - d^2$
10. $h^2 - 3h + 8 + 2$

Find the degree of each polynomial. *(Examples 6 & 7)*

11. $25x$
12. 5
13. $15m^2 + 4n$
14. $-6y^3z - 4y^2z$

15. **Geometry** To find the number of diagonals d in an n-sided figure, you can use the formula $d = 0.5n^2 - 1.5n$. *(Example 8)*
    a. Find the number of diagonals in an octagon.
    b. How many diagonals are in a hexagon?

Exercises

Practice

Homework Help	
For Exercises	See Examples
16–21	1, 2
22–33, 54, 55	3, 4, 5
33–45, 57	6, 7
56	8
Extra Practice	
See page 710.	

Determine whether each expression is a monomial. Explain why or why not.

16. $5z$
17. $7a + 2$
18. $\frac{5}{x}$
19. $8x^2y$
20. y^{-3}
21. $-10a^2b^2c$

State whether each expression is a polynomial. If it is a polynomial, identify it as a *monomial*, *binomial*, or *trinomial*.

22. y^3
23. $2y^2 + 5y - 7$
24. $a^3 - 3a$
25. $2r + 3s^{-3} - t$
26. $17 + 12x^3 - 3x^4$
27. $2m^2 + 5 - 1$
28. $-4j + 2k^2 - 3$
29. $x^2 - \frac{1}{2}x$
30. $\frac{x}{y}$
31. $-12a^2b^3cd^5e^3$
32. $2.5w^5 - v^3w^2$
33. $3x + 4x$

Find the degree of each polynomial.

34. 1
35. $8x^2$
36. $-14x^3$
37. $5p^2 + 3p + 5$
38. $4s^2 - 6t^8$
39. $p - 4q + 5r$
40. $25a + a^2$
41. $a^3 + a^2 + a$
42. $15g^3h^4 - 10gh^5$
43. $10v^4w^2 + vw^5$
44. $3rs^4 - 2r^2 + 7$
45. $5cd^4 - 2bcd$

Arrange the terms of each polynomial so that the powers of x are in descending order.

46. $2x^2 + x^4 - 6$
47. $x^2 - x + 25 - x^5$
48. $3x - 4x^2y + 5 - 3x^5$
49. $6w^3x + 3wx^2 - 10x^6 + 5x^7$

Arrange the terms of each polynomial so that the powers of x are in ascending order.

50. $2 + x^4 + x^5 + x^2$
51. $5x^3 - x^2 + 7 + 2x$
52. $a^3x - 6a^3x^3 + 0.5x^5 + 4x^2$
53. $3x^3y + 5xy^4 - x^2y^3 + y^6$

54. *True* or *false*: Every monomial is a polynomial.
55. *True* or *false*: Every polynomial is a monomial.

Applications and Problem Solving

56. **Geometry** The surface area of a right cylinder is given by the polynomial $2\pi rh + 2\pi r^2$, where r is the radius of the base of the cylinder and h is the height of the cylinder.

   a. Classify the polynomial.

   b. Find the surface area of a right circular cylinder with a height of 5 feet and a radius of 3 feet. Round to the nearest tenth.

   c. The volume of a cylinder is given by the expression $\pi r^2 h$. Using the same height and radius as in part b, find the volume of the cylinder. Round to the nearest tenth.

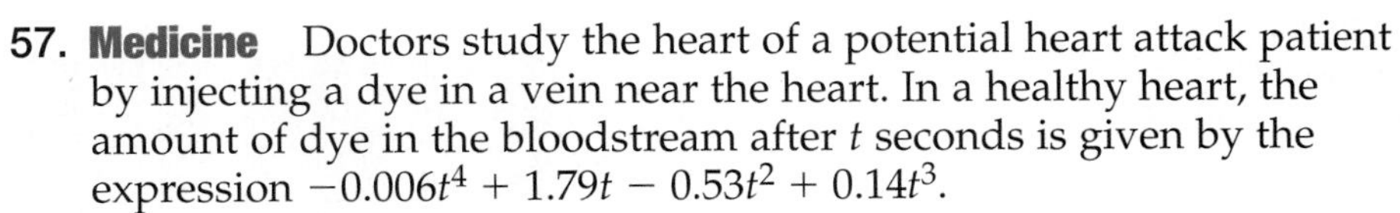

57. **Medicine** Doctors study the heart of a potential heart attack patient by injecting a dye in a vein near the heart. In a healthy heart, the amount of dye in the bloodstream after t seconds is given by the expression $-0.006t^4 + 1.79t - 0.53t^2 + 0.14t^3$.

   a. Arrange the terms of the polynomial so that the powers of t are in descending order.

   b. Find the degree of the polynomial.

58. **Critical Thinking** A number in base 10 can be written in polynomial form. For example, $3247 = 3(10)^3 + 2(10)^2 + 4(10) + 7(10)^0$.

   a. Write the year of your birth in polynomial form.

   b. Suppose 2356 is a number in base a. Write 2356 in polynomial form.

   c. What is the degree of the polynomial in part b?

Mixed Review

The lengths of three sides of a triangle are given. Determine whether each triangle is a right triangle. *(Lesson 8–7)*

59. 5 cm, 12 cm, 13 cm 60. 2 in., 3 in., 4 in. 61. 8 ft, 15 ft, 17 ft

Estimate each square root to the nearest whole number. *(Lesson 8–6)*

62. $\sqrt{20}$ 63. $\sqrt{50}$ 64. $\sqrt{75}$ 65. $\sqrt{120}$

66. **Population** In 2050, Earth's population is expected to be 8,900,000,000. Write this number in scientific notation. *(Lesson 8–4)*

Simplify each expression. *(Lesson 8–3)*

67. xy^{-3} 68. $\frac{k^{-3}}{k^4}$

Standardized Test Practice

69. **Multiple Choice** The graph of $y = 3x + 6$ is shown at the right. Which is the slope of a line parallel to the graph of $y = 3x + 6$? *(Lesson 7–7)*

   A -3 B 3

   C $-\frac{1}{3}$ D $\frac{1}{6}$

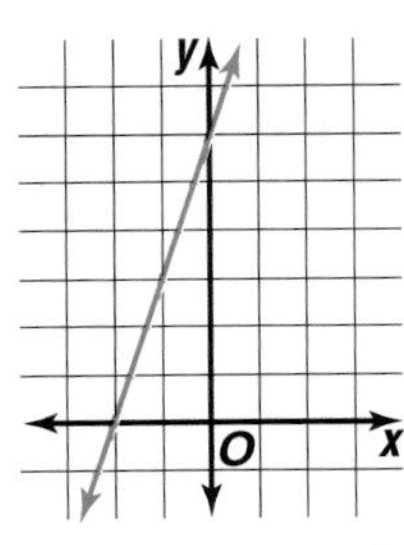

9–2 Adding and Subtracting Polynomials

What You'll Learn
You'll learn to add and subtract polynomials.

Why It's Important
Framing Addition of polynomials can be used to find the size of a picture.
See Exercise 43.

You can use algebra tiles like the ones below to model polynomials.

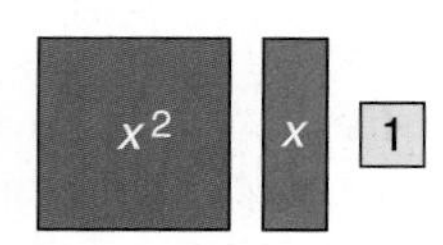

A model of the polynomial $2x^2 + 3x + 1$ is shown below.

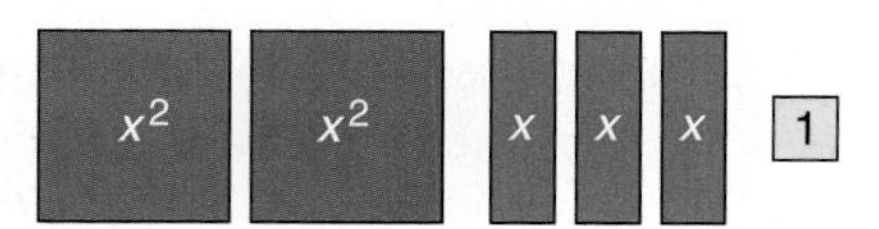

Red tiles are used to represent -1, $-x$, and $-x^2$. The model of the polynomial $-x^2 - 2x - 3$ is shown below.

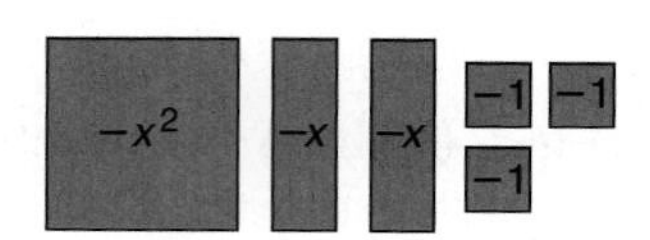

Once you know how to model polynomials using algebra tiles, you can use algebra tiles to add and subtract polynomials.

Hands-On Algebra
Algebra Tiles

Materials: algebra tiles

Add $x^2 + 2x - 3$ and $x^2 - x + 4$.

Step 1 Model each polynomial.

$x^2 + 2x - 3$ $\qquad$ $x^2 - x + 4$

Step 2 Combine like shapes and remove all zero pairs. Recall that a zero pair is formed by pairing one tile with its opposite.

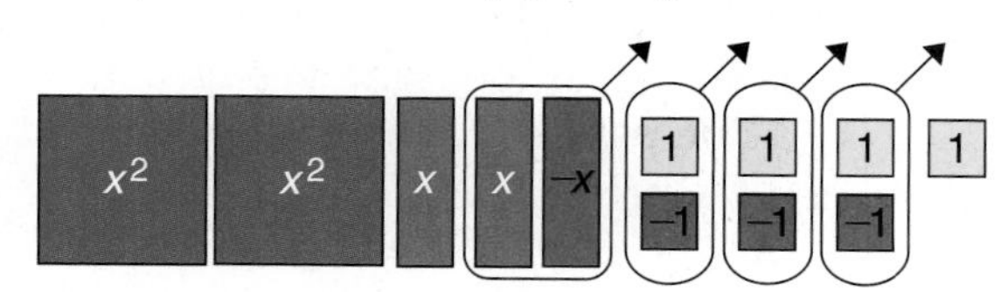

Therefore, $(x^2 + 2x - 3) + (x^2 - x + 4) = 2x^2 + x + 1$.

Try These

Use algebra tiles to find each sum.

1. $(3x + 5) + (2x - 3)$
2. $(-2x + 3) + (4x - 3)$
3. $(2x^2 + 2x - 4) + (x^2 + 3x + 7)$
4. $(3x^2 + x - 4) + (x^2 - 2x - 1)$
5. $(x^2 - 1) + (2x + 3)$
6. $(2x^2 + 3) + (x^2 - 2x - 1)$
7. **Write** a rule for adding polynomials without models.

You can add polynomials by grouping the like terms together and then finding their sum.

Examples

Find each sum.

1 $(4x - 3) + (2x + 5)$

Look Back

Like Terms: Lesson 1–4

Method 1

Group the like terms together.

$$\begin{aligned} (4x - 3) + (2x + 5) &= (4x + 2x) + (-3 + 5) && \textit{Group the like terms.} \\ &= (4 + 2)x + (-3 + 5) && \textit{Distributive Property} \\ &= 6x + 2 \end{aligned}$$

Method 2

Add in column form.

$$\begin{array}{rr} & 4x - 3 \\ (+) & 2x + 5 \\ \hline & 6x + 2 \end{array} \quad \textit{Align the like terms.}$$

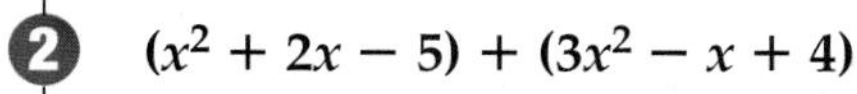

2 $(x^2 + 2x - 5) + (3x^2 - x + 4)$

$$\begin{aligned} (x^2 + 2x - 5) + (3x^2 - x + 4) &= (x^2 + 3x^2) + (2x - x) + (-5 + 4) \\ &= (1 + 3)x^2 + (2 - 1)x + (-5 + 4) \\ &= 4x^2 + 1x - 1 \text{ or } 4x^2 + x - 1 \end{aligned}$$

Check your answer by adding in column form.

3 $(2x^2 + 5xy + 3y^2) + (8x^2 - 7y^2)$

$$\begin{array}{rrrr} & 2x^2 & + 5xy & + 3y^2 \\ (+) & 8x^2 & & - 7y^2 \\ \hline & 10x^2 & + 5xy & - 4y^2 \end{array}$$

Your Turn

a. $(3x + 9) + (5x + 3)$

b. $(-2x^2 + x + 5) + (x^2 - 3x + 2)$

c. $\begin{array}{rrrr} & a^2 & - 2ab & + 4b^2 \\ (+) & 7a^2 & & - 2b^2 \\ \hline \end{array}$

d. $(7m^2 - 6) + (5m - 2)$

Recall that you can subtract an integer by adding its additive inverse or opposite.

$2 - 3 = 2 + (-3)$ *The additive inverse of 3 is −3.*

$5 - (-4) = 5 + 4$ *The additive inverse of −4 is 4.*

Similarly, you can subtract a polynomial by adding its additive inverse. To find the additive inverse of a polynomial, replace each term with its additive inverse.

Polynomial	Additive Inverse
$a + 2$	$-a - 2$
$x^2 + 3x - 1$	$-x^2 - 3x + 1$
$2x^2 - 5xy + y^2$	$-2x^2 + 5xy - y^2$

$-(a + 2) = -a - 2$
$-(x^2 + 3x - 1) = -x^2 - 3x + 1$
$-(2x^2 - 5xy + y^2) = -2x^2 + 5xy - y^2$

Examples

Find each difference.

4 **$(6x + 5) - (3x + 1)$**

Method 1

Find the additive inverse of $3x + 1$. Then group the like terms together and add.

The additive inverse of $3x + 1$ is $-(3x + 1)$ or $-3x - 1$.

$$\begin{aligned}(6x + 5) - (3x + 1) &= (6x + 5) + (-3x - 1) && \text{Add the additive inverse.}\\ &= (6x - 3x) + (5 - 1) && \text{Group the like terms.}\\ &= (6 - 3)x + (5 - 1) && \text{Distributive Property}\\ &= 3x + 4\end{aligned}$$

Method 2

Arrange like terms in column form.

$$\begin{array}{rr} & 6x + 5\\ (-) & 3x + 1\\ \hline \end{array} \qquad \textit{Add the additive inverse.} \qquad \begin{array}{rr} & 6x + 5\\ (+) & -3x - 1\\ \hline & 3x + 4\end{array}$$

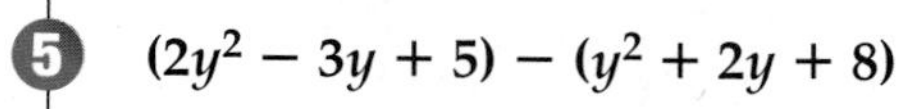

5 **$(2y^2 - 3y + 5) - (y^2 + 2y + 8)$**

The additive inverse of $y^2 + 2y + 8$ is $-(y^2 + 2y + 8)$ or $-y^2 - 2y - 8$.

$$\begin{aligned}(2y^2 - 3y + 5) - (y^2 + 2y + 8) &= (2y^2 - 3y + 5) + (-y^2 - 2y - 8)\\ &= (2y^2 - 1y^2) + (-3y - 2y) + (5 - 8)\\ &= (2 - 1)y^2 + (-3 - 2)y + (5 - 8)\\ &= 1y^2 + (-5y) + (-3) \text{ or } y^2 - 5y - 3\end{aligned}$$

To check the answer, add $y^2 - 5y - 3$ and $y^2 + 2y + 8$. The sum should be $2y^2 - 3y + 5$.

6 $(3x^2 + 5) - (-4x + 2x^2 + 3)$

Reorder the terms of the second polynomial so that the powers of x are in descending order.

$-4x + 2x^2 + 3 = 2x^2 - 4x + 3$

Then arrange like terms in column form.

$$\begin{array}{rrrr} & 3x^2 & & +\ 5 \\ (-) & 2x^2 & -\ 4x & +\ 3 \\ \hline \end{array} \quad \textit{Add the additive inverse.} \quad \begin{array}{rrrr} & 3x^2 & & +\ 5 \\ (+) & -2x^2 & +\ 4x & -\ 3 \\ \hline & x^2 & +\ 4x & +\ 2 \end{array}$$

Your Turn

e. $(3x - 2) - (5x - 4)$

f. $(10x^2 + 8x - 6) - (3x^2 + 2x - 9)$

g. $$\begin{array}{rrrr} & 6m^2 & & +\ 7 \\ (-) & -2m^2 & +\ 2m & -\ 3 \\ \hline \end{array}$$

h. $(5x^2 - 4x) - (2 - 3x)$

Polynomials can be used to represent measures of geometric figures.

Example 7

Geometry Link

The perimeter of triangle ABC is $7x + 2y$. Find the measure of the third side of the triangle.

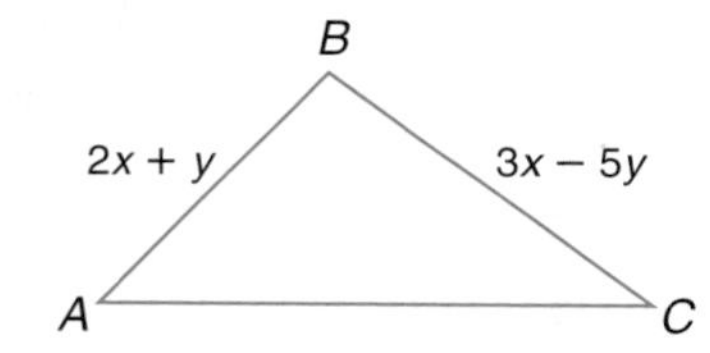

Explore You know the perimeter of the triangle and the measures of two sides. You need to find AC, the measure of the third side.

Plan The perimeter of a triangle is the sum of the measures of the three sides. To find AC, subtract the two given measures from the perimeter.

Solve

$$AC = \underbrace{\text{Perimeter}} - \underbrace{AB} - \underbrace{BC}$$

$$\begin{aligned} AC &= (7x + 2y) - (2x + y) - (3x - 5y) \\ &= (7x + 2y) + (-2x - y) + (-3x + 5y) \\ &= [7x + (-2x) + (-3x)] + [2y + (-y) + 5y] \\ &= [7 + (-2) + (-3)]x + [2 + (-1) + 5]y \\ &= 2x + 6y \end{aligned}$$

The measure of the third side of the triangle, AC, is $2x + 6y$.

Examine Check by adding the measures of the three sides.

$$\begin{array}{rrr} & 2x & +\ \ y \\ & 3x & -\ 5y \\ (+) & 2x & +\ 6y \\ \hline & 7x & +\ 2y \end{array}$$

The perimeter is $7x + 2y$. The answer checks.

Check for Understanding

Communicating Mathematics

1. **Describe** the first step you take when you add or subtract polynomials in column form.
2. **Explain** how addition and subtraction of polynomials are related.

Guided Practice

Getting Ready **Find the additive inverse of each polynomial.**

Sample 1: $6y - 3z$
Solution: $-6y + 3z$

Sample 2: $-a^2 + 2ab - 3b^2$
Solution: $a^2 - 2ab + 3b^2$

3. $2a + 9b$
4. $-4m + 6n$
5. $x^2 + 8x + 5$
6. $-3h^2 - 2h - 3$
7. $4xy^2 + 6x^2y - y^3$
8. $-2c^3 + c^2 - 3c$

Find each sum. *(Examples 1–3)*

9. $\begin{array}{r} 6y - 5 \\ (+)\ 2y + 7 \\ \hline \end{array}$
10. $\begin{array}{r} x^2 - 6x + 5 \\ (+)\ x^2 + 4x - 7 \\ \hline \end{array}$
11. $(2x^2 + 4x + 5) + (x^2 - 5x - 3)$
12. $(2x^2 - 5x + 4) + (3x^2 - 1)$

Find each difference. *(Examples 4–6)*

13. $\begin{array}{r} 3x + 4 \\ (-)\ x + 2 \\ \hline \end{array}$
14. $\begin{array}{r} 6m^2 - 5m + 3 \\ (-)\ 5m^2 + 2m - 7 \\ \hline \end{array}$
15. $(5x^2 + 4x - 1) - (4x^2 + x + 2)$
16. $(5x^2 - 4x) - (-3x + 2)$

17. **Geometry** The perimeter of triangle DEF is $12x^2 - 7x + 9$. Find the measure of the third side of the triangle. *(Example 7)*

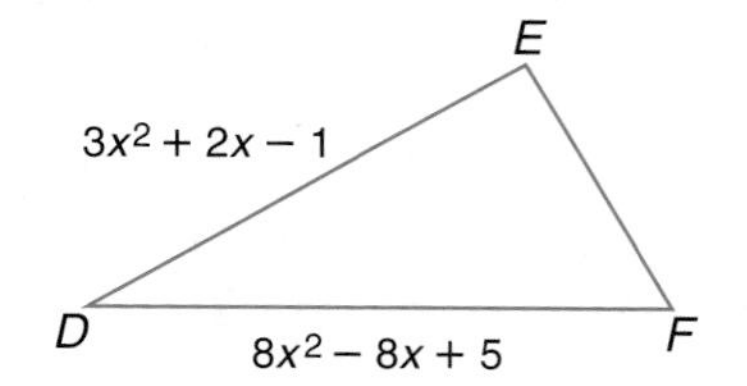

Exercises

Practice

Find each sum.

18. $\begin{array}{r} 2x + 3 \\ (+)\ x - 1 \\ \hline \end{array}$
19. $\begin{array}{r} 2x^2 - 5x + 4 \\ (+)\ 2x^2 + 8x - 1 \\ \hline \end{array}$
20. $\begin{array}{r} 2x^2 + 3xy - 4y^2 \\ (+)\ 4x^2 \qquad + 4y^2 \\ \hline \end{array}$
21. $(12x + 8y) + (2x - 7y)$
22. $(-7y + 3x) + (4x + 3y)$
23. $(n^2 + 5n + 3) + (2n^2 + 8n + 8)$
24. $(5x^2 - 7x + 9) + (3x^2 + 4x - 6)$
25. $(5n^2 - 4) + (2n^2 - 8n + 9)$
26. $(-2x^2 + 3xy - 3y^2) + (2x^2 - 5xy)$

Homework Help

For Exercises	See Examples
18–26, 37, 38, 40	1, 2, 3
27–36, 39, 41, 44	4, 5, 6
42	7

Extra Practice
See page 710.

Find each difference.

27. $\begin{array}{r} 9x + 5 \\ (-)\ 4x - 3 \\ \hline \end{array}$
28. $\begin{array}{r} 5a^2 + 7a + 9 \\ (-)\ 3a^2 + 4a + 1 \\ \hline \end{array}$
29. $\begin{array}{r} 3x^2 + 4x - 1 \\ (-)\ 4x^2 + \quad x + 2 \\ \hline \end{array}$

30. $(3x - 4) - (5x + 2)$

31. $(9x - 4y) - (12x - 5y)$

32. $(2x^2 + 5x + 3) - (x^2 - 2x + 3)$

33. $(3a^2 + 2a - 5) - (2a^2 - a - 4)$

34. $(3x^2 + 5x + 4) - (-1 + x^2)$

35. $(5x^2 - 4xy) - (2y^2 - 3xy)$

Find each sum or difference.

36. $(6x^3 - 10) - (2x^3 + 5x + 7)$

37. $(2pq + 3qr - 4pr) + (5pr - 3qr)$

38. $(3 + 2a - a^2) + (a^2 + 8a + 5)$

39. $(-x^2y - 4x^2y^2) - (x^2y + 3x^2y^2)$

40. Find the sum of $(x^2 + x + 5)$, $(3x^2 - 4x - 2)$, and $(2x^2 + 2x - 1)$.

41. What is $x^2 + 6x$ minus $3x^2 + 7$?

Applications and Problem Solving

Real World

42. **Geometry** The measure of each side of a triangle is $3x + 1$. Find its perimeter.

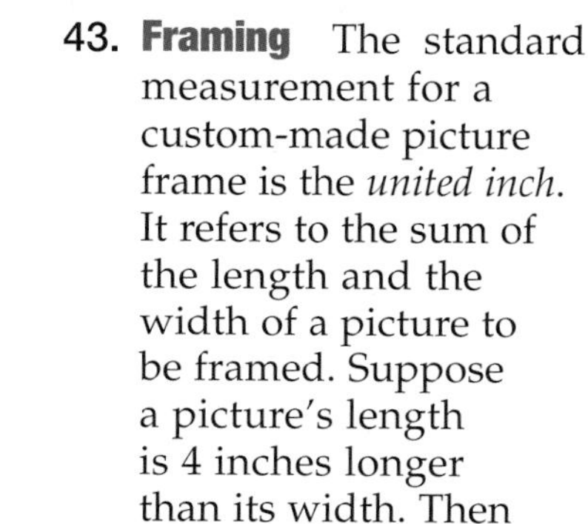

43. **Framing** The standard measurement for a custom-made picture frame is the *united inch.* It refers to the sum of the length and the width of a picture to be framed. Suppose a picture's length is 4 inches longer than its width. Then the width is w and the length would be $w + 4$.

a. Write a polynomial that represents the size of the picture in united inches.

b. If the width is 15 inches, use the polynomial from part a to find the size of the picture in united inches.

44. **Critical Thinking** One polynomial is subtracted from a second polynomial and the difference is $3m^2 + m - 5$. What is the difference when the second polynomial is subtracted from the first?

Mixed Review

Find the degree of each polynomial. *(Lesson 9–1)*

45. $y^2 + 3y + 2$

46. $n^6 + 3n^3 + 2$

47. $3x^2 + 2x^2y$

48. **Sailing** A rope from the top of a mast on a sailboat extends to a point 6 feet from the base of the mast. Suppose the rope is 24 feet long. To the nearest foot, how high is the mast? *(Lesson 8–7)*

Simplify each expression. *(Lesson 8–2)*

49. $(x^2y)(x^3y^2)$

50. $(4a^2)(2a^4)$

51. $(-2y)(-5xy)$

Standardized Test Practice

A B C D

52. **Multiple Choice** Which relation is a function? *(Lesson 6–4)*

A $\{(3, 1), (4, 2), (2, 3), (1, 2)\}$

B $\{(1, 1), (-1, 5), (3, 3), (-1, 2)\}$

C $\{(-1, 3), (1, 4), (3, 2), (1, 2)\}$

D $\{(-2, 1), (4, 2), (3, 3), (3, 2)\}$

9–3 Multiplying a Polynomial by a Monomial

What You'll Learn

You'll learn to multiply a polynomial by a monomial.

Why It's Important

Recreation You can use monomials and polynomials to solve problems involving recreation. *See Exercise 65.*

Suppose you have a square whose length and width are x units. If you increase the length by 3 units, what is the area of the new figure?

You can model this problem by using algebra tiles. The figures show how to make a rectangle whose length is $x + 3$ units and whose width is x units.

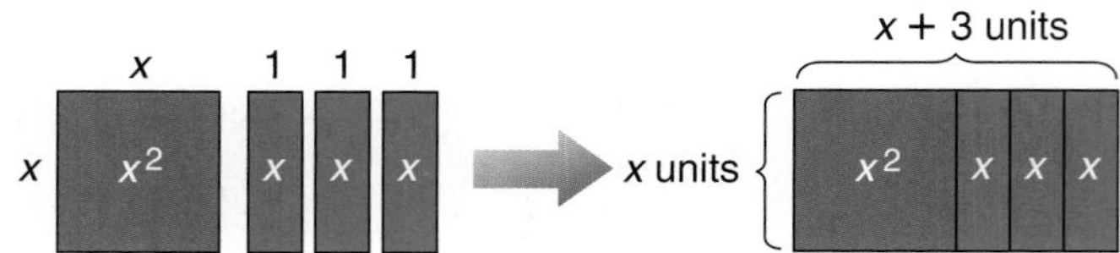

The area of any rectangle is the product of its length and its width. The area can also be found by adding the areas of the tiles.

Formula

$A = \ell w$

$= (x + 3)x$ or $x(x + 3)$

$= x^2 + 3x$

Algebra Tiles

$A = x^2 + x + x + x$

$= x^2 + 3x$

Since the areas are equal, $x(x + 3) = x^2 + 3x$. Another expression for the same area is $x^2 + 3x$ square units.

The example above shows how the Distributive Property can be used to multiply a polynomial by a monomial.

Multiplying a Polynomial by a Monomial

Words: To multiply a polynomial by a monomial, use the Distributive Property.

Symbols: $a(b + c) = ab + ac$

Model:

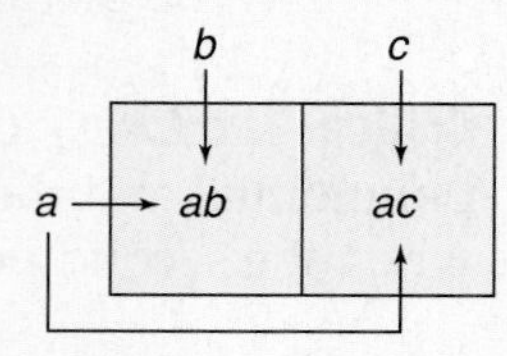

Examples

Find each product.

1 $y(y + 5)$

$y(y + 5) = y(y) + y(5)$

$= y^2 + 5y$

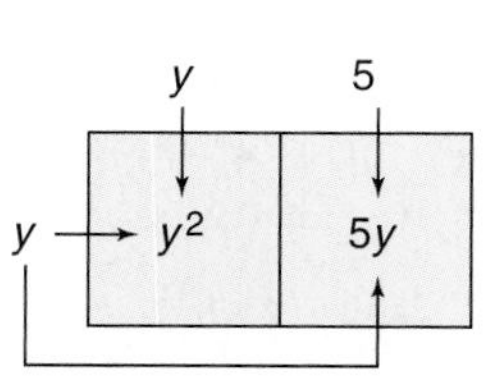

Multiplying Powers: Lesson 8–2

2 $b(2b^2 + 3)$

$b(2b^2 + 3) = b(2b^2) + b(3)$

$= 2b^3 + 3b$

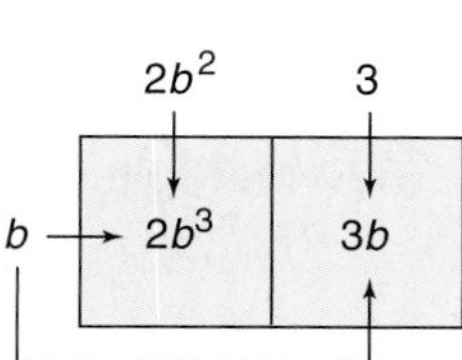

3 $-2n(7 - 5n^2)$

$$-2n(7 - 5n^2) = -2n(7) + (-2n)(-5n^2)$$
$$= -14n + 10n^3$$

	7	$-5n^2$
$-2n$	$-14n$	$10n^3$

4 $3x^3(2x^2 - 5x + 8)$

$$3x^3(2x^2 - 5x + 8) = 3x^3(2x^2) + 3x^3(-5x) + 3x^3(8)$$
$$= 6x^5 - 15x^4 + 24x^3$$

Your Turn

a. $7(2x + 5)$

b. $4x(3x^2 - 7)$

c. $-5a(6 - 3a^2)$

d. $2m^2(5m^2 - 7m + 8)$

Many equations contain polynomials that must be multiplied.

Examples

Solve each equation.

Look Back Solving Equations with Grouping Symbols: Lesson 4–7

5 $11(y - 3) + 5 = 2(y + 22)$

$11(y - 3) + 5 = 2(y + 22)$	
$11y - 33 + 5 = 2y + 44$	*Distributive Property*
$11y - 28 = 2y + 44$	*Combine like terms.*
$11y - 28 - 2y = 2y + 44 - 2y$	*Subtract 2y from each side.*
$9y - 28 = 44$	
$9y - 28 + 28 = 44 + 28$	*Add 28 to each side.*
$9y = 72$	
$\frac{9y}{9} = \frac{72}{9}$	*Divide each side by 9.*
$y = 8$	The solution is 8.

6 $w(w + 12) = w(w + 14) + 12$

$w(w + 12) = w(w + 14) + 12$	
$w^2 + 12w = w^2 + 14w + 12$	*Distributive Property*
$w^2 + 12w - w^2 = w^2 + 14w + 12 - w^2$	*Subtract w^2 from each side.*
$12w = 14w + 12$	
$12w - 14w = 14w + 12 - 14w$	*Subtract 14w from each side.*
$-2w = 12$	
$\frac{-2w}{-2} = \frac{12}{-2}$	*Divide each side by −2.*
$w = -6$	The solution is −6.

Your Turn

e. $2(5x - 12) = 6(-2x + 3) + 2$

f. $a(a + 2) + 3a = a(a - 3) + 8$

You can apply multiplication of a polynomial by a monomial to problems involving area.

Example 7
Geometry Link

Find the area of the shaded region in simplest form.

Subtract the area of the smaller rectangle from the area of the larger rectangle.

area of larger rectangle:	$3x(x + 2)$	*$A = \ell w$*
area of smaller rectangle:	$2x(x)$	
area of shaded region:	$3x(x + 2) - 2x(x)$	

$A = 3x(x + 2) - 2x(x)$

$= 3x(x) + 3x(2) - 2x(x)$ *Distributive Property*

$= 3x^2 + 6x - 2x^2$

$= 1x^2 + 6x$ or $x^2 + 6x$ *Combine like terms.*

The area of the shaded region is $x^2 + 6x$.

Check for Understanding

Communicating Mathematics

1. **Name** the property used to express $3n(2n - 6)$ as $6n^2 - 18n$.

2. **Write** an expression for the area of the rectangle at the right in the following two ways.

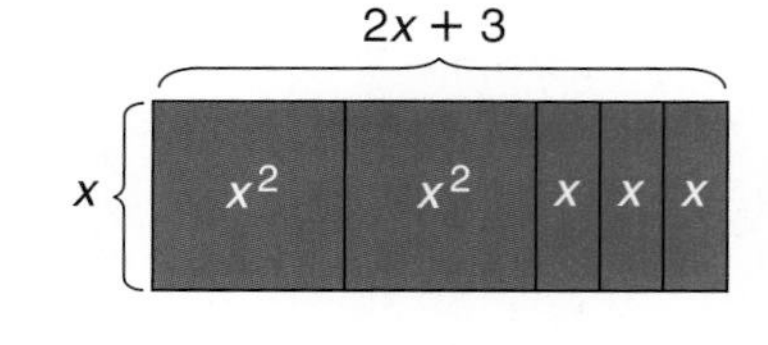

   a. a product of a monomial and a polynomial
   b. a sum of monomials

3. **You Decide?** Consuelo says that $2x(3x + 4) = 6x^2 + 8x$ is a true statement. Shawn says that $2x(3x + 4) = 6x^2 + 4$ is a true statement. Who is correct? Explain.

Guided Practice

Find each product. *(Examples 1–4)*

4. $2(y + 5)$
5. $x(x + 2)$
6. $3a(a - 1)$
7. $-4(x - 2)$
8. $-3n(5 - 2n)$
9. $4z(z^2 - 2z)$
10. $2(2d^2 + 3d + 8)$
11. $3(8y^2 + 3y - 5)$
12. $-2a(4a^2 - 3a - 1)$

Solve each equation. *(Examples 5 & 6)*

13. $7(x + 2) = 42$
14. $4(y - 8) + 10 = 2y + 12$
15. $-3(2a - 4) + 9 = 3(a + 1)$
16. $x(x + 3) + 5x = x(x + 5) + 9$

17. **Geometry** Find the area of the shaded region in simplest form. *(Example 7)*

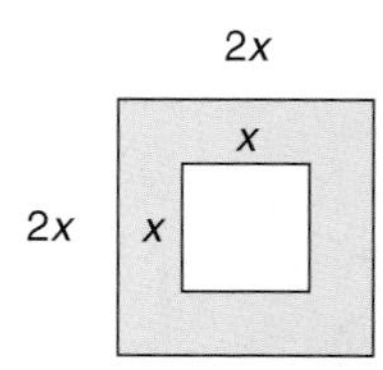

Exercises

Practice

Homework Help	
For Exercises	See Examples
18–26, 62, 65–66	1, 2
27–44, 57–61	3, 4
45–58	5, 6
63, 64	7

Extra Practice

See page 710.

Find each product.

18. $2(x + 6)$
19. $-3(y + 3)$
20. $7(2a + 3)$
21. $x(x - 5)$
22. $n(n + 4)$
23. $z(3z - 2)$
24. $2m(m + 4)$
25. $4x(2x - 3)$
26. $5y(y + 1)$
27. $-2a(a - 2)$
28. $-3x(x - 5)$
29. $-5y(6 - 2y)$
30. $3s(4s^2 - 7)$
31. $5d(d^2 + 3)$
32. $-7p(-3p - 6)$
33. $-2a(5a^2 - 7a + 2)$
34. $4x(-2x + 7x^2)$
35. $5n(8n^3 + 7n^2 - 3n)$
36. $-3y(6 - 9y + 4y^2)$
37. $7(-2a^2 + 5a - 11)$
38. $-5x^2(3x^2 - 8x - 12)$
39. $1.2(c^2 - 10)$
40. $0.1(4x^2 - 7x)$
41. $0.25x(4x - 8)$
42. $\frac{1}{2}(2x^2 - 6x)$
43. $\frac{2}{3}(6y^2 - 9y + 3)$
44. $\frac{1}{4}x(12x^2 + 8x)$

Solve each equation.

45. $8(x - 3) = 16$
46. $4(y - 3) = -8$
47. $40 = 5(a + 10)$
48. $2(x + 4) + 9 = 5x - 1$
49. $6(a + 2) - 5 = 2a + 3$
50. $8x + 14 = 3(x - 2) + 10$
51. $2(5a - 13) = 6(-2a + 3)$
52. $3(2y - 4) = -2(2y - 9)$
53. $5(n + 7) = 3(-2n + 1) - 1$
54. $b(b + 12) = b(b + 14) + 12$
55. $c(c + 8) - c(c + 3) - 23 = 3c + 11$
56. $m(m - 8) + 3m = -2 + m(m - 9)$

57. What is the product of $4x$ and $x^2 - 2x + 1$?

58. Multiply $-2n$ and $3 - 5n$.

Simplify.

59. $2a(a^2 - 5a + 4) - 6(a^3 + 3a - 11)$
60. $5t^2(t + 2) - 5t(4t^2 - 3t + 6) + 3(t^2 - 6)$
61. $3n^2(n - 4) + 6n(3n^2 + n - 7) - 4(n - 7)$

Applications and Problem Solving

62. **Sports** The length of a pool table is twice as long as its width. Suppose the width of a pool table is $4x - 1$. What is the length of the table?

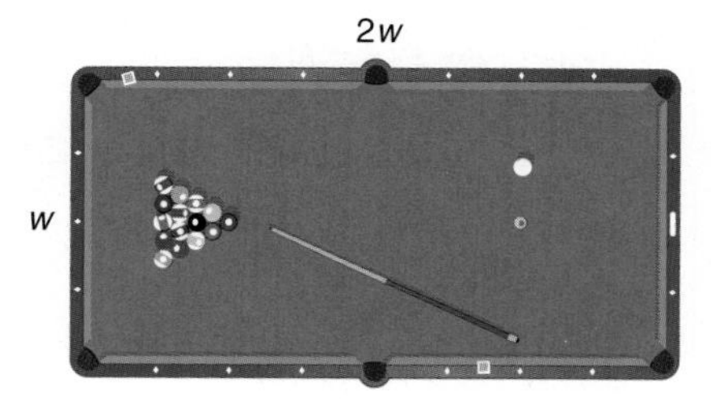

Geometry **Find the area of the shaded region for each figure.**

63.

3t
t − 1
3t
t

64.

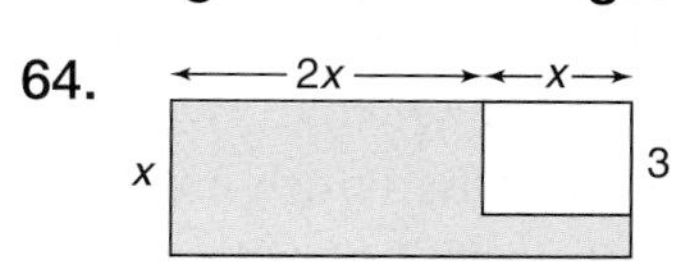

65. **Recreation** On the Caribbean island of Trinidad, children play a form of hopscotch called Jumby. The pattern for this game is shown at the right. Suppose the length of each rectangle is $y + 5$ units long and y units wide.

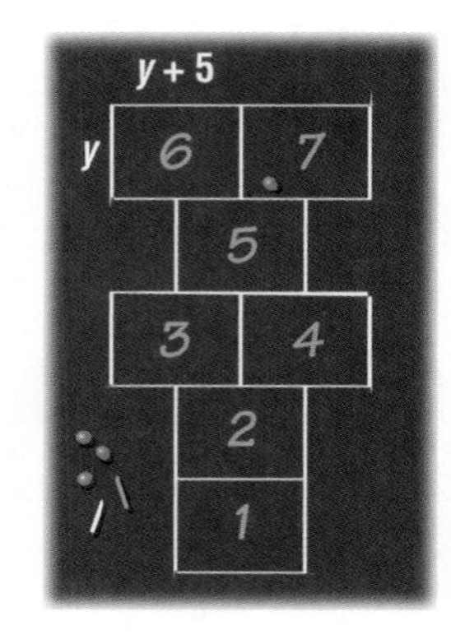

a. Write an expression in simplest form for the area of the pattern.

b. If y represents 10 inches, find the area of the pattern.

66. **Critical Thinking** Write three different multiplication expressions whose product is $6a^2 + 8a$.

Mixed Review

Find each sum or difference. *(Lesson 9–2)*

67. $(2x + 1) + (5x - 3)$

68. $(3a - 2) - (a + 4)$

69. $(x^2 + 2x - 3) + (2x^2 - 4x)$

70. $(y^2 + 8y) - (2y^2 + 5)$

71. **Ecology** The deer population of the Kaibab Plateau in Arizona from 1905 to 1930 can be estimated by the polynomial $-0.13x^5 + 3.13x^4 + 4000$, where x is the number of years after 1900. Find the degree of the polynomial. *(Lesson 9–1)*

Simplify. *(Lesson 8–5)*

72. $\sqrt{81}$

73. $-\sqrt{121}$

74. $\sqrt{256}$

75. $\sqrt{\frac{9}{100}}$

Find the odds of each outcome if a die is rolled. *(Lesson 5–6)*

76. a number less than 5

77. an odd number

Standardized Test Practice

78. **Multiple Choice** Which equation is equivalent to $12 + (x - 7) = 21$? *(Lesson 4–7)*

A $12x - 7 = 21$

B $12x - 84 = 21$

C $x + 5 = 21$

D $19 + x = 21$

Quiz 1 — Lessons 9–1 through 9–3

1. Find the degree of $3y^4 + 2y^3 - y^2 + 5y - 3$. *(Lesson 9–1)*
2. Arrange the terms of $2xy + x^2 - y^2$ so that the powers of x are in descending order. *(Lesson 9–1)*

Find each sum or difference. *(Lesson 9–2)*

3. $(4x^2 + 3x) - (6x^2 - 5x + 2)$

4. $(2w^2 - 4w - 12) + (15 - 3w^2 + 2w)$

5. **Sports** In the National Football League, the length of the playing field is 40 feet longer than twice its width. *(Lesson 9–3)*

a. Express the width and length of the playing field as polynomials.

b. Find the polynomial that represents the area of the playing field.

9–4 Multiplying Binomials

What You'll Learn
You'll learn to multiply two binomials.

Why It's Important
Packaging The FOIL method can be used to find the dimensions of a cereal box.
See Exercise 55.

Katie has a square herb garden in which she grows parsley. The measure of each side is x feet. She wants to increase the length by 2 feet and the width by 1 foot so she can grow sage, rosemary, and thyme. A plan for the garden is shown at the right. Find two different expressions for the area of the new garden.

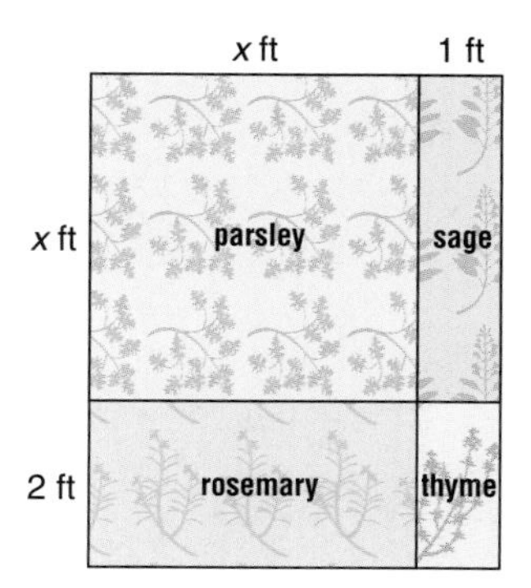

One way to find the area of the garden is to use the formula for area. Find the product of the length and width of the new garden.

Formula

$$A = \ell w$$
$$= (x + 2)(x + 1)$$

You can also find the area by adding the areas of the smaller regions.

Sum of Regions

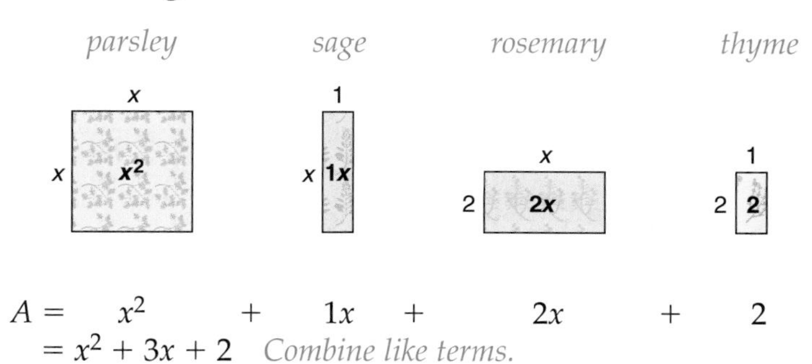

$$A = x^2 + 1x + 2x + 2$$
$$= x^2 + 3x + 2 \quad \textit{Combine like terms.}$$

Since the areas are equal, both expressions are equal. Therefore, $(x + 2)(x + 1) = x^2 + 3x + 2$.

The multiplication expression can also be shown in the model below. Notice that the Distributive Property is used twice.

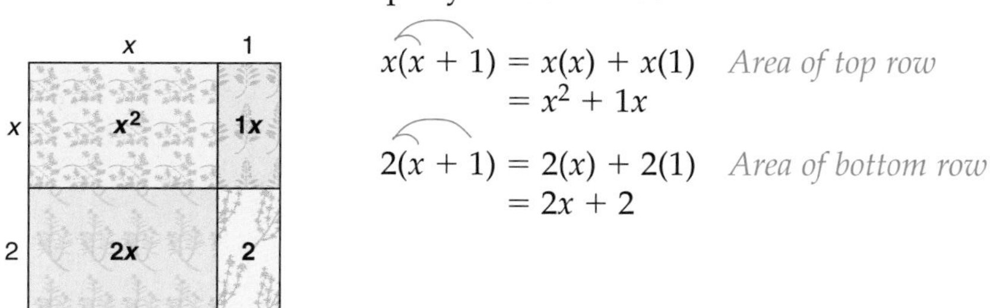

$x(x + 1) = x(x) + x(1)$ *Area of top row*
$= x^2 + 1x$

$2(x + 1) = 2(x) + 2(1)$ *Area of bottom row*
$= 2x + 2$

So, $(x + 2)(x + 1) = x^2 + 1x + 2x + 2$ or $x^2 + 3x + 2$.

Hands-On Algebra

Materials: straightedge

Use a model to find the product of $(x + 2)$ and $(x - 1)$.

Step 1 Put the $x + 2$ and $x - 1$ outside the box as shown. *Note that $x - 1 = x + (-1)$.*

Step 2 Use the Distributive Property to multiply x by $x - 1$ and place the products inside the boxes.

	x	-1
x	x^2	$-1x$
2		

Step 3 Use the Distributive Property to multiply 2 by $x - 1$ and place the products inside the boxes.

	x	-1
x	x^2	$-1x$
2	$2x$	-2

Step 4 Find the sum of the terms inside the boxes: $x^2 - 1x + 2x - 2 = x^2 + 1x - 2$ or $x^2 + x - 2$.

Therefore, $(x + 2)(x - 1) = x^2 + x - 2$.

Try These **Use a model to find each product.**

1. $(x + 2)(x + 3)$ **2.** $(x + 3)(x + 4)$ **3.** $(2x + 1)(x + 1)$

4. $(x - 2)(x + 1)$ **5.** $(x - 3)(x - 2)$ **6.** $(x - 1)(x + 1)$

You can also use the Distributive Property to multiply binomials.

Examples

Find each product.

1 $(x + 3)(x - 4)$

$$
\begin{aligned}
(x + 3)(x - 4) &= x(x - 4) + 3(x - 4) && \textit{Distributive Property} \\
&= x(x) + x(-4) + 3(x) + 3(-4) && \textit{Distributive Property} \\
&= x^2 - 4x + 3x - 12 && \textit{Simplify.} \\
&= x^2 - 1x - 12 \text{ or } x^2 - x - 12 && \textit{Combine like terms.}
\end{aligned}
$$

2 $(2y - 1)(y - 3)$

$(2y - 1)(y - 3)$

$$
\begin{aligned}
&= 2y(y - 3) + (-1)(y - 3) && \textit{Distributive Property} \\
&= 2y(y) + 2y(-3) + (-1)(y) + (-1)(-3) && \textit{Distributive Property} \\
&= 2y^2 - 6y - 1y + 3 && \textit{Simplify.} \\
&= 2y^2 - 7y + 3 && \textit{Combine like terms.}
\end{aligned}
$$

Your Turn

a. $(y + 4)(y - 2)$ **b.** $(m - 3)(m + 1)$

c. $(x - 5)(x - 2)$ **d.** $(2a - 3)(a - 2)$

Two binomials can always be multiplied using the Distributive Property. However, the following shortcut can also be used. It is called the **FOIL method**.

$$(3x + 1)(x + 2) = \underset{\substack{F \\ \text{product of} \\ \text{FIRST terms}}}{(3x)(x)} + \underset{\substack{O \\ \text{product of} \\ \text{OUTER terms}}}{(3x)(2)} + \underset{\substack{I \\ \text{product of} \\ \text{INNER terms}}}{(1)(1x)} + \underset{\substack{L \\ \text{product of} \\ \text{LAST terms}}}{(1)(2)}$$

$$= 3x^2 + 6x + 1x + 2$$
$$= 3x^2 + 7x + 2$$

FOIL Method for Multiplying Two Binomials

To multiply two binomials, find the sum of the products of

- **F** the First terms,
- **O** the Outer terms,
- **I** the Inner terms, and
- **L** the Last terms.

Examples

Find each product.

3 $(y + 4)(y + 6)$

$$(y + 4)(y + 6) = \overset{F}{(y)(y)} + \overset{O}{(y)(6)} + \overset{I}{(4)(y)} + \overset{L}{(4)(6)}$$
$$= y^2 + 6y + 4y + 24$$
$$= y^2 + 10y + 24 \quad \textit{Combine like terms.}$$

4 $(2x - 3)(2x + 2)$

$$(2x - 3)(2x + 2) = \overset{F}{(2x)(2x)} + \overset{O}{(2x)(2)} + \overset{I}{(-3)(2x)} + \overset{L}{(-3)(2)}$$
$$= 4x^2 + 4x - 6x - 6$$
$$= 4x^2 - 2x - 6 \quad \textit{Combine like terms.}$$

5 $(3a - b)(2a + 4b)$

$$(3a - b)(2a + 4b) = (3a)(2a) + (3a)(4b) + (-b)(2a) + (-b)(4b)$$
$$= 6a^2 + 12ab - 2ab - 4b^2$$
$$= 6a^2 + 10ab - 4b^2$$

Your Turn

e. $(n + 3)(n + 5)$ **f.** $(x - 4)(2x + 3)$ **g.** $(2x + y)(3x - 2y)$

Sometimes it is not possible to simplify the product of two binomials.

Example 6 **Find the product of $x^2 - 4$ and $x + 3$.**

$$(x^2 - 4)(x + 3) = (x^2)(x) + (x^2)(3) + (-4)(x) + (-4)(3)$$
$$= x^3 + 3x^2 - 4x - 12 \quad \textit{There are no like terms.}$$

Your Turn

h. Find the product of $x - 2$ and $x^2 + 3$.

You can use the FOIL method to solve problems involving volume.

Example 7 **Geometry Link**

The volume V of a rectangular prism is equal to the area of the base B times the height h. Express the volume of the prism as a polynomial. Use $V = Bh$.

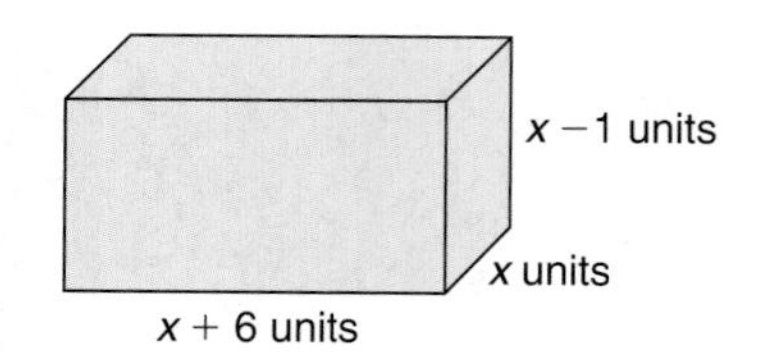

First, find the area of the base. The base is a rectangle.

$B = \ell w$

$= (x + 6)(x)$ *Replace ℓ with $x + 6$ and w with x.*

$= x^2 + 6x$ *Use the Distributive Property.*

To find the volume, multiply the area of the base by the height.

$V = Bh$

$= (x^2 + 6x)(x - 1)$ *Replace B with $x^2 + 6x$ and h with $x - 1$.*

$= (x^2)(x) + (x^2)(-1) + (6x)(x) + (6x)(-1)$ *Use the FOIL method.*

$= x^3 - x^2 + 6x^2 - 6x$

$= x^3 + 5x^2 - 6x$

The volume of the prism is $x^3 + 5x^2 - 6x$ cubic units.

Check for Understanding

Vocabulary
FOIL method

Communicating Mathematics

1. **Draw** a model that represents the product of $x - 2$ and $x + 3$. Then find the product.
2. **Write** the multiplication expression represented by each model.

a.

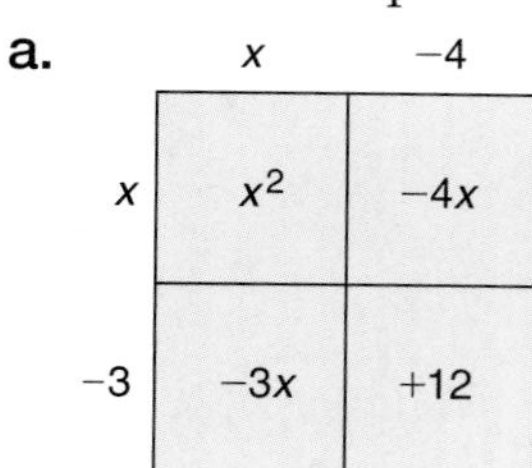

b.

	x	-4
$2x$	$2x^2$	$-8x$
5	$5x$	-20

3. **Compare and contrast** the procedure for multiplying two binomials and the procedure for multiplying a binomial and a monomial.

Guided Practice

Getting Ready **Find the sum of the products of the inner terms and the outer terms.**

Sample: $(x - 6)(x + 2)$

Solution: $(x - 6)(x + 2)$ (I = inner, O = outer)

$-6x + 2x = -4x$

4. $(a - 5)(a - 3)$
5. $(x + 1)(x + 5)$
6. $(g + 1)(g - 9)$
7. $(2m + 3)(3m + 2)$
8. $(2j + 1)(j - 3)$
9. $(y + 5)(2y + 4)$

Find each product. Use the Distributive Property or the FOIL method. *(Examples 1–5)*

10. $(x + 2)(x + 4)$
11. $(w - 7)(w + 5)$
12. $(a - 3)(a - 6)$
13. $(2y + 3)(y - 1)$
14. $(3x - 4)(2x - 3)$
15. $(2y + 7)(3y - 5)$
16. $(x + 2y)(x + 3y)$
17. $(2a - 5b)(a + 2b)$
18. $(y^2 + 3)(y - 2)$
19. $(m^3 - 2m)(m + 3)$

20. **Geometry** Express the volume of the rectangular prism as a polynomial. Use the formula $V = Bh$, where B is the area of the base and h is the height of the prism. *(Example 6)*

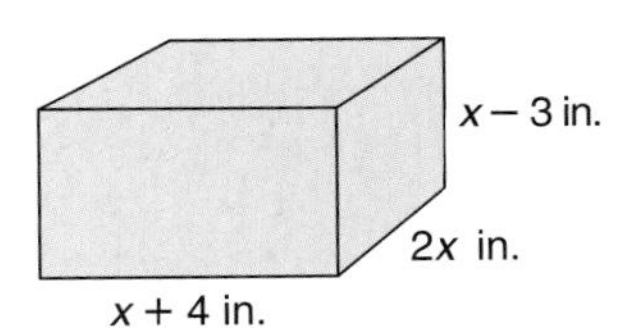

Exercises

Practice

Homework Help	
For Exercises	See Examples
21–39	1, 2, 3
40–52	4, 5, 6
53–56	7

Extra Practice

See page 711.

Find each product. Use the Distributive Property or the FOIL method.

21. $(x + 4)(x + 8)$
22. $(r + 7)(r + 2)$
23. $(a - 3)(a + 7)$
24. $(x - 3)(x - 7)$
25. $(n - 11)(n - 5)$
26. $(y - 4)(y + 15)$
27. $(z + 6)(z - 4)$
28. $(s - 11)(s + 5)$
29. $(x - 4)(3x + 2)$
30. $(2a + 5)(a - 7)$
31. $(z - 4)(3z - 5)$
32. $(y - 3)(2y - 5)$
33. $(5n + 2)(n - 3)$
34. $(x + 7)(2x - 3)$
35. $(3a + 1)(3a + 1)$
36. $(2x + 5)(5x + 3)$
37. $(4h - 3)(3h + 2)$
38. $(7a - 1)(2a - 3)$
39. $(2x + 7)(x - 3)$
40. $(8m + 2n)(6m + 5n)$
41. $(2y - 4z)(3y - 6z)$
42. $(3a - b)(2a + b)$
43. $(2x + 5y)(3x - y)$
44. $(5n - 2p)(5n + 2p)$
45. $(x^2 + 1)(x - 2)$
46. $(y + 3)(y^2 - 4)$
47. $(x^2 + 3)(3x^2 - 1)$
48. $(2y^2 + 1)(y + 1)$
49. $(x^2 + 2x)(x - 3)$
50. $(x + a)(x + b)$

51. Find the product of $(x + 3)$ and $(x - 3)$.
52. What is the product of x, $(2x + 1)$, and $(x - 3)$?

Applications and Problem Solving

53. **Geometry** Find the area of each shaded region in simplest form.

a.

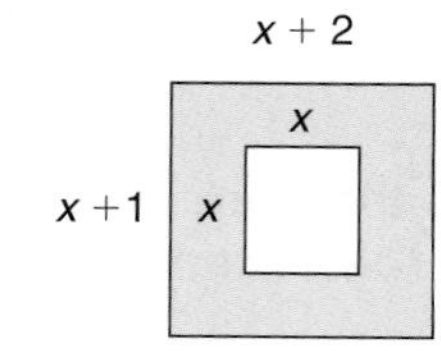

b.

$2x + 1$

$x - 1$

$2x$ $x - 1$

54. **Number Theory** Use the FOIL method to find the product of 16 and 38. (*Hint*: Write 16 as $10 + 6$ and 38 as $30 + 8$.)

55. **Packaging** A cereal box has a length of $2x$ inches, a width of $x - 2$ inches, and a height of $2x + 5$ inches.
    a. Express the volume of the package as a polynomial.
    b. Find the volume of the box if $x = 5$.

56. **Critical Thinking** Use the Distributive Property to find the product of $x + 2$ and $x^2 - 3x + 2$.

Mixed Review

57. **Clocks** Before mechanical clocks were invented, candles were used to keep track of time. One formula that was used was $c = 2(5 - t)$, where c is the height of the candle in inches and t is the time in hours that the candle burns. *(Lesson 9–3)*
    a. Use the Distributive Property to multiply 2 and $5 - t$.
    b. Find the height of the candle when $t = 2$.

58. Find the sum of $x^2 + 5x$ and $-3x - 7$. *(Lesson 9–2)*

Write each expression using exponents. *(Lesson 8–1)*

59. $2 \cdot 2 \cdot 2 \cdot x \cdot x \cdot x$ 60. 10 61. $(3)(3)(-2)(-2)(-2)$

Graph each equation. *(Lesson 7–5)*

62. $y = 2x + 3$ 63. $y = -x + 2$ 64. $4x + 5y = 20$

Data Update For the latest information on breakfast foods, visit: www.algconcepts.com

65. **Food** The graph shows the favorite breakfast food for students ages 9–12. Suppose you interview 200 students who are in this age range. How many would you expect to like cereal for breakfast? *(Lesson 5–4)*

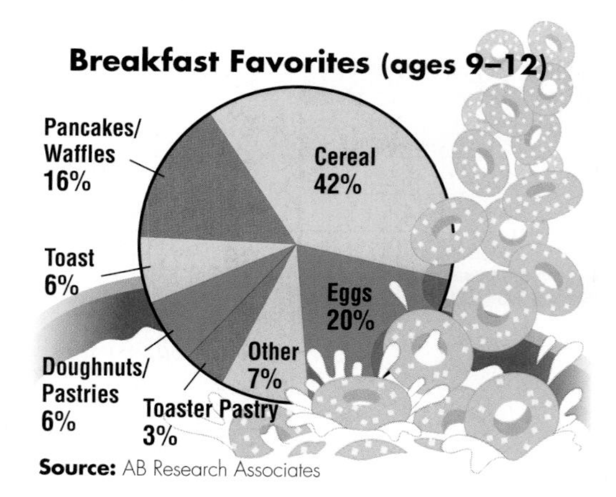

Source: AB Research Associates

Standardized Test Practice

66. **Extended Response** Draw and label a rectangle whose perimeter is 20 centimeters. *(Lesson 1–5)*

9–5 Special Products

What You'll Learn

You'll learn to develop and use the patterns for $(a + b)^2$, $(a - b)^2$, and $(a + b)(a - b)$.

Why It's Important

Biology Geneticists use a technique similar to finding $(a + b)^2$ to predict the characteristics of a population. *See Example 5.*

Some products of polynomials appear frequently in real-life problems. Expressions like $(a + b)^2$, $(a - b)^2$, and $(a + b)(a - b)$ occur so often that it is helpful to develop patterns for their products.

To develop these patterns, we will model a simpler expression, $(x + 1)^2$, geometrically.

The area of a square with a side length of $x + 1$ is $(x + 1)(x + 1)$ or $(x + 1)^2$. The total area can also be found by adding the areas of the inner regions together. The right side of the equation below represents the area of the square as the sum of the areas of the four small squares.

	x	1
x	x^2	$1x$
1	$1x$	1

$$(x + 1)^2 = x^2 + 1x + 1x + 1$$
$$= x^2 + 2x + 1$$

twice the product of 1 and x ($2x$)

You can use a similar model to find $(x - 1)^2$. $x - 1 = x + (-1)$

	x	-1
x	x^2	$-1x$
-1	$-1x$	1

$$(x - 1)^2 = x^2 - 1x - 1x + 1$$
$$= x^2 - 2x + 1$$

twice the product of −1 and x ($-2x$)

The square of a sum and the square of a difference can be found by using the following rules.

Square of a Sum and Square of a Difference

Symbols: $(a + b)^2 = a^2 + 2ab + b^2$
$(a - b)^2 = a^2 - 2ab + b^2$

Models:

	a	b
a	a^2	ab
b	ab	b^2

	a	$-b$
a	a^2	$-ab$
$-b$	$-ab$	b^2

Examples

Find each product.

1 $(x + 4)^2$

$(a + b)^2 = a^2 + 2ab + b^2$ *Square of a Sum*

$(x + 4)^2 = x^2 + 2(x)(4) + 4^2$ *Replace a with x and b with 4.*

$= x^2 + 8x + 16$

2 $(4m + n)^2$

$(a + b)^2 = a^2 + 2ab + b^2$ *Square of a Sum*

$(4m + n)^2 = (4m)^2 + 2(4m)(n) + n^2$ *Replace a with 4m and b with n.*

$= 16m^2 + 8mn + n^2$

3 $(w - 5)^2$

$(a - b)^2 = a^2 - 2ab + b^2$ *Square of a Difference*

$(w - 5)^2 = w^2 - 2(w)(5) + 5^2$ *Replace a with w and b with 5.*

$= w^2 - 10w + 25$

4 $(3p - 2q)^2$

$(a - b)^2 = a^2 - 2ab + b^2$ *Square of a Difference*

$(3p - 2q)^2 = (3p)^2 - 2(3p)(2q) + (2q)^2$ *Replace a with 3p and b with 2q.*

$= 9p^2 - 12pq + 4q^2$

Your Turn

a. $(y + 3)^2$

b. $(3g + h)^2$

c. $(d - 2)^2$

d. $(4x - 3y)^2$

Biologists use a method that is similar to squaring a sum to find the characteristics of offspring based on genetic information.

Example

Biology Link

5 **In a certain population, a parent has a 10% chance of passing the gene for brown eyes to its offspring. If an offspring receives one eye-color gene from its mother and one from its father, what is the probability that an offspring will receive at least one gene for brown eyes?**

There is a 10% chance of passing the gene for brown eyes. Therefore, there is a 90% chance of *not* passing the gene.

Look Back

Probability of Independent Events: Lesson 5–7

Use the model at the right to show all possible combinations. In the model, B represents the gene for brown eyes and b represents the gene for *not* brown eyes. *Note that the percents are written as decimals.*

		Father	
		B = 0.1	b = 0.9
Mother	B = 0.1	BB (0.1)(0.1) = 0.01	Bb (0.1)(0.9) = 0.09
	b = 0.9	bB (0.9)(0.1) = 0.09	bb (0.9)(0.9) = 0.81

www.algconcepts.com/extra_examples

Three of the four small squares in the model contain a B. Add their probabilities.

$0.01 + 0.09 + 0.09 = 0.19$

So, the probability of an offspring receiving at least one gene for brown eyes is 0.19 or 19%.

You can use the FOIL method to find product of the sum and difference of the same two terms. Consider $(x + 3)(x - 3)$.

	x	-3
x	x^2	$-3x$
3	$3x$	-9

$$(x + 3)(x - 3) = x^2 - 3x + 3x - 9$$
$$= x^2 - 9$$

square of the second term

square of the first term

Note that the product is the difference of the squares of the terms. The product of a sum and a difference can be found by using the following rule.

Product of a Sum and a Difference

Symbols: $(a + b)(a - b) = a^2 - b^2$

Model:

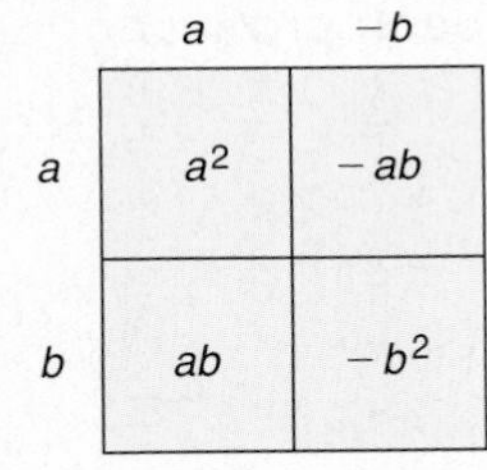

Examples

Find each product.

6 $(y + 2)(y - 2)$

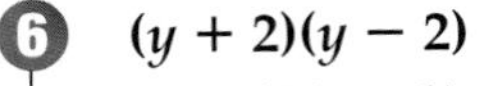

$(a + b)(a - b) = a^2 - b^2$ *Product of a Sum and a Difference*

$(y + 2)(y - 2) = y^2 - (2)^2$ *Replace a with y and b with 2.*

$= y^2 - 4$

7 $(2r + s)(2r - s)$

$(a + b)(a - b) = a^2 - b^2$ *Product of a Sum and a Difference*

$(2r + s)(2r - s) = (2r)^2 - s^2$ *Replace a with 2r and b with s.*

$= 4r^2 - s^2$

Your Turn

e. $(x + y)(x - y)$

f. $(5m - 6n)(5m + 6n)$

Check for Understanding

Communicating Mathematics

1. **Match** each description with an expression.
   a. square of a sum — i. $(r + t)(r - t)$
   b. square of a difference — ii. $(z + 1)^2$
   c. product of a sum and difference — iii. $(6 - a)^2$

2. 

Jessica says that the product of two binomials is always a trinomial. Hector disagrees. Who is correct? Explain your reasoning.

Guided Practice

Find each product. *(Examples 1–4, 6, & 7)*

3. $(y + 2)^2$
4. $(2a + 4b)^2$
5. $(m - 9)^2$
6. $(j - 7k)^2$
7. $(x + 7)(x - 7)$
8. $(5r + 9s)(5r - 9s)$

9. **Biology** In a certain population, a parent has a 5% chance of passing the gene for cystic fibrosis to its offspring. Use a model to find the probability that an offspring will *not* receive the gene for cystic fibrosis from either of its parents. *(Example 5)*

Exercises

Practice

Find each product.

10. $(r + 2)^2$
11. $(x + 5)^2$
12. $(a + 2b)^2$
13. $(w - 8)^2$
14. $(2x - y)^2$
15. $(m - 3n)^2$
16. $(p + 2)(p - 2)$
17. $(x + 4y)(x - 4y)$
18. $(2x - 3)(2x + 3)$
19. $(5 + k)^2$
20. $(3x - 2y)^2$
21. $(y + 3z)(y - 3z)$
22. $(6 - 2m)^2$
23. $(3x + 5)(3x - 5)$
24. $(5 + 2p)(5 + 2p)$
25. $(4 + 2x)^2$
26. $(5a - 2b)(5a + 2b)$
27. $(a - 3b)(a - 3b)$
28. $y(y + 1)(y - 1)$
29. $x(x + 1)^2$
30. $(x + 2)(x - 2)(x - 3)$

31. Find the product of $x - 2y$ and $x + 2y$.
32. What is the product of n, $n + 1$, and $n + 1$?

Homework Help

For Exercises	See Examples
10–12, 19, 24, 25, 29, 32, 34	1, 2
13–15, 20, 22, 27	3, 4
16–18, 21, 23, 26, 28, 30, 31, 33, 35	6, 7

Extra Practice

See page 711.

Applications and Problem Solving

33. **Number Theory** Explain how to find the product of 39 and 41 mentally. (*Hint*: Write 39 as 40 − 1 and 41 as 40 + 1.)

34. **Photography** Kareem cut off a 1-inch strip from each side of a square photograph so that it would fit into a square picture frame. He removed a total of 20 square inches.
    a. Draw and label a diagram that represents this situation. Let x represent the side length of the original photograph.
    b. Write an equation that could be used to find the dimensions of the original photograph.
    c. Find the original dimensions of the photograph.

35. **Geometry** The area of a triangle is given by the expression $\frac{1}{2}bh$, where b represents the length of the base and h represents the height. Suppose a right triangle has a base that measures $x - 3$ units and a height of $x + 3$ units.

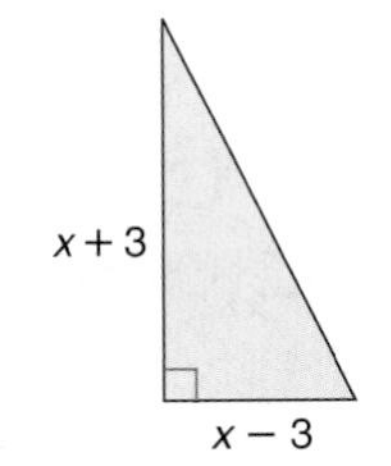

a. Express the area of the triangle as a sum of two monomials.

b. Find the area of the triangle if $x = 5$.

c. What is the length of the hypotenuse if $x = 6$?

36. **Critical Thinking** Use a model to find $(a + b + c)^2$.

Mixed Review

Geometry Find the area of each shaded region in simplest form. *(Lessons 9–3 & 9–4)*

37.

2x + 1

x

2x − 3 x

38.

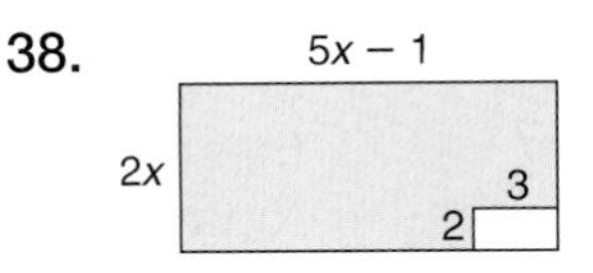

Write each equation in slope-intercept form. *(Lesson 7–3)*

39. $2y = 6x + 8$

40. $x + y = -4$

41. $2x - 5y = 15$

Standardized Test Practice

42. **Grid In** Enrique has 3 dimes, 2 quarters, and 5 nickels in his pocket. He takes one coin from his pocket at random. What is the probability that the coin is either a dime or a nickel? *(Lesson 5–7)*

43. **Multiple Choice** For what value(s) of b is the equation $3 + 2|b| = 7$ true? *(Lesson 3–7)*

A -5 and 5 B -2 and 2 C -2 only D 2 only

Quiz 2 Lessons 9–4 and 9–5

Find each product. *(Lessons 9–4 & 9–5)*

1. $(c + 2)(c + 8)$

2. $(5y - 3)(y + 2)$

3. $(c - 1)^2$

4. $(3x - 8)(3x + 8)$

5. **Geometry** Find the area of the shaded region if the length of a side of the large square is $x + 3$ centimeters and the length of a side of the smaller square is $x - 3$ centimeters. *(Lesson 9–5)*

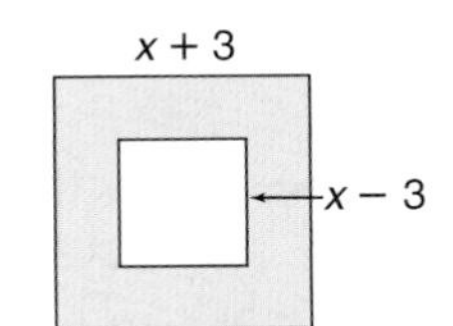

Chapter 9 Investigation

It's Greek to Me!

Areas and Ratio

Materials

 ruler

 calculator

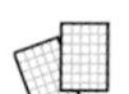 grid paper

Early Greek mathematicians enjoyed solving problems that required a lot of thought and investigation. Some of these problems still fascinate people today. For example, the Greeks wanted to construct a cube whose volume was twice the volume of a given cube. First, let's look at similar problems. What happens to the area of a square when you double the length of the sides of that square? Can you construct a square whose area is twice as great as a given square?

Temple of Poseidon

Investigate

Draw squares with different lengths on grid paper and find the ratios of their areas. Use a table or spreadsheet to record the lengths, areas, and ratios.

1. Draw a 3-centimeter by 3-centimeter square. Label it A. Find the area.
2. Draw a second square that has a side length twice that of square A. Label it B. Find the area.
3. Record your information in a table like the one below.

Square A		Square B		Ratios	
Length of Side	Area	Length of Side	Area	$\frac{\text{Length B}}{\text{Length A}}$	$\frac{\text{Area B}}{\text{Area A}}$
3	9	6	36	2:1	4:1

4. Draw three more pairs of squares and label them A and B. Find the areas and ratios for each pair of squares.
5. Make a conjecture about the ratio of areas of squares if the ratio of the lengths of their sides is 3:1.

Extending the Investigation

In this extension, you will investigate the ratios of the areas of other squares. You will also try to find the length of the sides of two squares so that the area of one square is twice the other.

- Draw pairs of squares so that the ratio of the side length of one square to the side length of the second square is 3:1. Make a conjecture about the ratio of their areas.
- Draw pairs of squares so that the ratio of the side length of one square to the side length of the second square is *not* 2:1 or 3:1. Find the ratio of the lengths of their sides. Make a conjecture about the ratio of their areas.
- Draw pairs of squares so that the lengths of their sides differ by 1. For example, one square might have a side length of 4 units and the second a side length of 5 units. Find the ratio of the lengths of their sides. Make a conjecture about the ratios of their areas.
- Draw pairs of squares so that the lengths of their sides differ by a number greater than 1. Find the ratio of the lengths of their sides. Make a conjecture about the ratios of their areas.
- Draw a 2-centimeter by 2-centimeter square. Then sketch a square whose area is twice as great as the area of the first square. Find the side length of the second square.

Presenting Your Conclusions

Here are some ideas to help you present your conclusions to the class.

- Make a bulletin board showing the results of this investigation.
- Write a report about your conjectures. Include diagrams or computations that help to explain your findings.
- Make a video showing the ratios that you discovered. Present your conjectures in a creative way.

Investigation For more information on three classical problems in Greek mathematics, visit: www.algconcepts.com

CHAPTER 9

Study Guide and Assessment

Understanding and Using the Vocabulary

After completing this chapter, you should be able to define each term, property, or phrase and give an example or two of each.

interNET CONNECTION Review Activities
For more review activities, visit: www.algconcepts.com

binomial *(p. 383)*
degree *(p. 384)*
FOIL method *(p. 401)*
monomial *(p. 382)*
polynomial *(p. 383)*
trinomial *(p. 383)*

Choose the letter of the term that best matches each statement or phrase. Each letter is used once.

1. $(x - y)^2$
2. a polynomial with two terms
3. a polynomial with three terms
4. a monomial or a sum of monomials
5. the sum of the exponents of the variables
6. a number, a variable, or a product of numbers and variables
7. Subtract polynomials by adding this.
8. Add polynomials by grouping these together.
9. Use this to multiply any two binomials.
10. Use this to multiply a polynomial by a monomial.

a. additive inverse
b. binomial
c. degree of a polynomial
d. Distributive Property
e. FOIL method
f. like terms
g. monomial
h. polynomial
i. square of a difference
j. trinomial

Skills and Concepts

Objectives and Examples

- **Lesson 9–1** Identify and classify polynomials and find their degree.

Identify $3x^3 - 2x^2$ as a *monomial*, *binomial*, or *trinomial*.

$3x^3 - 2x^2$ can be written as a sum of two monomials, $3x^3$ and $-2x^2$. It is a binomial.

Review Exercises

State whether each expression is a polynomial. If it is a polynomial, identify it as either a *monomial*, *binomial*, or *trinomial*.

11. $7m$
12. $5g^{-2}$
13. $2x^2 + 3x - 4$
14. $3y + 5y + 8$

Find the degree of each polynomial.

15. 7
16. $-4m^5$
17. $10x^2 - x^3y$
18. $2abc + 9a^5b - 4bc^3$
19. Arrange $3d^3 - d^2 + 7cd$ so that the powers of d are in ascending order.

Objectives and Examples

• **Lesson 9–2** Add and subtract polynomials.

$(2x + 4) + (5x - 7)$
$= (2x + 5x) + (4 - 7)$
$= (2 + 5)x + (4 - 7)$
$= 7x - 3$

$(7s^2 + 3s - 4) - (3s^2 - 2s + 5)$

The additive inverse of $(3s^2 - 2s + 5)$ is $(-3s^2 + 2s - 5)$.

$(7s^2 + 3s - 4) - (3s^2 - 2s + 5)$
$= (7s^2 + 3s - 4) + (-3s^2 + 2s - 5)$
$= (7s^2 - 3s^2) + (3s + 2s) + (-4 - 5)$
$= (7 - 3)s^2 + (3 + 2)s + (-4 - 5)$
$= 4s^2 + 5s - 9$

Review Exercises

Find each sum or difference.

20. $\begin{array}{r} 8x - 7 \\ \underline{(+)\ 3x + 5} \end{array}$

21. $\begin{array}{r} 6x^2 \qquad\ + 4 \\ \underline{(+)\ 3x^2 - 3x + 2} \end{array}$

22. $\begin{array}{r} 9m^2 + 6m - 6 \\ \underline{(-)\ 7m^2 - 3m + 3} \end{array}$

23. $\begin{array}{r} 15s^2 \qquad\ - 6 \\ \underline{(-)\ 8s^2 + 6s - 4} \end{array}$

24. $(17x - 3y) + (-2x + 5y)$
25. $(10g^2 + 5g) - (6g^2 - 3g)$
26. $(14x^2 + 3x - 6) - (8 - 2x)$
27. $(-3s^2t^2 - 4st^2) + (-7s^2t^2 + 6st^2)$

28. What is $5n^2 + 3$ minus $2n^2 - 4$?

• **Lesson 9–3** Multiply a polynomial by a monomial.

$2x^2(3x^3 + 2x^2 - x)$
$= 2x^2(3x^3) + 2x^2(2x^2) + 2x^2(-x)$
$= 6x^5 + 4x^4 - 2x^3$

Find each product.

29. $3(x - 5)$
30. $n(n + 3)$
31. $2m(m^2 + 4m)$
32. $x^4(3x^2 - 2x)$
33. $5(2x^2 + 3x - 2)$
34. $-3m^2(m^2 - 2m + 1)$

Solve each equation.

35. $x(x + 8) = x(x + 11) + 9$
36. $9(w - 4) + 10 = 2w + 16$

• **Lesson 9–4** Multiply two binomials.

Find the product of $x + 4$ and $x - 3$.

F L I O
$(x + 4)(x - 3)$

$= (x)(x) + (x)(-3) + (4)(x) + (4)(-3)$
$= x^2 - 3x + 4x - 12$
$= x^2 + x - 12$

Find each product.

37. $(y - 2)(y + 6)$
38. $(m + 5)(m + 7)$
39. $(x - 1)(x - 3)$
40. $(a + 11)(a - 4)$
41. $(2m - 1)(m - 4)$
42. $(x + 4)(3x - 2)$
43. $(4y + 2)(3y + 1)$
44. $(6x - 3)(2x + 5)$

45. Find the product of $5x + 4$ and $2x - 3$.

• Extra Practice
See pages 710–711.

Objectives and Examples	Review Exercises

• **Lesson 9–5** Develop and use the patterns for $(a + b)^2$, $(a - b)^2$, and $(a + b)(a - b)$.

$(x + 3)^2 = x^2 + 2(x)(3) + 3^2$ *Square of a sum*
$= x^2 + 6x + 9$

$(y - 5)^2 = y^2 - 2(y)(5) + 5^2$ *Square of a difference*
$= y^2 - 10y + 25$

$(s + 4)(s - 4) = s^2 - 4^2$ *Product of a sum and difference*
$= s^2 - 16$

Find each product.

46. $(w + 6)^2$ **47.** $(2x + 3)^2$

48. $(8 + g)^2$ **49.** $(x - 2)^2$

50. $(3y - 1)^2$ **51.** $(5m - 3n)^2$

52. $(y + 4)(y - 4)$ **53.** $(2a + 3)(2a - 3)$

54. $(2p - 3q)(2p + 3q)$ **55.** $(5a + 1)(5a + 1)$

Applications and Problem Solving

56. Recreation Bocce is a game similar to lawn bowling, but it is played on a rectangular dirt court. If the length of the rectangle is $12x$ and the width is $2x + 2$, find the area of the court in terms of x. Write your answer as a polynomial in simplest form. *(Lesson 9–3)*

57. Geometry The area of a triangle is given by $\frac{1}{2}bh$, where b is the length of the base of the triangle and h is the measure of the height of the triangle. In triangle ABC, $b = 3x + 8$ and $h = 7x - 4$. Write the polynomial representing the measure of the area of triangle ABC. *(Lesson 9–4)*

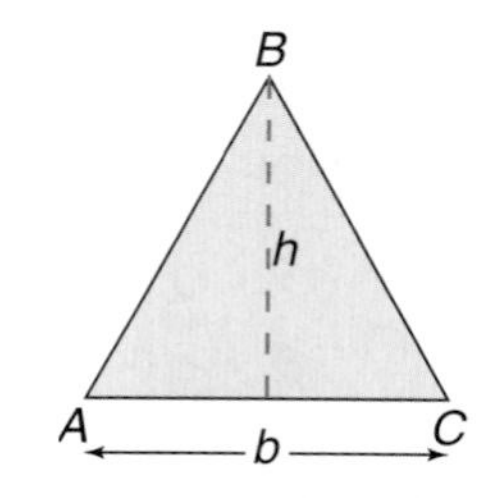

58. Gardening A rectangular garden is 10 feet longer than it is wide. It is surrounded by a brick walkway 3 feet wide, as shown at the right. Suppose the total area of the walkway is 396 square feet. *(Lesson 9–4)*

a. Find the area of the garden.

b. What are the dimensions of the garden?

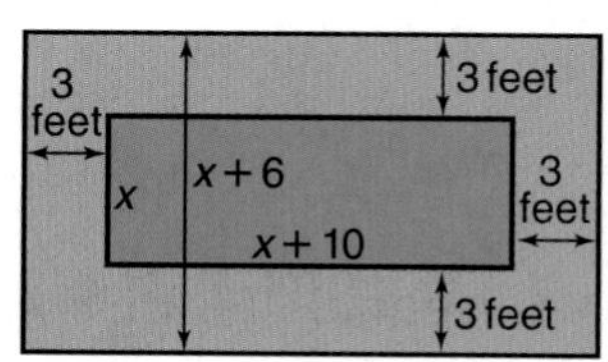

59. Quilting Kirsten is making a quilt to enter in the arts festival. The diagram gives the dimensions of each block of the quilt. *(Lesson 9–5)*

a. Find the area of each block.

b. The completed quilt is a square of 36 blocks. Write a product of binomials that could be used to determine the area of the entire quilt.

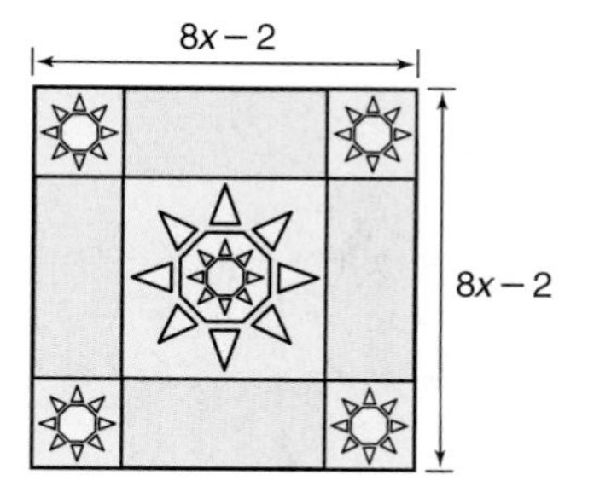

CHAPTER 9

Test

1. **Explain** how to use the FOIL method to multiply two binomials.
2. **Explain** how to subtract polynomials.
3. You Decide? Mark says that $-4wx^{-2}y^3$ is a monomial of degree 2. Linda disagrees. Who is correct? Explain your reasoning.

State whether each expression is a polynomial. If it is a polynomial, identify it as either a *monomial*, *binomial*, or *trinomial*, and state its degree.

4. $4x^3 + 3x^2$
5. $12r^{-2} - 3r + 6$
6. m^0
7. $5w^3yz^2$

Find each sum or difference.

8. $\begin{array}{r} 3x^2 - 5x + 4 \\ \underline{(+)\ 5x^2 + 7x + 8} \end{array}$
9. $\begin{array}{r} 8z^2 - 5z \\ \underline{(-)\ 6z^2 - 3z + 4} \end{array}$
10. $(y^3 + 6y^2 + 4y) - (2y^3 - 7y)$
11. $(-9h^2 - 5g) + (2g + 7h^2)$

Find each product.

12. $3x(2x^2 + 4x - 5)$
13. $5t^2(-2t + 3t^3 + 4)$
14. $(x + 2)(x + 3)$
15. $(y - 4)(y - 5)$
16. $(x - 2)(x + 5)$
17. $(3n + 4)(2n + 3)$
18. $(x + 5)^2$
19. $(2w - 3)^2$
20. $(7r - 4)(7r + 4)$
21. $(m - 3n)(m + 3n)$

Solve.

22. $2(y - 3) + 9 = 5y - 6$
23. $x(x + 4) + 5x = x(x - 1) + 20$
24. **Geometry** Find the area of the shaded region.

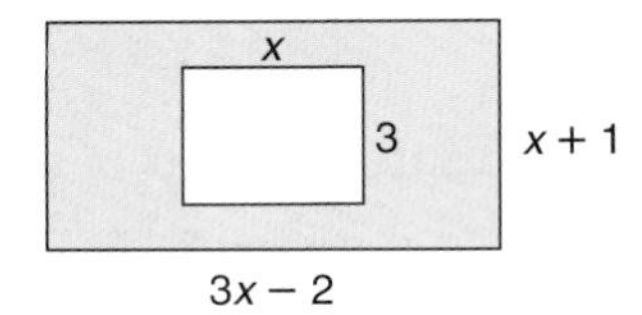

Exercise 24

25. **Family Life** Mrs. Douglas wants her five children, Aaron, Briana, Casey, Danielle, and Eddie, to spend a total of 35 hours on chores this week. Aaron, the oldest, works twice as many hours as the others. Danielle has earned an hour off from chores by getting an A on her algebra test. If x is the number of hours Danielle will work, then Briana, Casey, and Eddie will each work $x + 1$ hours, and Aaron will work twice as long, 2 + 2 hours.
    a. Write an equation to represent the number of hours the children will work.
    b. Use the equation to determine the number of hours each child will work on chores this week.

Preparing for Standardized Tests

Percent Problems

Standardized tests contain several types of percent problems. They may be numeric problems, word problems, or data analysis problems.

Familiarize yourself with common fractions, decimals, and percents. Below are some examples of common equivalents.

$0.01 = \frac{1}{100} = 1\%$ $\quad 0.1 = \frac{1}{10} = 10\%$ $\quad 0.333\ldots = \frac{1}{3} \approx 33.3\%$

$0.25 = \frac{1}{4} = 25\%$ $\quad 0.5 = \frac{1}{2} = 50\%$ $\quad 0.75 = \frac{3}{4} = 75\%$

The Princeton Review

A percent is a fraction whose denominator is 100.

State Test Example

The table shows recorded music sales. What percent of total sales in 1996 was on compact disc (CD)? Round to the nearest percent.

Sales of Recorded Music (millions)			
Year	Format		
	CD	Cassette	Record
1988	149.7	450.1	72.4
1992	407.5	366.4	2.3
1996	778.9	225.3	2.9

Hint Write a ratio, and then write the fraction as a percent.

Solution First, find the total sales in 1996.

$$778.9 + 225.3 + 2.9 = 1007.1$$

Next, write the ratio of compact disc sales to total sales.

$$\frac{778.9}{1007.1}$$

Then write this fraction as a decimal and as a percent. Use your calculator.

$$\frac{778.9}{1007.1} = 0.7734 \text{ or } 77\% \text{ to the nearest percent}$$

Compact disc sales were about 77% of total recorded music sales in 1996.

ACT Example

A bus company charges $5 each way to shuttle passengers between the hotel and a shopping mall. On a given day, the bus company has a total capacity of 250 people on the way to the mall and back. If the bus company runs at 90% of capacity, how much money would it collect that day?

A $1147.50 **B** $1250 **C** $2250
D $2500 **E** $2625

Hint Avoid partial answers. Be sure you answer the question that is asked.

Solution First determine how much money the bus company makes when it runs at total capacity. The 250 passengers would pay $10 each, because the charge is $5 each way.

$$250(\$10) = \$2500$$

Notice that this total amount is answer choice D, but it is a wrong answer. The question asks for the amount when the bus runs at 90% of capacity.

The word *of* is a clue to multiply. Find 90% of $2500.

$$(90\%)(\$2500) = (0.90)(2500) \text{ or } 2250$$

The bus company would collect $2250. Therefore, the answer is C.

After you work each problem, record your answer on the answer sheet provided or on a sheet of paper.

Multiple Choice

1. The areas of two rooms are 150 square feet and 135 square feet. If the total area of the home is 2000 square feet, what percent of the total area is the area of the two rooms?

   A $6\frac{3}{4}\%$ **B** $7\frac{1}{2}\%$
   C $14\frac{1}{4}\%$ **D** $85\frac{3}{4}\%$

2. A $9.95 calendar is marked down 40%. Before tax, how much is saved on the purchase of one calendar?

   A $1.99 **B** $3.98
   C $4.00 **D** $5.97

3. On a 16-question quiz, Tom answered 2 questions incorrectly. If each question is worth the same number of points, what percent of the total points is his point total?

   A 12.5% **B** 16% **C** 85%
   D 87.5% **E** 94%

4. An increase in prices has made the cost of remodeling an office 12% more than the original cost. The original cost was $7145. What is the best estimate of the new cost?

   A $700 **B** $7700
   C $10,000 **D** $70,000

5. The graph shows the attendance at a park. Predict the attendance in the year 2005.

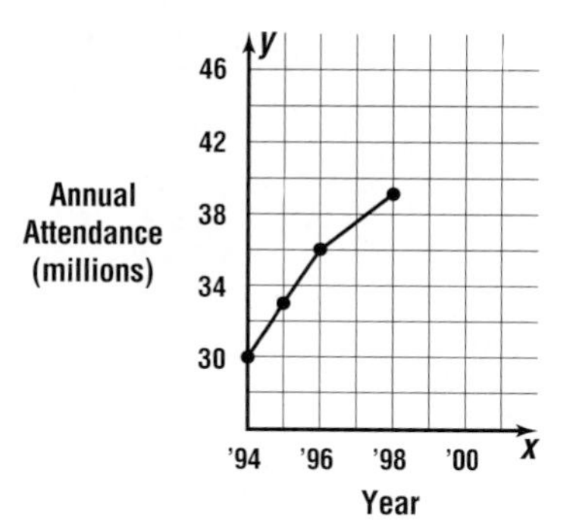

   A 38 million **B** 48 million
   C 60 million **D** 100 billion

6. A pair of shoes that regularly sold for $44 is now on sale for $33. What is the discount rate?

   A 11% **B** 25% **C** 75%
   D $33\frac{1}{3}\%$ **E** $66\frac{2}{3}\%$

7. There are 60 students in the band. How many play a percussion instrument?

Instruments in Bay City Band

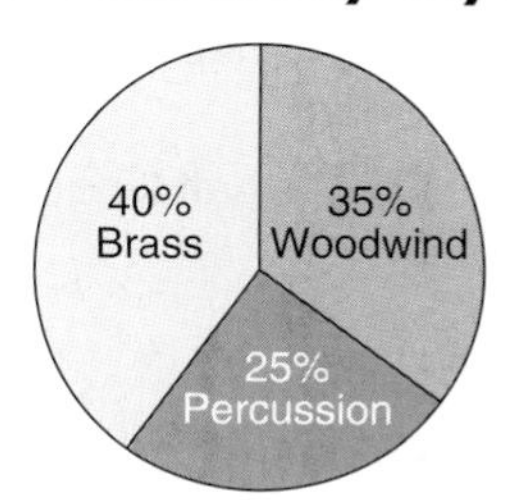

   A 15 **B** 21 **C** 24 **D** 25

8. Which expression can be used to find the value of y?

x	y
1	2
2	5
3	8
4	11

   A $2x + 1$ **B** $1 - 3x$
   C $3x - 1$ **D** $3x + 1$

Grid In

9. A CD player is on sale for $250. If there is a 6% sales tax, what is the total cost?

Extended Response

10. The total land area of a state is 23,159,000 acres. Of that, 13,513,000 acres are cropland, 1,866,000 acres are pastureland, and 3,626,000 acres are forest.

    Part A To the nearest percent, what percent of the state's area is cropland?

    Part B What percent of the state's area is *not* cropland, pastureland, or forest?

CHAPTER 10

Factoring

FOLDABLES™

Study Organizer

Make this Foldable to help you organize your notes on factoring. Begin with a sheet of plain $8\frac{1}{2}$" by 11" paper.

1. **Fold** in half lengthwise.

2. **Fold** again in thirds.

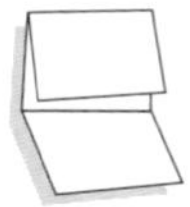

3. **Open** Cut along the second fold to make three tabs.

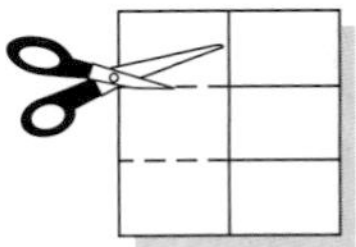

4. **Label** each tab as shown.

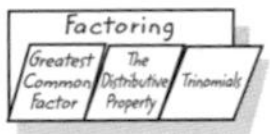

Factoring
Greatest Common Factor | The Distributive Property | Trinomials

Reading and Writing As you read and study the chapter, unfold each page and fill the journal with notes and examples.

Problem-Solving Workshop

Project

In 1964, there were about 7600 shopping malls in the United States. By 1997, this number had increased to 42,874. In this project, you will design a new shopping mall for a city near you. Your mall should occupy at least 750,000 square feet. Include a scale drawing of your mall and a one-page proposal promoting your design.

Working on the Project

Work with a partner and choose a strategy to help analyze and complete this project. Develop a plan. Brainstorm with your partner on possible dimensions for the mall. Here are some questions to help you get started.

- Suppose your mall design has three levels and each level is shaped like a square. What is the length of one side of the building?
- Suppose the mall has two levels, each shaped like a rectangle. What are two possible dimensions for the building?

Strategies

Look for a pattern.
Draw a diagram.
Make a table.
Work backward.
Use an equation.
Make a graph.
Guess and check.

Technology Tools

- Use a **calculator** to find possible dimensions for the mall.
- Use **word processing software** to write your proposal.
- Use **drawing software** to make your scale drawing.

*inter***NET** CONNECTION **Research** For more information about shopping malls, visit: www.algconcepts.com

Presenting the Project

Prepare a portfolio of your scale drawings. Write a one-page proposal that highlights the features of your mall. Make sure that your drawings and proposal include:

- your calculations for the total number of square feet occupied by the mall,
- labels for all dimensions in your scale drawings, and
- the scale that you used for your drawings.

10–1 Factors

What You'll Learn
You'll learn to find the greatest common factor of a set of numbers or monomials.

Why It's Important
Crafts Quilters use greatest common factors when they cut fabric.
See Exercise 57.

Recall that when two or more numbers are multiplied, each number is a *factor* of the product. For example, 12 can be expressed as the product of different pairs of whole numbers. Factors can be shown geometrically.

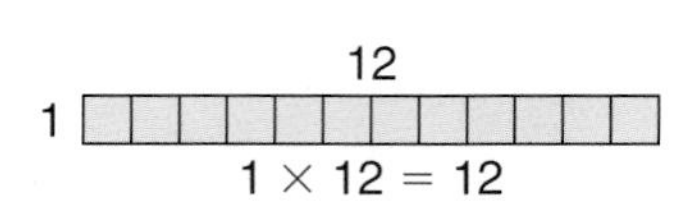

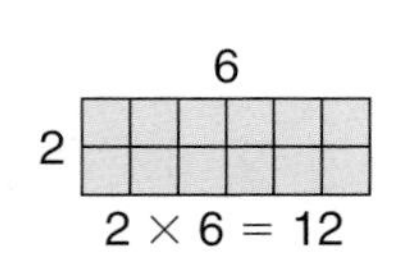

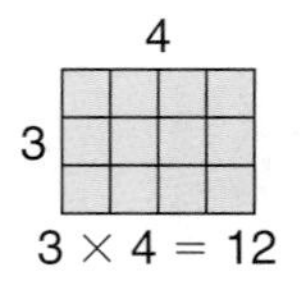

The whole numbers 1, 12, 2, 6, 3, and 4 are the factors of 12.

Some whole numbers have exactly two factors, the number itself and 1. Recall that these numbers are called *prime numbers*. Whole numbers that have more than two factors, such as 12, are called *composite numbers*.

Prime Numbers Less Than 20	2, 3, 5, 7, 11, 13, 17, 19
Composite Numbers Less Than 20	4, 6, 8, 9, 10, 12, 14, 15, 16, 18
Neither Prime nor Composite	0, 1

Examples

1 **Find the factors of each number. Then classify each number as *prime* or *composite*.**

72

To find the factors of 72, list all pairs of whole numbers whose product is 72.

1×72 2×36 3×24 4×18 6×12 8×9

The factors of 72 are 1, 2, 3, 4, 6, 8, 9, 12, 18, 24, 36, and 72. Since 72 has more than two factors, it is a composite number.

interNET CONNECTION
Data Update For the latest information on prime numbers, visit: www.algconcepts.com

2 **37**

There is only one pair of whole numbers whose product is 37.
1×37

The factors of 37 are 1 and 37. Therefore, 37 is a prime number.

Your Turn

a. 25 **b.** 23 **c.** 79 **d.** 51

You can use a graphing calculator to investigate factor patterns.

Graphing Calculator Exploration

Graphing Calculator Tutorial
See pp. 724–727.

The table below shows the numbers 2 through 12 and their factors arranged by the number of factors.

2 Factors	3 Factors	4 Factors	5 Factors	6 Factors
2: 1, 2 **3:** 1, 3 **5:** 1, 5 **7:** 1, 7 **11:** 1, 11	**4:** 1, 2, 4 **9:** 1, 3, 9	**6:** 1, 2, 3, 6 **8:** 1, 2, 4, 8 **10:** 1, 2, 5, 10	none	**12:** 1, 2, 3, 4, 6, 12

Step 1: Copy the table above.

Step 2: Use the graphing calculator program below to find the factors of the numbers 13 through 20.

```
PROGRAM:FACTOR
:Input "ENTER NUMBER", N
:For (D, 1, N)
:If iPart (N/D) = (N/D)
:Disp D
:END
```

Try These

1. Place the numbers 13 through 20 in the correct column of the table.
2. Predict a number from 21 through 100 for each column. Check your prediction by using the calculator program.
3. Explain the pattern in each column.

Since $4 \cdot 3 = 12$, 4 is a factor of 12. However, it is not a prime factor of 12 because 4 is not a prime number. Recall that when a number is expressed as a product of prime factors, the expression is called the *prime factorization* of the number.

You can use a *factor tree* to find the prime factorization of a number. Two different factor trees are shown for the prime factorization of 12.

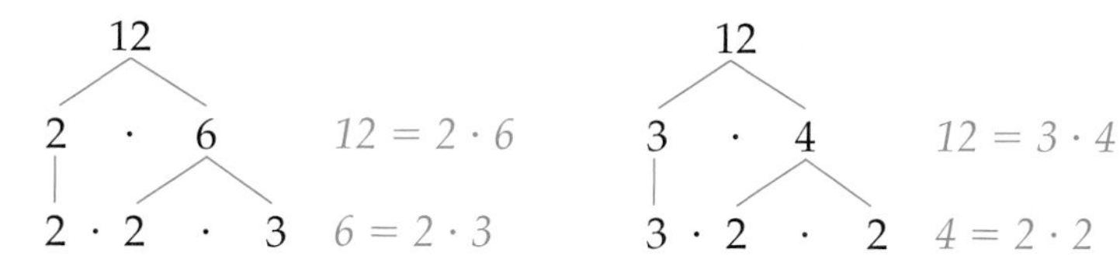

All of the factors in the last row are prime numbers. The factors are in a different order, but the result is the same. Except for the order of the factors, there is only one prime factorization of a number. Thus, the prime factorization of 12 is $2 \cdot 2 \cdot 3$ or $2^2 \cdot 3$.

You can use prime factorization to factor monomials. A monomial is in *factored form* when it is expressed as the product of prime numbers and variables and no variable has an exponent greater than 1.

Examples

Look Back

Monomials: Lesson 9–1

Factor each monomial.

3 $\mathbf{12a^2b}$

$12a^2b = 2 \cdot 2 \cdot 3 \cdot a \cdot a \cdot b$ *$12 = 2 \cdot 2 \cdot 3,\ a^2 = a \cdot a$*

4 $\mathbf{100mn^3}$

$100mn^3 = 2 \cdot 2 \cdot 5 \cdot 5 \cdot m \cdot n \cdot n \cdot n$ *$100 = 2 \cdot 2 \cdot 5 \cdot 5,\ n^3 = n \cdot n \cdot n$*

5 $\mathbf{-25x^2}$

To factor a negative integer, first express it as the product of a whole number and −1. Then find the prime factorization.

$-25x^2 = -1 \cdot 25x^2$ *$-25 = -1 \cdot 25$*

$= -1 \cdot 5 \cdot 5 \cdot x \cdot x$ *$25 = 5 \cdot 5$*

Your Turn

e. $15ab^2$ **f.** $84yz^2$ **g.** $-36b^3$

Two or more numbers may have some common prime factors. Consider the prime factorization of 36 and 42.

$$36 = 2 \cdot 2 \cdot 3 \cdot 3$$
$$42 = 2 \cdot 3 \cdot 7$$

Line up the common factors.

The integers 36 and 42 have 2 and 3 as common prime factors. The product of these prime factors, $2 \cdot 3$ or 6, is called the **greatest common factor (GCF)** of 36 and 42. The GCF is the greatest number that is a factor of both original numbers.

Greatest Common Factor

The greatest common factor of two or more integers is the product of the prime factors common to the integers.

The GCF of two or more monomials is the product of their common factors when each monomial is expressed in factored form.

Examples

Find the GCF of each set of numbers or monomials.

6 **24, 60, and 72**

$24 = 2 \cdot 2 \cdot 2 \cdot 3$ — *Find the prime factorization of each number.*

$60 = 2 \cdot 2 \cdot 3 \cdot 5$ — *Line up as many factors as possible.*

$72 = 2 \cdot 2 \cdot 2 \cdot 3 \cdot 3$ — *Circle the common factors.*

The GCF of 24, 60, and 72 is $2 \cdot 2 \cdot 3$ or 12.

7 **15 and 8**

$15 = 3 \cdot 5$

$8 = 2 \cdot 2 \cdot 2$

There are no common prime factors. The only common factor is 1. So, the GCF of 15 and 8 is 1.

Reading Algebra

Two numbers or monomials whose greatest common factor is 1 are called *relatively prime*. So, 15 and 8 are relatively prime.

8 **$15a^2b$ and $18ab$**

$15a^2b = 3 \cdot 5 \cdot a \cdot a \cdot b$

$18ab = 2 \cdot 3 \cdot 3 \cdot a \cdot b$

The GCF of $15a^2b$ and $18ab$ is $3 \cdot a \cdot b$ or $3ab$.

Your Turn

h. 75, 100, and 150 **i.** $5a$ and $8b$ **j.** $24ab^2c$ and $60a^2bc$

Knowing the factors of a number can help you with geometry.

Example
Geometry Link

9 **The area of a rectangle is 18 square inches. Find the length and width so that the rectangle has the least perimeter. Assume that the length and width are both whole numbers.**

Explore You know that the area of the rectangle is 18 square inches. You want to find the length and width so that the rectangle has the least perimeter.

Plan Find the factors of 18 and draw rectangles with each length and width. Then find each perimeter.

Solve

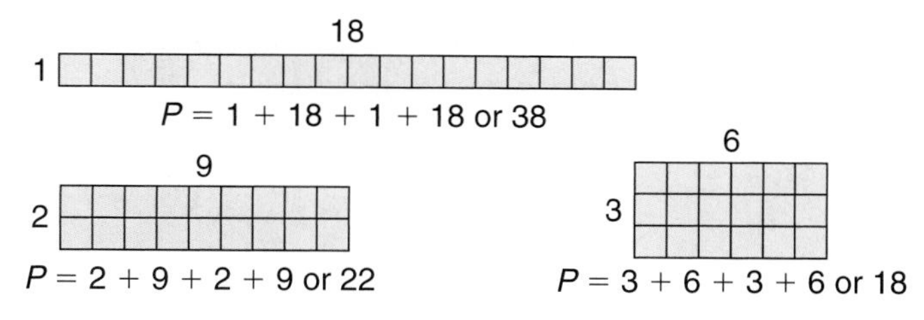

The least perimeter is 18 inches. The rectangle has a length of 6 inches and a width of 3 inches. *Examine this solution.*

Check for Understanding

Communicating Mathematics

1. **List** the prime numbers between 20 and 50.
2. **Name** two numbers whose GCF is 4.
3. **Explain** how to find the GCF of $8x^2$ and $16x$.
4. 

Jennifer believes that $2 \cdot 3 \cdot 4 \cdot 5$ is the prime factorization of 120, but Arturo disagrees. Who is correct? Explain.

Vocabulary
greatest common factor (GCF)

Guided Practice

Getting Ready **Find the prime factorization of each number.**

Sample 1: 28
Solution: $28 = 2 \cdot 2 \cdot 7$

Sample 2: 60
Solution: $60 = 2 \cdot 2 \cdot 3 \cdot 5$

5. 21
6. 72
7. 51
8. 150
9. 108
10. 110

Find the factors of each number. Then classify each number as *prime* or *composite*. *(Examples 1 & 2)*

11. 42
12. 47

Factor each monomial. *(Examples 3–5)*

13. $24x^2y$
14. $-16ab^2c$

Find the GCF of each set of numbers or monomials. *(Examples 6–8)*

15. 15, 70
16. 16, 24, 28
17. 20, 21
18. $2x$, $5y$
19. $7y^2$, $14y^3$
20. $-12ab$, $4a^2b^3$

21. **Geometry** The area of a rectangle is 72 square centimeters. Find the length and width so that the rectangle has the greatest perimeter. Assume that the length and width are both whole numbers. *(Example 9)*

Exercises

Practice

Find the factors of each number. Then classify each number as *prime* or *composite*.

22. 19
23. 20
24. 61
25. 45
26. 49
27. 91

Factor each monomial.

28. $20x^2$
29. $-15a^2b$
30. $-24c^3$
31. $50m^2n^2$
32. $44r^2s$
33. $90yz^2$

Homework Help

For Exercises	See Examples
22–27, 52, 54, 57	1, 2
28–33	3, 4, 5
34–51, 53	6, 7, 8
56	1, 2, 6, 7, 8

Extra Practice

See page 711.

Find the GCF of each set of numbers or monomials.

34. 24, 40
35. 12, 8
36. 17, 21
37. 18, 36
38. 20, 30
39. 45, 72
40. $3x^2$, $3x$
41. $18y^2$, $3y$
42. $-5ab$, $6b^2$
43. -18, $45mn$
44. $24a^2$, $60ab$
45. $9x^2y$, $10m^2n$
46. 6, 8, 12
47. 20, 21, 25
48. 18, 30, 54
49. $5m^2$, $15n^2$, $25mn$
50. $6ax^2$, $18ay^2$, $9az^3$
51. $15r^2$, $35s^2$, $70rs$

52. What is the greatest prime number less than 90?

53. Find the greatest common factor of $5x^2$, $5y^2$, and $10xy$.

54. *Twin primes* are prime numbers that differ by 2, such as 5 and 7. Find two other sets of twin primes that are between 25 and 45.

Applications and Problem Solving

55. **Crafts** Ashley wants to make a quilt from two different kinds of fabric. One is 60 inches wide, and the other is 48 inches wide. What are the dimensions of the largest squares she can cut from both fabrics so that no fabric is wasted?

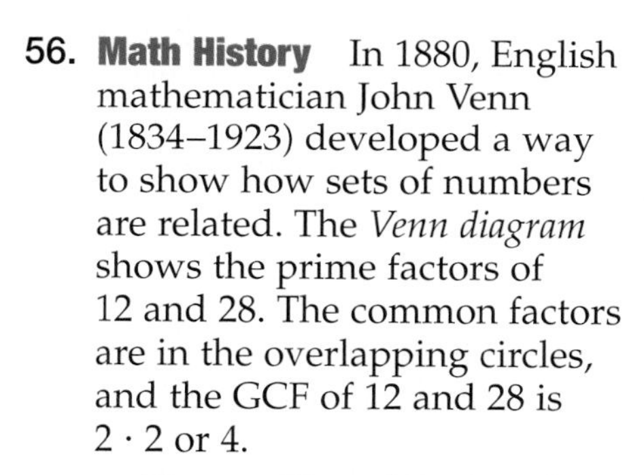

56. **Math History** In 1880, English mathematician John Venn (1834–1923) developed a way to show how sets of numbers are related. The *Venn diagram* shows the prime factors of 12 and 28. The common factors are in the overlapping circles, and the GCF of 12 and 28 is $2 \cdot 2$ or 4.

Prime Factors of 12 and 18

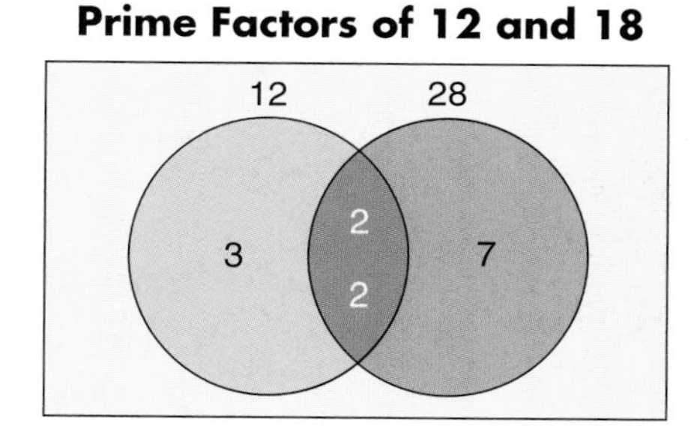

a. Draw a Venn diagram showing the prime factors of 36 and 45.

b. Find the GCF of 36 and 45.

57. **Critical Thinking** Explain why 2 is the only even prime number.

Mixed Review

Find each product. *(Lessons 9–3, 9–4, and 9–5)*

58. $(x + 3)(x - 3)$
59. $(2y + 1)(2y + 1)$
60. $(3a - 2)^2$
61. $(z + 4)(z + 3)$
62. $(x - 5)(x - 4)$
63. $(2n + 1)(n + 4)$
64. $3(x - 5)$
65. $2a(3 - a^2)$
66. $4x^2y(3x - 2y)$

Standardized Test Practice

67. **Short Response** In 2025, there are expected to be 817,000,000 cars in use worldwide. Write 817,000,000 in scientific notation. *(Lesson 8–4)*

68. **Multiple Choice** Which graph below is *not* the graph of a function? *(Lesson 6–4)*

A
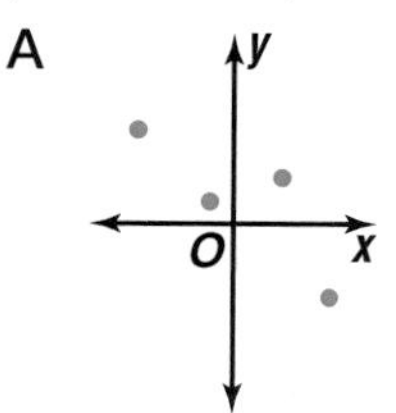

B
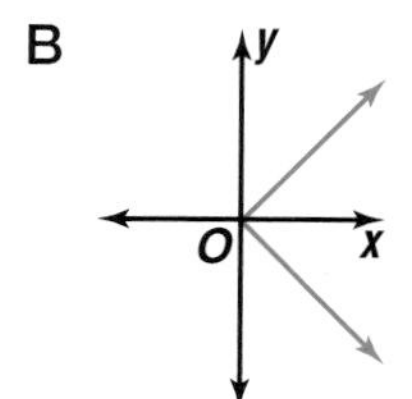

C
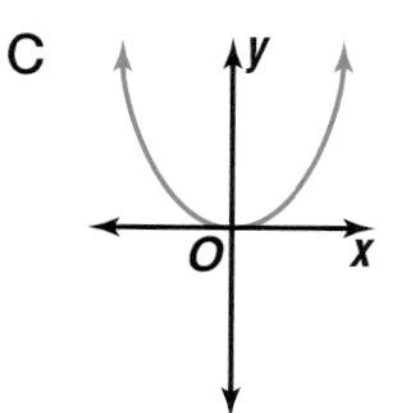

D
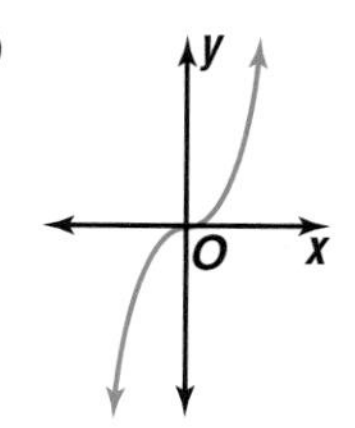

Chapter 10 Investigation

A Puzzling Perimeter Problem

Materials

Perimeter and Area

You know that the perimeter of a rectangle is the distance around the outside of the figure. It is measured in units like inches, centimeters, or feet. The area of a rectangle is the number of square units needed to cover the surface. It is measured in units like square inches or square centimeters. Are there any rectangles in which the measure of the perimeter is equal to the measure of the area? Let's investigate.

Investigate

1. Use the 1-tiles from a set of algebra tiles.

a. Make a table like the one below with 21 rows.

Area of Rectangle	Possible Dimensions of Rectangle	Perimeter of Rectangle
1		
2		

A 1-by-2 rectangle and a 2-by-1 rectangle have the same perimeter. So, they are listed only once in the table.

b. Select one tile. The measure of its area is 1. The dimension of this rectangle is 1 by 1. Find the measure of the perimeter of the rectangle and enter it in column 3.

c. Select two tiles. You can form a 1-by-2 rectangle with the tiles. This rectangle has an area of 2. Write 1 by 2 in column 2. What is the perimeter of this rectangle? Enter it in column 3.

d. Repeat this process using three tiles.

e. When you select four tiles, you have two options: a 1-by-4 rectangle or a 2-by-2 rectangle.

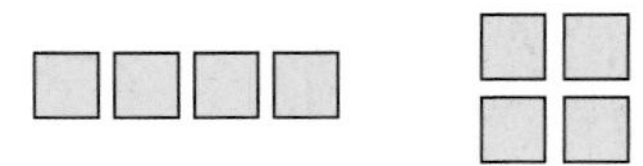

Write 1 by 4 and 2 by 2 in column 2 as the dimensions of the rectangles. Find the perimeters of the two rectangles and enter the results in column 3.

Your table should look like this.

Area of Rectangle	Possible Dimensions of Rectangle	Perimeters of Rectangle
1	1 by 1	4
2	1 by 2	6
3	1 by 3	8
4	1 by 4, 2 by 2	10, 8

2. Use tiles to form all possible rectangles with areas from 5 through 20. Write the dimensions in column 2 and the perimeters in column 3.
3. Analyze the data. Which rectangle(s) has the same numerical values for perimeter and area?
4. Make a conjecture about whether there are other rectangles that have the same numerical values for perimeter and area.

Extending the Investigation

In this extension, you will continue to investigate the area and perimeter of rectangles. Extend your table through at least 30 squares.

- Study the perimeters for only the rectangles whose dimensions are 1 by n. Make an ordered list of the perimeters. Describe the pattern of the perimeters.
- Study the perimeters for only the rectangles whose dimensions are 2 by n, but not 2 by 2. Make an ordered list of the perimeters. Describe the pattern of the perimeters.
- Study the perimeters for only the rectangles whose dimensions are 3 by n, but not 3 by 3. Make an ordered list of the perimeters. Describe the pattern of the perimeters.
- Study the perimeters for rectangles that are also squares. Make an ordered list of these perimeters. Describe the pattern of the perimeters.

Presenting Your Conclusions

Here are some ideas to help you present your conclusions to the class.

- Make a brochure describing your findings. Include figures and tables to illustrate your results.
- Make a video showing the patterns you discovered. You may want to have your classmates take on the roles of rectangles with various dimensions.

Investigation For more information on perimeter and area, visit: www.algconcepts.com

10–2 Factoring Using the Distributive Property

What You'll Learn
You'll learn to use the GCF and the Distributive Property to factor polynomials.

Why It's Important
Marine Biology You can find the height a dolphin jumps out of the water by evaluating an expression that is written in factored form.
See Exercise 46.

Sometimes, you know the product and are asked to find the factors. This process is called **factoring**.

For example, suppose you want to paint a rectangle on a wall and you only have enough paint to cover 20 square feet. If the length of each side must be an integer, what are the dimensions of all the possible rectangles you could paint?

Recall that the formula for the area of a rectangle is $A = \ell w$. If $A = 20$ square feet, then the measures of the length and width of the painted rectangle must be factor pairs of 20. The factor pairs of 20 are 1 and 20, 2 and 10, and 4 and 5. The figures below show rectangles with these factors as measures of length and width.

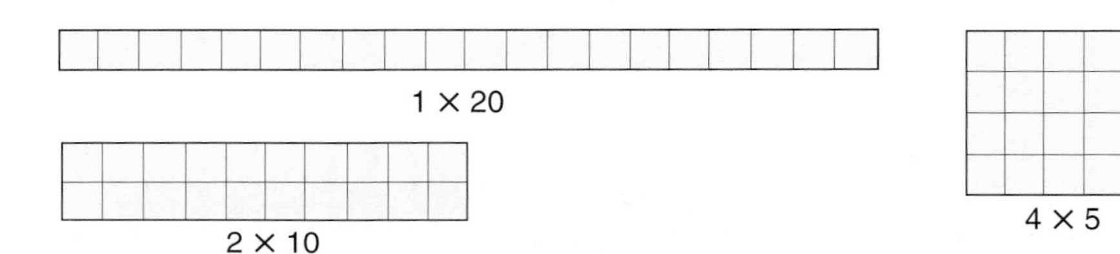

Hands-On Algebra
Algebra Tiles

Use algebra tiles to factor $2x + 8$.

Step 1 Model the polynomial $2x + 8$.

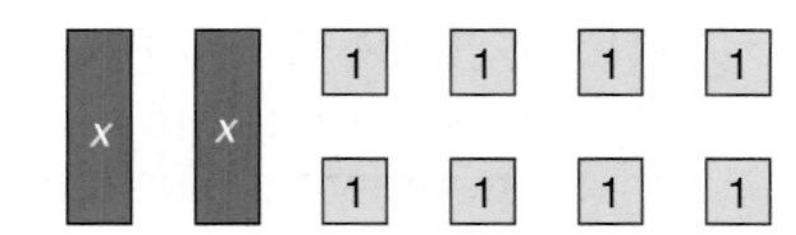

Step 2 Arrange the tiles into a rectangle. The total area of the tiles represents the product. Its length and width represent the factors. The rectangle has a width of 2 and a length of $x + 4$. So, $2x + 8 = 2(x + 4)$.

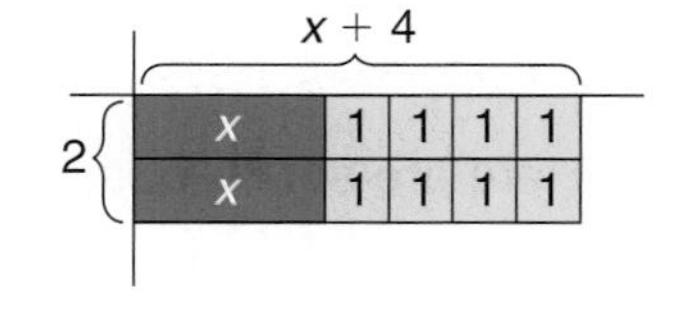

Try These

Use algebra tiles to factor each binomial.

1. $3x + 9$ **2.** $4x + 10$ **3.** $x^2 + 5x$ **4.** $3x^2 + 4x$

In Chapter 9, you used the Distributive Property to multiply a polynomial by a monomial.

$$2y(4y + 5) = 2y(4y) + 2y(5)$$
$$= 8y^2 + 10y$$

	$4y$	5
$2y$	$8y^2$	$10y$

You can reverse this process to express a polynomial in *factored form*. A polynomial is in factored form when it is expressed as the product of polynomials. For example, to factor $8y^2 + 10y$, find the greatest common factor of $8y^2$ and $10y$.

$$8y^2 = 2 \cdot 2 \cdot \textcircled{2} \cdot \textcircled{y} \cdot y$$
$$10y = \textcircled{2} \cdot 5 \cdot \textcircled{y}$$

The GCF of $8y^2$ and $10y$ is $2y$. Write each term as a product of the GCF and its remaining factors. Then use the Distributive Property.

$$8y^2 + 10y = 2y(4y) + 2y(5)$$
$$= 2y(4y + 5) \quad \textit{Distributive Property}$$

$8y^2 + 10y$ written in factored form is $2y(4y + 5)$.

Examples

Factor each polynomial.

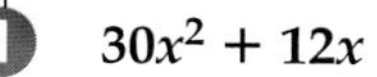

1 $\mathbf{30x^2 + 12x}$

First, find the GCF of $30x^2$ and $12x$.

$$30x^2 = \textcircled{2} \cdot \textcircled{3} \cdot 5 \cdot \textcircled{x} \cdot x$$
$$12x = 2 \cdot \textcircled{2} \cdot \textcircled{3} \cdot \textcircled{x}$$

	$5x$	2
$6x$	$30x^2$	$12x$

The GCF of $30x^2$ and $12x$ is $6x$. Write each term as a product of the GCF and its remaining factors.

$$30x^2 + 12x = 6x(5x) + 6x(2)$$
$$= 6x(5x + 2) \quad \textit{Distributive Property}$$

2 $\mathbf{15ab^2 - 25abc}$

$$15ab^2 = 3 \cdot \textcircled{5} \cdot \textcircled{a} \cdot \textcircled{b} \cdot b$$
$$25abc = \textcircled{5} \cdot 5 \cdot \textcircled{a} \cdot \textcircled{b} \cdot c$$

	$3b$	$-5c$
$5ab$	$15ab^2$	$-25abc$

The GCF is $5ab$.

$$15ab^2 - 25abc = 5ab(3b) - 5ab(5c)$$
$$= 5ab(3b - 5c) \quad \textit{Distributive Property}$$

Examples

Factor each polynomial.

3 $\mathbf{18x^2y + 12xy^2 + 6xy}$

$$18x^2y = 2 \cdot 3 \cdot 3 \cdot x \cdot x \cdot y$$
$$12xy^2 = 2 \cdot 2 \cdot 3 \cdot x \cdot y \cdot y$$
$$6xy = 2 \cdot 3 \cdot x \cdot y$$

	$3x$	$2y$	1
$6xy$	$18x^2y$	$12xy^2$	$6xy$

The GCF is $6xy$. When $6xy$ is factored from $6xy$ the remaining factor is 1.

$$18x^2y + 12xy^2 + 6xy = 6xy(3x) + 6xy(2y) + 6xy(1)$$
$$= 6xy(3x + 2y + 1) \quad \textit{Distributive Property}$$

4 $\mathbf{7x^2 + 9yz}$

$$7x^2 = 7 \cdot x \cdot x$$
$$9yz = 3 \cdot 3 \cdot y \cdot z$$

There are no common factors of $7x^2$ and $9yz$ other than 1. Therefore, $7x^2 + 9yz$ cannot be factored using the GCF. It is a prime polynomial.

Reading Algebra

A polynomial that cannot be written as a product of two polynomials with integral coefficients is called a *prime polynomial*.

Your Turn

a. $12n^2 - 8n$

b. $16a^2b + 10ab^2$

c. $20rs^2 - 15r^2s + 5rs$

d. $21x + 5y + 16z$

If you know a product and one of its factors, you can use division to find the other factor. To divide a polynomial by a monomial, divide each term of the polynomial by the monomial.

Example

5 **Divide $\mathbf{15x^3 + 12x^2}$ by $\mathbf{3x}$.**

$$(15x^3 + 12x^2) \div 3x = \frac{15x^3}{3x} + \frac{12x^2}{3x} \quad \textit{Divide each term by 3x.}$$

$$= \frac{\overset{5}{\cancel{15}} \cdot \overset{1}{\cancel{x}} \cdot x \cdot x}{\underset{1}{\cancel{3}} \cdot \underset{1}{\cancel{x}}} + \frac{\overset{4}{\cancel{12}} \cdot \overset{1}{\cancel{x}} \cdot x}{\underset{1}{\cancel{3}} \cdot \underset{1}{\cancel{x}}} \quad \textit{Simplify.}$$

$$= 5x^2 + 4x$$

Therefore, $(15x^3 + 12x^2) \div 3x = 5x^2 + 4x$.

Look Back

Dividing Powers: Lesson 8–2

Your Turn

Find each quotient.

e. $(9b^2 - 15) \div 3$

f. $(10x^2y^2 + 5xy) \div 5xy$

Factoring a polynomial can help simplify computations.

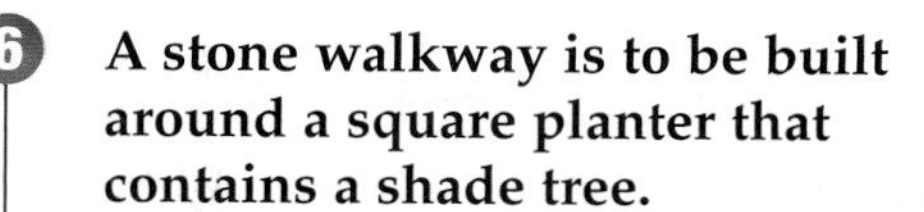

Example 6

Landscaping Link

A stone walkway is to be built around a square planter that contains a shade tree.

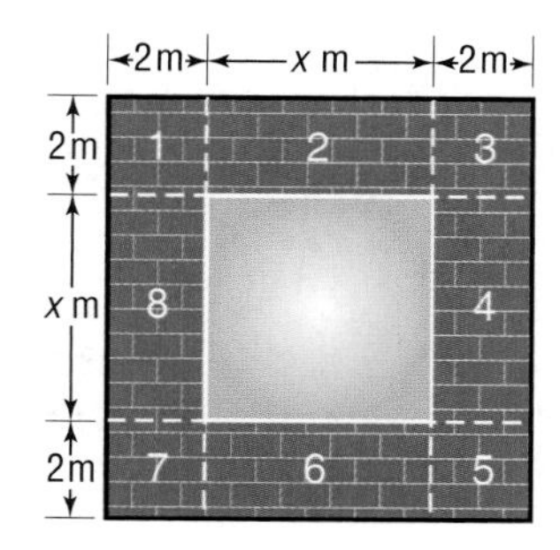

A. If the walkway is 2 meters wide, write an expression in factored form that represents the area of the walkway.

Let x represent the length and width of the planter. You can find the area of the walkway by finding the sum of the areas of the 8 rectangular sections shown in the figure.

The resulting expression can be simplified by using the Distributive Property to combine like terms and then factoring.

Regions

$$A = \underbrace{2 \cdot 2}_{1} + \underbrace{2 \cdot x}_{2} + \underbrace{2 \cdot 2}_{3} + \underbrace{2 \cdot x}_{4} + \underbrace{2 \cdot 2}_{5} + \underbrace{2 \cdot x}_{6} + \underbrace{2 \cdot 2}_{7} + \underbrace{2 \cdot x}_{8}$$

$$= 4 + 2x + 4 + 2x + 4 + 2x + 4 + 2x$$

$= 16 + 8x$ *$4 + 4 + 4 + 4 = 16$ and $2x + 2x + 2x + 2x = 8x$*

$= 8(2) + 8(x)$ *The GCF of 16 and $8x$ is 8.*

$= 8(2 + x)$

B. If the dimensions of the square planter are 1.5 meters by 1.5 meters, find the area of the walkway.

$A = 8(2 + x)$

$= 8(2 + 1.5)$ *Replace x with 1.5.*

$= 8(3.5)$ or 28

The area of the walkway is 28 square meters.

Check for Understanding

Communicating Mathematics

1. **Illustrate** with algebra tiles or a drawing how to factor $x^2 + 2x$.
2. **Explain** what it means to factor a polynomial.
3. **Write** a few sentences explaining how the Distributive Property is used to factor polynomials. Include at least two examples.

Vocabulary
factoring

Guided Practice

Getting Ready **Find the GCF of the terms in each expression.**

Sample: $9a^2 + 3a$ **Solution:** $9a^2 = 3 \cdot 3 \cdot a$
$3a = 3 \cdot a$ The GCF is $3a$.

4. $8xy + 12mn$ **5.** $x^2 + 2x$ **6.** $5xy + y^2$
7. $18y^2 + 30y$ **8.** $3ab - 2a^2b$ **9.** $15m^2n - 20m^2n$

Factor each polynomial. If the polynomial cannot be factored, write *prime*. *(Examples 1–4)*

10. $3x + 6$ **11.** $2x^2 + 4x$ **12.** $12a^2b + 6a$
13. $7mn - 13yz$ **14.** $3x^2y + 6xy + 9y^2$ **15.** $2a^3b^2 + 8ab + 16a^2b^3$

Find each quotient. *(Example 5)*

16. $(12m^2 - 15m) \div 3m$ **17.** $(3c^2d + 9cd) \div 3cd$

18. **Landscaping** Kiyoshi is planning to build a walkway around her square koi pond. The walkway is 6 feet wide.

a. If x represents the measure of one side of the pond, write an expression in factored form that represents the area of the walkway. *(Example 6)*

b. If the dimensions of the pond are 8 feet by 8 feet, find the area of the walkway. *(Example 7)*

Exercises

Practice

Homework Help	
For Exercises	See Examples
19–30, 46	1, 2
31–36	3, 4
37–43, 45	5
47	6
Extra Practice	
See page 712.	

Factor each polynomial. If the polynomial cannot be factored, write *prime*.

19. $9x + 15$ **20.** $6x + 3x^2$ **21.** $8x + 2x^2y$
22. $7a^2b^2 + 3ab^3$ **23.** $3c^2d - 6c^2d^2$ **24.** $7x - 3y$
25. $36mn - 11mn^2$ **26.** $18xy^2 + 24x^2y$ **27.** $19ab + 21xy$
28. $14mn^2 - 2mn$ **29.** $12xy^3 + y^4$ **30.** $3a^2b - 6a^2b^2$
31. $24xy + 18xy^2 - 3y$ **32.** $3x^3y + 9xy + 36xy^2$
33. $x + x^2y^3 + x^3y^2$ **34.** $6x^2 + 9xy + 24x^2y^2$
35. $12axy - 14ay + 20ax$ **36.** $42xyz - 12x^2y^2 + 3x^3y^3$

Find each quotient.

37. $(27x^2 - 21y^2) \div 3$ **38.** $(5abc + c) \div c$
39. $(14ab + 28b) \div 14b$ **40.** $(16x + 24xy) \div 8x$
41. $(4x^2y^2z + 6xz^2) \div 2xz$ **42.** $(3x^2y + 12xyz^2) \div 3xy$

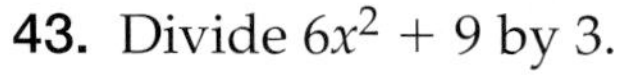

43. Divide $6x^2 + 9$ by 3.
44. What is the GCF of $14abc^2$ and $18c$?

Applications and Problem Solving

45. **Geometry** The area of a rectangle is $(16x + 4y)$ square feet. If the width is 4 feet, find the length.

46. **Marine Biology** In a pool at a water park, a dolphin jumps out of the water traveling at 24 feet per second. Its height h, in feet, above the water after t seconds is given by the formula $h = 24t - 16t^2$.

a. Factor the expression $24t - 16t^2$.

b. Find the height of a dolphin when $t = 0.75$ second.

47. **Geometry** Write an expression in factored form that represents the area of the shaded region.

a.

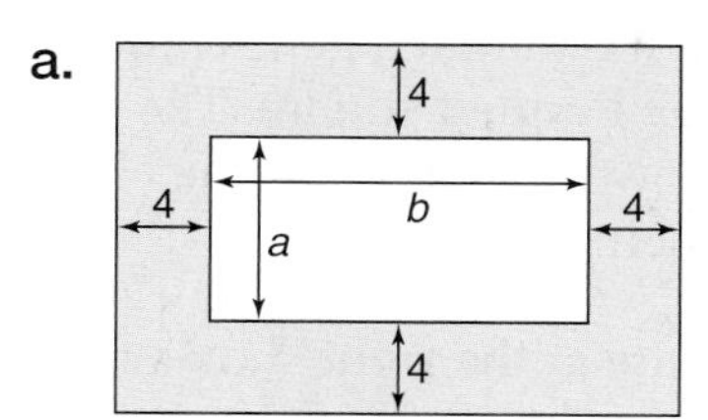

b.

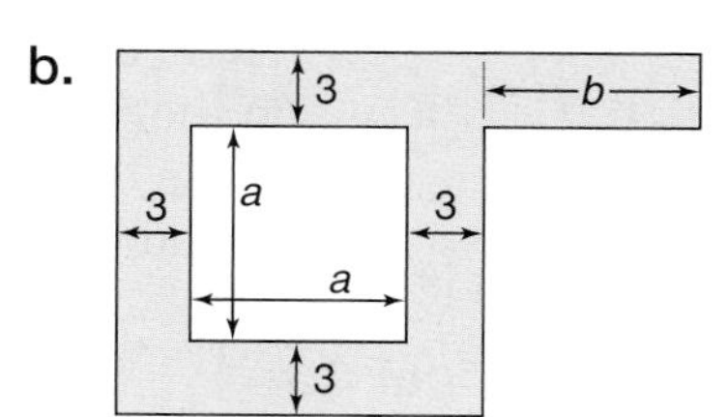

48. **Critical Thinking** The length and width of a rectangle are represented by $2x$ and $9 - 4x$. If x must be an integer, what are the possible measures for the area of this rectangle?

Mixed Review

Classify each number as *prime* or *composite*. *(Lesson 10–1)*

49. 2 50. 21 51. 49 52. 53 53. 90

54. **Geometry** The length of a side of a square is $3x + 5$ units. What is the area of the square? *(Lesson 9–5)*

Add or subtract. *(Lesson 9–2)*

55. $(x^2 + 4x - 3) + (2x^2 - 6x - 9)$ 56. $(2y^2 - 5y + 3) - (5y^2 - 4)$

Standardized Test Practice

A B C D

57. **Short Response** Write a second degree polynomial. *(Lesson 9–1)*

Quiz 1 Lessons 10–1 and 10–2

1. Find the prime factorization of 24. *(Lesson 10–1)*

Find the GCF of the terms in each expression. Then factor the expression. *(Lessons 10–1, 10–2)*

2. $20s + 40s^2$ 3. $ax^3 + 7bx^3 + 11cx^3$ 4. $6x^3 + 12x^2 + 6x$

5. **Landscaping** A 2-foot wide stone path is to be built along each side of a rectangular flower garden. The length of the garden is twice the width. If the flower garden is bordered on one side by a house, write an expression in factored form to represent the area of the path. *(Lesson 10–2)*

10–3 Factoring Trinomials: $x^2 + bx + c$

What You'll Learn

You'll learn to factor trinomials of the form $x^2 + bx + c$.

Why It's Important

Biology Geneticists use Punnett squares, which are similar to the models for factoring trinomials.
See Exercise 51.

In biology, *Punnett squares* are used to show possible ways that traits can be passed from parents to their offspring.

Each parent has two genes for each trait. The letters representing the parent's genes are placed on the outside of the Punnett square. The letters inside the boxes show the possible gene combinations for their offspring.

	G	*g*
G	*GG*	*Gg*
g	*Gg*	*gg*

The Punnett square at the right shows the gene combinations for fur color in rabbits.

- *G* represents the dominant gene for gray fur.
- *g* represents the recessive gene for white fur.

Notice that the Punnett square is similar to the model for multiplying binomials. The model below shows the product of $(x + 1)$ and $(x + 3)$.

$$(x + 1)(x + 3) = x^2 + 3x + 1x + 3$$
$$= x^2 + 4x + 3$$

In this lesson, you will factor a trinomial into the product of two binomials.

	x	3
x	x^2	$3x$
1	$1x$	3

Hands-On Algebra

Materials:  straightedge

Use a model to factor $x^2 + 5x + 4$.

Step 1 Draw a square with four sections. Put the first and last terms into the boxes as shown.

x^2	
	4

Step 2 Factor x^2 as $x \cdot x$ and place the factors outside the box.

Now, think of factors of 4 to place outside the box.

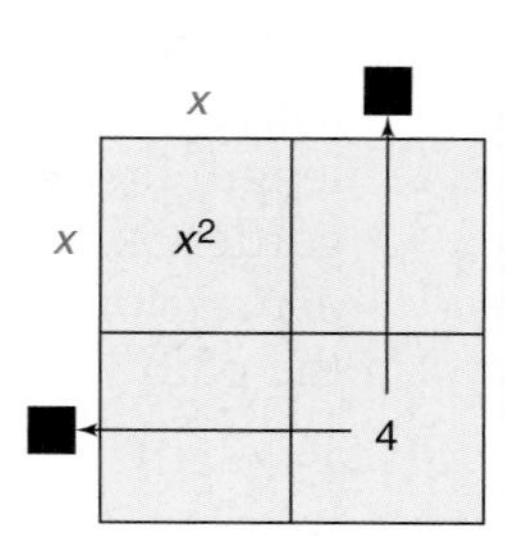

Step 3 The number 4 has two different factor pairs, 2 and 2, and 4 and 1. Try the factor pairs until you find the one that results in a middle term of $5x$.

Try 2 and 2.

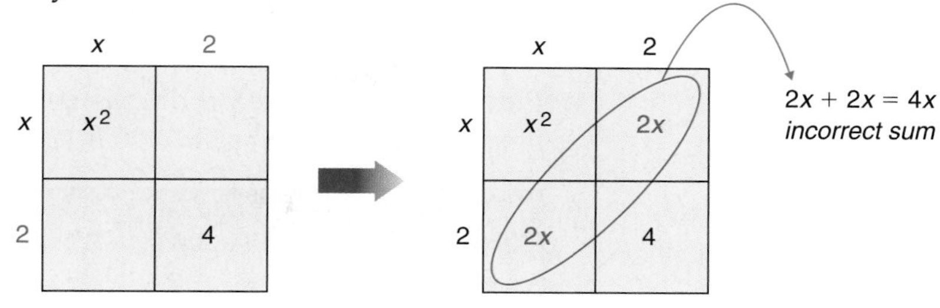

Try 4 and 1.

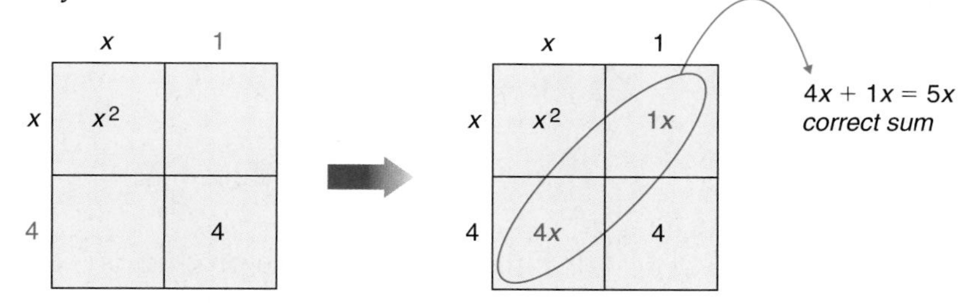

Step 4 The integers 4 and 1 result in the correct middle term, $5x$. Therefore, $x^2 + 5x + 4 = (x + 4)(x + 1)$.

Try These

Use a model to factor each trinomial.

1. $x^2 + 6x + 5$ **2.** $x^2 + 7x + 6$ **3.** $x^2 + 8x + 12$

4. $x^2 - 3x + 2$ **5.** $x^2 - 6x + 8$ **6.** $x^2 - 6x + 9$

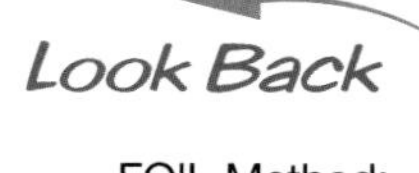

FOIL Method: Lesson 9–4

The FOIL method will help you factor trinomials without models. Use the following method to factor $x^2 + 6x + 8$.

Step 1 x^2 is the product of the **F**irst terms, and 8 is the product of the **L**ast terms.

$$x^2 + 6x + 8 = (x + \blacksquare)(x + \blacksquare)$$

Step 2 Try several factor pairs of 8 until the sum of the products of the **O**uter and **I**nner terms is $6x$. Check by using FOIL.

Try 1 and 8. $(x + 1)(x + 8) = x^2 + 8x + 1x + 8$
$= x^2 + 9x + 8$ *9x is not the correct term.*

Try 2 and 4. $(x + 2)(x + 4) = x^2 + 4x + 2x + 8$
$= x^2 + 6x + 8$ ✓

Therefore, $x^2 + 6x + 8 = (x + 2)(x + 4)$.

Examples

Factor each trinomial.

1 $x^2 - 7x + 10$

$$x^2 - 7x + 10 = (x + \blacksquare)(x + \blacksquare)$$

Find integers whose product is 10 and whose sum is -7. *Recall that the product of two negative integers is positive.*

Product	Integers	Sum
10	−1, −10	$-1 + (-10) = -11$
10	−2, −5	$-2 + (-5) = -7$ ✓

Therefore, $x^2 - 7x + 10 = (x - 2)(x - 5)$.

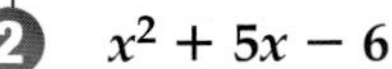

2 $x^2 + 5x - 6$

$$x^2 + 5x - 6 = (x + \blacksquare)(x + \blacksquare)$$

Find integers whose product is -6 and whose sum is 5. *Recall that the product of a positive integer and a negative integer is negative.*

Product	Integers	Sum
−6	−2, 3	$-2 + 3 = 1$
−6	2, −3	$2 + (-3) = -1$
−6	−1, 6	$-1 + 6 = 5$ ✓

You can stop listing factors when you find a pair that works.

Therefore, $x^2 + 5x - 6 = (x - 1)(x + 6)$.

3 $x^2 - 7 - 3x$

First, write the trinomial as $x^2 - 3x - 7$.

$$x^2 - 3x - 7 = (x + \blacksquare)(x + \blacksquare)$$

Find two integers whose product is -7 and whose sum is -3.

Product	Integers	Sum
−7	−1, 7	$-1 + 7 = 6$
−7	1, −7	$1 + (-7) = -6$

There are no factors of -7 whose sum is -3. Therefore, $x^2 - 3x - 7$ is a prime polynomial.

Your Turn

a. $x^2 + 3x + 2$ **b.** $a^2 + 4a + 3$ **c.** $b^2 + 4b + 4$

d. $y^2 - 7y + 12$ **e.** $n^2 - 5n - 14$ **f.** $m^2 - m + 1$

In the previous lesson, you learned that the terms of a polynomial might have a GCF that can be factored using the Distributive Property. When you factor trinomials, always check for a GCF first.

Example 4

Factor $2x^2 - 20x - 22$.

First, check for a GCF.

$2x^2 - 20x - 22 = 2(x^2 - 10x - 11)$ *The GCF is 2.*

Now, factor $x^2 - 10x - 11$. *Find two integers whose product is -11 and whose sum is -10.*

Product	Integers	Sum
−11	−1, 11	−1 + 11 = 10
−11	1, −11	1 + (−11) = −10 ✓

So, $x^2 - 10x - 11 = (x + 1)(x - 11)$.

Therefore, $2x^2 - 20x - 22 = 2(x + 1)(x - 11)$. *Check by using FOIL.*

Your Turn

Factor each polynomial.

g. $3y^2 - 9y - 54$

h. $5m^2 + 45m + 100$

The area of a figure can often be expressed as a trinomial.

Example 5

Tammy is planning a rectangular garden in which the width will be 4 feet less than its length. She has decided to put a birdbath within the garden, occupying a space 3 feet by 4 feet. How many square feet are now left for planting? Express the answer in factored form.

Let ℓ = the length of the original rectangle.

Let $\ell - 4$ = the width of the original rectangle.

ℓ

3

$\ell - 4$

4

Find the area of the original rectangle.

$A = \ell w$ *Area = length × width*

$A = \ell(\ell - 4)$ *Replace w with $\ell - 4$.*

$A = \ell^2 - 4\ell$ *Distributive Property*

Find the area of the small rectangle.

$A = 4(3)$ or 12

Remaining area = area of original rectangle − area of small rectangle

$= \ell^2 - 4\ell - 12$

The remaining area is $\ell^2 - 4\ell - 12$ or $(\ell - 6)(\ell + 2)$.

Check for Understanding

Communicating Mathematics

1. **Illustrate** how to factor $x^2 + 7x + 6$ using a model.
2. **Explain** why the trinomial $x^2 + x + 5$ cannot be factored.
3. **Complete** the following sentence.
When you factor $m^2 - 3m - 10$, you want to find two integers whose product is ___?___ and whose sum is ___?___.

Guided Practice

Getting Ready **Find two integers whose product is the first number and whose sum is the second number.**

Sample: 10, 7 **Solution:** $2 \times 5 = 10$, $2 + 5 = 7$

4. 30, 11 **5.** 12, −7 **6.** −10, 3 **7.** −6, −5 **8.** −30, −7

Factor each trinomial. If the trinomial cannot be factored, write *prime*. *(Examples 1–4)*

9. $x^2 + 5x + 6$ **10.** $y^2 + 9y + 20$ **11.** $a^2 - 5a + 4$
12. $z^2 - 8z + 16$ **13.** $x^2 + 3x - 10$ **14.** $m^2 - 4m - 21$
15. $w^2 + w + 2$ **16.** $3a^2 + 15a + 12$ **17.** $2c^2 - 12c - 14$

18. Geometry Find the area of the shaded region. Express the area in factored form. *(Example 5)*

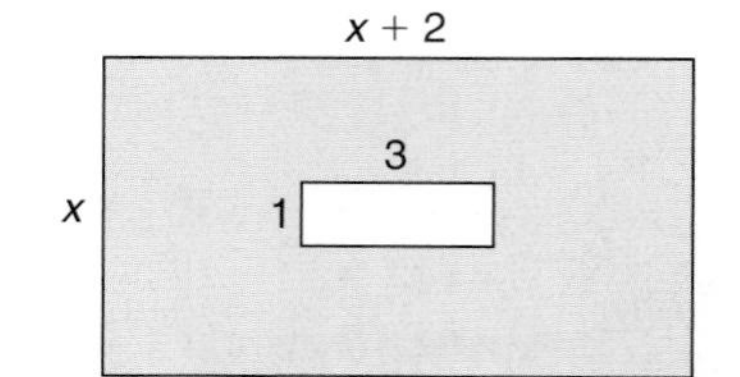

Exercises

Practice

Homework Help	
For Exercises	See Examples
19–39, 46, 48, 50, 52	1–3
40–45	4
49	5
Extra Practice	
See page 712.	

Factor each trinomial. If the trinomial cannot be factored, write *prime*.

19. $b^2 + 5b + 4$ **20.** $x^2 + 10x + 25$ **21.** $a^2 + 7a + 12$
22. $a^2 + 3a + 5$ **23.** $y^2 + 12y + 27$ **24.** $z^2 + 13z + 40$
25. $x^2 - 8x + 15$ **26.** $a^2 - 4a + 4$ **27.** $c^2 - 13c + 36$
28. $d^2 - 11d + 28$ **29.** $m^2 - 5m + 1$ **30.** $y^2 - 12y + 32$
31. $c^2 + 2c - 3$ **32.** $x^2 - 5x - 24$ **33.** $r^2 - 3r - 18$
34. $m^2 - 2m - 24$ **35.** $n^2 + 13n - 30$ **36.** $m^2 + 11m - 12$
37. $x^2 - 17x + 72$ **38.** $a^2 - a - 90$ **39.** $r^2 + 22r - 48$
40. $4x^2 + 28x + 40$ **41.** $3y^2 - 21y + 36$ **42.** $z^3 + z^2 - 12z$
43. $m^3 + 3m^2 + 2m$ **44.** $3y^3 - 24y^2 + 36y$ **45.** $2a^3 + 14a^2 - 16a$

46. Express $x^2 + 24x + 95$ as the product of two binomials.
47. Write a trinomial that cannot be factored.
48. Complete the trinomial $x^2 + 6x +$ ___?___ with a positive integer so that the resulting trinomial can be factored.

Applications and Problem Solving

49. **Geometry** Refer to the figure at the right.

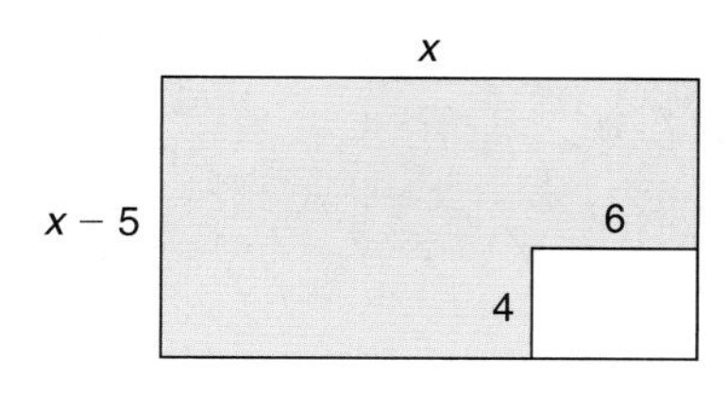

a. Express the area of the shaded region as a polynomial.

b. Express the area in factored form.

50. **Geometry** The volume of a rectangular prism is $x^3 + 4x^2 + 3x$. Find the length, width, and height of the prism if each dimension can be written as a monomial or binomial with integral coefficients. (*Hint*: Use the formula $V = \ell wh$.)

51. **Genetics** In guinea pigs, a black coat is a dominant trait over a white coat. Let C represent a black coat and c represent a white coat in the Punnett squares below. Find the missing genes or gene pair.

a.

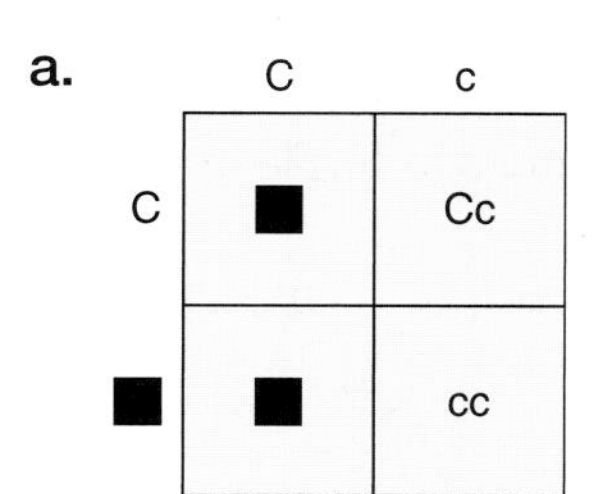

b.

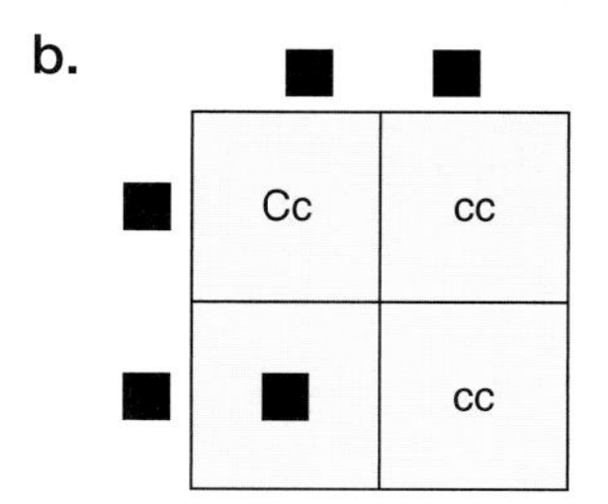

52. **Critical Thinking** Find all values of k so that the trinomial $x^2 + kx + 10$ can be factored.

Mixed Review

Find each quotient. *(Lesson 10–2)*

53. $(10x^2 + 25y^2) \div 5$

54. $(2y^2 + 4y) \div y$

55. $(6a^2 + 8ab - 6b^2) \div 2$

56. $(3x^2y^2 + 9x^3y^2z) \div 3x^2y^2$

Find the GCF of each set of numbers or monomials. *(Lesson 10–1)*

57. $12a, 16b$

58. $6a^2b, 9ab^2$

59. $15x, 7y$

60. 15, 60, 75

Standardized Test Practice

61. **Extended Response** The graph shows the value of riding lawn mower shipments in 1997 and 2000. *(Lesson 7–4)*

a. Write an equation of the line in slope-intercept form.

b. What does the slope represent?

c. Use the equation to predict the value of riding lawn mower shipments in 2005.

Riding Lawn Mower Shipments (billion dollars)

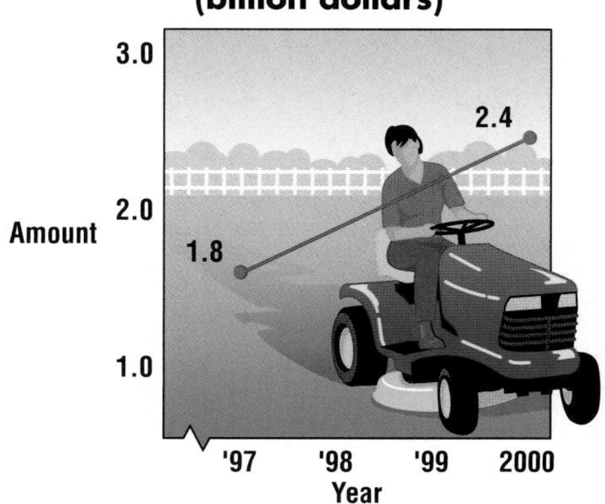

Source: Freedonia Group

62. **Multiple Choice** Evaluate $8x + 3y$ if $x = 9$ and $y = -2$. *(Lesson 2–5)*

A 66 B 57 C 78 D 11

10–4 Factoring Trinomials: $ax^2 + bx + c$

What You'll Learn
You'll learn to factor trinomials of the form $ax^2 + bx + c$.

Why It's Important
Manufacturing The volume of a rectangular crate can be expressed in factored form. *See Example 4.*

In this lesson, you will learn to factor trinomials in which the coefficient of x^2 is a number other than 1.

Hands-On Algebra

Materials:  straightedge

Use a model to factor $2x^2 + 7x + 6$.

Step 1 Draw a square with four sections. Put the first and last terms into the boxes as shown.

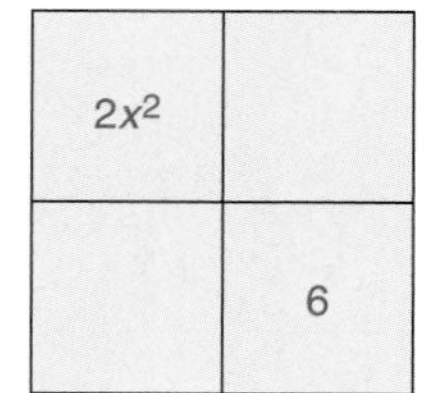

Step 2 Factor $2x^2$ as $2x \cdot x$ and place the factors outside the box.

Think of factors of 6 to place outside the box.

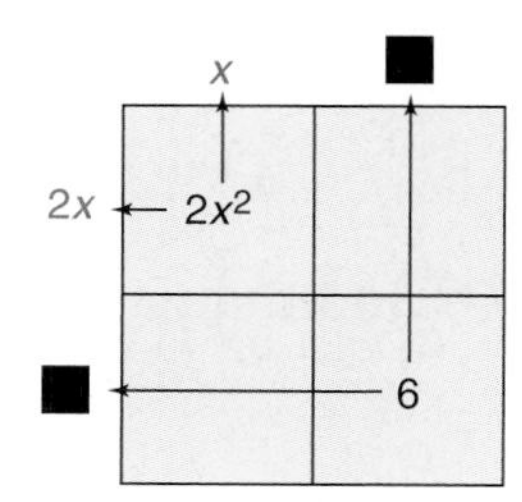

Step 3 The number 6 has two different factor pairs, 2 and 3, and 1 and 6. Try the factor pairs until you find the one that results in a middle term of $7x$. First, try 2 and 3. Note that there are two different ways of placing the 2 and 3 outside of the box.

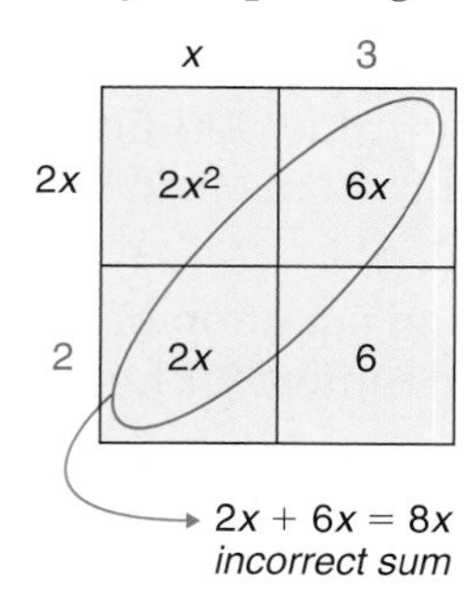

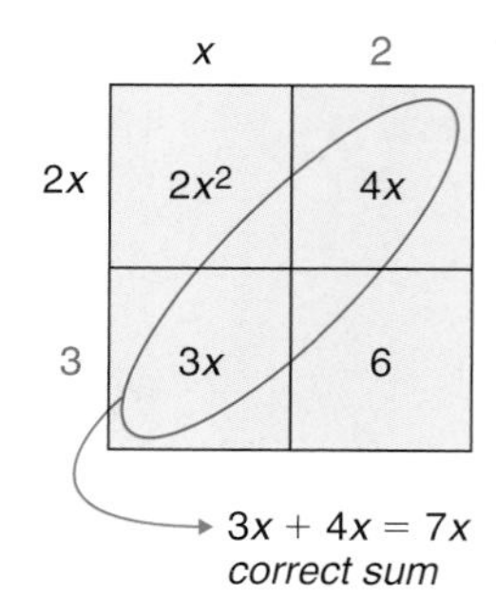

Step 4 The second model results in the correct middle term, $7x$. Therefore, $2x^2 + 7x + 6 = (2x + 3)(x + 2)$.

Try These **Use a model to factor each trinomial.**

1. $2x^2 + 7x + 3$
2. $2x^2 + 5x + 3$
3. $3x^2 + 7x + 2$
4. $3x^2 + 8x + 5$
5. $4x^2 + 8x + 3$
6. $4x^2 + 13x + 3$

The FOIL method will help you factor trinomials without models.

Examples

Factor each trinomial.

1 $2x^2 - 7x + 3$

$2x^2$ is the product of the **F**irst terms, and 3 is the product of the **L**ast terms.

$$2x^2 - 7x + 3 = (2x + \blacksquare)(x + \blacksquare)$$

The last term, 3, is positive. The sum of the inside and outside terms, −7, is negative. So, both factors of 3 must be negative. Try factor pairs of 3 until the sum of the products of the **O**uter and **I**nner terms is $-7x$.

Try −3 and −1. $(2x - 3)(x - 1) = 2x^2 - 2x - 3x + 3$
$= 2x^2 - 5x + 3$ *$-5x$ is not the correct middle term.*
$(2x - 1)(x - 3) = 2x^2 - 6x - 1x + 3$
$= 2x^2 - 7x + 3$ ✓

Therefore, $2x^2 - 7x + 3 = (2x - 1)(x - 3)$.

2 $3y^2 + 2y - 5$

$3y^2$ is the product of the **F**irst terms, and −5 is the product of the **L**ast terms.

$$3y^2 + 2y - 5 = (3y + \blacksquare)(y + \blacksquare)$$

Find integers whose product is −5. Try factor pairs of −5 until the sum of the products of the **O**uter and **I**nner terms is $2y$.

Try −5 and 1. $(3y - 5)(y + 1) = 3y^2 + 3y - 5y - 5$
$= 3y^2 - 2y - 5$ *$-2y$ is not the correct middle term.*
$(3y + 1)(y - 5) = 3y^2 - 15y + 1y - 5$
$= 3y^2 - 14y - 5$ *$-14y$ is not the correct middle term.*

Try 5 and −1. $(3y + 5)(y - 1) = 3y^2 - 3y + 5y - 5$
$= 3y^2 + 2y - 5$ ✓

Therefore, $3y^2 + 2y - 5 = (3y + 5)(y - 1)$.

Your Turn

a. $2x^2 + 3x + 1$ **b.** $5y^2 + 2y - 3$ **c.** $3z^2 - 8z + 4$

Sometimes the coefficient of x^2 can be factored into more than one pair of integers.

Example 

Factor $4x^2 + 12x + 5$.

Number	Factor Pairs
4	4 and 1, 2 and 2
5	5 and 1

Reading Algebra

When factoring this kind of trinomial, it is important to keep an organized list of the factors.

Try 4 and 1. $(4x + 5)(1x + 1) = 4x^2 + 4x + 5x + 5$
$= 4x^2 + 9x + 5$ *9x is not the correct middle term.*

$(4x + 1)(1x + 5) = 4x^2 + 20x + 1x + 5$
$= 4x^2 + 21x + 5$ *21x is not the correct middle term.*

Try 2 and 2. $(2x + 5)(2x + 1) = 4x^2 + 2x + 10x + 5$
$= 4x^2 + 12x + 5$ ✓

Therefore, $4x^2 + 12x + 5 = (2x + 5)(2x + 1)$.

Your Turn

d. $6x^2 + 17x + 5$

e. $4x^2 - 8x - 5$

Recall that the first step in factoring any polynomial is to factor out any GCF other than 1.

Real World

Example 4

Manufacturing Link

The volume of a rectangular shipping crate is $6x^3 - 15x^2 - 36x$. Find possible dimensions for the crate.

The formula for the volume of a rectangular prism is $V = \ell wh$. Find three factors of $6x^3 - 15x^2 - 36x$. First, look for a GCF.

$6x^3 - 15x^2 - 36x = 3x(2x^2 - 5x - 12)$ *The GCF is 3x.*

$3x$ is one factor of $6x^3 - 15x^2 - 36x$. Factor $2x^2 - 5x - 12$ to find the other two factors.

$2x^2 - 5x - 12 = (2x + \blacksquare)(x + \blacksquare)$

The factors of -12 are -3 and 4, 3 and -4, -2 and 6, 2 and -6, -1 and 12, and 1 and -12. Check several combinations; the correct factors are 3 and -4.

$2x^2 - 5x - 12 = (2x + 3)(x - 4)$

So, $6x^3 - 15x^2 - 36x = 3x(2x + 3)(x - 4)$. Therefore, the dimensions can be $3x$, $2x + 3$, and $x - 4$.

Check for Understanding

Communicating Mathematics

1. **Write** the trinomial and its binomial factors shown by each model.

a.

	$2x$	3
x	$2x^2$	$3x$
-4	$-8x$	-12

b.

	$3x$	-1
$2x$	$6x^2$	$-2x$
-3	$-9x$	3

c.

	x	5
$5x$	$5x^2$	$25x$
3	$3x$	15

2. **You Decide?** Jamal factored the trinomial $18k^2 - 24k + 8$ as $(3k - 2)(6k - 4)$. Jacqui disagrees with Jamal's answer. She says that he did not factor the trinomial completely. Who is correct? Explain your reasoning.

Guided Practice

Factor each trinomial. If the trinomial cannot be factored, write *prime*. *(Examples 1–4)*

3. $2a^2 + 5a + 3$
4. $3y^2 + 7y + 2$
5. $5x^2 + 13x + 6$
6. $2x^2 + x - 3$
7. $2x^2 + x - 21$
8. $2n^2 - 11n + 7$
9. $10a^2 - 9a + 2$
10. $6y^2 - 11y + 4$
11. $6x^2 + 16x + 10$

12. **Geometry** The measure of the volume of a rectangular prism is $2x^3 + x^2 - 15x$. Find possible dimensions for the prism. *(Example 4)*

Exercises

Practice

Factor each trinomial. If the trinomial cannot be factored, write *prime*.

Homework Help	
For Exercises	See Examples
22, 26–35, 37–38, 40–41	1, 2
13–21, 23–25, 36, 39, 44	3, 4
42	4
Extra Practice	
See page 712.	

13. $2y^2 + 7y + 3$
14. $2x^2 + 11x + 5$
15. $4a^2 + 8a + 3$
16. $2x^2 - 9x - 5$
17. $2q^2 - 9q - 18$
18. $5x^2 - 13x - 6$
19. $7a^2 + 22a + 3$
20. $3y^2 + 7y + 15$
21. $3x^2 + 14x + 8$
22. $2z^2 - 11z + 15$
23. $3x^2 + 14x + 15$
24. $3m^2 + 10m + 8$
25. $3x^2 + 5x + 1$
26. $4x^2 - 8x + 3$
27. $14x^2 + 33x - 5$
28. $6y^2 - 11y + 4$
29. $8m^2 - 10m + 3$
30. $6r^2 + 9r - 42$
31. $6x^2 + 3x - 30$
32. $4x^2 + 10x - 6$
33. $2x^3 + 5x^2 - 12x$
34. $7x - 5 + 6x^2$
35. $11y + 6y^2 - 2$
36. $15x^3 - 11x^2 - 12x$
37. $2a^2 + 5ab - 3b^2$
38. $15x^2 - 13xy + 2y^2$
39. $9k^2 + 30km + 25m^2$

40. Factor $2x^2 + 5x - 25$.
41. What are the factors of the trinomial $6x^3 + 15x^2 - 9x$?

Real World

Applications and Problem Solving

42. **Measurement** The volume of a rectangular prism is 60 cubic feet. If the measure of the length, width, and height are consecutive integers, find the dimensions.

43. **Manufacturing** The dimensions of a rectangular piece of metal are shown at the right.

2y in.

$y - 7$ in.

a. If a 1-inch by 1-inch square is removed from each corner, write an expression that represents the area of the remaining piece of metal. Express the area in factored form.

b. If the metal is folded along the dashed lines, an open box is formed. Write an expression that represents the volume of the box.

c. If $y = 10$ inches, find the area of the metal and the volume of the box.

44. **Critical Thinking** Find all values of k so that the trinomial $4y^2 + ky + 5$ can be factored.

Mixed Review

Factor each polynomial. *(Lessons 10–2 & 10–3)*

45. $x^2 + 14x - 32$

46. $y^2 - 7x + 12$

47. $3a^3 - 15a^2 + 6a$

48. $2n^2 + 2n - 24$

49. **Geometry** Find the length of the diagonal of a rectangle whose length is 24 feet and whose height is 7 feet. *(Lesson 8–7)*

Solve. Assume that y varies directly as x. *(Lesson 6–5)*

50. If $y = 28$ when $x = 7$, find x when $y = 52$.

51. Find x when $y = 45$, if $y = 27$ when $x = 6$.

Standardized Test Practice

52. **Multiple Choice** What is the solution of $10 - 3(x + 4) = 16$? *(Lesson 4–7)*

A -6 B $-\frac{12}{7}$ C 2 D 6

Quiz 2 Lessons 10–3 and 10–4

Factor each trinomial.

1. $x^2 + 3x - 10$

2. $x^2 - 5x - 24$ *(Lesson 10–3)*

3. $2x^2 + 9x + 7$

4. $8x^2 - 16x - 10$ *(Lesson 10–4)*

5. **Geometry** Find the area of the shaded region. Express the area in factored form. *(Lesson 10–3)*

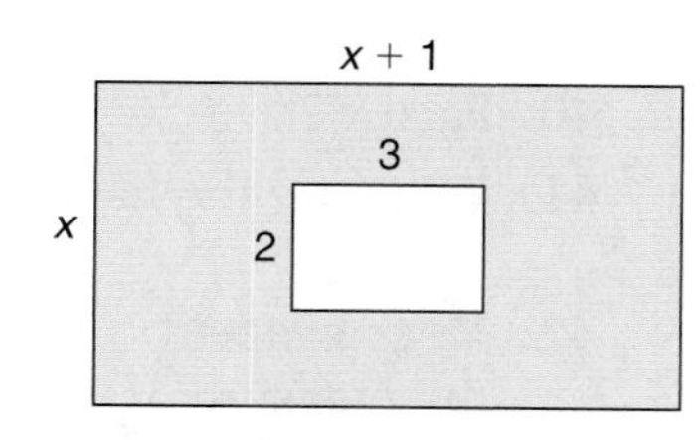

10–5 Special Factors

What You'll Learn

You'll learn to recognize and factor the differences of squares and perfect square trinomials.

Why It's Important

Manufacturing You can find the area of a washer by using the difference of squares. *See Exercise 52.*

In this lesson, you will learn to recognize and factor polynomials that are **perfect square trinomials**. They have two equal binomial factors.

The model below shows the product $(x + 3)^2$.

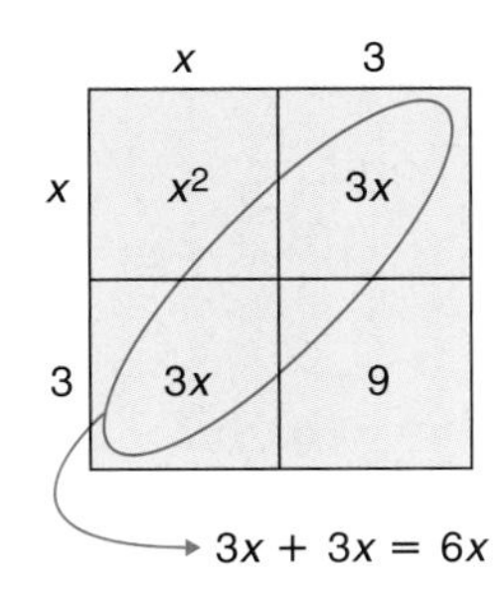

You can also use the FOIL method to find the product.

$$(x + 3)(x + 3) = x^2 + 3x + 3x + 9$$
$$= x^2 + 6x + 9$$

perfect squares

twice the product of x and 3

Square of a Sum: Lesson 9–5

The square of $(x + 3)$ is the sum of

- the square of the first term of the binomial,
- the square of the last term of the binomial, and
- twice the product of the terms of the binomial.

These observations will help you recognize when a trinomial is a perfect square trinomial. They can be factored as shown.

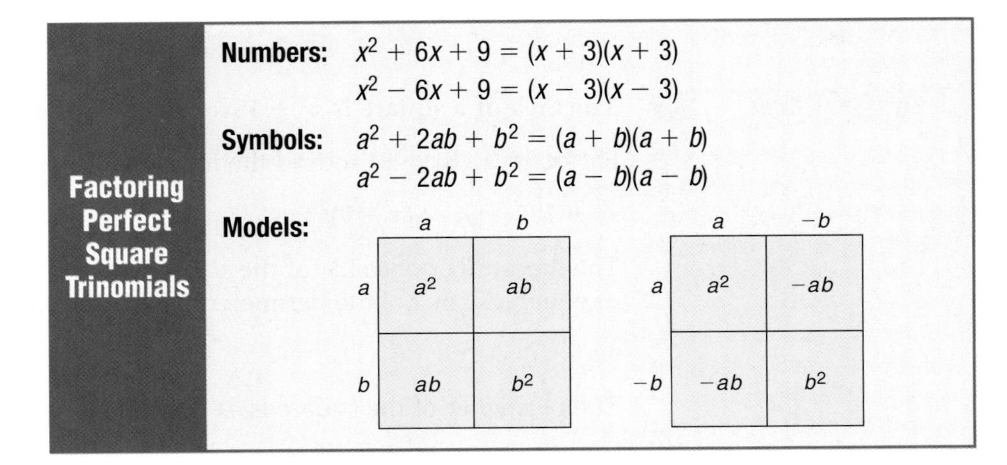

Factoring Perfect Square Trinomials

Numbers: $x^2 + 6x + 9 = (x + 3)(x + 3)$
$x^2 - 6x + 9 = (x - 3)(x - 3)$

Symbols: $a^2 + 2ab + b^2 = (a + b)(a + b)$
$a^2 - 2ab + b^2 = (a - b)(a - b)$

Models:

Examples

Determine whether each trinomial is a perfect square trinomial. If so, factor it.

1 $x^2 + 10x + 25$

To determine whether $x^2 + 10x + 25$ is a perfect square trinomial, answer each question.

- Is the first term a perfect square? Yes, x^2 is the square of x.
- Is the last term a perfect square? Yes, 25 is the square of 5.
- Is the middle term twice the product of x and 5? Yes, $10x = 2(5x)$.

Therefore, $x^2 + 10x + 25$ is a perfect square trinomial.
$x^2 + 10x + 25 = (x + 5)^2$

2 $4n^2 - 4n + 1$

- Is the first term a perfect square? Yes, $4n^2$ is the square of $2n$.
- Is the last term a perfect square? Yes, 1 is the square of 1 and -1.
- Is the middle term twice the product of $2n$ and -1? Yes, $2(-2n) = -4n$.

Therefore, $4n^2 - 4n + 1$ is a perfect square trinomial.
$4n^2 - 4n + 1 = (2n - 1)^2$

3 $4p^2 - 12p + 36$

- Is the first term a perfect square? Yes, $4p^2$ is the square of $2p$.
- Is the last term a perfect square? Yes, 36 is the square of 6 and -6.
- Is the middle term twice the product of $2p$ and -6? No, $2(-12p) \neq -12p$.

Therefore, $4p^2 - 12p + 36$ is *not* a perfect square trinomial.

Your Turn

a. $a^2 + 2a + 1$ b. $16x^2 + 20x + 25$ c. $49x^2 - 14x + 1$

Geometry Link

4 **The area of a square is $x^2 + 18x + 81$. Find the perimeter.**

Factor $x^2 + 18x + 81$ to find the measure of one side of the square.

$x^2 + 18x + 81 = (x + 9)^2$

The measure of one side of the square is $x + 9$. A square has four sides of equal length. So, the perimeter is four times the length of a side.

$4(x + 9) = 4x + 36$ *Distributive Property*

The perimeter of the square is $4x + 36$.

A polynomial like $x^2 - 9$ is called the **difference of squares**. Although this is not a trinomial, it *can* be factored into two binomials. The model shows how to factor $x^2 - 9$.

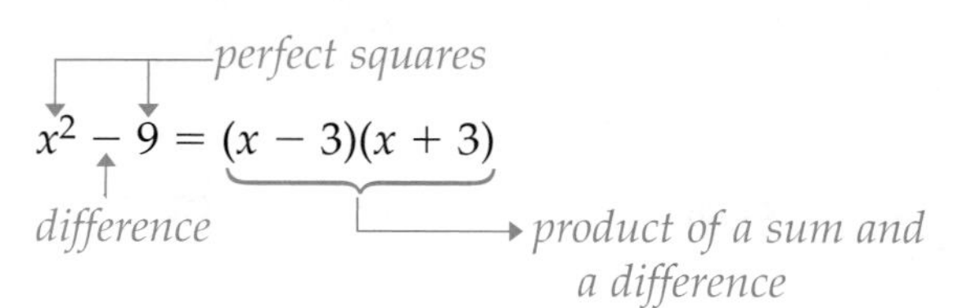

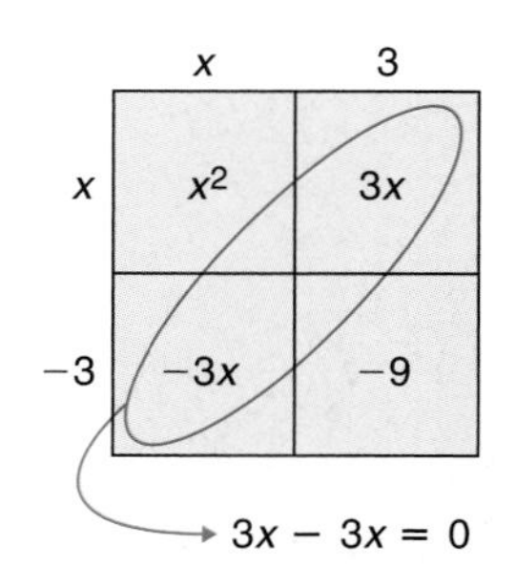

A difference of squares can be factored as shown.

Factoring a Difference of Squares	
Numbers:	$x^2 - 9 = (x - 3)(x + 3)$
Symbols:	$a^2 - b^2 = (a - b)(a + b)$
Model:	

	a	b
a	a^2	ab
$-b$	$-ab$	$-b^2$

Examples

Determine whether each binomial is the difference of squares. If so, factor it.

5 $a^2 - 25$

a^2 and 25 are both perfect squares, and $a^2 - 25$ is a difference.

$a^2 - 25 = (a)^2 - (5)^2$ *$a \cdot a = a^2$, $5 \cdot 5 = 25$*

$= (a - 5)(a + 5)$ *Difference of Squares*

6 $y^2 + 100$

y^2 and 100 are both perfect squares. But $y^2 + 100$ is a sum, not a difference. Therefore $y^2 + 100$ is *not* a difference of squares. It is a prime polynomial.

7 $3n^2 - 48$

First, look for a GCF. Then, determine whether the remaining factor is a difference of squares.

$3n^2 - 48 = 3(n^2 - 16)$ *The GCF of $3n^2$ and 48 is 3.*

$= 3[(n)^2 - (4)^2]$ *$n \cdot n = n^2$, $4 \cdot 4 = 16$*

$= 3(n - 4)(n + 4)$ *Difference of Squares*

Your Turn

d. $121 - p^2$ **e.** $25x^3 - 100x$ **f.** $4a^2 + 49$

The following chart summarizes factoring methods.

Concept Summary	Factoring Method	Number of Terms		
		Two	Three	Four or more
	greatest common factor	√	√	√
	difference of squares	√		
	perfect square trinomials		√	
	trinomial with two binomial factors		√	

Check for Understanding

Communicating Mathematics

Vocabulary
perfect square trinomials
difference of squares

1. **State** whether $4c^2 - 7$ can be factored as a difference of squares. Explain.
2. **Copy** the chart shown above into your math journal. Then write a polynomial that can be factored by each method.

Guided Practice

Determine whether each trinomial is a perfect square trinomial. If so, factor it. *(Examples 1–3)*

3. $y^2 + 14y + 49$
4. $x^2 - 10x + 100$
5. $a^2 - 10a + 25$

Determine whether each binomial is the difference of squares. If so, factor it. *(Examples 5–7)*

6. $16x^2 - 25$
7. $8x^2 - 50y^2$
8. $49m^2 + 16$

Factor each polynomial. If the polynomial cannot be factored, write *prime*.

9. $3x^2 + 15$
10. $y^2 + 6y - 9$
11. $3y^2 + 21y - 24$

12. **Geometry** The area of a square is $4x^2 + 20xy + 25y^2$. Find the perimeter. *(Example 4)*

Exercises

Practice

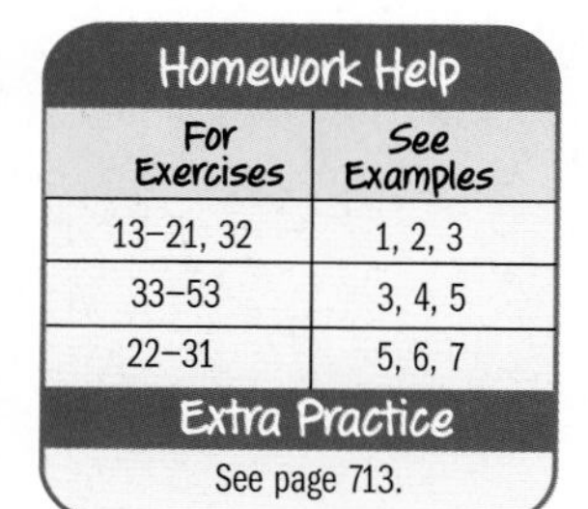
Homework Help

For Exercises	See Examples
13–21, 32	1, 2, 3
33–53	3, 4, 5
22–31	5, 6, 7

Extra Practice
See page 713.

Determine whether each trinomial is a perfect square trinomial. If so, factor it.

13. $r^2 + 8r + 16$
14. $x^2 - 16x + 64$
15. $a^2 + 2a + 1$
16. $4a^2 + 4a + 1$
17. $4z^2 - 20z + 25$
18. $9m^2 + 15m + 25$
19. $9a^2 + 24a + 16$
20. $d^2 - 22d + 121$
21. $49 + 14z + z^2$

Determine whether each binomial is the difference of squares. If so, factor it.

22. $x^2 - 16$
23. $a^2 - 36$
24. $y^2 - 20$
25. $1 - 9m^2$
26. $16m^2 - 25n^2$
27. $y^2 + z^2$
28. $8a^2 - 18$
29. $2z^2 - 98$
30. $49 - a^2b^2$

31. Write a polynomial that is the difference of two squares. Then factor it.

32. Is $x^2 + x - 1$ a perfect square trinomial? Explain.

Factor each polynomial. If the polynomial cannot be factored, write *prime*.

33. $5x^2 + 25$	34. $a^2 - 16b^2$	35. $y^2 - 5y + 6$
36. $m^2 + 8m + 16$	37. $2x^2 - 72$	38. $3a^2b + 6ab + 9ab^2$
39. $8xy^2 - 13x^2y$	40. $2r^2 + 3r + 1$	41. $x^2 - 6x - 9$
42. $z^3 + 6z^2 + 9z$	43. $20n^2 + 34n + 6$	44. $b^2 + 6 - 7b$
45. $8w^2 + 14w - 15$	46. $a^3 - 17a^2 + 72a$	47. $5x^2 + 15x + 10$
48. $7a^2 - 21a$	49. $2x^3 - 32x$	50. $2x^2 - 11x - 21$

Applications and Problem Solving

51. **Number Theory** The difference of two numbers is 2. The difference of their squares is 12. Find the numbers.

52. **Manufacturing** A metal washer is manufactured by stamping out a circular hole from a metal disk. In the figure, r represents the radius of the metal disk. The radius of the hole is 1 centimeter.

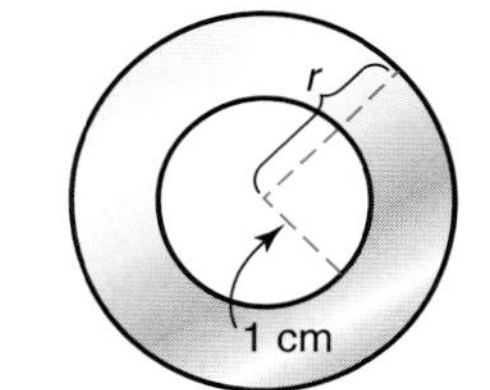

a. Write an expression in factored form for the area of the washer. (*Hint*: Use $A = \pi r^2$.)

b. If $r = 2$ centimeters, find the area of the washer to the nearest hundredth.

53. **Critical Thinking** The area of a square is $81 - 90x + 25x^2$. If x is a positive integer, what is the least possible measure for the square's perimeter?

Mixed Review

Factor each polynomial. *(Lessons 10–3 & 10–4)*

54. $4y^2 + 16y + 15$	55. $4x^2 + 11x - 3$
56. $a^3 - 7a^2 + 12a$	57. $m^2 - 5m - 14$

Simplify each expression. *(Lesson 8–3)*

58. n^{-2} 59. $a^5(a^{-3})$ 60. $\frac{1}{r^{-3}}$ 61. $\frac{3c^2d^3f^4}{9c^4d^2f^4}$

Standardized Test Practice

A B C D

62. **Extended Response** Describe the difference between the graphs of $y = 4x$ and $y = 4x - 5$. *(Lesson 7–6)*

63. **Multiple Choice** Which graph is the best example of data that exhibit a linear relationship between the variables x and y? *(Lesson 6–3)*

A 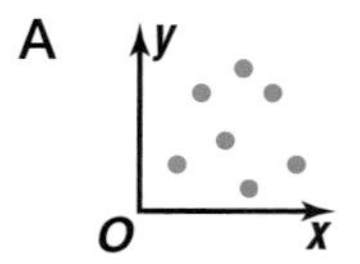

B 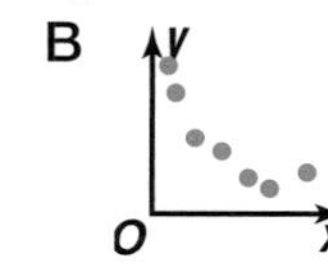

C 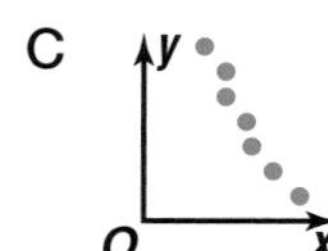

D 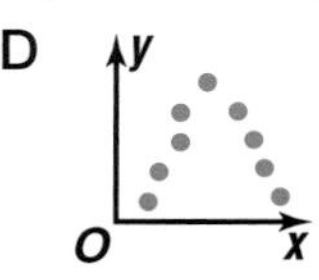

CHAPTER 10

Study Guide and Assessment

Understanding and Using the Vocabulary

After completing this chapter, you should be able to define each term, property, or phrase and give an example or two of each.

difference of squares *(p. 447)*
factoring *(p. 428)*
greatest common factor (GCF) *(p. 422)*
perfect square trinomials *(p. 445)*
prime polynomial *(p. 430)*

*inter*NET CONNECTION **Review Activities**
For more review activities, visit: www.algconcepts.com

State whether each sentence is *true* or *false*. If false, replace the underlined word or number to make a true sentence.

1. The prime factorization of 12 is $\underline{3 \cdot 4}$.
2. $\underline{3x}$ is the greatest common factor of $6x^2$ and $9x$.
3. When two or more numbers are multiplied, each number is a <u>factor</u> of the product.
4. $2y$ and $(y + 3)$ are factors of $\underline{2y^2 + 3}$.
5. When you factor trinomials, always check for a <u>GCF</u> first.
6. The number <u>51</u> is an example of a prime number.
7. $(x - 3)(x + 3)$ is the factored form of $\underline{x^2 + 9}$.
8. Whole numbers that have more than two factors are called <u>composite numbers</u>.
9. A polynomial is in <u>factored form</u> when it is expressed as the product of polynomials.
10. $4a^2 - b^2$ is an example of a <u>perfect square trinomial</u>.

Skills and Concepts

Objectives and Examples

- **Lesson 10–1** Find the greatest common factor of a set of numbers or monomials.

Find the GCF of $12x^2y$ and $30xy^2$.

$$12x^2y = 2 \cdot (2) \cdot (3) \cdot (x) \cdot x \cdot (y)$$
$$30xy^2 = (2) \cdot (3) \cdot 5 \cdot (x) \cdot (y) \cdot y$$

The GCF of $12x^2y$ and $30xy^2$ is $2 \cdot 3 \cdot x \cdot y$ or $6xy$.

Review Exercises

Find the GCF of each set of numbers or monomials.

11. 20, 25
12. 12, 18, 42
13. 20, 25, 28
14. $5xy$, $10x$
15. $9x^2$, $9x$
16. $6a^2b$, $18a^2b^2$, $9ab^2$

Objectives and Examples

• **Lesson 10–2** Use the GCF and the Distributive Property to factor polynomials.

Factor $12x^2 - 8xy$.

The GCF of $12x^2$ and $8xy$ is $4x$. Write each term as a product of the GCF and its remaining factors.

$$12x^2 - 8xy = 4x(3x) - 4x(2y)$$
$$= 4x(3x - 2y) \quad \text{Distributive Property}$$

Review Exercises

Factor each polynomial. If the polynomial cannot be factored, write *prime*.

17. $5x + 30y$
18. $16a^2 + 32b^2$
19. $12ab - 18a^2$
20. $5mn^2 + 10mn$
21. $3xy + 12x^2y^2$

Find each quotient.

22. $(20x^3 + 15x^2) \div 5x$
23. $(40a^2b^2 - 8ab) \div 8ab$

• **Lesson 10–3** Factor trinomials of the form $x^2 + bx + c$.

Factor $x^2 + 3x - 10$.

Find integers whose product is -10 and whose sum is 3.

Product	Integers	Sum
−10	2, −5	2 + (−5) = −3
−10	−2, 5	−2 + 5 = 3 ✓

Therefore, $x^2 + 3x - 10 = (x - 2)(x + 5)$.

Factor each trinomial. If the trinomial cannot be factored, write *prime*.

24. $y^2 + 9y + 14$
25. $x^2 - 8x + 15$
26. $a^2 + 5a - 7$
27. $x^2 - 2x - 8$
28. $y^2 + 7y + 12$
29. $x^2 + 2x - 35$
30. $a^2 - a - 1$
31. $2n^2 - 8n - 24$

• **Lesson 10–4** Factor trinomials of the form $ax^2 + bx + c$.

Factor $2x^2 + 5x + 3$.

$$2x^2 + 5x + 3 = (2x + \blacksquare)(x + \blacksquare)$$

$$(2x + 3)(x + 1) = 2x^2 + 2x + 3x + 3$$
$$= 2x^2 + 5x + 3$$

Therefore, $2x^2 + 5x + 3 = (2x + 3)(x + 1)$.

Factor each trinomial. If the trinomial cannot be factored, write *prime*.

32. $2z^2 + 7z + 5$
33. $3x^2 + 8x + 5$
34. $3a^2 + 8a + 4$
35. $6a^2 - a - 2$
36. $3x^2 - 7x - 6$
37. $2y^2 - 9y - 18$
38. $2x^2 + 5x + 6$
39. $15a^2 - 20a + 5$

● Extra Practice
See pages 711–713.

Objectives and Examples

- **Lesson 10–5** Recognize and factor the differences of squares and perfect square trinomials.

Factor $a^2 + 6a + 9$.

$a^2 + 6a + 9 = a^2 + 2(3a) + 3^2$
$= (a + 3)^2$

Factor $x^2 - 25$.

$x^2 - 25 = x^2 - 5^2$
$= (x + 5)(x - 5)$

Review Exercises

Factor each polynomial. If the polynomial cannot be factored, write *prime*.

40. $y^2 + 8y + 16$

41. $a^2 - 12a + 36$

42. $n^2 + 2n - 1$

43. $25x^2 + 20x + 4$

44. $y^2 - 81$

45. $4x^2 - 9$

46. $a^2 + 49$

47. $12c^2 - 12$

Applications and Problem Solving

48. Physics If a flare is launched into the air, its height h feet above the ground after t seconds is given by the formula $h = vt - 16t^2$. In the formula, v represents the initial velocity in feet per second. *(Lesson 10–2)*

a. Factor the expression $vt - 16t^2$.

b. If the flare is launched with an initial velocity of 144 feet per second, find the height after 2 seconds.

49. Genetics The Punnett square below represents the possible gene combinations for hair length in dogs. H represents long hair, and h represents short hair. Find the missing genes for the parents. *(Lesson 10–3)*

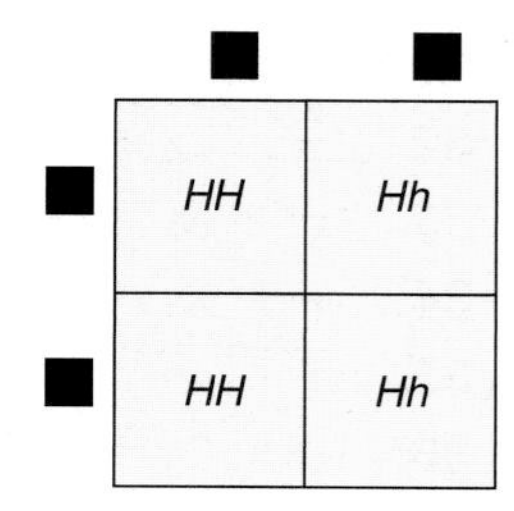

50. Geometry The area of a square is $(25x^2 + 30x + 9)$ square units. Find the perimeter. *(Lesson 10–5)*

CHAPTER 10 Test

1. **Explain** what it means to *factor* a polynomial.
2. **Write** two monomials whose GCF is 1.
3. **Write** the trinomial and its binomial factors shown by the model.
4. **List** two different methods of factoring polynomials.
5. **Classify** the number 15 as *prime* or *composite*. Explain your reasoning.

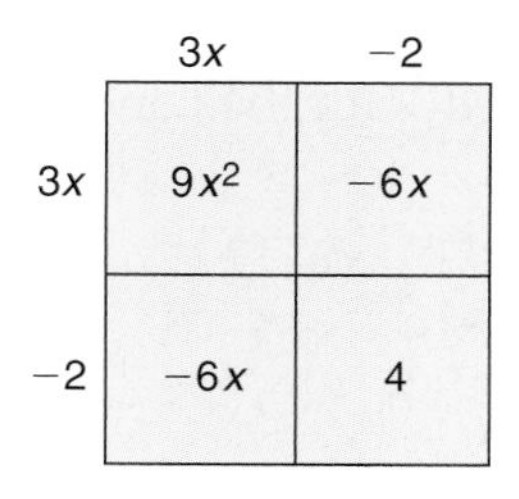

Exercise 3

Factor each monomial.

6. $25x^2y^2$
7. $-15b^3$
8. $24a^2b$

Find the GCF of each set of numbers or monomials.

9. 24, 60
10. $16a^2, 30a^3$
11. $20a^2b, 25a^2b^2$

Factor each polynomial. If the polynomial cannot be factored, write *prime*.

12. $12x^2 + 18x$
13. $3x^2y - 12xy^2$
14. $6a^3 + 8a^2 + 2a$
15. $x^2 + 9x + 8$
16. $m^2 - 10m + 24$
17. $y^2 - 3y - 18$
18. $3x^2 + x - 14$
19. $3m^2 + 17m + 10$
20. $2x^2 - 18$
21. $n^2 - 8n - 16$
22. $y^2 + 10y + 25$
23. $25m^2 - 16$
24. $3r^2 + r + 1$
25. $6x^3 + 15x^2 - 9x$

26. **Geometry** Find the area of the shaded region. Express the area in factored form.

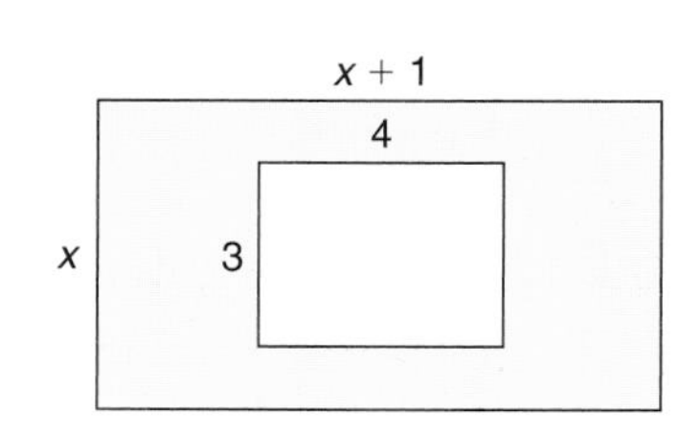

CHAPTER 10

Preparing for Standardized Tests

Function and Graph Problems

All standardized tests include problems with functions and graphs.

Familiarize yourself with the concepts below.

function	equation of a line
table of values	y-intercept, x-intercept
graph of a line	slope

Slope is the ratio of the change in y to the change in x. A line that slopes upward from left to right has a positive slope.

State Test Example

Martin is paid by the hour for babysitting. His hourly wage is a fixed amount plus an additional amount for each child. The graph shows his hourly wage for up to 5 children. If x represents the number of children, which expression can be used to find Martin's hourly wage, y?

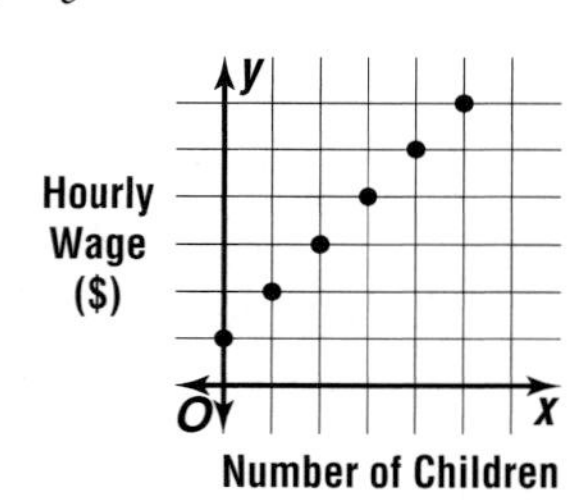

A x **B** $x - 1$ **C** $2x + 1$ **D** $x + 1$

Hint Study the graph. The points lie on a line. Find the y-intercept and the slope.

Solution The fixed amount is represented by the y-intercept. It is the point that represents 0 children. The y-intercept is 1.

The slope of the line shows the amount for each child. Moving left to right, each point is one unit higher than the previous point. So the slope is 1.

Martin's hourly wage is the fixed amount, \$1, plus the amount per child, \$1, times the number of children, x. The expression is $1x + 1$ or $x + 1$. The answer is D.

SAT Example

What is the equation of a line that is parallel to the line whose equation is $y = \frac{2}{3}x + 5$ and passes through the point at $(-6, 2)$?

A $y = \frac{2}{3}x + 5$ **B** $y = \frac{2}{3}x - 2$ **C** $y = \frac{2}{3}x + 6$

D $y = \frac{2}{3}x - \frac{22}{3}$ **E** $y = -\frac{3}{2}x - 7$

Hint Memorize the slope-intercept and point-slope forms of linear equations.

Solution The equation of the given line is in slope-intercept form. The slope is $\frac{2}{3}$. Parallel lines have the same slope. So, the slope of the parallel line must also be $\frac{2}{3}$. This eliminates answer choice E.

The parallel line must pass through $(-6, 2)$. Write the equation of the line in point-slope form.

$$y - 2 = \frac{2}{3}[x - (-6)]$$

$$y - 2 = \frac{2}{3}(x + 6)$$

$$y - 2 = \frac{2}{3}x + \frac{2}{3}(6) \quad \textit{Distributive Property}$$

$$y - 2 = \frac{2}{3}x + 4$$

$$y = \frac{2}{3}x + 6$$

The answer is C.

After you work each problem, record your answer on the answer sheet provided or on a sheet of paper.

Multiple Choice

1. Use the function table to find the value of y when $x = 5$.

x	y
0	3
1	17
2	31
3	45

A 59 **B** 60

C 73 **D** 75

2. What is the y-intercept of the line determined by the equation $5x + 2 = 7y - 3$?

A -1 **B** $-\frac{1}{7}$ **C** $\frac{1}{7}$ **D** $\frac{5}{7}$ **E** 5

3. Which expression can be used to find the value of y in the graph?

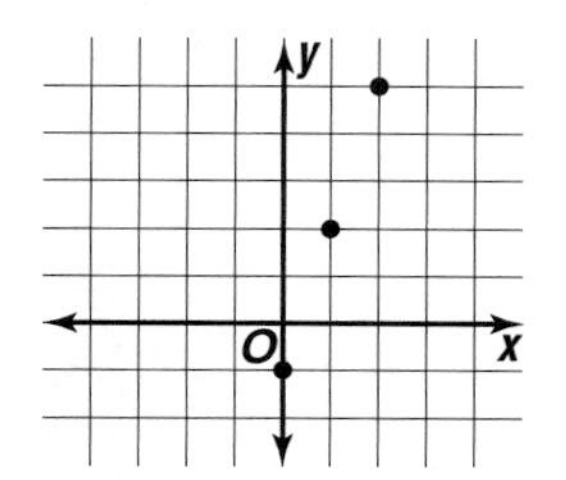

A $1 - 2x$ **B** $1 - 3x$

C $2x + 1$ **D** $3x - 1$

4. The charge to enter a nature reserve is a fixed amount per vehicle plus a fee for each person in it. The table shows some charges. What would the charge be for a vehicle with 8 people?

People	Charge
1	\$1.50
2	\$2.00
3	\$2.50
4	\$3.00

A \$3.50 **B** \$4.00 **C** \$5.00 **D** \$6.00

5. At what point does the line MN cross the y-axis?

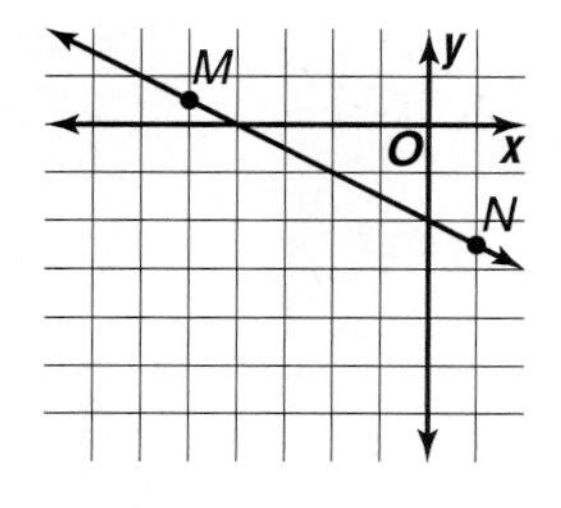

A $(-4, 0)$ **B** $(0, -4)$

C $(-2, 0)$ **D** $(0, -2)$

6. The average of two numbers x and y is A. Which of the following is an expression for y?

A $\frac{A + x}{2}$ **B** $\frac{A}{2} - x$ **C** $2A - x$

D $A - x$ **E** $x - A$

7. Write 4^{-4} without using an exponent.

A 0.00039 **B** 0.0039

C 0.016 **D** 256

8. Which expression should come next in the pattern $2x, 4x^2, 8x^3, 16x^4, \ldots$?

A $24x^5$ **B** $32x^5$

C $24x^6$ **D** $32x^6$

Grid In

9. The graph of $y = 4x - 2$ is shown. What is the x-intercept?

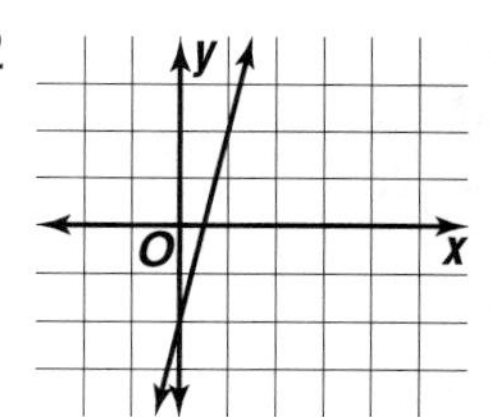

Extended Response

10. The graph shows the distance traveled by an African elephant.

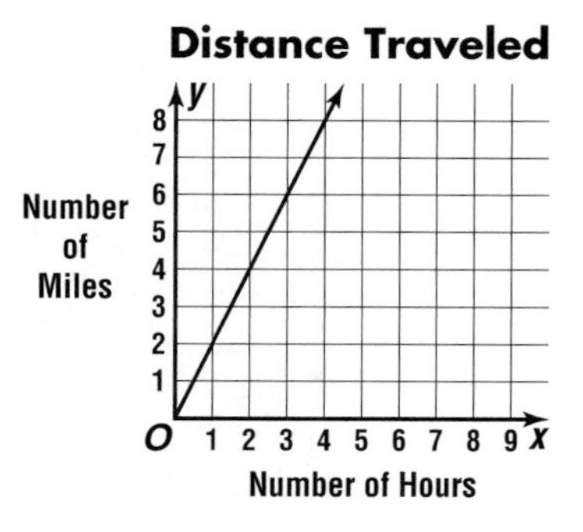

Part A What is the slope of the line?

Part B Explain what the slope represents.

CHAPTER 11 Quadratic and Exponential Functions

FOLDABLES™ Study Organizer

Make this Foldable to help you organize information about quadratic and exponential functions. Begin with four sheets of plain paper.

1. **Fold** each sheet in half along the width.

2. **Unfold** each sheet and tape to form one long piece.

3. **Label** each page with the lesson number as shown.

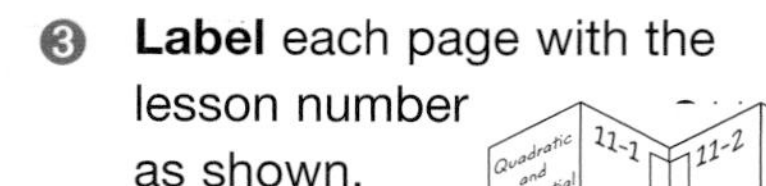

4. **Refold** to form a booklet. Label the front cover "Quadratic and Exponential Functions."

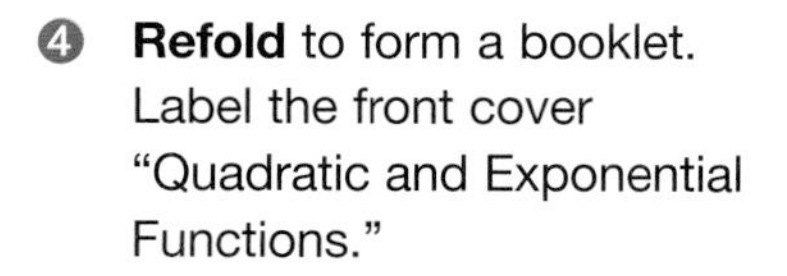

Reading and Writing As you read and study the chapter, write notes and examples for each lesson on each page of the journal.

Problem-Solving Workshop

Project

Do you want to be a millionaire? In this project, you will make a plan for saving money. Your goal is to reach $1,000,000 at the end of a 40-year period, following the guidelines listed below.

1. Your initial deposit is any amount of your choice.
2. The money is invested for exactly 40 years.
3. The annual growth rate of your investment is 5%.
4. Interest is calculated at the end of each year.
5. Choose an additional amount of money to deposit after interest is calculated each year. This amount is fixed, or the same, for all 40 years.
6. After 40 years, you should have as close to $1,000,000 as possible.

Working on the Project

Work with a partner and choose a strategy to help analyze and solve the problem. Here are some questions to help you get started.

- At the beginning of year 1, you decide to invest $5000. How much money will you have at the end of year 1, including interest?
- At the end of year 1, you invest an additional $2000. How much money will you have at the end of year 2, including interest?

Strategies

Look for a pattern.

Draw a diagram.

Make a table.

Work backward.

Use an equation.

Make a graph.

Guess and check.

Technology Tools

- Use a **spreadsheet** to calculate each year-end balance.
- Use **graphing software** to make a graph of your investment over 40 years.

*inter*NET CONNECTION **Research** For more information about investing, visit: www.algconcepts.com

Presenting the Project

Write a one-page paper describing your investment plan. Include:

- successful and unsuccessful strategies used to plan,
- a table or spreadsheet showing each year-end balance, and
- a graph of your savings over the 40-year period.

11–1 Graphing Quadratic Functions

What You'll Learn

You'll learn to graph quadratic functions.

Why It's Important

Architecture Architects graph quadratic functions when designing arches. *See Exercise 15.*

The shape of a parabola is like an arch. **Parabolas** are modeled by **quadratic functions** of the form $y = ax^2 + bx + c$. The first term cannot equal zero because it will make the function linear, or a straight line. Therefore, the coefficient a cannot be zero.

Quadratic Function

Words: A quadratic function is a function that can be described by an equation of the form $y = ax^2 + bx + c$, where $a \neq 0$.

Models:

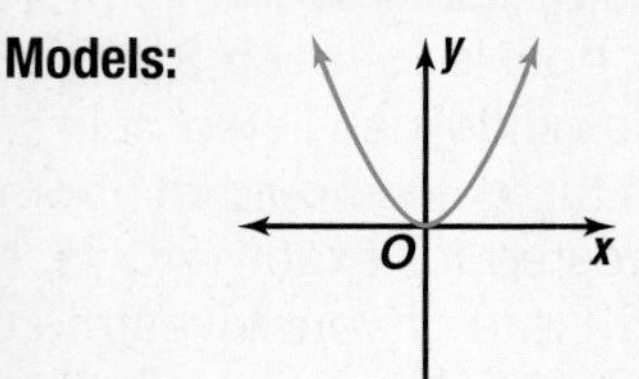

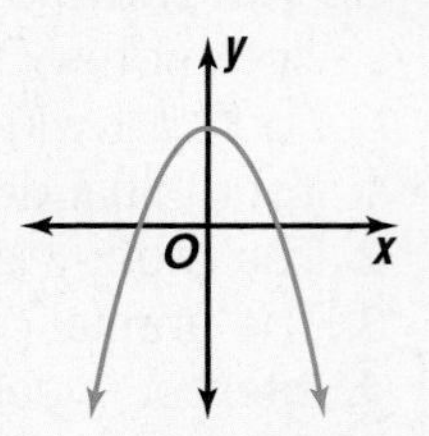

Graphs of all quadratic functions have the shape of a parabola.

Examples

Graph each quadratic function by making a table of values.

1 $y = x^2 + 1$

First, choose integer values for x. Evaluate the function for each x-value. Graph the points and connect them with a smooth curve.

Look Back

Graphing Relations: Lesson 6–3

x	$x^2 + 1$	y	(x, y)
−2	$(-2)^2 + 1$	5	(−2, 5)
−1	$(-1)^2 + 1$	2	(−1, 2)
0	$0^2 + 1$	1	(0, 1)
1	$1^2 + 1$	2	(1, 2)
2	$2^2 + 1$	5	(2, 5)

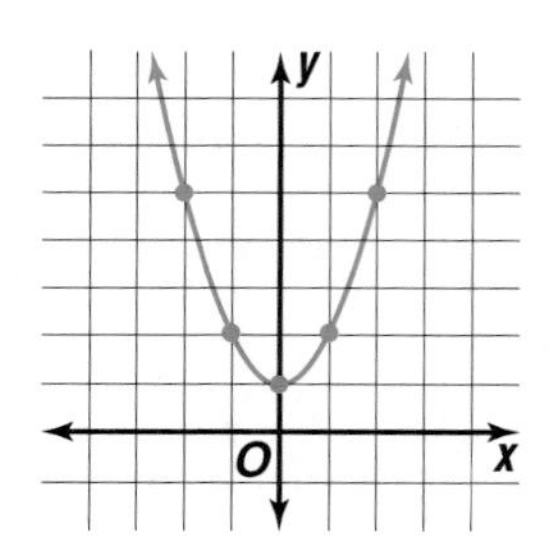

2 $y = -x^2$

x	$-x^2$	y	(x, y)
−2	$-(-2)^2$	−4	(−2, −4)
−1	$-(-1)^2$	−1	(−1, −1)
0	$-(0)^2$	0	(0, 0)
1	$-(1)^2$	−1	(1, −1)
2	$-(2)^2$	−4	(2, −4)

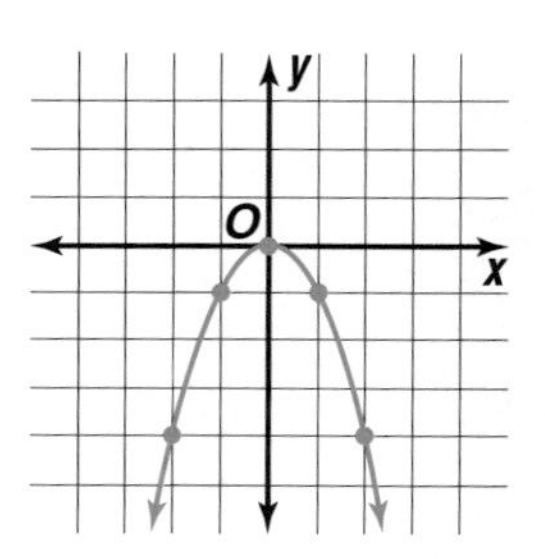

Your Turn

a. $y = x^2 + 6$ b. $y = -2x^2 - 1$

In Example 1, the *lowest point*, or **minimum**, of the graph of $y = x^2 + 1$ is at (0, 1). Since the coefficient of x^2 is positive, the graph opens *upward*. In Example 2, the *highest point*, or **maximum**, of the graph of $y = -x^2$ is at (0, 0). Since the coefficient of x^2 is negative, the graph opens *downward*.

As with linear functions, in a quadratic function x is the independent variable and y is the dependent variable. Since the graph of a quadratic function extends forever to the left and to the right, the domain (x values) of a quadratic function is the set of all real numbers. For a quadratic function with a graph that opens upward, the range (y values) is all real numbers greater than or equal to the minimum value. For a quadratic function with a graph that opens downward, the range is all real numbers less than or equal to the maximum value.

The maximum or minimum point of a parabola is called the **vertex**. The vertical line containing the vertex of a parabola is called the **axis of symmetry**. If you fold the graph of $y = x^2 + 1$ or $y = -x^2$ along the axis of symmetry, the two halves of each graph will coincide. In both examples, the axis of symmetry is the line $x = 0$.

You can use the rule below to find the equation of the axis of symmetry.

Equation of the Axis of Symmetry

Words: The equation of the axis of symmetry for the graph of $y = ax^2 + bx + c$, where $a \neq 0$, is $x = -\frac{b}{2a}$.

Model:

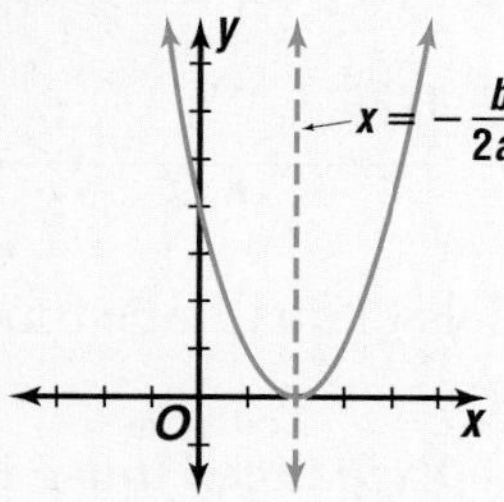

Example 3 Use characteristics of quadratic functions to graph $y = x^2 - 4x - 1$.

A. Find the equation of the axis of symmetry.

B. Find the coordinates of the vertex of the parabola.

C. Graph the function.

A. First identify a, b, and c.

$y = ax^2 + bx + c$

$y = 1x^2 - 4x - 1$ $x^2 = 1x^2$

So, $a = 1$, $b = -4$, and $c = -1$.

Now, find the equation of the axis of symmetry.

$x = -\frac{b}{2a}$ *Equation of axis of symmetry*

$x = -\frac{-4}{2(1)}$ $a = 1, b = -4$

$x = 2$ *Simplify.*

(continued on the next page)

B. Next, find the vertex. Since the equation of the axis of symmetry is $x = 2$, the x-coordinate of the vertex must be 2. Substitute 2 for x in the equation $y = x^2 - 4x - 1$ to solve for y.

$$\begin{aligned} y &= x^2 - 4x - 1 \\ &= (2)^2 - 4(2) - 1 \\ &= 4 - 8 - 1 \text{ or } -5 \end{aligned}$$

The point at $(2, -5)$ is the vertex. *This point is a minimum.*

C. Construct a table. Choose some values for x that are less than 2 and some that are greater than 2. This ensures that points on each side of the axis of symmetry are graphed.

x	$x^2 - 4x - 1$	y	(x, y)
0	$0^2 - 4(0) - 1$	-1	$(0, -1)$
1	$1^2 - 4(1) - 1$	-4	$(1, -4)$
2	$2^2 - 4(2) - 1$	-5	$(2, -5)$
3	$3^2 - 4(3) - 1$	-4	$(3, -4)$
4	$4^2 - 4(4) - 1$	-1	$(4, -1)$

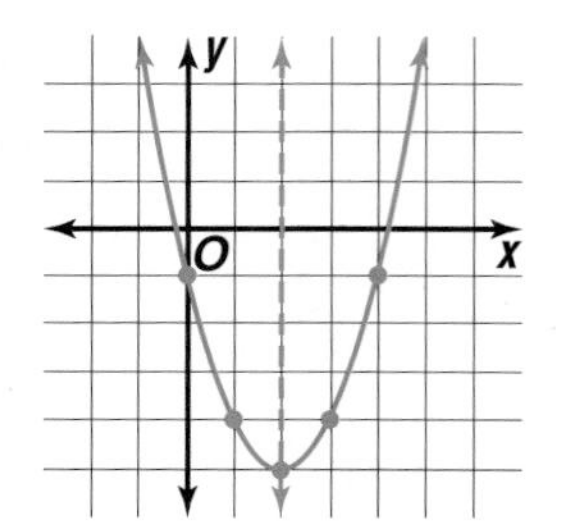

Your Turn

Find the coordinates of the vertex and the equation of the axis of symmetry for the graph of each equation. Then graph the function.

c. $y = x^2 + x$

d. $y = -x^2 + 2x - 3$

Models of quadratic functions can be found in the real world.

Architecture Link

4 **The Exchange House in London, England, is supported by a steel arch shaped like a parabola. This parabola can be modeled by the quadratic function $y = -0.025x^2 + 2x$, where y represents the height of the arch and x represents the horizontal distance from one end of the base in meters. What is the highest point of the parabolic arch?**

The highest point of the arch is the y-coordinate of the vertex. Find the equation of the axis of symmetry for $h(x) = -0.025x^2 + 2x$.

$$\begin{aligned} x &= -\frac{b}{2a} \\ &= -\frac{2}{2(-0.025)} \quad \textit{a = -0.025, b = 2} \\ &= -\frac{2}{-0.05} \text{ or } 40 \quad \textit{Simplify.} \end{aligned}$$

Prerequisite Skills Review
Operations with Decimals, p. 684

Next, find the vertex. Since the equation of the axis of symmetry is $x = 40$, the x-coordinate of the vertex must be 40. Substitute 40 for x in the function $h(x) = -0.025x^2 + 2x$. Then solve for h.

$h(x) = -0.025x^2 + 2x$

$h(40) = -0.025(40)^2 + 2(40)$ *Replace x with 40.*

$= -40 + 80$ or 40 *Simplify.*

The point (40, 40) is the vertex. So, the maximum height is 40 meters.

Check for Understanding

Communicating Mathematics

1. **Compare and contrast** quadratic and linear functions.
2. **Explain** what effect a negative coefficient of x^2 has on the orientation of a parabola.
3. **Name** the point on the graph of a quadratic function that has a unique y-coordinate.

Vocabulary

parabola
quadratic function
minimum
maximum
vertex
axis of symmetry

Guided Practice

Identify the values for a, b, and c for each quadratic function in the form $y = ax^2 + bx + c$.

Sample 1: $y = x^2 - 9$
Solution: $a = 1, b = 0, c = -9$

Sample 2: $y = -4x^2 - 9x + 13$
Solution: $a = -4, b = -9, c = 13$

4. $y = 6x^2 - 3x + 1$
5. $y = x^2 + 4$
6. $y = -x^2 - x - 12$
7. $y = 3x^2 + 4x$

Graph each quadratic function by making a table of values. *(Examples 1 & 2)*

8. $y = -2x^2$
9. $y = x^2 + 3x$
10. $y = -x^2 - 2x + 5$

Write the equation of the axis of symmetry and the coordinates of the vertex of the graph of each quadratic function. Then graph the function. *(Example 3)*

11. $y = x^2 + 2$
12. $y = x^2 + 8x + 12$
13. $y = -x^2 + 4x + 3$
14. $y = -x^2 + 10x$

15. **Architecture** Mr. Kwan is drafting the windows for a building. Their shape is a parabola modeled by the equation $h = -w^2 + 9$, where h is the height of the window and w is the width in feet. *(Example 4)*
    a. Graph the function.
    b. Find the maximum height of each window.
    c. Find the width of each window at its base.

Exercises

Practice

Homework Help	
For Exercises	See Examples
16–24	1, 2
25–44	3, 4
Extra Practice	
See page 713.	

Graph each quadratic function by making a table of values.

16. $y = 2x^2$ **17.** $y = 3x^2$ **18.** $y = -5x^2$

19. $y = x^2 + 4$ **20.** $y = 2x^2 - 1$ **21.** $y = -x^2 + 8x$

22. $y = -x^2 + 4x + 1$ **23.** $y = 3x^2 - 6x + 1$ **24.** $y = \frac{1}{2}x^2 - 6x + 5$

Write the equation of the axis of symmetry and the coordinates of the vertex of the graph of each quadratic function. Then graph the function.

25. $y = x^2$ **26.** $y = -2x^2$ **27.** $y = 4x^2$

28. $y = 2x^2 + 3$ **29.** $y = -x^2 + 1$ **30.** $y = x^2 - 6x$

31. $y = x^2 - 4x + 2$ **32.** $y = x^2 - 4x + 10$ **33.** $y = -3x^2 - 6x + 4$

34. $y = -\frac{1}{4}x^2 + x - 2$ **35.** $y = x^2 + 5x$ **36.** $y = 3x^2 - 3x - 2$

Match each function with its graph.

37. $y = x^2 + 2x + 1$ **38.** $y = x^2 - 2x - 1$ **39.** $y = -x^2 + 4x - 3$

A
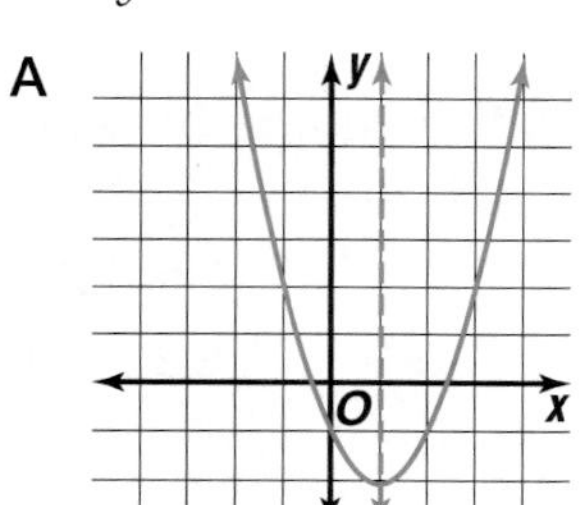

B
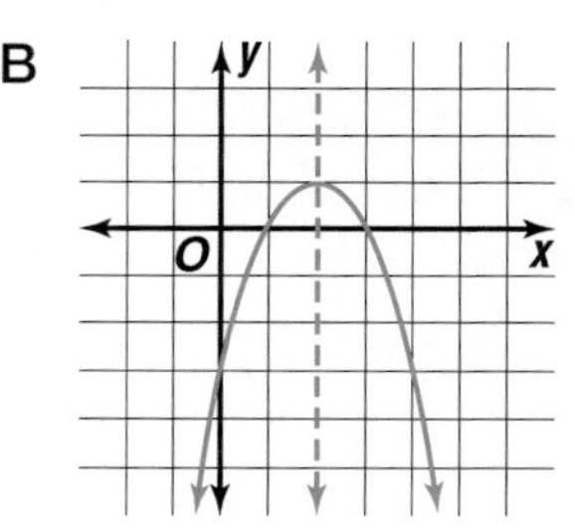

C
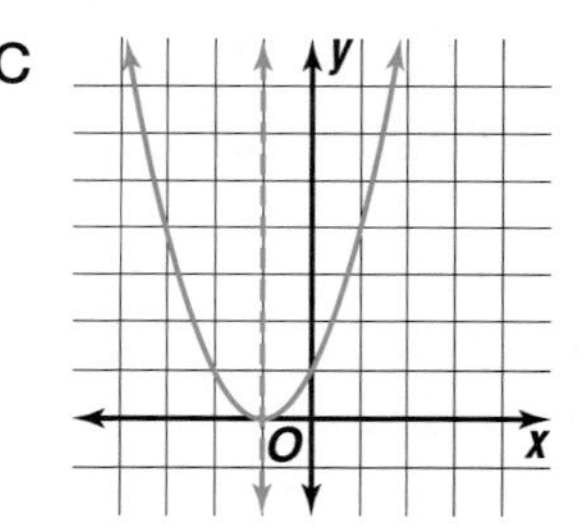

40. Suppose the equation of the axis of symmetry for a quadratic function is $x = -3$ and one of the x-intercepts is -8. What is the other x-intercept?

41. Suppose the points at $(-8, 5)$ and $(6, 5)$ are on the graph of a parabola. What is the equation of the axis of symmetry?

Applications and Problem Solving

42. Sports Mark wanted to know the angle at which he should kick a football for maximum distance. He used a device that kicked a football at a constant velocity at varying angles. He recorded his results in the table.

Angle	Distance (ft)
30°	140
35°	152
40°	160
45°	162
50°	160
55°	152

a. Graph the data in the table.

b. Which angle gives the maximum distance?

c. Predict how far the football will go if it is kicked at a 60° angle.

43. **Business** The profit function of a small business can be expressed as $P(x) = -x^2 + 300x$, where x represents the number of employees.

a. How many employees will yield the maximum profit?

b. What is the maximum profit?

44. **Critical Thinking** Graph $y = x^2$. Then graph $y = x^2 - 4$ on the same axes. Describe the difference between the graphs.

Mixed Review

Factor each polynomial. *(Lesson 10–5)*

45. $x^2 - 49$

46. $a^2 + 10a + 25$

47. $2g^2 - 16g + 32$

48. **Geometry** The volume of a rectangular prism is $3x^3 - 6x^2 - 24x$. Find dimensions for the prism in terms of x. *(Lesson 10–4)*

49. Arrange the terms of the polynomial $-4x^2 + 5 - x^3 - 8x$ so that the powers of x are in descending order. Then state the degree of the polynomial. *(Lesson 9–1)*

Use the map for Exercises 50–51.

50. You can use the letters and numbers on the map to form ordered pairs and name square areas. State the locations of Fancyburg and Northam Parks as ordered pairs. *(Lesson 2–2)*

51. Find the straight-line distance in miles between Fancyburg and Northam Parks. *(Lesson 5–2)*

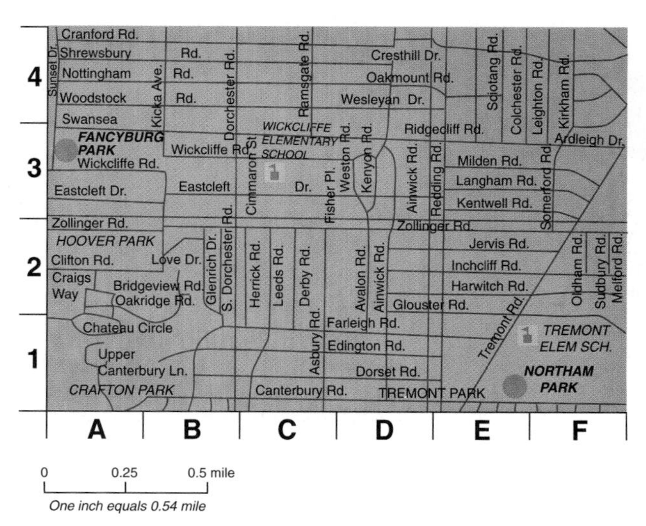

Standardized Test Practice

A B C D

52. **Multiple Choice** The line graph shows the amount of soda an average American drinks over a five-year period. During which of the following periods was the greatest change in consumption? *(Lesson 1–7)*

A 1993–1994

B 1994–1995

C 1995–1996

D 1996–1997

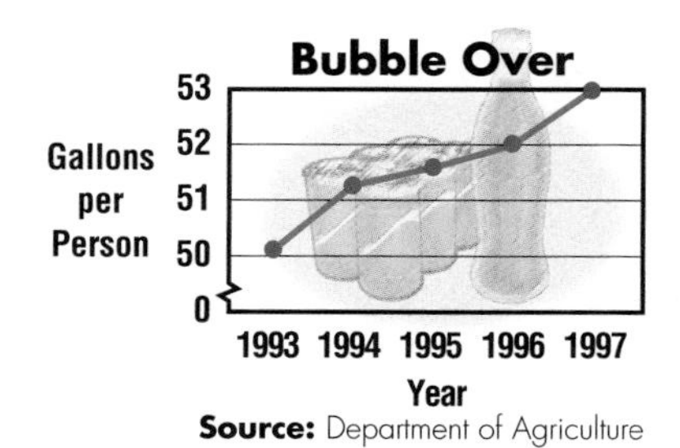

11–2 Families of Quadratic Functions

What You'll Learn
You'll learn the characteristics of families of parabolas.

Why It's Important
Animation Digital artists use families of parabolas to create the illusion of motion. *See Example 5.*

In families of linear graphs, lines either have the same slope or the same y-intercept. However, in families of parabolas, graphs either share a vertex or an axis of symmetry, or both. Also, a family can consist of parabolas of the same shape.

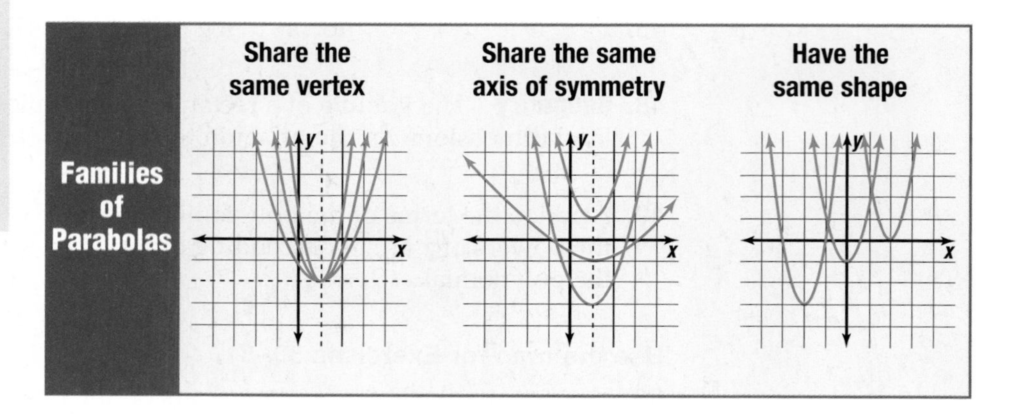

Examples

Graph each group of equations on the same screen. Compare and contrast the graphs. What conclusions can be drawn?

1 $y = x^2, y = 0.2x^2, y = 3x^2$

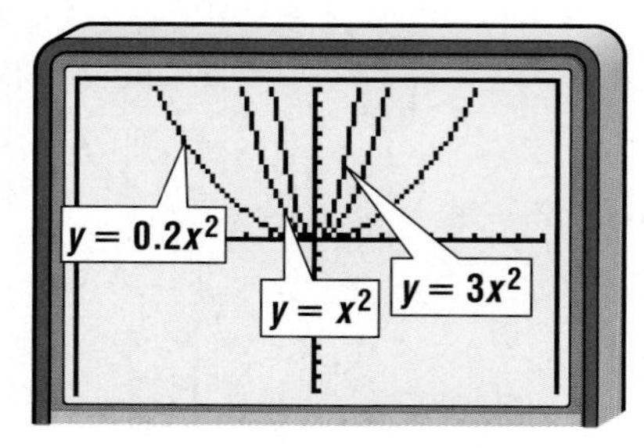

Each graph opens upward and has its vertex at the origin. Therefore, these equations are a family of parabolas. The graph of $y = 0.2x^2$ is wider than the graph of $y = x^2$. The graph of $y = 3x^2$ is more narrow than the graph of $y = x^2$.

Technology Tip
Use the [Y=] screen to enter functions into the graphing calculator.

The parent graph for each family is $y = x^2$.

The shape of the parabola narrows as the coefficient of x^2 becomes greater. The shape widens as the coefficient of x^2 becomes smaller.

2 $y = x^2, y = x^2 - 6, y = x^2 + 3$

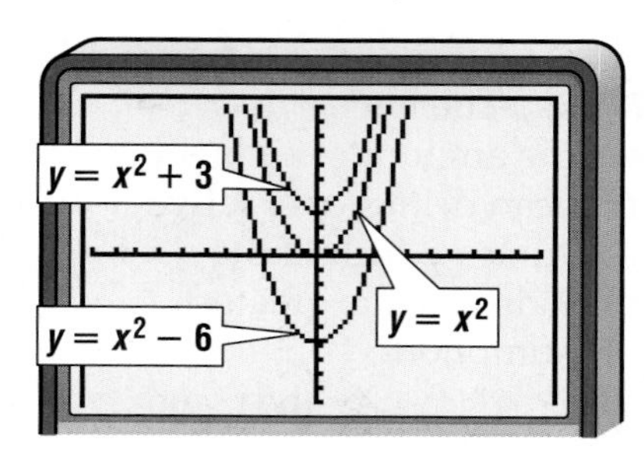

Each graph opens upward and has the same shape as $y = x^2$ so they form a family. Yet, each parabola has a different vertex located along the y-axis.

A constant greater than 0 shifts the graph upward, and a constant less than 0 shifts the graph downward along the axis of symmetry.

3 $y = x^2, y = (x + 2)^2, y = (x - 4)^2$

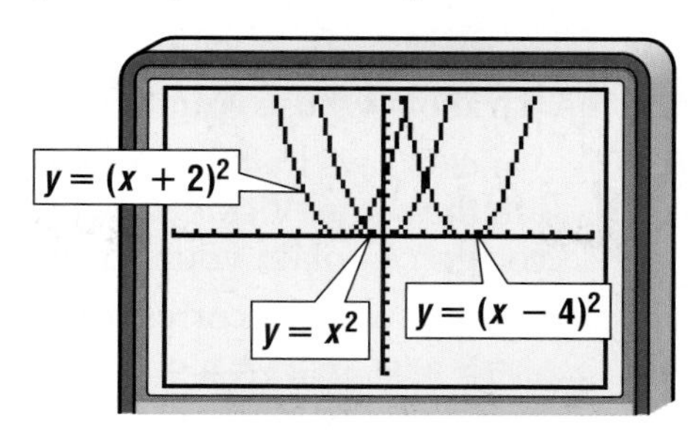

Each graph opens upward and has the same shape as $y = x^2$. However, each parabola has a different vertex located along the x-axis.

Find the number for x that results in 0 inside the parentheses. The graph shifts this number of units to the left or right.

4 $y = x^2, y = (x - 7)^2 + 2$

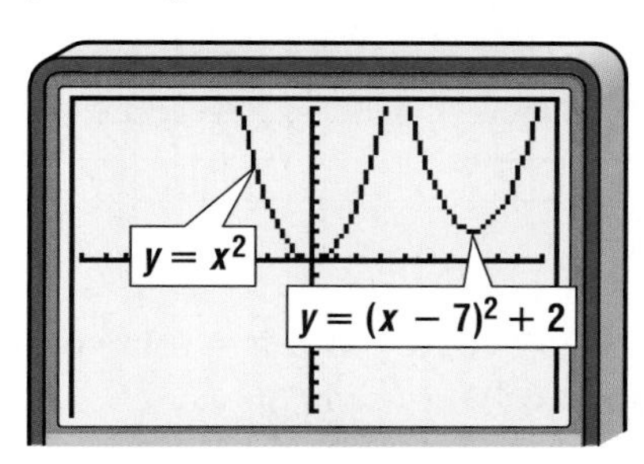

The graph of $y = (x - 7)^2 + 2$ has the same shape as the graph of $y = x^2$. However, it shifts to the right 7 units because a positive 7 will result in zero inside the parentheses. It also shifts upward 2 units because of the constant 2 outside the parentheses.

Your Turn

a. $y = x^2, 2x^2, 4x^2$

b. $y = x^2, x^2 - 1, x^2 - 8$

c. $y = x^2, -x^2$

d. $y = -x^2, y = -(x + 2)^2, y = -(x + 4)^2$

Sometimes computers are used to generate families of graphs.

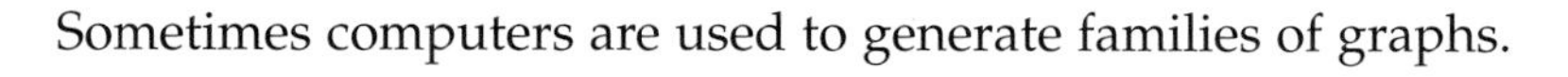

Example

Computer Animation Link

5 **In a computer game, a player dodges space shuttles that are shaped like parabolas. Suppose the vertex of one shuttle is at the origin. The shuttle's initial shape and position are given by the equation $y = 0.5x^2$. It leaves the screen with its vertex at (6, 5). Find an equation to model the final shape and position of the shuttle.**

The shape of the shuttle remains the same. However, the vertex shifts from (0, 0) to the right 6 units and up 5 units.

Begin with the original equation.

$y = 0.5x^2$

$y = 0.5(x - 6)^2$

If $x = 6$, then $x - 6 = 0$. Shift the vertex to the right 6 units.

$y = 0.5(x - 6)^2 + 5$

The 5 outside the parentheses shifts the entire parabola up 5 units.

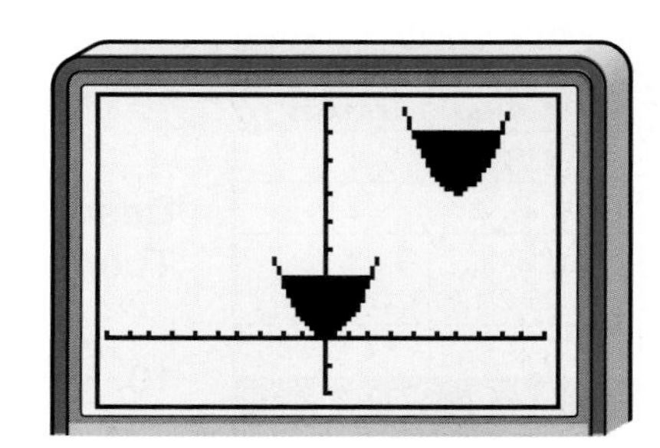

So, the final shape and position of the shuttle can be described by the equation $y = 0.5(x - 6)^2 + 5$.

interNET CONNECTION

Data Update For the latest information on computer animation, visit: www.algconcepts.com

Check for Understanding

Communicating Mathematics

1. **Describe** the parabola whose equation is $y = 100x^2$.
2. 

Vanessa says that the graphs of $y = (x - 2)^2$ and $y = x^2 - 2$ are the same. Vickie says that they are different. Who is correct? Explain your answer and sketch the graphs.
3. **Match** each equation with its corresponding graph.

$y = -3x^2$ $\quad$ $y = x^2 + 3$ $\quad$ $y = (x - 3)^2$

A
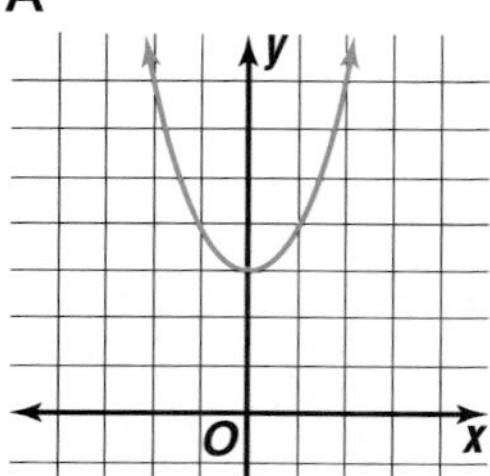

B
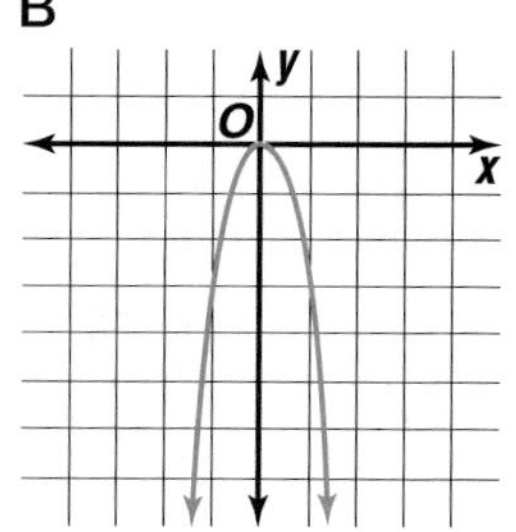

C
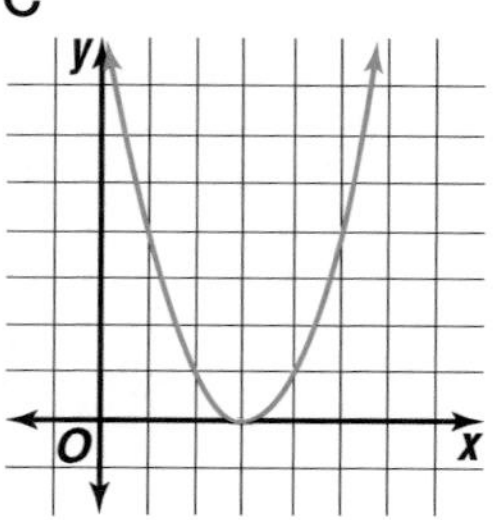

Guided Practice

4. Graph $y = x^2$, $y = 0.5x^2$, and $y = 0.1x^2$ on the same axes. Compare and contrast the graphs. *(Example 1)*

Describe how each graph changes from the parent graph of $y = x^2$. Then name the vertex of each graph. *(Examples 1–4)*

5. $y = 0.7x^2$
6. $y = x^2 + 10$
7. $y = (x + 4)^2$
8. $y = (x + 2)^2 - 9$

9. **Computer Animation** Refer to Example 5. Suppose the shuttle is programmed to move to point (4, 3) before leaving the screen. Write the equation that describes its location. *(Example 5)*

Exercises

Practice

Graph each group of equations on the same screen. Compare and contrast the graphs.

10. $y = -x^2$
    $y = -4x^2$
    $y = -6x^2$
11. $y = (x + 1)^2$
    $y = (x + 2)^2$
    $y = (x + 3)^2$
12. $y = -x^2 - 1$
    $y = -x^2 - 3$
    $y = -x^2 - 5$

Homework Help	
For Exercises	See Examples
10, 13, 17, 18	1
12, 14, 26	2
11, 15, 16, 25, 27	3
23, 28, 29	4, 5

Extra Practice
See page 713.

Describe how each graph changes from the parent graph of $y = x^2$. Then name the vertex of each graph.

13. $y = 5x^2$
14. $y = x^2 - 8$
15. $y = (x - 7)^2$
16. $y = (x - 3)^2$
17. $y = -2x^2$
18. $y = -0.9x^2$
19. $y = 2x^2 + 1$
20. $y = -(x + 4)^2$
21. $y = 0.4x^2 - 8$
22. $y = -x^2 + 6$
23. $y = (x + 1)^2 - 5$
24. $y = [x - (-2)]^2 + 3$

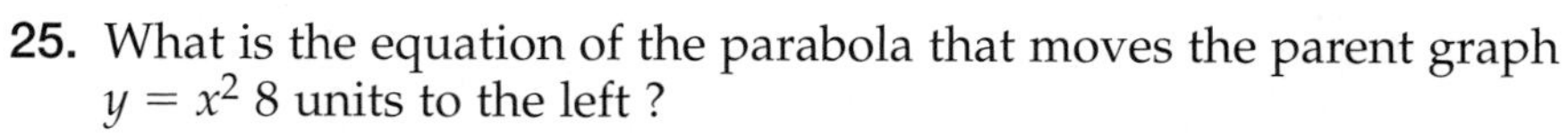

25. What is the equation of the parabola that moves the parent graph $y = x^2$ 8 units to the left ?

26. Suppose the parent graph is $y = x^2 + 3$. Write the equation of the parabola that would move it down 5 units.

Applications and Problem Solving

Real World

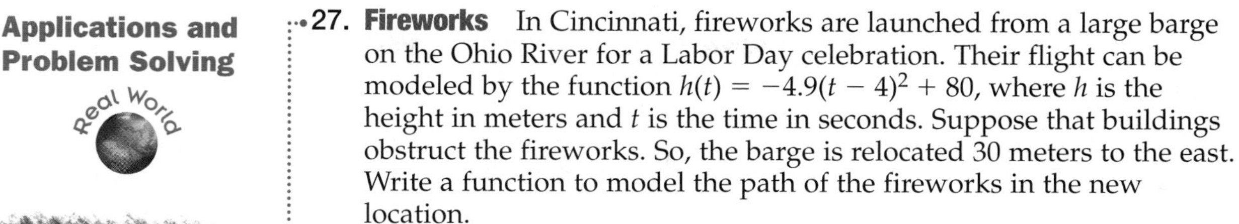

27. **Fireworks** In Cincinnati, fireworks are launched from a large barge on the Ohio River for a Labor Day celebration. Their flight can be modeled by the function $h(t) = -4.9(t - 4)^2 + 80$, where h is the height in meters and t is the time in seconds. Suppose that buildings obstruct the fireworks. So, the barge is relocated 30 meters to the east. Write a function to model the path of the fireworks in the new location.

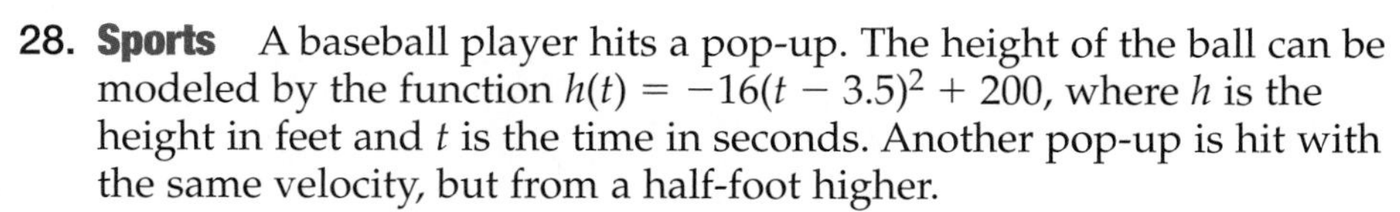

28. **Sports** A baseball player hits a pop-up. The height of the ball can be modeled by the function $h(t) = -16(t - 3.5)^2 + 200$, where h is the height in feet and t is the time in seconds. Another pop-up is hit with the same velocity, but from a half-foot higher.
    a. Write equations to model the height of the ball for each hit.
    b. Graph both equations and find each vertex.
    c. Does the higher height affect the maximum height? Does the higher height affect the time the ball takes to reach its maximum height?

Fireworks display

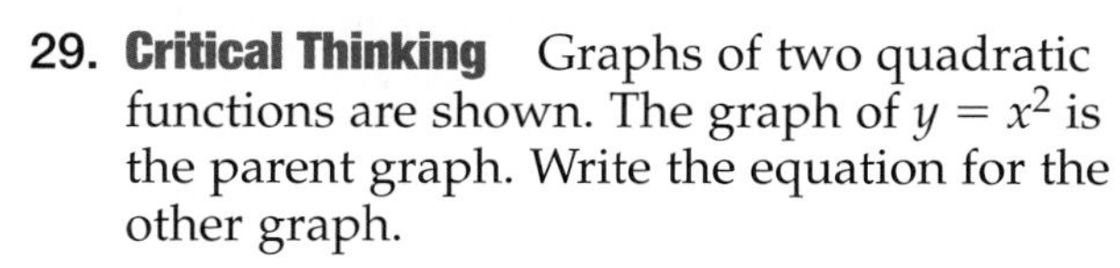

29. **Critical Thinking** Graphs of two quadratic functions are shown. The graph of $y = x^2$ is the parent graph. Write the equation for the other graph.

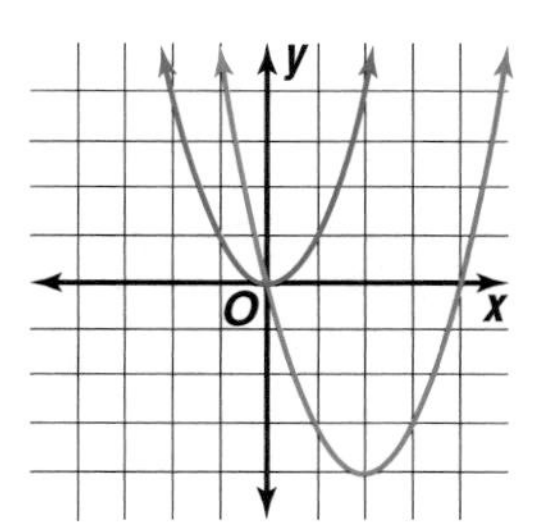

Mixed Review

Find the coordinates of the vertex for each quadratic function. *(Lesson 11–1)*

30. $y = -4x^2 + 8x + 13$

31. $y = 2x^2 + 6x + 3$

32. **Finance** Tracy invested \$500, earning an annual interest rate r. The value of her investment at the end of two years can be represented by the polynomial $500r^2 + 1000r + 500$. *(Lesson 10–5)*
    a. Write an expression in factored form for the value of her investment at the end of two years.
    b. If $r = 6\%$, find the value of her investment at the end of two years.

Find each product. *(Lesson 9–4)*

33. $(a + 7)(a - 4)$

34. $(3n + 6)(n - 4)$

Standardized Test Practice

A B C D

35. **Multiple Choice** Nine tickets, numbered 3 through 11, are placed in an empty hat. If one ticket is drawn at random from the hat, what is the probability that a prime number will be on the ticket? *(Lesson 5–6)*

A $\frac{1}{9}$ B $\frac{4}{9}$ C $\frac{5}{9}$ D $\frac{1}{3}$

11–3 Solving Quadratic Equations by Graphing

What You'll Learn
You'll learn to locate the roots of quadratic equations by graphing the related functions.

Why It's Important
Manufacturing
You can use quadratic equations to solve problems with production and profit. *See Exercise 24.*

In a **quadratic equation,** the value of the related quadratic function is 0. For example, if you substitute 0 for y in the quadratic function, the result is the quadratic equation $0 = ax^2 + by + c$. The solutions of a quadratic equation are called the **roots** of the equation. The roots of a quadratic equation can be found by finding the x-intercepts or **zeros** of the related quadratic function.

Examples

Landmark Link

1 **The Buckingham Fountain in Chicago has 133 jets through which water flows. The path of water streaming from a jet is in the shape of a parabola. Find the distance from the jet where the water hits the ground by graphing. Use the function $h(d) = -2d^2 + 4d + 6$, where $h(d)$ represents the height of a stream of water at any distance d from its jet in feet.**

Explore Graph the parabola to determine where the stream of water will hit the ground.

Plan Find the solution of the equation by looking at the values of d where $h(d)$ is 0.

Solve Make a table of values to graph the related function $h(d) = -2d^2 + 4d + 6$.

d	$-2d^2 + 4d + 6$	$h(d)$
−1	$-2(-1)^2 + 4(-1) + 6$	0
0	$-2(0)^2 + 4(0) + 6$	6
1	$-2(1)^2 + 4(1) + 6$	8
2	$-2(2)^2 + 4(2) + 6$	6
3	$-2(3)^2 + 4(3) + 6$	0

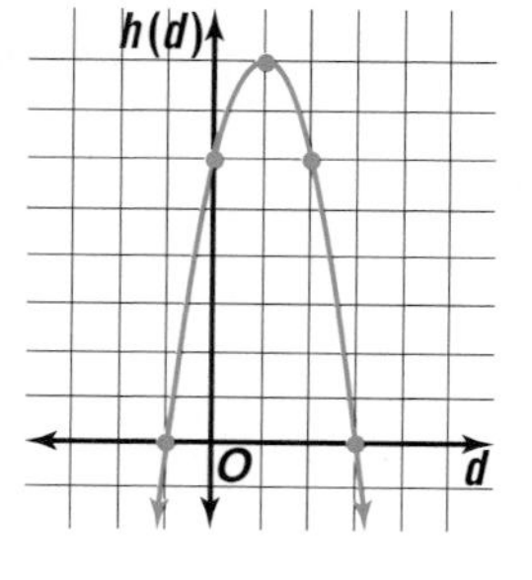

Buckingham Fountain, Chicago

The roots of $0 = -2d^2 + 4d + 6$ are −1 and 3.

Examine Since d represents distance, it cannot be negative. Therefore, $d = -1$ is not a solution, and the only reasonable solution is 3. The water will hit the ground 3 feet from the jet.

Press 2nd [TABLE] 2 to find the roots of an equation graphed on the calculator screen.

2 **Find the roots of $x^2 - 10x + 16 = 0$ by graphing the related function.**

Graph the related function $f(x) = x^2 - 10x + 16$. Before making a table of values, find the equation of the axis of symmetry. This will make selecting x-values for your table easier.

$x = -\frac{b}{2a}$ *Equation of the axis of symmetry*

$x = -\frac{-10}{2(1)}$ or 5 *$a = 1$ and $b = -10$*

The equation of the axis of symmetry is $x = 5$. Now, make a table using x-values around 5. Graph each point on a coordinate plane.

x	$x^2 - 10x + 16$	$f(x)$
2	$2^2 - 10(2) + 16$	0
4	$4^2 - 10(4) + 16$	−8
5	$5^2 - 10(5) + 16$	−9
6	$6^2 - 10(6) + 16$	−8
8	$8^2 - 10(8) + 16$	0

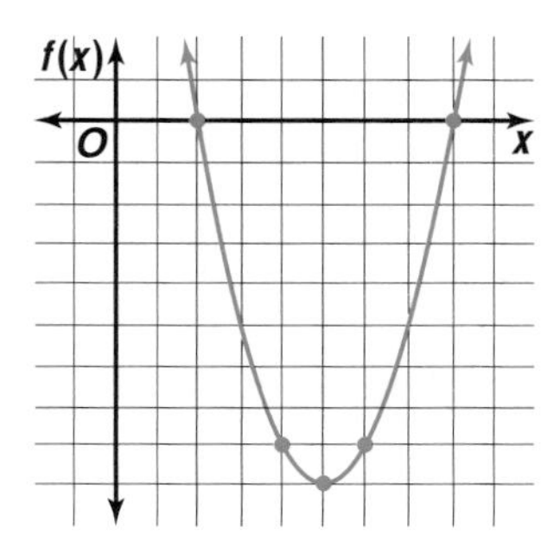

The zeros of the function appear to be 2 and 8. So, the roots are 2 and 8.

Check: Substitute 2 and 8 for x in the equation $x^2 - 10x + 16 = 0$.

$$x^2 - 10x + 16 = 0 \qquad x^2 - 10x + 16 = 0$$
$$2^2 - 10(2) + 16 \stackrel{?}{=} 0 \qquad 8^2 - 10(8) + 16 \stackrel{?}{=} 0$$
$$4 - 20 + 16 \stackrel{?}{=} 0 \qquad 64 - 80 + 16 \stackrel{?}{=} 0$$
$$0 = 0 \ \checkmark \qquad 0 = 0 \ \checkmark$$

Your Turn

a. Find the roots of $0 = x^2 - 5x + 4$ by graphing the related function.

Sometimes exact roots cannot be found by graphing. In this case, estimate solutions by stating the consecutive integers between which the roots are located.

Example **3** **Estimate the roots of $-x^2 + 2x + 1 = 0$.**

Find the equation of the axis of symmetry.

$x = -\frac{b}{2a}$ *Equation of the axis of symmetry*

$x = -\frac{2}{2(-1)}$ or 1 *$a = -1$ and $b = 2$*

The equation of the axis of symmetry is $x = 1$. Now, make a table using x-values around 1. Graph each point on a coordinate plane.

(continued on the next page)

x	$-x^2 + 2x + 1 = 0$	$f(x)$
-1	$-(-1)^2 + 2(-1) + 1$	-2
0	$-0^2 + 2(0) + 1$	1
1	$-(1)^2 + 2(1) + 1$	2
2	$-(2)^2 + 2(2) + 1$	1
3	$-(3)^2 + 2(3) + 1$	-2

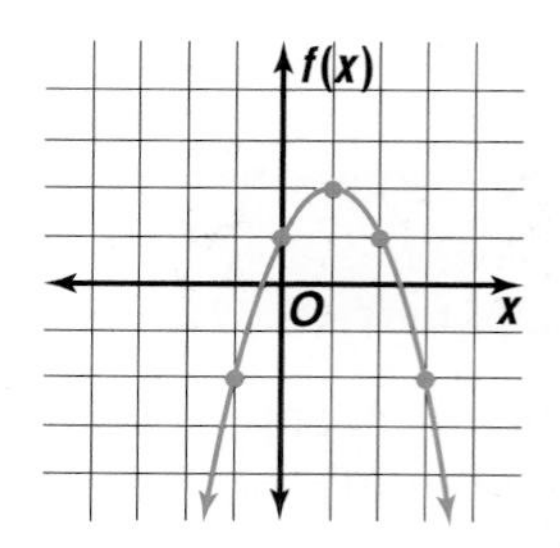

The x-intercepts of the graph are between -1 and 0 and between 2 and 3. So, one root of the equation is between -1 and 0, and the other root is between 2 and 3.

Your Turn

b. Estimate the roots of $y = x^2 - 2x - 9$.

Example 4
Number Theory Link

Find two numbers whose sum is 4 and whose product is 5.

Explore Let x = one of the numbers.
Then $4 - x$ = the other number.

Plan Since the product of the two numbers is 5, you know that $5 = x(4 - x)$.

$5 = x(4 - x)$

$5 = 4x - x^2$ *Distributive Property*

$0 = -x^2 + 4x - 5$ *Subtract 5 from each side.*

Solve You can solve $0 = -x^2 + 4x - 5$ by graphing the related function $f(x) = -x^2 + 4x - 5$. The equation of the axis of symmetry is $x = -\frac{4}{2(-1)}$ or 2. So, choose x-values around 2 for your table.

x	$-x^2 + 4x - 5$	$f(x)$
0	$-0^2 + 4(0) - 5$	-5
1	$-(1)^2 + 4(1) - 5$	-2
2	$-(2)^2 + 4(2) - 5$	-1
3	$-(3)^2 + 4(3) - 5$	-2
4	$-(4)^2 + 4(4) - 5$	-5

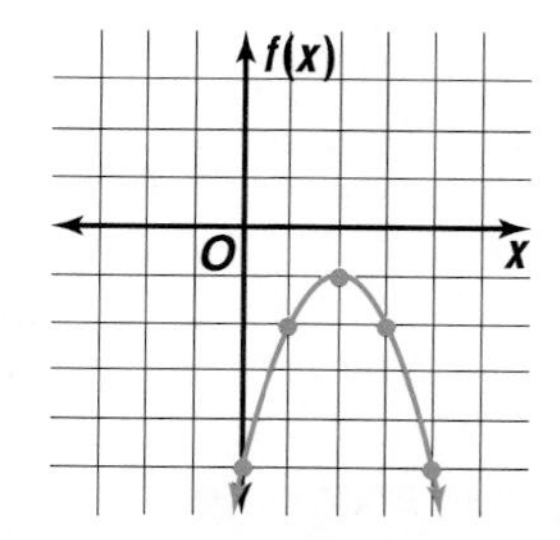

The graph has no x-intercepts since it does not cross the x-axis. This means the equation $x^2 - 4x + 5 = 0$ has no real roots. Thus, it is *not* possible for two numbers to have a sum of 4 and a product of 5. *Examine this solution by testing several pairs of numbers.*

The graphing calculator is a useful tool when making tables in order to solve functions.

Graphing Calculator Exploration

Graphing Calculator Tutorial
See pp. 724–727.

Solve $x^2 + x - 2 = 0$ by making a table.

Enter the equation in the [Y=] screen. Then press [2nd] [TABLE]. A table with two columns labeled X and Y1 will appear on your screen.

X	Y1
-3	4
-2	0
-1	-2
0	-2
1	0

X=

Begin entering values for x. For each x-value entered, a corresponding y-value is calculated using the equation $y = x^2 + x - 2$.

Since $y = 0$ when $x = -2$ and $x = 1$, the roots of the equation are -2 and 1.

Try These

Make a table to find the roots of each equation.

1. $x^2 - 18x + 81 = 0$
2. $x^2 - 2x - 15 = 0$

Check for Understanding

Communicating Mathematics

1. **Explain** why finding the x-intercepts of the graph of a quadratic function can be used to solve the related quadratic equation.
2. **Write** the related function for $4 = x^2 - 3x$.
3. **Sketch** a parabola that has one root.

Vocabulary
quadratic equation
roots
zeros

Guided Practice

Getting Ready **State the roots of each quadratic equation.**

Sample:

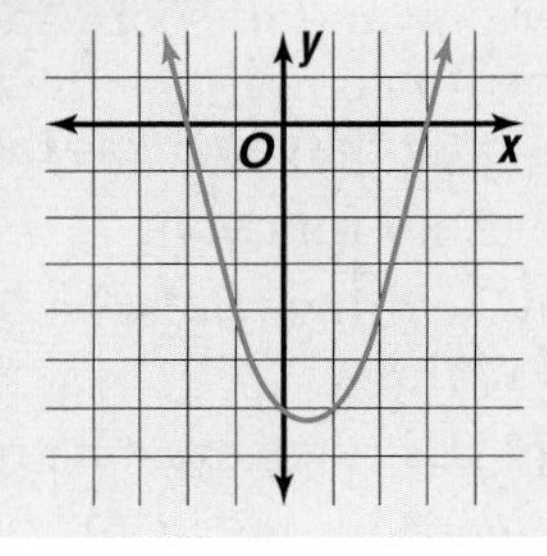

Solution: The x-intercepts of the graph are -2 and 3. So, the roots are -2 and 3.

4. 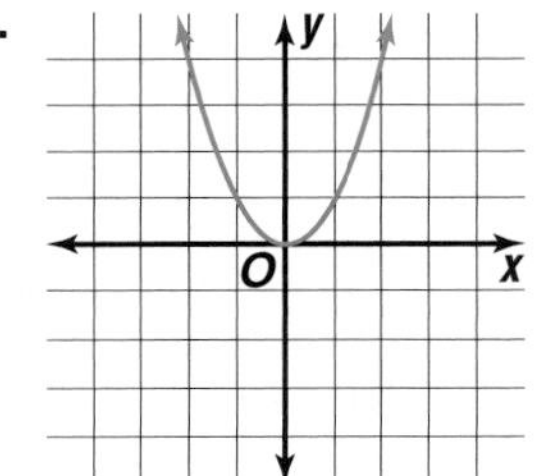

5. 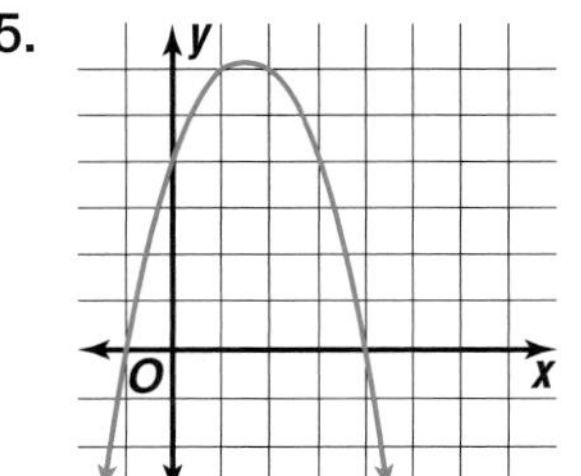

6. 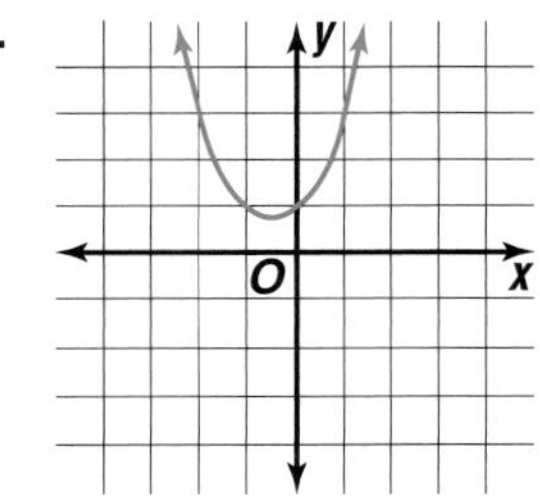

Solve each equation by graphing the related function. If exact roots cannot be found, state the consecutive integers between which the roots are located. *(Examples 1–3)*

7. $x^2 - 5x + 6 = 0$
8. $x^2 + 2x - 15 = 0$
9. $2x^2 - 3x - 7 = 0$

10. **Number Theory** Use a quadratic equation to find two numbers whose sum is 4 and whose product is -12. *(Example 4)*

Exercises

Practice

Homework Help	
For Exercises	See Examples
11–19	1, 2
27	2, 3
20, 21, 24, 25	3, 4
Extra Practice	
See page 714.	

Solve each equation by graphing the related function. If exact roots cannot be found, state the consecutive integers between which the roots are located.

11. $x^2 + 2x + 1 = 0$
12. $x^2 - 4x + 3 = 0$
13. $x^2 - 5x - 14 = 0$
14. $-x^2 + 6x + 7 = 0$
15. $x^2 - 10x + 25 = 0$
16. $x^2 + 3x - 2 = 0$
17. $x^2 + 4x - 2 = 0$
18. $x^2 - 2x + 2 = 0$
19. $-2x^2 + 3x + 4 = 0$

Use a quadratic equation to determine the two numbers that satisfy each situation.

20. Their sum is 18 and their product is 81.
21. Their difference is 4 and their product is 32.

Use the given roots and vertex of a quadratic equation to graph the related quadratic function.

22. roots: -2, 4
    vertex: (1, 9)
23. roots: -7, -3
    vertex: $(-5, -4)$

Applications and Problem Solving

24. **Business** Mr. Jamison owns a manufacturing company that produces key rings. Last year, he collected data about the number of key rings produced per day and the corresponding profit. He then modeled the data using the function $P(k) = -2k^2 + 12k - 10$, where P is the profit in thousands of dollars and k is the number of key rings in thousands.
    a. Graph the profit function.
    b. How many key rings must be produced per day so that there is no profit and no loss?
    c. How many key rings must be produced for the maximum profit?
    d. What is the maximum profit?

25. **Skydiving** In a recent year, Adrian Nicholas broke two world records by flying 10 miles in 4 minutes 55 seconds without a plane. He jumped from an airplane at 35,000 feet and did not activate his parachute until he was 500 feet above the ground. If air resistance is ignored, how long did Nicholas free-fall? Use the formula $h(t) = -16t^2 + h_0$, where t is the time in seconds and h_0 is the initial height in feet.

26. **Geometry** Mrs. Parker wants to enclose a rectangular running yard at her dog kennel with 200 feet of fencing.

a. Make a drawing of the enclosed yard. Then write a formula for the perimeter and solve for the width.

b. Write a quadratic equation for the area A of the yard. Use the expression for the width from part a.

c. Show graphically how the area of the yard changes when the length changes. Can the yard have a length of 100 feet? Explain.

27. **Critical Thinking** Suppose the value of a quadratic function is negative when $x = 1$ and positive when $x = 2$. Explain why it is reasonable to assume that the related equation has a root between 1 and 2.

Mixed Review

Describe how each graph changes from its parent graph of $y = x^2$. Then name the vertex. *(Lesson 11–2)*

28. $y = -x^2$ 29. $y = x^2 + 4$ 30. $y = (x + 6)^2$

Standardized Test Practice

A B C D

31. **Grid In** If the equation of the axis of symmetry of a quadratic function is $x = 0$, and one of the x-intercepts is -4, what is the other x-intercept? *(Lesson 11–1)*

32. **Multiple Choice** Evaluate $4^x + (x^2)^3$ if $x = 2$. *(Lesson 8–1)*

A 6 B 20 C 32 D 80

Quiz 1 Lessons 11–1 through 11–3

Write the equation of the axis of symmetry and the coordinates of the vertex for each quadratic function. Then graph the function. *(Lesson 11–1)*

1. $y = x^2 + 6x - 5$ 2. $y = -2x^2 - 9$ 3. $y = -3x^2 - 6x + 4$

Describe how each graph changes from its parent graph of $y = x^2$. Then name the vertex of each graph. *(Lesson 11–2)*

4. $y = x^2 + 7$ 5. $y = (x - 6)^2$ 6. $y = -0.8x^2$

Solve each equation by graphing the related function. If exact roots cannot be found, state the consecutive integers between which the roots are located. *(Lesson 11–3)*

7. $x^2 - 5x + 6 = 0$ 8. $x^2 + 6x + 10 = 0$ 9. $x^2 - 2x - 1 = 0$

10. **Sports** Anna is doing a back dive from a 10-meter platform. Her path of descent is given by the graph of $h(d) = -d^2 + 2d + 10$, where the height h and distance from the platform d are both in meters. *(Lessons 11–1 & 11–3)*

a. Graph the function.

b. At her maximum height, how far away from the platform is she?

c. About how far away from the platform will she enter the water?

11–4 Solving Quadratic Equations by Factoring

What You'll Learn
You'll learn to solve quadratic equations by factoring and by using the Zero Product Property.

Why It's Important
Photography You can use quadratic equations to solve problems involving photography.
See Exercise 34.

In the previous lesson you learned to solve quadratic equations by graphing. Quadratic equations can also be solved by factoring. For example, the equation $8x^2 - 4x = 0$ can be solved by finding factors.

$8x^2 - 4x = 0$

$4x(2x - 1) = 0$ *Factor $8x^2 - 4x$.*

To solve this equation, find values of x that make the product $4x(2x - 1)$ equal to 0. Since the product of 0 and any number is 0, *at least one* of the factors in the expression must be zero.

$4x = 0$ or $2x - 1 = 0$

$x = 0$ $\qquad x = \frac{1}{2}$

The solutions of $8x^2 - 4x = 0$ are 0 and $\frac{1}{2}$.

This method of solving quadratic equations uses the **Zero Product Property**.

Zero Product Property	For all numbers a and b, if $ab = 0$, then $a = 0$, $b = 0$ or both a and b equal 0.

We can use this property to solve any equation that is written in the form $ab = 0$.

Example 1 **Solve $3x(x - 1) = 0$. Check your solution.**

If $3x(x - 1) = 0$, then $3x = 0$ or $x - 1 = 0$. *Zero Product Property*

$3x = 0$ or $x - 1 = 0$ *Solve each equation.*

$x = 0$ $\qquad x = 1$

Check: Substitute 0 and 1 for x in the original equation.

$$3x(x - 1) = 0 \qquad \text{or} \qquad 3x(x - 1) = 0$$
$$3(0)(0 - 1) \stackrel{?}{=} 0 \qquad\qquad 3(1)(1 - 1) \stackrel{?}{=} 0$$
$$0(-1) \stackrel{?}{=} 0 \qquad\qquad 3(0) \stackrel{?}{=} 0$$
$$0 = 0 \ \checkmark \qquad\qquad 0 = 0 \ \checkmark$$

The solutions are 0 and 1.

Your Turn **Solve each equation. Check your solution.**

a. $z(z - 8) = 0$

b. $(a - 4)(4a + 3) = 0$

Example 2 — Recreation Link

At an adventure park, you can jump off a 26-foot cliff into a pool of water below. The equation $h = -16t^2 + 4t + 26$ describes your height h in feet t seconds after your jump. If you jump up and off the cliff, what time will you pass the height of 26 feet again?

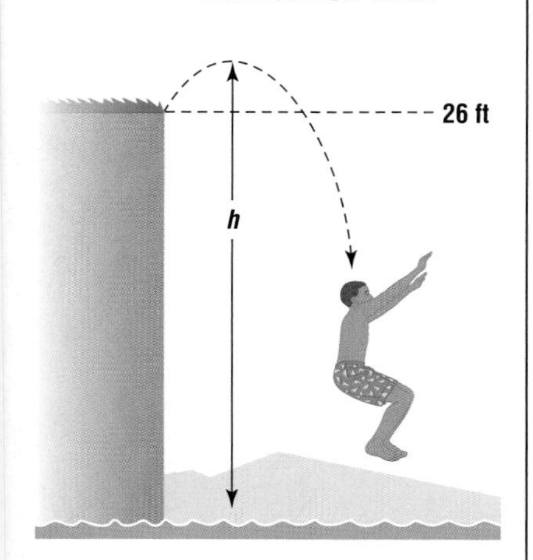

To find the time at which this occurs, let $h = 26$ and solve for t.

$26 = -16t^2 + 4t + 26$

$0 = -16t^2 + 4t$ *Subtract 26 from each side.*

$0 = 4t(-4t + 1)$ *Factor –16t – 8t.*

If $4t(-4t + 1) = 0$, then $4t = 0$ or $-4t + 1 = 0$. *Zero Product Property*

$4t = 0$ or $-4t + 1 = 0$ *Solve each equation.*

$t = 0$ $\qquad -4t = -1$

$t = \frac{1}{4}$

Check: Graph the equation and find the x-intercepts. We will graph the equation $h = 4t(-4t + 1)$ on a graphing calculator.

Enter the equation in the [Y=] screen. Press [GRAPH]. To find the zeros, press [2nd] [CALC] [▼] 2. Use the arrow keys to move the cursor to the left of one of the x-intercepts. Press [ENTER]. Move the cursor to the right of the same x-intercept. Press [ENTER] twice. Coordinates of the root will appear. Repeat for the other root.

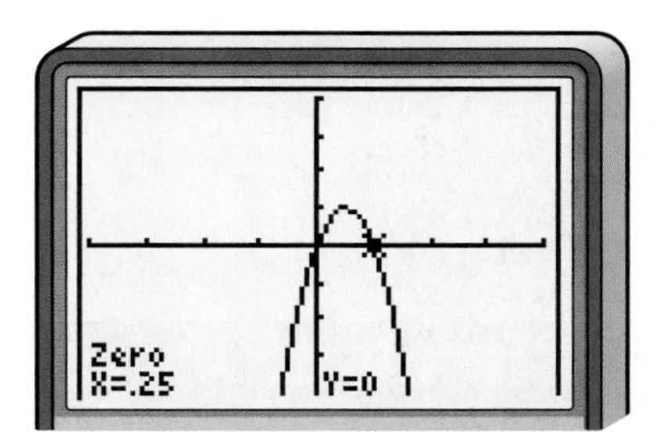

Set viewing window for x: [−1, 1] by 0.25 and y: [−1, 1] by 0.25.

The solutions are 0 and $\frac{1}{4}$. The solution 0 represents the beginning of the jump. So, you would return to your original location after a fourth of a second.

You will often need to factor an equation before using the Zero Product Property to solve the equation.

Example 3

Solve $x^2 + 4x - 12 = 0$. Check your solution.

$x^2 + 4x - 12 = 0$

$(x - 2)(x + 6) = 0$ *Factor.*

$x - 2 = 0$ or $x + 6 = 0$ *Zero Product Property*

$x = 2$ $\qquad x = -6$ *Check this solution.*

The solutions are -6 and 2.

Factoring Trinomials: Lessons 10–3 & 10–4

Your Turn

c. Solve $x^2 - 2x = 3$. Check your solution.

Example 4

Geometry Link

The length ℓ of a rectangle is 2 feet more than twice its width w. The area of the rectangle is 144 square feet. Find the measures of the sides.

Explore The formula for the area of a rectangle is $A = \ell w$ and $\ell = 2w + 2$.

Plan

$$\underbrace{144}_{\text{Area}} \underbrace{=}_{\text{equals}} \underbrace{2w + 2}_{\text{length}} \underbrace{\cdot}_{\text{times}} \underbrace{w}_{\text{width.}}$$

Solve

$144 = 2w^2 + 2w$ *Distributive Property*

$0 = 2w^2 + 2w - 144$ *Subtract 144 from each side.*

$0 = 2(w^2 + w - 72)$ *Factor out the GCF, 2.*

$0 = w^2 + w - 72$ *Divide each side by 2.*

$0 = (w + 9)(w - 8)$ *Factor.*

$w + 9 = 0$ or $w - 8 = 0$ *Zero Product Property*

$w = -9$ $\quad w = 8$ *Solve each equation.*

Examine Since width cannot be negative, the width must be equal to 8 feet. Substituting 8 for w, the length of the rectangle is $2(8) + 2$ or 18 feet.

Check for Understanding

Communicating Mathematics

1. **Define** the Zero Product Property in your own words.
2. **Explain** why you should check your answers by substituting values into the original equation rather than into a simplified version.
3. **You Decide?** Lashonda said that she should solve $2x^2 - 3x - 35 = 0$ by graphing. Jerome disagreed. He said that she should solve the equation by factoring. Who is correct, and why?

Guided Practice

Solve each equation. Check your solution.

4. $(x + 1)(x - 7) = 0$
5. $3b(b - 5) = 0$
6. $(2m - 1)(m + 4) = 0$
7. $x^2 - 6x + 9 = 0$
8. $x^2 = x$
9. $7x^2 - 14x = 56$
10. **Geometry** The height of a triangle measures 5 centimeters more than its base. The area of the triangle is 18 square centimeters. Find the measures of the base and the height of the triangle.

Exercises

Practice

Solve each equation. Check your solution.

11. $3m(m + 2) = 0$
12. $5z(z + 8) = 0$
13. $(p + 7)(p - 6) = 0$
14. $(s - 2)(s + 11) = 0$
15. $(r + 1)(2r - 8) = 0$
16. $(x - 2)(3x + 4) = 0$
17. $x^2 + 10x + 16 = 0$
18. $z^2 + 5z - 6 = 0$
19. $p^2 - 11p + 24 = 0$
20. $n^2 + 9n + 18 = 0$
21. $m^2 - m - 12 = 0$
22. $r^2 + 14r = 0$

Homework Help	
For Exercises	See Examples
11–16	1, 2
17–34	3, 4
Extra Practice	
See page 714.	

23. $3w^2 - 9w = 0$ **24.** $15 = x^2 - 2x$ **25.** $2x^2 - 8x + 8 = 0$

26. $2a^2 - 70 = 4a$ **27.** $3b^2 - 12b - 15 = 0$ **28.** $2v^2 - 17v = 9$

For each problem, define a variable. Then use an equation to solve the problem.

29. The length of Luis' house is ten feet longer than it is wide. The area in square feet is 875. Find the dimensions of the house.

30. Find two consecutive even integers whose product is 120.

31. Find two integers whose difference is 3 and whose product is 88.

Applications and Problem Solving

32. Physics A flare is launched from a life raft with an initial upward velocity of 192 feet per second. How many seconds will it take for the flare to return to the sea? Use the formula $h = 192t - 16t^2$, where h is the height of the flare in feet and t is the time in seconds.

33. Geometry Four corners are cut from a rectangular piece of cardboard that measures 3 feet by 5 feet. The cuts are x feet from the corners as shown in the figure at the right. After the cuts are made, the sides are folded up to form an open box. The area of the bottom of the box is 3 square feet. Find the dimensions of the box.

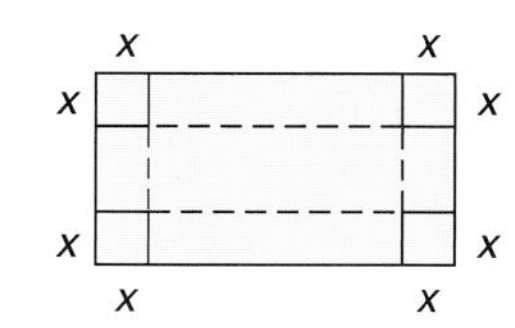

34. Photography A rectangular photograph is 4 inches wide and 6 inches long. The photograph is enlarged by increasing the length and width by an equal amount. If the area of the new photograph is twice as large as the original area, what are the dimensions of the new photograph?

35. Critical Thinking The graph of a quadratic function is shown.

a. What are the zeros of the function?

b. Write an equation for the graph.

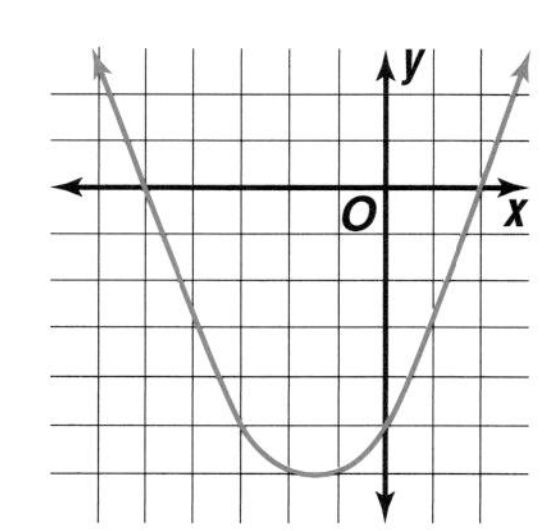

Mixed Review

Solve each equation by graphing the related function. *(Lesson 11–3)*

36. $x^2 - 10x + 21 = 0$ **37.** $x^2 - 2x + 5 = 0$ **38.** $x^2 + 4x = 12$

39. Graph $y = 0.25x^2$, $y = x^2$, and $y = 4x^2$ on the same set of axes. Compare and contrast the graphs. *(Lesson 11–2)*

Factor. *(Lesson 10–3)*

40. $y^2 + 7y + 6$ **41.** $b^2 + b - 42$

Standardized Test Practice

42. Multiple Choice Which of the following expressions is equivalent to $2^3 \times 4$? *(Lesson 8–2)*

A 8^4 **B** 8^3 **C** 6^4 **D** 2^6 **E** 2^5

11–5 Solving Quadratic Equations by Completing the Square

What You'll Learn

You'll learn to solve quadratic equations by completing the square.

Why It's Important

Construction You can use quadratic equations to determine the dimensions of a room.
See Exercise 13.

Carmen planted daylilies on a square piece of land with an area of 49 square feet. She wants to plant petunias around the daylilies to form a border whose area is 72 square feet. What length should she make the sides of the outer square?

Let x represent the length of a side of the outer square. Find a relationship between the variable and given numerical information to write and solve an equation.

Total area	*minus*	*the daylily area*	*is*	*the petunia area.*
x^2	$-$	49	$=$	72

$x^2 = 121$ *Add 49 to each side.*

$\sqrt{x^2} = \pm\sqrt{121}$ *Find the square root of each side.*

$x = \pm 11$

Length cannot be negative, so the sides of the outer square measure 11 feet.

Look Back

Perfect Square Trinomials: Lesson 10–5

We were able to solve the problem easily because 121 is a perfect square. However, few quadratic expressions are perfect squares. To make *any* quadratic expression a perfect square, a method called **completing the square** can be used.

When given an equation, you can complete the square by writing one side of the equation as a perfect square. This is modeled below.

Hands-On Algebra
Algebra Tiles

Materials: algebra tiles, equation mat

Use algebra tiles to complete the square for $x^2 + 2x - 4 = 0$.

Step 1 Model $x^2 + 2x - 4 = 0$ on the mat.

$x^2 + 2x - 4 = 0$

Step 2 Add 4 to each side of the mat. Remove the zero pairs.

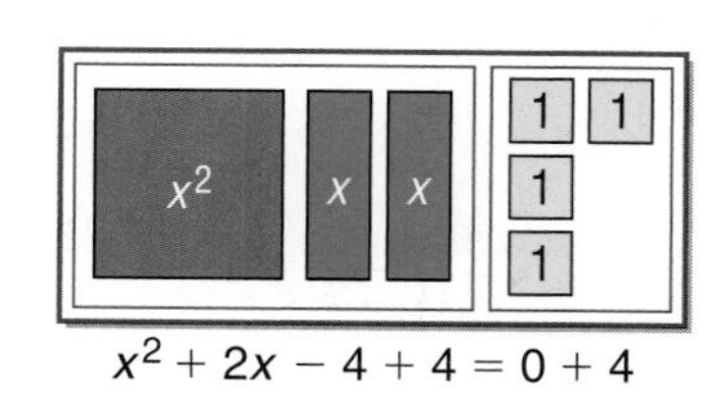

$x^2 + 2x - 4 + 4 = 0 + 4$

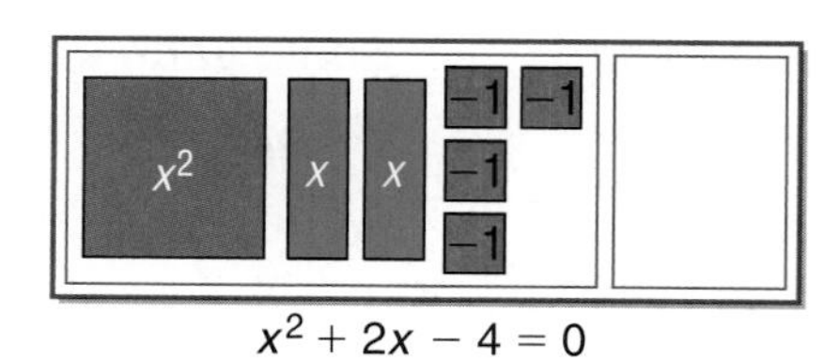

Step 3 Begin to arrange the x^2 and x-tiles into a square.

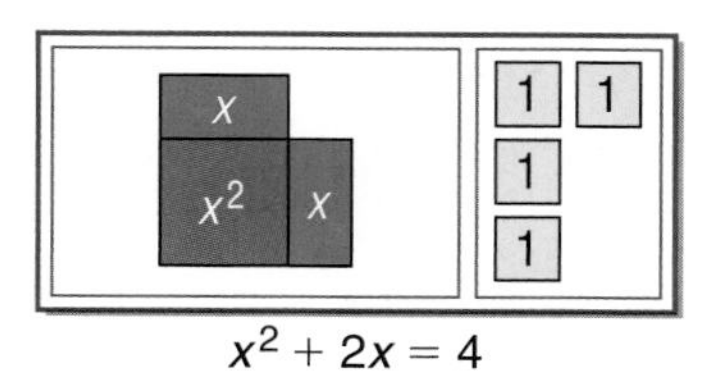

$x^2 + 2x = 4$

Step 4 To complete the square, add 1 yellow tile to each side of the mat. So, the equation is $x^2 + 2x + 1 = 5$ or $(x + 1)^2 = 5$.

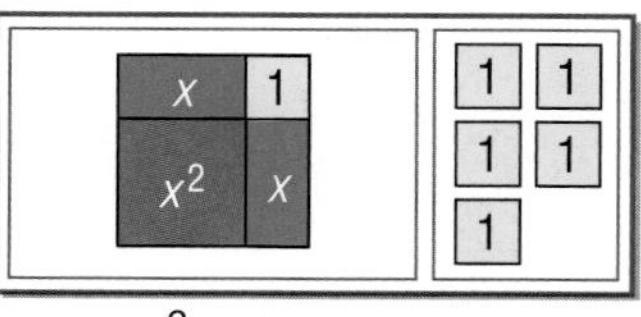

$x^2 + 2x + 1 = 5$

Try These

Use algebra tiles to complete the square of each equation.

1. $x^2 + 4x + 1 = 0$

2. $x^2 - 6x = -5$

To complete the square for any quadratic expression of the form $x^2 + bx$, you can follow the steps below.

Step 1 Take $\frac{1}{2}$ of b, the coefficient of x.

Step 2 Square the result of Step 1.

Step 3 Add the result of Step 2, $\left(\frac{b}{2}\right)^2$, to $x^2 + bx$.

Example 1

Find the value of c that makes $x^2 + 14x + c$ a perfect square.

Step 1 Find one half of 14. $\quad \frac{14}{2} = 7$

Step 2 Square the result of Step 1. $\quad 7^2 = 49$

Step 3 Add the result of Step 2 to $x^2 + 14x$. $\quad x^2 + 14x + 49$

Thus, $c = 49$. Notice that $x^2 + 14x + 49 = (x + 7)^2$.

Your Turn

a. Find the value of c that makes $x^2 - 6x + c$ a perfect square.

Once a perfect square is found, we can solve the equation by taking the square root of each side.

Example 2

Solve $x^2 + 12x - 13 = 0$ by completing the square.

Reading Algebra

Read $\pm\sqrt{49}$ as *plus or minus the square root of 49.*

$x^2 + 12x - 13 = 0$	*$x^2 + 12x - 13$ is not a perfect square.*
$x^2 + 12x = 13$	*Add 13 to each side.*
$x^2 + 12x + 36 = 13 + 36$	*Since $\left(\frac{12}{2}\right)^2 = 36$, add 36 to each side.*
$(x + 6)^2 = 49$	*Factor $x^2 + 12x + 36$.*
$\sqrt{(x + 6)^2} = \pm\sqrt{49}$	*Take the square root of each side.*

(continued on the next page)

$$x + 6 = \pm 7$$

$$x + 6 - 6 = \pm 7 - 6 \quad \textit{Subtract 6 from each side.}$$

$$x = 7 - 6 \quad \text{or} \quad x = -7 - 6 \quad \textit{Simplify each equation.}$$

$$x = 1 \qquad\qquad x = -13$$

The solutions are -13 and 1. *Check the solution by using the graphing calculator or by substituting -13 and 1 for x in the original equation.*

Your Turn

b. Solve $x^2 + 6x - 16 = 0$ by completing the square.

You can complete the square only if the coefficient of the first term is 1. If the coefficient is *not* 1, first divide each term by the coefficient.

3 **A school wants to redesign its nurse's station. Because of building codes, the maximum length of the rectangular room is 20 meters less than twice its width. Find the dimensions, to the nearest tenth of a meter, for the widest possible nurse's station if its area is to be 60 square meters.**

Explore Draw a picture of the room.
Let w = the maximum width.
Then $2w - 20$ = the maximum length.

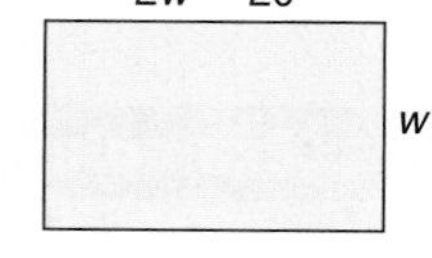

Plan

$$\underbrace{\textit{Area}}_{60} \quad \underbrace{\textit{equals}}_{=} \quad \underbrace{\textit{length}}_{2w - 20} \quad \underbrace{\textit{times}}_{\cdot} \quad \underbrace{\textit{width.}}_{w}$$

Solve

$$60 = w(2w - 20)$$

$$60 = 2w^2 - 20w \quad \textit{Distributive Property}$$

$$30 = w^2 - 10w \quad \textit{Divide each side by 2.}$$

$$30 + 25 = w^2 - 10w + 25 \quad \left(\frac{-10}{2}\right)^2 = 25. \textit{ Add 25 to each side.}$$

$$55 = (w - 5)^2 \quad \textit{Factor } w^2 - 10w + 25.$$

$$\sqrt{55} = \sqrt{(w - 5)^2} \quad \textit{Take the square root of each side.}$$

$$\pm\sqrt{55} = w - 5$$

$$5 \pm \sqrt{55} = w - 5 + 5 \quad \textit{Add 5 to each side.}$$

$$5 \pm \sqrt{55} = w$$

The roots are $5 + \sqrt{55}$ and $5 - \sqrt{55}$. You can use a calculator to find decimal approximations for these numbers.

$$5 + \sqrt{55} \approx 12.4 \quad \text{and} \quad 5 - \sqrt{55} \approx -2.4$$

The solution of -2.4 is not reasonable since there cannot be negative width. The dimensions of the nurse's station should be about 12.4 meters by 2(12.4) − 20 or 4.8 meters.

Examine Since 12.4(4.8) = 59.52, the answer seems reasonable.

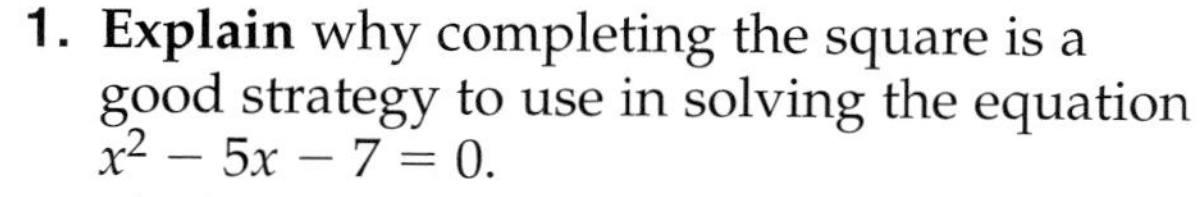

Check for Understanding

Communicating Mathematics

1. **Explain** why completing the square is a good strategy to use in solving the equation $x^2 - 5x - 7 = 0$.

Vocabulary

completing the square

Math Journal

2. **List** the steps necessary to complete the square of the expression $x^2 - 10x$.

Guided Practice

Getting Ready **State whether each trinomial is a perfect square.**

Sample 1: $x^2 - 4x - 4$
Solution: No, it is not factorable.

Sample 2: $x^2 + 6x + 9$
Solution: Yes, $x^2 + 6x + 9 = (x + 3)^2$.

3. $x^2 - 2x + 1$
4. $x^2 - 18x - 9$
5. $x^2 + 7x - 14$
6. $x^2 + 8x + 16$

Find the value of c that makes each trinomial a perfect square. *(Example 1)*

7. $x^2 + 2x + c$
8. $z^2 - 18z + c$
9. $m^2 + 3m + c$

Solve each equation by completing the square. *(Example 2)*

10. $p^2 - 6p = 0$
11. $r^2 + 7r + 12 = 0$
12. $x^2 - 4x - 9 = 0$

13. **Construction** The Thompsons' rectangular bathroom measures 6 feet by 9 feet. They want to double the area of the bathroom by increasing the length and width by the same amount. Find the dimensions, to the nearest foot, of the new bathroom. *(Example 3)*

Exercises

Practice

Find the value of c that makes each trinomial a perfect square.

14. $z^2 - 10z + c$
15. $f^2 + 12f + c$
16. $h^2 - 20h + c$
17. $s^2 - 2s + c$
18. $q^2 + 14q + c$
19. $x^2 - 24x + c$
20. $z^2 + 16z + c$
21. $k^2 + 5k + c$
22. $r^2 + 9r + c$

Homework Help

For Exercises	See Examples
14–22	1
23–34, 35	2
36	3

Extra Practice

See page 714.

Solve each equation by completing the square.

23. $x^2 + 2x - 3 = 0$
24. $x^2 - 8x + 7 = 0$
25. $p^2 + 6p = -5$
26. $r^2 - 16r = 0$
27. $s^2 - 14s = 0$
28. $x(x + 10) - 1 = 0$
29. $q^2 - 2q = 30$
30. $4x^2 + 16x + 24 = 0$
31. $3z^2 - 18z = 30$
32. $m^2 - 2m = 6$
33. $d^2 + 3d - 10 = 0$
34. $a^2 - \frac{7}{2}a + \frac{3}{2} = 0$

Applications and Problem Solving

35. **Physical Science** Omar set off a rocket in a field. The height h of the rocket in feet can be roughly estimated using the formula $h = -16t^2 + 128t$, where t is the time in seconds. How long will it take the rocket to reach a height of 240 feet?

36. **Photography** Sheila places a 54 square inch photo behind a 12-inch-by-12-inch piece of matting. The photograph is positioned so that the matting is twice as wide at the top and bottom as the sides.

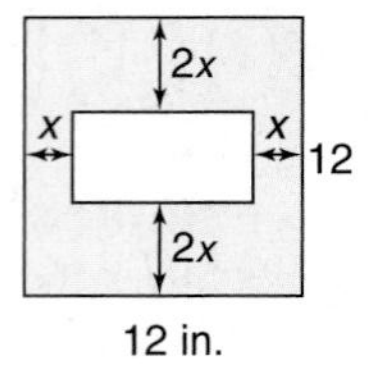

a. Write an equation for the area of the photo in terms of x.

b. Find the dimensions of the photo.

37. **Critical Thinking** Find the values of b that make $x^2 + bx + 81$ a perfect square.

Mixed Review

Solve each quadratic equation by factoring. *(Lesson 11–4)*

38. $a^2 + 3a - 28 = 0$

39. $x^2 - 7x + 10 = 0$

40. $z^2 + 5z = -4$

41. $2b^2 + 12b + 18 = 0$

42. **Rockets** Jon is launching rockets in an open field. The path of the rocket can be modeled by the quadratic function $h(t) = -16t^2 + 96t$, where $h(t)$ is the height in feet any time t in seconds. *(Lesson 11–3)*

a. Graph the quadratic function.

b. After how many seconds will the rocket reach a height of 80 feet for the *second* time?

c. After how many seconds will the rocket hit the ground?

d. What is the rocket's maximum height?

Standardized Test Practice

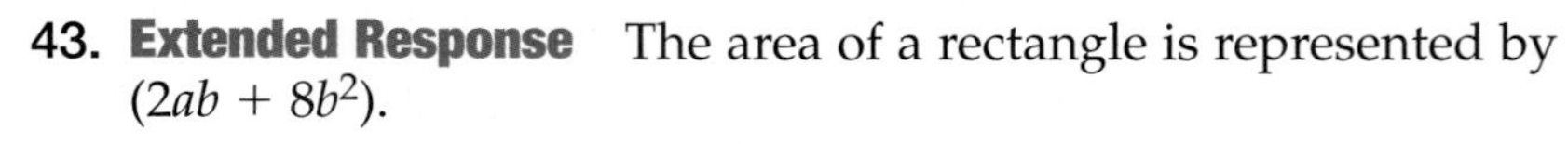

43. **Extended Response** The area of a rectangle is represented by $(2ab + 8b^2)$.

a. Factor the expression to find the dimensions of the rectangle. *(Lesson 10–2)*

b. Using the dimensions from part a, write an expression for the perimeter of the rectangle in simplest form. *(Lessons 9–2 & 9–3)*

c. Find the dimensions, area, and perimeter of the rectangle if $a = 5$ inches and $b = 1$ inch. *(Lesson 1–2)*

44. **Multiple Choice** The graph shows how each Ohio tax dollar was divided among programs and services in a recent year. If Maria paid $32 in state taxes on her last paycheck, what amount of her money went towards education? *(Lessons 5–3 & 5–4)*

A $5.12 B $10.24

C $15.36 D $24.96

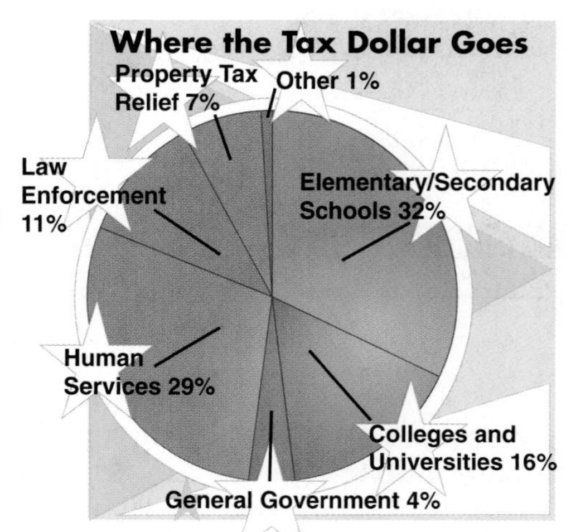

Source: Ohio Dept. of Taxation

11–6 The Quadratic Formula

What You'll Learn

You'll learn to solve quadratic equations by using the Quadratic Formula.

Why It's Important

Science You can use the Quadratic Formula to solve problems dealing with physics. *See Exercise 10.*

The table below summarizes the ways you have learned to solve quadratic equations. Although these methods can be used to solve all quadratic equations, each method is most useful in particular situations.

Method	When Is the Method Useful?
Graphing	Use only to estimate solutions.
Factoring	Use when the quadratic expression is easy to factor.
Completing the Square	Use when the coefficient of x^2 is 1 and all other coefficients are fairly small.

An alternative method is to develop a general formula for solving *any* quadratic equation. Begin with the general form of a quadratic equation, $ax^2 + bx + c = 0$, where $a \neq 0$, and complete the square.

$$ax^2 + bx + c = 0$$

$$x^2 + \frac{b}{a}x + \frac{c}{a} = 0 \quad \textit{Divide by a so the coefficient of } x^2 \textit{ is 1.}$$

$$x^2 + \frac{b}{a}x = -\frac{c}{a} \quad \textit{Subtract } \frac{c}{a} \textit{ from each side.}$$

Now complete the square.

$$x^2 + \frac{b}{a}x + \left(\frac{b}{2a}\right)^2 = -\frac{c}{a} + \left(\frac{b}{2a}\right)^2 \quad \frac{1}{2} \cdot \frac{b}{a} = \frac{b}{2a}$$

$$\left(x + \frac{b}{2a}\right)^2 = -\frac{c}{a} + \frac{b^2}{4a^2} \quad \textit{Factor the left side of the equation.}$$

$$\left(x + \frac{b}{2a}\right)^2 = \frac{4a}{4a}\left(-\frac{c}{a}\right) + \frac{b^2}{4a^2} \quad \textit{The common denominator is } 4a^2.$$

$$\left(x + \frac{b}{2a}\right)^2 = \frac{b^2 - 4ac}{4a^2} \quad \textit{Simplify the right side of the equation.}$$

Finally, take the square root of each side and solve for x.

$$\sqrt{\left(x + \frac{b}{2a}\right)^2} = \pm\sqrt{\frac{b^2 - 4ac}{4a^2}} \quad \textit{Take the square root of each side.}$$

$$x + \frac{b}{2a} = \frac{\pm\sqrt{b^2 - 4ac}}{2a} \quad \textit{Simplify: } \sqrt{4a^2} = 2a.$$

$$x = \frac{\pm\sqrt{b^2 - 4ac}}{2a} - \frac{b}{2a} \quad \textit{Subtract } \frac{b}{2a} \textit{ from each side.}$$

$$x = \frac{-b \pm \sqrt{b^2 - 4ac}}{2a} \quad \textit{Combine the terms.}$$

The result is called the **Quadratic Formula** and can be used to solve *any* quadratic equation.

The Quadratic Formula	$x = \frac{-b \pm \sqrt{b^2 - 4ac}}{2a}, a \neq 0$

Examples

Use the Quadratic Formula to solve each equation.

1 $x^2 - 4x + 3 = 0$

$$x = \frac{-b \pm \sqrt{b^2 - 4ac}}{2a}$$

$$= \frac{-(-4) \pm \sqrt{(-4)^2 - 4(1)(3)}}{2(1)} \quad a = 1, b = -4, \text{ and } c = 3$$

$$= \frac{4 \pm \sqrt{16 - 12}}{2} \quad (-4)^2 = 16 \text{ and } 4(1)(3) = 12$$

$$= \frac{4 \pm \sqrt{4}}{2}$$

$$= \frac{4 \pm 2}{2}$$

$$= \frac{4 + 2}{2} \quad \text{or} \quad x = \frac{4 - 2}{2}$$

$$= \frac{6}{2} \text{ or } 3 \qquad x = \frac{2}{2} \text{ or } 1$$

Check: Substitute values into the original equation.

$$x^2 - 4x + 3 = 0 \quad \text{or} \quad x^2 - 4x + 3 = 0$$
$$3^2 - 4(3) + 3 \stackrel{?}{=} 0 \qquad 1^2 - 4(1) + 3 \stackrel{?}{=} 0$$
$$9 - 12 + 3 \stackrel{?}{=} 0 \qquad 1 - 4 + 3 \stackrel{?}{=} 0$$
$$0 = 0 \ \checkmark \qquad 0 = 0 \ \checkmark$$

The roots are 3 and 1.

2 $-2x^2 + 3x - 5 = 0$

$$x = \frac{-b \pm \sqrt{b^2 - 4ac}}{2a}$$

$$= \frac{-3 \pm \sqrt{3^2 - 4(-2)(-5)}}{2(-2)} \quad a = -2, b = 3, \text{ and } c = -5$$

$$= \frac{-3 \pm \sqrt{9 - 40}}{-4} \quad 3^2 = 9 \text{ and } 4(-2)(-5) = 40$$

$$= \frac{-3 \pm \sqrt{-31}}{-4}$$

The square root of a negative number, such as $\sqrt{-31}$, is not a real number. So, there are no real solutions for x.

Check: Use a graphing calculator. It is clear that the graph of $f(x) = -2x^2 + 3x - 5$ never crosses the x-axis. Therefore, there are no real roots for the equation $-2x^2 + 3x - 5 = 0$.

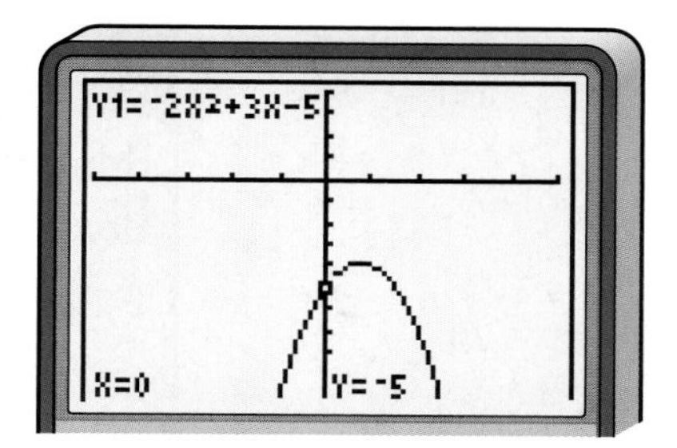

Your Turn

a. $-3x^2 + 6x + 9 = 0$

b. $x^2 + 4x + 2 = 0$

When the solutions are irrational numbers, use a calculator to estimate.

Example 3
Sports Link

"Hang time" is the total amount of time a ball stays in the air. Manuel can kick a football with an upward velocity of 64 ft/s. His foot meets the ball 2 feet off the ground. What is Manuel's hang time if the ball is not caught? Use the formula $h(t) = -16t^2 + 64t + 2$, where $h(t)$ is the ball's height in feet for any time t, in seconds, after the ball is kicked.

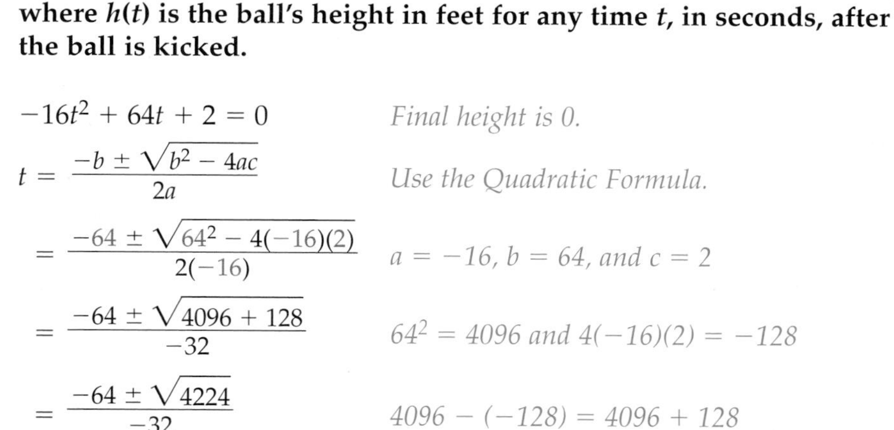

$-16t^2 + 64t + 2 = 0$ *Final height is 0.*

$t = \dfrac{-b \pm \sqrt{b^2 - 4ac}}{2a}$ *Use the Quadratic Formula.*

$= \dfrac{-64 \pm \sqrt{64^2 - 4(-16)(2)}}{2(-16)}$ *$a = -16$, $b = 64$, and $c = 2$*

$= \dfrac{-64 \pm \sqrt{4096 + 128}}{-32}$ *$64^2 = 4096$ and $4(-16)(2) = -128$*

$= \dfrac{-64 \pm \sqrt{4224}}{-32}$ *$4096 - (-128) = 4096 + 128$*

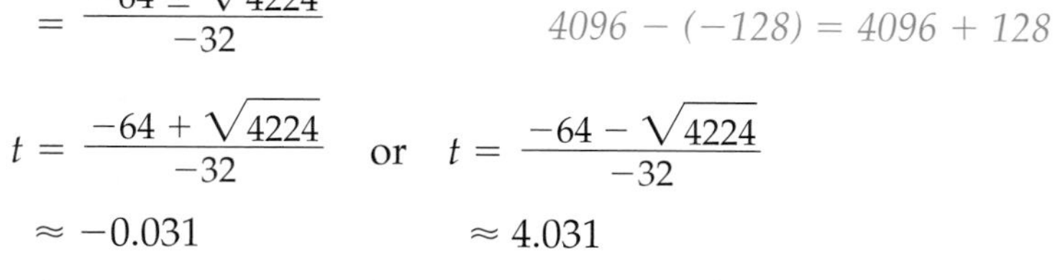

$t = \dfrac{-64 + \sqrt{4224}}{-32}$ or $t = \dfrac{-64 - \sqrt{4224}}{-32}$

≈ -0.031 ≈ 4.031

Since time cannot be negative, the only approximate solution is 4.031. The football has a hang time of about 4 seconds.

Check for Understanding

Communicating Mathematics

Vocabulary
Quadratic Formula

1. **Choose** the correct term and justify your answer. Quadratic equations (*sometimes, always, never*) have at least one real solution.

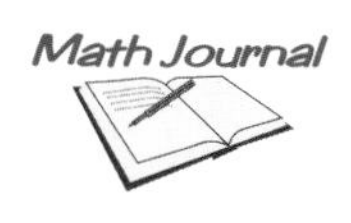

2. **List** the steps you take when you use the Quadratic Formula to solve a quadratic equation.

Guided Practice

Getting Ready **Find the value of $b^2 - 4ac$ for each equation.**

Sample 1: $-3z^2 + 8z + 5 = 0$
Solution: $8^2 - 4(-3)(5)$
$= 64 + 60$
$= 124$

Sample 2: $6a^2 = 3$
Solution: Rewrite $6a^2 = 3$ as
$6a^2 - 3 = 0.$
$0^2 - 4(6)(-3) = 72$

3. $x^2 + 10x + 13 = 0$
4. $3x^2 - 9x + 2 = 0$
5. $4x^2 = 12$
6. $-2x^2 - 7x = -5$

Use the Quadratic Formula to solve each equation. *(Examples 1 & 2)*

7. $z^2 - 25 = 0$
8. $r^2 - 5r + 7 = 0$
9. $d^2 + 4d = -3$

10. **Physical Science** Josefina tosses a ball upward with an initial velocity of 76 feet per second. She tosses it from an initial height of 2 feet. Find the approximate time that the ball hits the ground. Use the function $h(t) = -16t^2 + 76t + 2$. *(Example 3)*

Exercises

Practice

Use the Quadratic Formula to solve each equation.

11. $x^2 + 7x + 6 = 0$
12. $r^2 + 10r + 9 = 0$
13. $-a^2 + 5a - 6 = 0$
14. $-b^2 + 6b + 7 = 0$
15. $9d^2 - 4d + 3 = 0$
16. $3v^2 + 5v - 8 = 0$
17. $x^2 - 8x = 0$
18. $-p^2 + 6p - 5 = 0$
19. $w^2 + w - 7 = 0$
20. $z^2 + 9z - 2 = 0$
21. $4g^2 + 8g - 3 = 0$
22. $-2t^2 + 4t = 3$
23. $2c^2 + 14c = 10$
24. $5z^2 + 12 = -6z$
25. $8k^2 - 5k = 7$

Homework Help

For Exercises	See Examples
11–25	1, 2
26–27	3

Extra Practice
See page 715.

Applications and Problem Solving

26. **Skiing** Brandi, an amateur skier, begins on the "bunny hill," which has a vertical drop of 200 feet. Her speed at the bottom of the hill is given by the equation $v^2 = 64h$, where the velocity v is in feet per second and the height h of the hill is in feet. Assuming there is no friction, approximate Brandi's velocity at the bottom of the hill.

27. **Space** Suppose an astronaut can jump vertically with an initial velocity of 5 m/s. The time that it takes him to touch the ground is given by the equation $0 = 5t - 0.5at^2$. The time t is in seconds and the acceleration due to gravity a is in m/s^2.

a. If the astronaut jumps on the moon where $a = 1.6\ m/s^2$, how long will it take him to reach the ground?

b. How long will it take him to reach the ground if he jumps on Earth where $a = 9.8\ m/s^2$?

28. **Critical Thinking** The expression $b^2 - 4ac$ is called the **discriminant** of a quadratic equation. It determines the number of real solutions you can expect when solving a quadratic equation.

a. Copy and complete the table below.

Equation	$x^2 - 4x + 1 = 0$	$x^2 + 6x + 11 = 0$	$x^2 - 4x + 4 = 0$
Value of the Discriminant			
Number of x-intercepts			
Number of Real Solutions			

b. Use the table above to describe the discriminant of a quadratic equation for each type of real solution.

i. two solutions ii. one solution iii. no solutions

Mixed Review

Find the value of c to make a perfect square. *(Lesson 11–5)*

29. $a^2 + 4a + c$ 30. $x^2 + 8x + c$ 31. $v^2 - 18v + c$

32. **Geometry** The area of a circle varies directly as the square of the radius r. If the radius of a circle is doubled, how many times as large is the area of the circle? *(Lesson 6–5)*

Standardized Test Practice

33. **Multiple Choice** If four times a number is equal to that number decreased by 5, what is the number? *(Lesson 4–6)*

A -5 B $-\frac{5}{3}$ C $-\frac{1}{4}$ D $\frac{1}{3}$ E 4

Quiz 2 Lessons 11–4 through 11–6

Solve each quadratic equation by factoring. *(Lesson 11–4)*

1. $x^2 + 4x - 12 = 0$ 2. $x^2 - 8x + 16 = 0$ 3. $x^2 - 10x = -21$

Solve each quadratic equation by completing the square. *(Lesson 11–5)*

4. $x^2 + 4x + 3 = 0$ 5. $x^2 - 6x + 7 = 0$ 6. $2x^2 - 8x = 4$

Solve each equation by using the Quadratic Formula. *(Lesson 11–6)*

7. $x^2 + 3x + 3 = 0$ 8. $x^2 + 8x + 15 = 0$ 9. $x^2 + 5x + 3 = 0$

10. **Construction** A company is designing a parabolic arch bridge to span a creek. The height of the bridge is given by the equation $h = -0.02d^2 + 0.8d$, where the height h and the distance d from one end of the base are in feet. Find the width of the bridge. *(Lesson 11–6)*

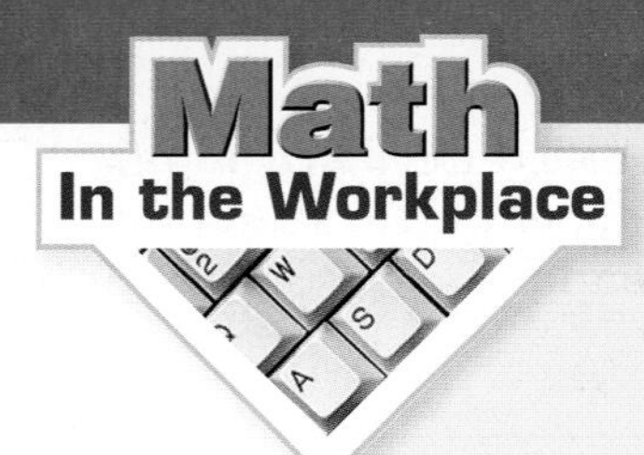

Civil Engineer

Civil engineers work in the oldest branch of engineering. They design and supervise the construction of buildings, bridges, roads, and sewer systems. One of the most innovative structures in bridges, the parabolic arch, was first used extensively by the Romans. Civil engineers continue to use the parabolic arch today for its simple beauty.

A parabolic arch supports the Hulme Arch Bridge in England. The Hulme Arch can be modeled by the quadratic function $h(x) = -0.037x^2 + 25$, where $h(x)$ represents the height of the arch at any point x to the left or right of the center. All measures are in meters.

1. Use the Quadratic Formula to approximate the zeros of the quadratic function $h(x) = -0.037x^2 + 25$ to the nearest whole number.
2. What do the zeros represent on the Hulme Arch?
3. Find the width of the Hulme Arch to the nearest meter.
4. What is the height of the Hulme Arch?
5. Graph the Hulme Arch on a coordinate plane.

FAST FACTS About Civil Engineers

Working Conditions

- usually work in office buildings, industrial plants, or outside on construction sites
- 40-hour work weeks, except around deadlines when more hours are expected

Education

- college degree in engineering, physical science, or mathematics
- creative and detail-oriented personalities are well-suited for the profession

Earnings

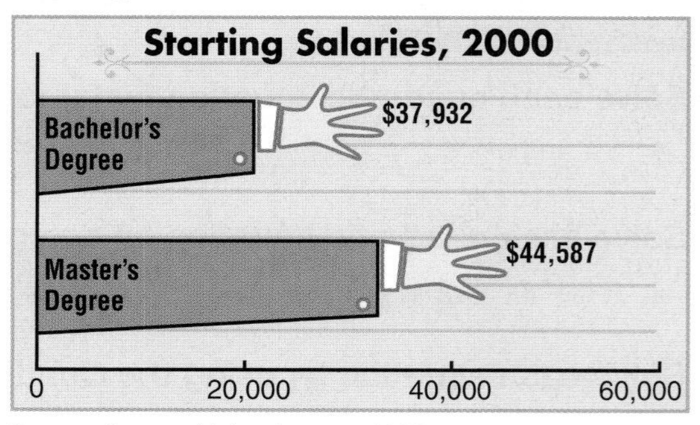

Source: Bureau of Labor Statistics, 2001

Career Data For the latest information on civil engineers, visit:
www.algconcepts.com

11–7 Exponential Functions

What You'll Learn

You'll learn to graph exponential functions.

Why It's Important

Finance You can use exponential functions to study the growth of bank accounts and populations. *See Example 4.*

Often, patterns can be used to describe and classify types of functions.

Hands-On Algebra

Materials: large sheet of paper

Step 1 Fold a large sheet of paper in half. Unfold it and record how many sections are formed by the creases. Refold the paper.

Step 2 Fold the paper in half again. Record how many sections are formed by the creases. Refold the paper.

Step 3 Continue folding in half and recording the number of sections until you can no longer fold the paper.

Try These

1. How many folds could you make?
2. How many sections were formed?

The data collected in the Hands-On Algebra activity can be represented in a table to see a pattern more easily. Each fold increases the number of sections by a factor of 2.

Folds	Sections	Pattern
1	2	$2 = 2^1$
2	4	$2\ (2) = 2^2$
3	8	$2(2)(2) = 2^3$
4	16	$2(2)(2)(2) = 2^4$
5	32	$2(2)(2)(2)(2) = 2^5$

Let x equal the number of folds and let y equal the number of sections. Then the function $y = 2^x$ represents the number of sections for any number of folds. This is an example of an **exponential function**. An exponential function is of the form $y = a^x$, where $a > 0$ and $a \neq 1$. Some people also refer to equations of the form $y = ab^x + c$ as exponential functions. For this form of the exponential function, the value a is called the *coefficient*.

You can make a table to help graph exponential functions.

Example 1 Graph $y = 2^x$.

x	2^x	y
−1	2^{-1}	0.5
0	2^0	1
1	2^1	2
2	2^2	4
3	2^3	8

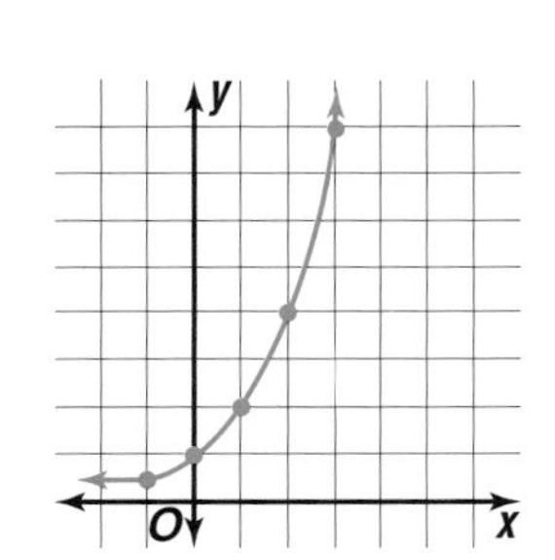

The graph of an exponential function changes little for small x values. However, as the values of x increase, the y values increase quickly.

The **initial value** of an exponential function is the value of the function when $x = 0$. This is the same as the y-intercept. Exponential functions of the form $y = a^x$ have an initial value of 1. Exponential functions of the form $y = ab^x + c$ have an initial value of c.

Examples

Graph each exponential function. Then state the y-intercept.

2 $y = 5^x$

x	5^x	y
−2	5^{-2}	0.04
−1	5^{-1}	0.2
0	5^0	1
1	5^1	5
2	5^2	25

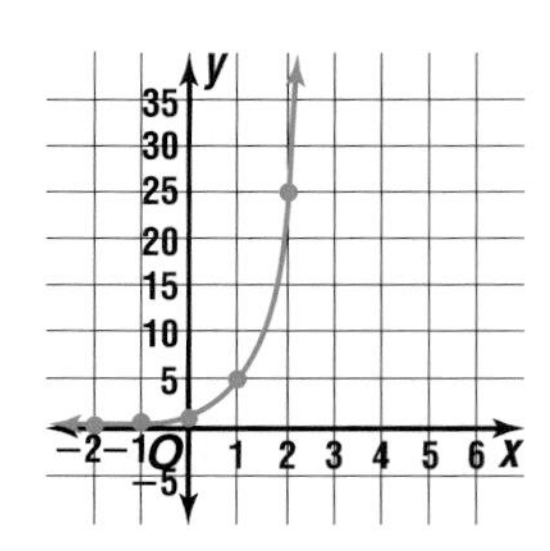

To find the y-intercept, let $x = 0$ and solve for y.

$$y = 5^0 \text{ or } 1$$

In this case, the y-intercept is 1.

3 $y = 2^x + 3$

Reading Algebra

Read $2^x + 3$ as *two to the x plus three.*

x	$2^x + 3$	y
−2	$2^{-2} + 3$	3.25
−1	$2^{-1} + 3$	3.5
0	$2^0 + 3$	4
1	$2^1 + 3$	5
2	$2^2 + 3$	7
3	$2^3 + 3$	11

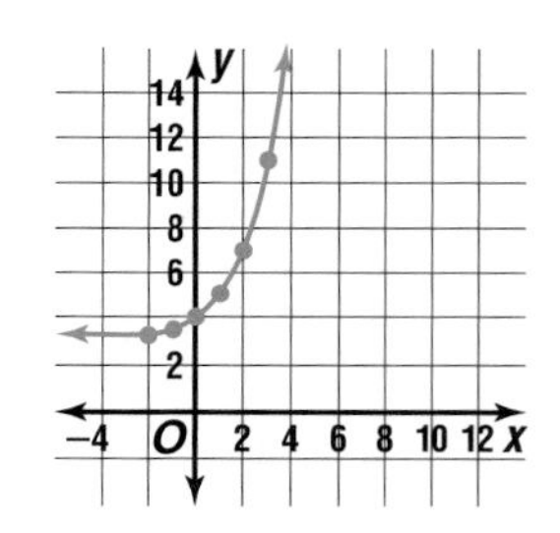

The y-intercept is 4. *The constant is 3: 1 + 3 = 4.*

Your Turn **Graph each function. Then state the y-intercept.**

a. $y = 2.5^x$

b. $y = 4^x + 3$

Quantities that increase rapidly have *exponential growth.* For instance, money in the bank may grow at an exponential rate.

Real World Example **Finance Link**

4 **When Taina was 10 years old, she received a certificate of deposit (CD) for $2000 with an annual interest rate of 5%. After eight years, how much money will she have in the account?**

After the first year, the CD is worth the initial deposit plus interest.

$2000 + 2000(0.05) = 2000(1 + 0.05)$ *Distributive Property*

$= 2000(1.05)$

Taina can make a table to look for patterns in the growth of her CD.

Year	Balance	Pattern
0	2000	$2000(1.05)^0$
1	2000(**1.05**) = 2100	$2000(1.05)^1$
2	(2000 · **1.05**)**1.05** = 2205.00	$2000(1.05)^2$
3	(2000 · **1.05** · **1.05**)**1.05** = 2315.25	$2000(1.05)^3$

Let x represent the number of years. Then the function that represents the balance in Taina's CD is $B(x) = 2000(1.05)^x$. After eight years, it will have a balance of $2954.91.

For exponential data, you can create a best-fit curve that passes through most of the data points. You can then use this curve to make predictions.

Graphing Calculator Tutorial
See pp. 724–727.

Graphing Calculator Exploration

A city Web site tracks the number of "hits". The table shows the average number of hits per week since the launch.

Years After Launch	"Hits" per Week
4	10,000
5	20,000
6	37,500
7	62,500
8	190,000

Step 1 Use the Edit option in the STAT menu to enter the year data in L1 and the hits data in L2.

Step 2 Find an equation for a best-fit curve.

Enter: [STAT] [▶] 0 [2nd] [L1] [,] [2nd] [L2] [,] [VARS] [▶] [ENTER] [ENTER] [ENTER]

The screen displays the equation $y = a \cdot b^x$ and gives values for a and b.

Step 3 The exponential equation has been stored in the [Y=] menu. To see a graph of the function in the proper window, press [ZOOM] 9.

Make sure that the STATPLOT is turned on.

Try This

Use the graph of the best-fit curve to predict the number of weekly hits during the 15th year after the launch.

Check for Understanding

Communicating Mathematics

Vocabulary
exponential function
initial value

1. **Describe** the general shape of an exponential function.
2. **Explain** why the y-intercept of an exponential function in the form $y = a^x$ is always 1.

Guided Practice

Getting Ready Use a calculator to find each value to the nearest hundredth.

Sample 1: 3^{-4}
Solution:
3 [^] [(–)] 4 [ENTER] 0.01

Sample 2: 1.25^5
Solution:
1.25 [^] 5 [ENTER] 3.05

3. 2^8
4. 4^{-1}
5. 1.08^3

Graph each exponential function. Then state the *y*-intercept. *(Examples 1–3)*

6. $y = 2^x + 2$
7. $y = 2^x - 6$
8. $y = 3^x + 5$

9. **Finance** When Lindsay was born, her parents opened a savings account for her with \$500. The account pays an annual interest of 2%. *(Example 4)*
   a. Make a table showing the account's growth for 3 years.
   b. Write a function that represents the amount of money in Lindsay's account after x number of years.
   c. How much money will be in Lindsay's account after 18 years?
   d. If the initial deposit had been \$1000, how much would Lindsay's account be worth after 18 years?

Exercises

Practice

Homework Help	
For Exercises	See Examples
10, 11, 16, 18, 25, 26	1, 2
12–15, 17	3
19–21, 24	4
Extra Practice	
See page 715.	

Graph each exponential function. Then state the *y*-intercept.

10. $y = 3^x$
11. $y = 4^x$
12. $y = 2^x - 5$
13. $y = 2^x + 1$
14. $y = 3^x + 2$
15. $y = 4^x - 3$
16. $y = 2^{2x}$
17. $y = 3^{2x} - 1$
18. $y = 2^{0.5x}$

Find the amount of money in a bank account given the following conditions.

19. initial deposit = \$5000, annual rate = 3%, time = 2 years
20. initial deposit = \$1500, annual rate = 10%, time = 10 years
21. initial deposit = \$3000, annual rate = 5.5%, time = 5 years

Applications and Problem Solving

22. **Taxes** In a recent year, the average tax refund had risen to about \$1700. The table shows the nest egg you would make for yourself by investing that refund at a 10% interest rate. Find your savings after 30 years.

Compounding Tax Refunds

Year	Savings
0	1700
1	1700(1.1)
2	1700(1.1)(1.1)

Source: National Tax Services

23. **Finance** Nuno bought a new car for \$18,000. Once the car is driven off the lot, the car depreciates in value 15% each year. How much will the car be worth in 5 years?

Downtown Atlanta

24. **Population** The Atlanta, Georgia, metropolitan area experienced population growth throughout the 1990s. If the population in 1990 was 2,960,000 and the annual rate of increase was 3%, predict the population in the year 2002. Use the function $P(t) = 2{,}960{,}000(1.03)^t$, where t is the number of years since 1990.

25. **Money** Suppose on the first day of the year, your parents offered you $500 for the month of January. If you did not take the $500, they would give you a penny on the first day and then continue to double the amount each following day. Which is the better deal?
   a. Copy and complete the table.
   b. Write a function that represents the amount of money you will receive after x days.
   c. Use the equation in part b to find the day when you receive a payment of more than $500.

Day	Cents	Pattern
1	1	
2	2 · 1	
3	2 · 2 · 1	
4		
5		

26. **Critical Thinking** Make a graph of each exponential function listed below. Then describe the shape of an exponential function when a number less than 1 is raised to the x power compared to a number greater than 1.
   a. $y = 2^x$ b. $y = 1.5^x$ c. $y = 0.5^x$ d. $y = 0.25^x$

Mixed Review

Use the Quadratic Formula to solve each equation. *(Lesson 11–6)*

27. $3p^2 - 12 = 0$

28. $x^2 + 4x = 10$

29. **Flooring** There are 30 square yards in a roll of carpet. If the carpet is unrolled, it forms a rectangle. The length is four yards more than twice its width. *(Lesson 11–5)*
   a. Write a quadratic equation to represent the carpet's area.
   b. Find the dimensions.

Write the point-slope form of an equation for each line passing through the given point and having the given slope. *(Lesson 7–2)*

30. $(4, -3), 2$

31. $(-2, 6), 0$

Standardized Test Practice

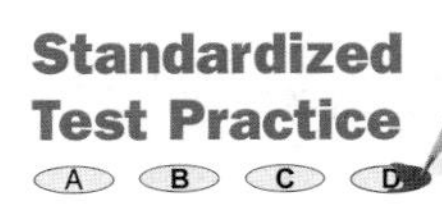

32. **Extended Response** During a road-trip, Javonte recorded the number of miles he had driven by the end of each hour. He graphed the number of miles driven on the number line shown. *(Lesson 7–1)*

Distance (mi)
270
240
210
180
150
120
90
60
30
O 1 2 3 4 5 6
Time (hr)

   a. Find the slope of the line between hours 0 and 2.
   b. Find the slope of the line between hours 2 and 3.
   c. Find the slope of the line between hours 3 and 6.
   d. What does the slope represent? (*Hint:* Look at the units.)

Chapter 11 Investigation

Sticky Note Sequences

Materials:

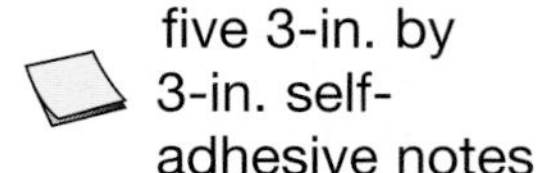

five 3-in. by 3-in. self-adhesive notes

scissors

Geometric Sequences

In the Chapter 3 Investigation, you examined arithmetic sequences. In this investigation, you will look at a different type of sequence.

Investigate

1. Use a self-adhesive note to represent 1 whole unit. Label it "1." Cut a second self-adhesive note in half to represent one half of a unit. Label one of the resulting pieces "$\frac{1}{2}$." Cut the other piece in half again to represent one-fourth of a unit. Label one of the resulting pieces "$\frac{1}{4}$." Continue this process. Arrange the self-adhesive notes as shown below.

1 $\quad \frac{1}{2} \quad \frac{1}{4} \quad \frac{1}{8} \quad \frac{1}{16} \quad \frac{1}{32}$

a. Make a table like the one shown. Write the numbers 1–10 in column 1. Record the fractions from each note in column 2.

b. Find the ratio of the new portion and the previous portion to complete column 3.

Term Number	Portion of Self-Adhesive Note	$\frac{\text{New Portion}}{\text{Previous Portion}}$
1	1	—
2	$\frac{1}{2}$	$\frac{1}{2} \div 1 = ?$
3	$\frac{1}{4}$	$\frac{1}{4} \div \frac{1}{2} = ?$

c. List the numbers in column 2 from greatest to least.

d. The sequence in part c is called a **geometric sequence** because the quotient between any two consecutive terms, called the **common ratio**, is always the same. What is the common ratio?

2. A geometric sequence can be written as a formula in several ways.
   a. Copy and complete a table like the one below for terms 1–10.

Term	Portion of Self-Adhesive Note	Form 1	Form 2	Form 3
1	1	1	1	$1 \cdot \left(\frac{1}{2}\right)^0$
2	$\frac{1}{2}$	$1 \cdot \frac{1}{2}$	$1 \cdot \frac{1}{2}$	$1 \cdot \left(\frac{1}{2}\right)^1$
3	$\frac{1}{4}$	$\frac{1}{2} \cdot \frac{1}{2}$	$1 \cdot \frac{1}{2} \cdot \frac{1}{2}$	$1 \cdot \left(\frac{1}{2}\right)^2$

   b. You can describe Form 1 by using the formula *portion = previous portion times one half*. Another way to write this formula is $f_{n+1} = f_n \times \frac{1}{2}$. f_{n+1} is the value you are trying to find. f_n is the value of the previous term. How could you describe Forms 2 and 3 using a formula like $f_{n+1} = f_n \times \frac{1}{2}$?

3. Some bacteria reproduce exponentially by *fission*. One cell splits to become two cells. There are now two bacteria instead of one.

Term Number	Number of Bacteria	New Number of Bacteria / Previous Number of Bacteria
1	3	—
2	6	$6 \div 3 = ?$

   a. Copy and complete a table like the one at the left for terms 1–10. Use 3 self-adhesive notes to represent 3 bacteria. Cut each in half to produce 6 bacteria. Continue this process.
   b. Write a sequence that models the bacteria's growth. Then write a recursive formula representing the sequence.

Extending the Investigation

In this extension, you will examine other sequences and their formulas.

1. Graph the sequence in Exercise 1. Label the *x*-axis "term number" and the *y*-axis "portion of self-adhesive note." What type of function would describe the graph?
2. Use the sequence in Exercise 1 to find the sum of the first 10 terms. Suppose the sequence continued for 100 terms. What would you expect the sum to be?
3. Repeat 1 and 2 for Exercise 3.

Presenting Your Investigation

Here are some ideas to help you present your conclusions to the class.

- Make a brochure showing the sequences you wrote in this investigation.
- Describe a situation that results in a sequence similar to those in this investigation.

*inter***NET** CONNECTION **Investigation** For more information on geometric sequences, visit: www.algconcepts.com

CHAPTER 11

Study Guide and Assessment

Understanding and Using the Vocabulary

After completing this chapter, you should be able to define each term, property, or phrase and give an example or two of each.

*inter*NET CONNECTION **Review Activities**
For more review activities, visit: www.algconcepts.com

axis of symmetry *(p. 459)*
completing the square *(p. 478)*
discriminant *(p. 487)*
exponential function *(p. 489)*
geometric sequence *(p. 494)*
initial value *(p. 490)*
maximum *(p. 459)*
minimum *(p. 459)*
parabola *(p. 458)*
quadratic equation *(p. 468)*
Quadratic Formula *(p. 484)*
quadratic function *(p. 458)*
roots *(p. 468)*
vertex *(p. 459)*
zeros *(p. 468)*

Choose the letter of the term that best matches each statement or phrase. Each letter is used once.

1. the maximum or minimum point of a parabola
2. Use this to find the exact roots of any quadratic equation.
3. the method of making a quadratic expression into a perfect square
4. the solutions of a quadratic equation
5. the shape of the graph of any quadratic function
6. a vertical line containing the vertex of a parabola
7. $f(x) = a^x$, $a > 0$ and $a \neq 1$
8. $f(x) = ax^2 + bx + c$, $a \neq 0$
9. the x-intercepts of a quadratic function
10. the highest point on the graph of a quadratic function

a. exponential function
b. Quadratic Formula
c. completing the square
d. zeros
e. maximum
f. vertex
g. roots
h. axis of symmetry
i. quadratic function
j. parabola

Skills and Concepts

Objectives and Examples

• **Lesson 11–1** Graph quadratic functions.

Write the equation of the axis of symmetry and the coordinates of the vertex of $f(x) = x^2 - 8x + 12$.

$$x = -\frac{-8}{2(1)} \text{ or } 4$$

Since $4^2 - 8(4) + 12 = -4$, the graph has a vertex at $(4, -4)$.

Review Exercises

Write the equation of the axis of symmetry and the coordinates of the vertex of each quadratic function.

11. $y = x^2 - 3$
12. $y = -x^2 + 4x + 5$
13. $y = 3x^2 + 6x - 17$
14. $y = x^2 - 3x - 4$

Chapter 11 Study Guide and Assessment

Objectives and Examples

• **Lesson 11–1** Graph quadratic functions.

Graph
$f(x) = x^2 - 8x + 12$.

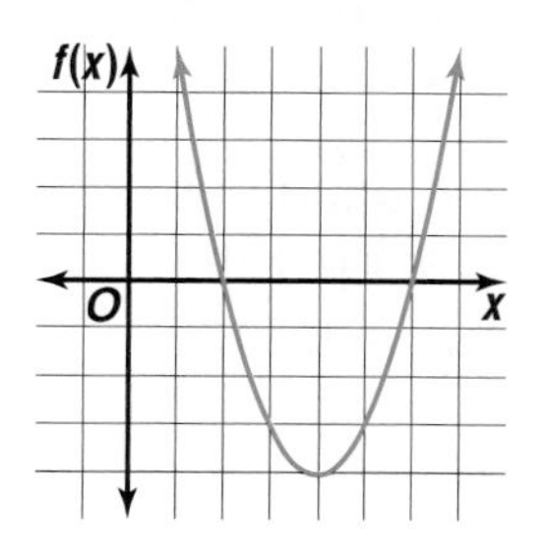

• **Lesson 11–2** Recognize characteristics of families of parabolas.

Describe how the graph of $y = (x + 12)^2 - 5$ changes from the graph of $y = x^2$.

If $x = -12$, then $x + 12 = 0$. The graph shifts 12 units to the left. The -5 outside the parentheses shifts the graph 5 units down. The vertex is at $(-12, -5)$.

• **Lesson 11–3** Locate the roots of quadratic equations by graphing.

Find the roots of $0 = x^2 - 8x + 12$.

Based on the graph of the related function shown above, the roots are 2 and 6.

• **Lesson 11–4** Solve quadratic equations by factoring.

Solve $x^2 - 8x - 20 = 0$.

$$x^2 - 8x - 20 = 0$$
$$(x - 10)(x + 2) = 0$$
$$x - 10 = 0 \quad \text{or} \quad x + 2 = 0$$
$$x = 10 \qquad\qquad x = -2$$

Review Exercises

Use the results from Exercises 11–14 and a table to graph each quadratic function.

15. $y = x^2 - 3$
16. $y = -x^2 + 4x + 5$
17. $y = 3x^2 + 6x - 17$
18. $y = x^2 - 3x - 4$

Describe how each graph changes from its parent graph of $y = x^2$. Then name the vertex.

19. $y = (x - 8)^2$
20. $y = x^2 - 10$
21. $y = 0.4x^2$
22. $y = (x + 3)^2$
23. $y = (x - 1)^2 + 7$
24. $y = -1.5x^2$

Solve each equation by graphing the related function.

25. $x^2 - x - 12 = 0$
26. $x^2 + 10x + 25 = 0$
27. $2x^2 - 5x + 4 = 0$
28. $x^2 + x - 4 = 0$
29. $6x^2 - 13x = 15$

Solve each equation. Check your solution.

30. $y(y + 11) = 0$
31. $(t + 4)(2t - 10) = 0$
32. $z^2 - 36 = 0$
33. $x^2 + 16x + 64 = 0$
34. $b^2 - 5b - 14 = 0$
35. $4g^2 - 4 = 0$
36. $2x^2 + 10x + 12 = 0$

• Extra Practice
See pages 713–715.

Objectives and Examples | **Review Exercises**

• **Lesson 11–5** Solve quadratic equations by completing the square.

$$x^2 + 6x + 2 = 0$$
$$x^2 + 6x = -2 \quad \textit{Subtract 2.}$$
$$x^2 + 6x + 9 = -2 + 9 \quad \textit{Complete the square.}$$
$$(x + 3)^2 = 7$$
$$x + 3 = \pm\sqrt{7} \quad \textit{Take the square root.}$$
$$x = -3 \pm \sqrt{7} \quad \textit{Subtract 3.}$$

Find the value of c that makes each trinomial a perfect square.

37. $r^2 + 4r + c$ **38.** $t^2 - 12t + c$

Solve each quadratic equation by completing the square.

39. $t^2 - 4t - 21 = 0$ **40.** $x^2 - 16x + 23 = 0$

• **Lesson 11–6** Solve equations by using the Quadratic Formula.

Solve $-x^2 + 3x + 40 = 0$.

$$x = \frac{-b \pm \sqrt{b^2 - 4ac}}{2a}$$
$$= \frac{-3 \pm \sqrt{3^2 - 4(-1)(40)}}{2(-1)}$$
$$= \frac{-3 \pm \sqrt{169}}{-2}$$
$$= \frac{-3 + 13}{-2} \quad \text{or} \quad x = \frac{-3 - 13}{-2}$$
$$= -5 \qquad x = 8$$

Solve each equation by using the Quadratic Formula. Check your solution.

41. $r^2 + 10r + 9 = 0$

42. $-2d^2 + 8d = 0$

43. $3n^2 - 2n - 1 = 0$

44. $s^2 + 2s + 2 = 0$

45. $m^2 + 5m = -3$

46. $-4x^2 + 8x + 3 = 0$

• **Lesson 11–7** Graph exponential functions.

Graph $y = 2^x - 1$.

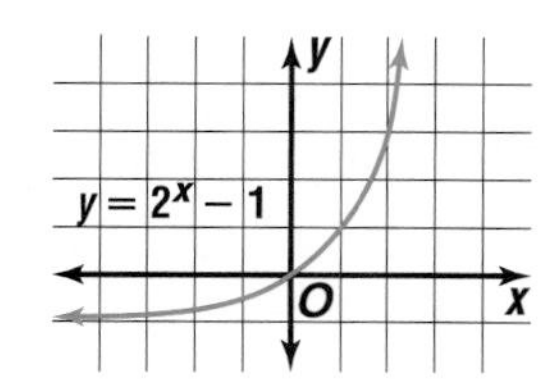

Graph each exponential function. Then state the y-intercept.

47. $y = 2^x$

48. $y = 2.7^x$

49. $y = 3^x + 3$

50. $y = 4^x - 4$

Applications and Problem Solving

51. Finance Game room tickets cost \$5. The function $I(x) = -50x^2 + 100x + 6000$ represents the weekly income when x is the number of 50¢ price increases. Graph the function to find the ticket price that results in the maximum income. *(Lesson 11–1)*

52. Tuition A college board voted to increase tuition prices a maximum of 4% annually. If tuition is \$6678, in how many years will it exceed \$10,000? Use the function $P(t) = 6678(1.04)^t$, where $P(t)$ is tuition and t is the number of years. *(Lesson 11–7)*

CHAPTER 11 Test

For each situation, state whether it is *always* true, *sometimes* true, or *never* true.

1. The graph of a quadratic function opens upward in a "u-shape."
2. You can use the Quadratic Formula to solve quadratic equations.
3. There are two real solutions to any quadratic equation.

Write the equation of axis of symmetry and the coordinates of the vertex of the graph of each quadratic function. Then graph the function.

4. $y = x^2 + 6x + 8$
5. $y = -x^2 + 3x$

Describe how each graph changes from its parent graph of $y = x^2$. Then name the vertex of each graph.

6. $y = x^2 - 10$
7. $y = (x - 9)^2$
8. $y = (x + 7)^2 + 3$

Solve each equation by graphing the related function. If exact roots cannot be found, state the consecutive integers between which the roots are located.

9. $y = x^2 - 8x + 16$
10. $y = x^2 - 5x - 1$

Solve each equation by factoring. Check your solution.

11. $(r - 3)(r - 8) = 0$
12. $s^2 - 5s + 4 = 0$
13. $x^2 + 7x + 10 = 0$

Find the value of c that makes each trinomial a perfect square.

14. $b^2 + 10b + c$
15. $x^2 + 8x + c$
16. $g^2 - 2g + c$

Solve each equation by completing the square.

17. $a^2 + 14a = -45$
18. $v^2 - 6v + 3 = 0$

Solve each equation by using the Quadratic Formula.

19. $2n^2 - n - 3 = 0$
20. $3x^2 + x + 5 = 0$

Graph each exponential function. Then state the y-intercept.

21. $y = 1.5^x$
22. $y = 2^x - 3$

23. **Physics** A plane flying 400 feet over Antarctica drops equipment needed by research scientists below. How long will it take the equipment to touch ground? Use the formula $h = -16t^2 + 400$, where the height h is in feet and the time t is in seconds.

24. **Geometry** The length of a rectangle is 4 inches more than its width. The area of the rectangle is 60 square inches. Find the dimensions of the rectangle.

25. **Finance** A school district has a $64 million budget. The school board has voted to increase the budget by only 1% each year. After how many years will the budget be over $70 million? Use the function $B(t) = 64(1.01)^t$, where $B(t)$ is the budget and t is the time in years.

CHAPTER 11

Preparing for Standardized Tests

Polynomial and Factoring Problems

State proficiency tests may include a few polynomial problems. In addition, many questions on the SAT and ACT ask you to factor or evaluate polynomials.

Memorize these polynomial relationships.

1. difference of squares $a^2 - b^2 = (a + b)(a - b)$
2. perfect square trinomials $a^2 + 2ab + b^2 = (a + b)^2$; $a^2 - 2ab + b^2 = (a - b)^2$

The Princeton Review

Look for factors in all problems that include polynomials. Factoring is often the quick way to find an answer.

State Test Example

Evaluate $2x^2y - 4y + x^4$ if $x = 2$ and $y = -1$.

A 12 **B** 16 **C** 20 **D** 28

Hint Apply the rules for operations carefully when negative numbers are involved.

Solution Substitute 2 for x and -1 for y. Then simplify the expression.

$$\begin{aligned} 2x^2y - 4y + x^4 &= 2(2)^2(-1) - 4(-1) + (2)^4 \\ &= 2(4)(-1) - 4(-1) + 16 \\ &= 8(-1) + 4 + 16 \\ &= -8 + 20 \\ &= 12 \end{aligned}$$

The answer is A.

SAT Example

If $y < 0$, which of the polynomials must be less than $y^2 + 150y + 75^2$?

A $y^2 + 160y + 75^2$

B $(y + 75)^2$

C $y^2 - 10y + 75^2$

D $(y - 75)^2$

Hint Consider the sign of y and the coefficients of each polynomial.

Solution First expand the binomials in choices B and D. Since $(y + 75)^2 = y^2 + 150y + 75^2$, the polynomial in choice B is equal to the original polynomial, not less than it. This eliminates answer choice B. The binomial in choice D expands to $y^2 - 150y + 75^2$.

The original polynomial and each of the three remaining choices, A, C, and D, have terms y^2 and 75^2. The only differences are in the y terms. The y terms for the original polynomial and the remaining answer choices are $150y$, $160y$, $-10y$, and $-150y$. Since y is negative, the least term is $160y$.

Therefore, the polynomial that must be less than $y^2 + 150y + 75^2$ is $y^2 + 160y + 75^2$.

The answer is A.

After you work each problem, record your answer on the answer sheet provided or on a sheet of paper.

Multiple Choice

1. Evaluate $h^2 - 5h$ if $h = -3$.
   - A -24
   - B -6
   - C 6
   - D 24

2. A swimming pool is shaped like a rectangular prism. Its volume is 1200 cubic feet. The dimensions of a smaller pool are half the size of the larger pool's dimensions. What is the volume, in cubic feet, of the smaller pool?
   - A 150
   - B 300
   - C 600
   - D 1200

3. For all $x \neq -3$, which of the following is equivalent to the expression $\frac{x^2 - x - 12}{x + 3}$?
   - A $x - 4$
   - B $x - 2$
   - C $x + 2$
   - D $x + 4$
   - E $x + 6$

4. If $x = -2$, then find $3x^2 - 5x - 6$.
   - A -8
   - B -4
   - C 10
   - D 16

5. The figures below represent a sequence of numbers. What is an expression for the nth term of this sequence?

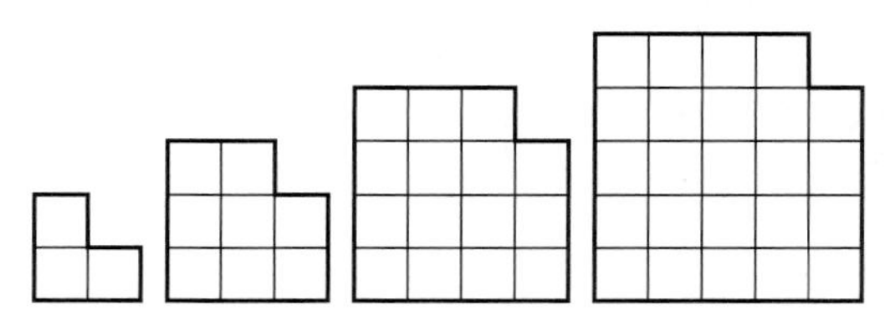

   - A $n^2 + 2$
   - B $n^2 - 1$
   - C $n(n + 2)$
   - D $2n^3 - 1$

6. Which quadratic equation has roots $\frac{1}{2}$ and $\frac{1}{3}$?
   - A $5x^2 - 5x - 2 = 0$
   - B $5x^2 - 5x + 1 = 0$
   - C $6x^2 + 5x + 1 = 0$
   - D $6x^2 - 6x + 1 = 0$
   - E $6x^2 - 5x + 1 = 0$

7. Use the function table to find the value of y when $x = 8$.
   - A 9
   - B 10.5
   - C 12
   - D 13.5

x	y
1	1.5
2	3.0
3	4.5
4	6.0
5	7.5

8. If $a > 0$ and $a \neq b$, which statement must be true?
   - A $\frac{a^2 - b^2}{a - b} - b$ is positive.
   - B $\frac{a^2 - b^2}{a - b} + b$ is positive.
   - C $\frac{a^2 - b^2}{a - b}$ is positive.
   - D $\frac{a^2 - b^2}{a - b}$ is negative.

Grid In

9. Solve for x if $\frac{x^2 + 7x + 12}{x + 4} = 5$.

Extended Response

10. Consider $2x + 3y = 15$.

    Part A Make a table of at least six pairs of values that satisfy the equation above.

    Part B Use your table from Part A to graph the equation.

CHAPTER 12

Inequalities

Make this Foldable to help you organize information about the material in this chapter. Begin with four sheets of graph paper.

❶ **Fold** each sheet in half from top to bottom.

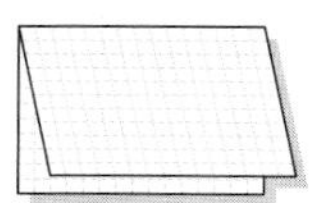

❷ **Cut** along fold. Staple the eight half-sheets together to form a booklet.

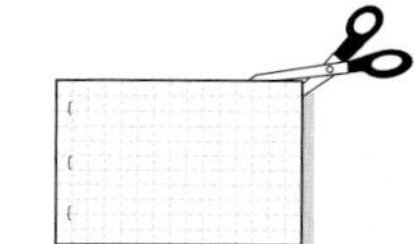

❸ **Label** each page with a lesson number and title.

Reading and Writing As you read and study the chapter, use each page to write notes and to graph examples.

Problem-Solving Workshop

Project

Plan a one-week menu of foods that require little or no preparation.

1. Total Calories must be between 1200 and 1800 per day.
2. Calories from fat must be at most 30% of the total Calories. Note that 1 gram of fat has 9 Calories.
3. Total sodium must be less than 2400 milligrams.
4. Total protein must be at least 46 grams.

Working on the Project

Work with a partner to solve the problem.

- Write inequalities to represent the Calories, fat (g), sodium (mg), and protein (g) allowed per day.
- Determine whether the sample menu in the table below satisfies the nutritional guidelines.

Technology Tools

- Use a **spreadsheet** to help write a menu that meets all guidelines.
- Use **publishing software** to illustrate your menu.

Strategies

Look for a pattern.
Draw a diagram.
Make a table.
Work backward.
Use an equation.
Make a graph.
Guess and check.

Food (one serving)	Calories	Fat (g)	Sodium (mg)	Protein (g)
canned chicken noodle soup	90	2	940	6
granola cereal and skim milk	270	9	20	5
canned tuna in water (2 ounces)	60	0.5	250	13
frozen dinner	450	20	1530	10
white bread (3 slices)	240	3	480	6

interNET CONNECTION **Research** For more information about nutrition, visit: www.algconcepts.com

Presenting the Project

Compile a portfolio of your work for this project. Include the menu for each of seven days, the nutritional value of each food, and a list of inequalities to show that all guidelines were met.

12–1 Inequalities and Their Graphs

What You'll Learn
You'll learn to graph inequalities on a number line.

Why It's Important
Postal Service Inequalities can be used to describe postal regulations. *See Exercise 12.*

You already know that equations are mathematical statements that describe two expressions with equal values. When the values of the two expressions are *not* equal, their relationship can be described in an inequality.

Verbal phrases like *greater than* or *less than* describe inequalities. For example, 6 is *greater than* 2. This is the same as saying 2 is *less than* 6.

The chart below lists other phrases that indicate inequalities and their corresponding symbols.

Inequalities			
$<$	$\leq$	$>$	$\geq$
• less than • fewer than	• less than or equal to • at most • no more than • a maximum of	• greater than • more than	• greater than or equal to • at least • no less than • a minimum of

Example 1

Suppose the minimum driving age in your state is 16. Write an inequality to describe people who are *not* of legal driving age in your state.

Let d represent the ages of people who are *not* of legal driving age.

The ages of all drivers → d *are greater than or equal to* → $\geq$ *16 years.* → 16

$d \geq 16$ is the same as $16 \leq d$.

Then d is *less than* 16, or $d < 16$.

Your Turn

a. Lisa can carry no more than 50 newspapers on her paper route. Express the number of papers that Lisa can carry as an inequality.

Not only can inequalities be expressed through words and symbols, but they can also be graphed.

On a graph, a circle means that the number is *not* included. A bullet means that the number *is* included.

Consider the inequality $d \geq 16$.

Step 1 Graph a bullet at 16 to show that d can *equal* 16.

Step 2 Graph all numbers *greater than* 16 by drawing a line and an arrow to the right.

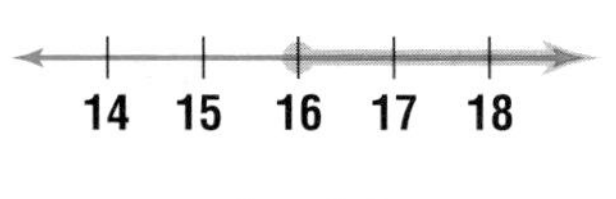

$d \geq 16$

The graph shows all values that are *greater than or equal to* 16.

Now suppose we want to graph all numbers that are *less than* 16. We want to include all numbers up to 16, but not including 16.

Step 1 Graph a circle at 16 to show that *d does not equal* 16.

Step 2 Graph all numbers *less than* 16 by drawing a line and an arrow to the left.

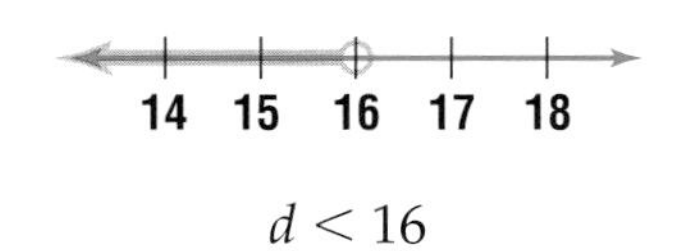

$d < 16$

The graph shows all values that are *less than* 16.

Examples

Graph each inequality on a number line.

2 $x > -4$

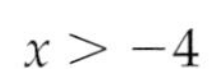

Since x cannot equal -4, graph a circle at -4 and shade to the right.

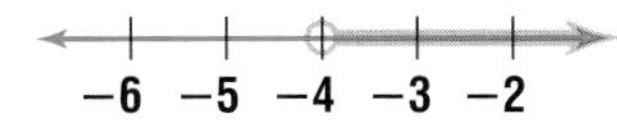

3 $n \geq 1.5$

Since n can equal 1.5, graph a bullet at 1.5 and shade to the right.

Your Turn

b. $y < 3$

c. $p \leq -\frac{1}{2}$

You can also write inequalities given their graphs.

Examples

Write an inequality for each graph.

4

−1 0 1 2 3

Locate where the graph begins. This graph begins at 1, but 1 is not included. Also note that the arrow is to the left. The graph describes values that are less than 1.

So, $x < 1$.

5

0 $\frac{1}{4}$ $\frac{2}{4}$ $\frac{3}{4}$ 1

Locate where the graph begins. This graph begins at $\frac{1}{4}$ and includes $\frac{1}{4}$. Note that the arrow is to the right. The graph describes values greater than or equal to $\frac{1}{4}$.

So, $x \geq \frac{1}{4}$.

Your Turn

d. −4 −3 −2 −1 0

e.

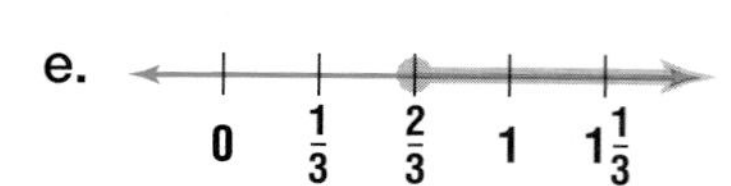

Inequalities are commonly used in the real world.

Example
Sports Link

6 **To play junior league soccer, you must be at least 14 years of age.**

A. Write an inequality to represent this situation.

Let a represent the ages of people who can play junior league soccer. Then write an inequality using ≥ since the soccer players have to be greater than or equal to 14 years of age.

$a \geq 14$

B. Graph the inequality on a number line.

To graph the inequality, first graph a bullet at 14. Then include all ages *greater than* 14 by drawing a line and an arrow to the right.

12 13 14 15 16

C. The Valdez children are 10, 13, 14, and 16 years old. Which of the children can play junior league soccer?

The set of the children's ages, {10, 13, 14, and 16}, can be called a *replacement set*. It includes possible values of the variable a. In this case, only 14 and 16 satisfy the inequality $a \geq 14$ and are members of the *solution set*. So, the two Valdez children that are 14 and 16 years old can play junior league soccer.

Check for Understanding

Communicating Mathematics

1. **Match** each inequality symbol with its description.

i. less than or equal to	**a.** ≥
ii. at least	**b.** ≤
iii. fewer than	**c.** >
iv. greater than	**d.** <

2. **Describe** the graph of $x \geq 12$.
3. **Explain** the difference between the graphs of $x \leq 4$ and $x < 4$.
4. **You Decide** Soto says that $x \geq 7$ is the same as $7 \leq x$. Darrell says that it is not. Who is correct? Explain.

Guided Practice

Write an inequality to describe each number. *(Example 1)*

5. a number less than 14
6. a number no less than 0

Graph each inequality on a number line. *(Examples 2 & 3)*

7. $x > 6$
8. $x \leq 2$
9. $5 > x$

Write an inequality for each graph. *(Examples 4 & 5)*

10. 6 7 8 9 10

11. 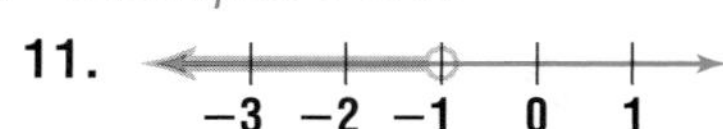

12. **Mail** All letters mailed with one first-class stamp can weigh no more than 1 ounce. Write an inequality to represent this situation. Then graph the inequality. *(Example 6)*

Exercises

Practice

Homework Help	
For Exercises	See Examples
13–18	1
19–30	2, 3
31–38	4, 5
39–42	6

Extra Practice
See page 715.

Write an inequality to describe each number.

13. a number greater than 4
14. a number less than or equal to 13
15. a number less than 9
16. a number no more than 7
17. a number that is at least −8
18. a number more than −6

Graph each inequality on a number line.

19. $x < 7$
20. $y > 3$
21. $b \leq 0$
22. $a \leq 5$
23. $x > -4$
24. $1 \leq z$
25. $x \geq 2.5$
26. $d < 1.9$
27. $x < -3.4$
28. $g > \frac{3}{4}$
29. $1\frac{1}{2} \geq x$
30. $h \leq -\frac{1}{3}$

Write an inequality for each graph.

31. −8 −7 −6 −5 −4

32. −2 −1 0 1 2

33. 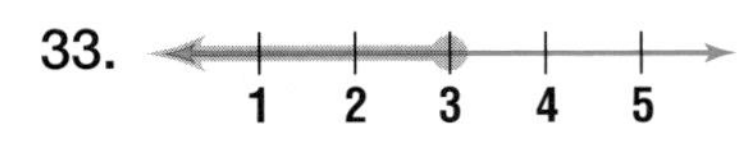

34. 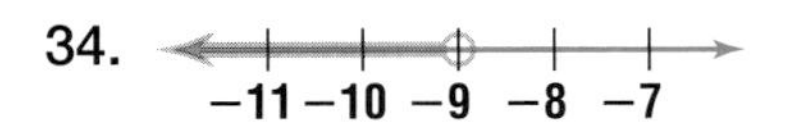

35. 3 4 5 6 7

36. 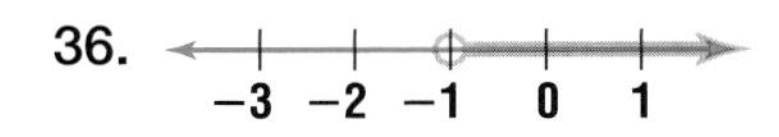

37. 2 3 4

38. 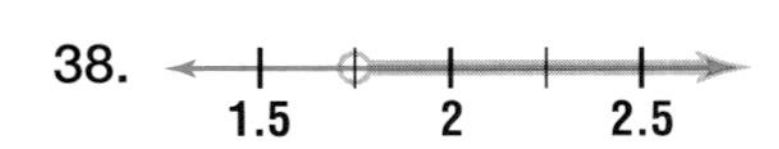

Applications and Problem Solving

For Exercises 39–41, write an inequality to represent each situation. Then graph the inequality on a number line.

39. **Conservation** At a pond, you must return any fish that you catch if it weighs less than 3 pounds.

40. **Entertainment** A roller coaster requires a person to be at least 42 inches tall to ride it.

41. **Safety** The elevators in an office building have been approved to hold a maximum of 3600 pounds.

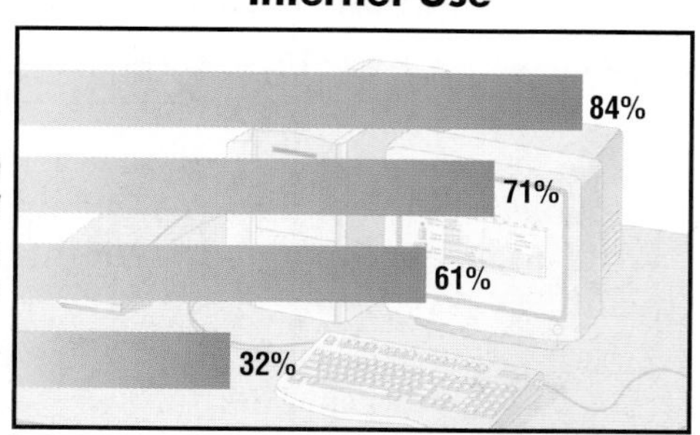

Source: *USA TODAY*

42. **Critical Thinking** The graph shows survey results on the Internet use of 100 college students. Write an inequality to represent the number of students that may spend at least 5 hours online each day.

Mixed Review

43. **Finance** Mr. Johnson invested \$10,000 at an annual interest rate of 6%. Graph the function $B(t) = 10{,}000(1.06)^t$, where t is the time in years. How many years will it take before the balance is over \$15,000? *(Lesson 11–7)*

Solve each equation by using the Quadratic Formula. *(Lesson 11–6)*

44. $t^2 - 3t - 40 = 0$
45. $3a^2 + a + 1 = 0$
46. $x^2 - 6x + 7 = 0$

47. Find the prime factorization of 24. How many distinct prime factors are there? *(Lesson 10–1)*

Simplify each expression.

48. $-\sqrt{16}$ *(Lesson 8–5)*
49. $\frac{-18x^3y}{3x}$ *(Lesson 8–2)*

Standardized Test Practice

50. **Multiple Choice** If two lines are perpendicular to each other, which of the following statements could be true? *(Lesson 7–7)*
    - **I.** The slopes of both lines are equal.
    - **II.** One slope is positive, and the other slope is negative.
    - **III.** The slope of one line is the reciprocal of the slope of the other line.

    A I only B II only C III only
    D I and II E II and III

12–2 Solving Addition and Subtraction Inequalities

What You'll Learn
You'll learn to solve inequalities involving addition and subtraction.

Why It's Important
Sports You can use inequalities to set goals in competitions. *See Example 2.*

The Iditarod, nicknamed "The Last Great Race on Earth," is a dogsled race in Alaska. The 12- to 16-dog teams cover over 1150 miles in subzero weather. The record time for finishing the race is 218 hours (9 days, 2 hours).

A dogsled race

Suppose a team takes 73 hours to reach the first checkpoint of the Iditarod and 98 hours to reach the second. How much time can be spent on the last leg of the race to beat the record time?

Let t represent the time for the last leg of the race. Write an inequality.

The sum of the times	*must be less than*	*the record time.*
$73 + 98 + t$	$<$	218

$$171 + t < 218$$

If this were an equation, we would subtract 171 from each side to solve for t. The same procedure can be used with inequalities, as explained by the properties below. *This problem will be solved in Example 2.*

These properties are also true for $\leq$ and $\geq$.

Addition and Subtraction Properties for Inequalities

Words: For any inequality, if the same quantity is added or subtracted to each side, the resulting inequality is true.

Symbols: For all numbers a, b, and c,
1. if $a > b$, then $a + c > b + c$ and $a - c > b - c$.
2. if $a < b$, then $a + c < b + c$ and $a - c < b - c$.

Numbers:

$-5 < 1$	$2 > -4$
$-5 + 2 < 1 + 2$	$2 - 3 > -4 - 3$
$-3 < 3$	$-1 > -7$

Example 1 **Solve $x + 14 \geq 5$. Check your solution.**

$$x + 14 \geq 5$$

$$x + 14 - 14 \geq 5 - 14 \quad \textit{Subtract 14 from each side.}$$

$$x \geq -9$$

(continued on the next page)

Check: Substitute a number less than −9, the number −9, and a number greater than −9 into the inequality.

Let $x = -10$.	Let $x = -9$.	Let $x = 0$.
$x + 4 \geq 5$	$x + 14 \geq 5$	$x + 14 \geq 5$
$-10 + 14 \stackrel{?}{\geq} 5$	$-9 + 14 \stackrel{?}{\geq} 5$	$0 + 14 \stackrel{?}{\geq} 5$
$4 \geq 5$ false	$5 \geq 5$ true	$14 \geq 5$ true

The solution is {all numbers greater than or equal to −9}.

Your Turn **Solve each inequality. Check your solution.**

a. $x + 2 < 7$ **b.** $x - 6 \geq 12$

A more concise way to express the solution to an inequality is to use **set-builder notation**. The solution in Example 1 in set-builder notation is $\{x \mid x \geq -9\}$.

$$\underbrace{\{x}_{\text{The set of all numbers } x} \; \underbrace{\mid}_{\text{such that}} \; \underbrace{x \geq -9\}}_{x \text{ is greater than or equal to } -9.}$$

In Lesson 12–1, you learned that you can show the solution to an inequality on a line graph. The solution, $\{x \mid x \geq -9\}$, is shown below.

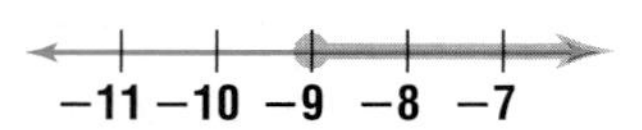

Real World
Examples
Sports Link

2 **Refer to the application at the beginning of the lesson. Solve $171 + t < 218$ to find the time needed to finish the last leg of the Iditarod and beat the record.**

$171 + t < 218$

$171 + t - 171 < 218 - 171$ *Subtract 171 from each side.*

$t < 47$ *This means all numbers less than 47.*

The solution can be written as $\{t \mid t < 47\}$. So any time less than 47 hours will beat the record.

3 **Solve $7y + 4 > 8y - 12$. Graph the solution.**

$7y + 4 > 8y - 12$

$7y + 4 - 7y > 8y - 12 - 7y$ *Subtract 7y from each side.*

$4 > y - 12$

$4 + 12 > y - 12 + 12$ *Add 12 to each side.*

$16 > y$

Since $16 > y$ is the same as $y < 16$, the solution is $\{y \mid y < 16\}$.

The graph of the solution has a circle at 16, since 16 is not included. The arrow points to the left.

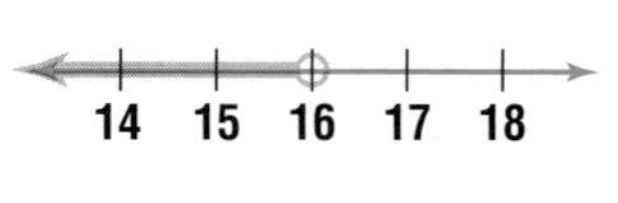

Your Turn **Solve each inequality. Graph the solution.**

c. $5y - 3 > 6y - 9$

d. $3r + 7 \leq 2r + 4$

You can also use inequalities to describe some geometry concepts.

Hands-On Algebra

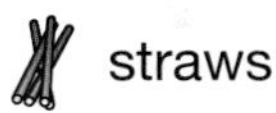

Materials: straws, scissors, pipe cleaners, ruler

Step 1 Cut one straw so that it is 3 inches long. Cut a second straw 4 inches long. Cut a third straw 5 inches long.

Step 2 Insert a pipe cleaner into each straw. Then form a triangle. Label each side like the one shown at the right.

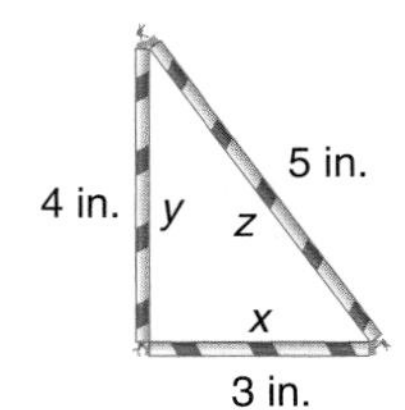

Try These

1. Repeat Steps 1 and 2 for each of the side lengths listed below. In each case, can a triangle be formed? Explain.
   a. $x = 4$ in., $y = 8$ in., $z = 6$ in.
   b. $x = 3$ in., $y = 5$ in., $z = 2$ in.
   c. $x = 1$ in., $y = 6$ in., $z = 3$ in.
2. Study the triangles in Exercise 1. Use an inequality to describe what must be true about the side lengths to form a triangle.
3. Can a triangle be formed with side lengths 9 cm, 12 cm, and 15 cm? Explain.

Check for Understanding

Communicating Mathematics

1. **Compare and contrast** solving inequalities by using addition and subtraction with solving equations by using addition and subtraction.
2. **Write** two inequalities that each have $\{y \mid y > 4\}$ as their solution.
3. **List** three situations in which solving an inequality may be helpful.

Vocabulary
set-builder notation

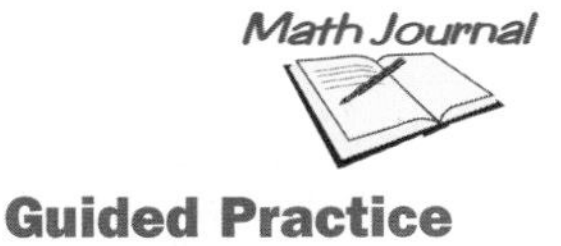

Guided Practice

Getting Ready **Write an inequality for each statement.**

Sample: Five more than a number is greater than sixteen.
Solution: $n + 5 > 16$

4. A number minus three is greater than or equal to ten.
5. The sum of 8 and a number is at most 12.

Solve each inequality. Check your solution. *(Examples 1 & 2)*

6. $x + 9 < 12$
7. $v - 12 \geq 5$
8. $a - 6 \leq -14$
9. $h - 7 \leq 14$
10. $4.6 + x > 2.1$
11. $y - \frac{2}{3} < 1\frac{1}{6}$

Data Update For the latest prices on computer hardware, visit: www.algconcepts.com

Solve each inequality. Graph the solution. *(Example 3)*

12. $z - 12 \geq 2z + 4$
13. $4x - 1 > 5x$

14. **Budgeting** Antonio can spend no more than $1000 on a new computer system. The hard drive he wants costs $220. The monitor costs $300. How much money does Antonio have to spend on other components? *(Example 2)*

Exercises

Practice

Solve each inequality. Check your solution.

15. $n + 6 > 9$
16. $x - 7 > -3$
17. $-3 + b < -8$
18. $g + 12 \leq 5$
19. $r - 8 < 11$
20. $x + 6 \geq 14$
21. $w - 9 > 13$
22. $4 + p \leq 1$
23. $t - 5 \leq -5$
24. $x + 3 \geq 19$
25. $-2 < c + 2$
26. $11 \leq -4 + m$
27. $d + 1.4 < 6.8$
28. $-3 + x > 11.9$
29. $-0.2 \geq 0.3 + z$
30. $s - (-2) > \frac{3}{4}$
31. $3\frac{3}{8} + v \leq 5\frac{7}{8}$
32. $\frac{1}{2} > x - \frac{2}{3}$

Solve each inequality. Graph the solution.

33. $7 > 8g - 7g - 6$
34. $5n < 4n + 10$
35. $3x + 12 \geq 2x - 4$
36. $6a \leq 5(a - 1)$
37. $-(-t + 9) \geq 0$
38. $3(v - 5) > 2(v - 2)$

Write an inequality for each statement. Then solve.

39. Eight less than a number is not greater than 12.
40. The sum of 5, 10, and a number is more than 25.

Applications and Problem Solving

41. **Fundraising** Westfield High School is having a raffle to raise money for Habitat for Humanity. Any homeroom that sells at least 150 tickets will get to help build a home. Ms. Martinez' homeroom is keeping a table of the number of tickets sold each day. How many more tickets do they need to sell to help build a home?

Day	Tickets
1	32
2	19
3	?
4	?
5	?

42. **Sports** Carlos weighs 175 pounds. At least how much weight must he lose to wrestle in the 171-pound weight class?

43. **Academics** Alissa must earn 475 out of 550 points to receive a grade of B. So far, she has earned 244 test points, 82 quiz points, and 50 homework points. How many points must she score on her final exam to earn at least a B in the class?

44. **Volunteerism** Each summer, 70 men bicycle in the Journey of Hope to raise money for people with disabilities. They begin their journey in San Francisco, California, and end in Washington, D.C. The map shows the miles they cycle in California. It takes them at most 285 miles to cycle to Nevada. How many miles is it from Jackson, California, to their first stop in Nevada?

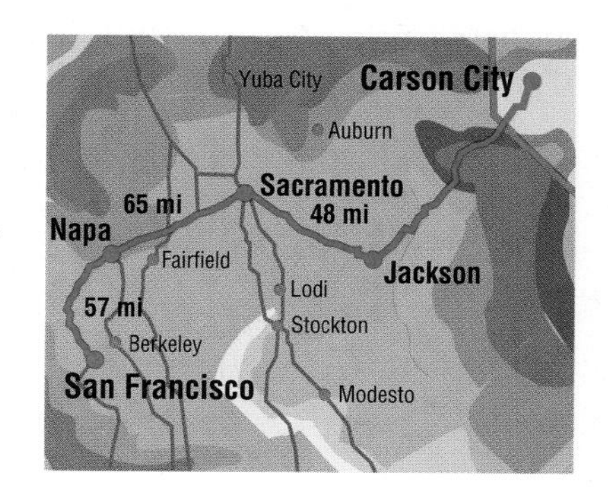

45. **Critical Thinking** Choose the correct term and justify your answer. The solution set of an inequality is (*sometimes, always, never*) the empty set $(\varnothing)$.

Mixed Review

Write an inequality to describe each number. *(Lesson 12–1)*

46. a number no more than 1

47. a number less than -8

48. a number greater than 3

49. a minimum number of 5

50. The graph of which of the following equations is shown at the right: $y = 2^x$, $y = 2^x + 1$, $y = 2^x - 1$, or $y = -2^x$? *(Lesson 11–7)*

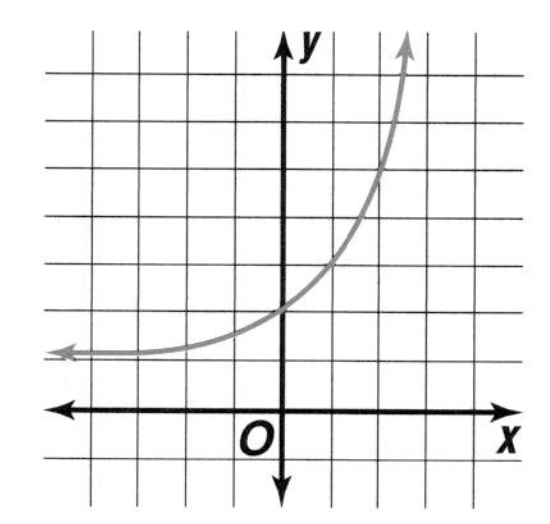

51. Factor the trinomial $3x^2 - 13x - 10$. *(Lesson 10–4)*

52. Write $3x + y = -8$ in slope-intercept form. *(Lesson 7–3)*

Standardized Test Practice

53. **Grid In** If y varies inversely as x and $y = 8$ when $x = 24$, find y when $x = 6$. *(Lesson 6–6)*

54. **Multiple Choice** Suppose $**x** = 4x - 2$. If $**x** = 10$, then what is the value of x? *(Lesson 4–4)*

A 2 B 3 C 38 D 40

12–3 Solving Multiplication and Division Inequalities

What You'll Learn
You'll learn to solve inequalities involving multiplication and division.

Why It's Important
Savings You can use inequalities when you are trying to budget your money.
See Exercise 37.

In the previous lesson, you learned that addition and subtraction inequalities are solved with the same procedures as addition and subtraction equations. Can you use the procedures for solving multiplication and division equations to solve inequalities as well?

For example, to solve $4x = 36$, we would divide each side by 4. Will this work when solving inequalities? Consider the inequality $4 < 36$.

$$4 < 36$$
$$\frac{4}{4} \stackrel{?}{<} \frac{36}{4} \quad \textit{Divide each side by 4.}$$
$$1 < 9 \quad \text{true}$$

So, it is possible to solve an inequality when dividing by a positive number. What happens when you divide by a negative number? Consider the inequality $-4 < 36$.

$$-4 < 36$$
$$\frac{-4}{-4} \stackrel{?}{<} \frac{36}{-4} \quad \textit{Divide each side by } -4.$$
$$1 < -9 \quad \text{false}$$

When dividing by a negative number, you must reverse the symbol for the inequality to remain true. This example leads us to the Division Property for Inequalities.

This property is also true for $\leq$ and $\geq$.

Division Property for Inequalities

Words: If you divide each side of an inequality by a positive number, the inequality remains true. If you divide each side of an inequality by a negative number, the inequality symbol must be reversed for the inequality to remain true.

Symbols: For all numbers a, b, and c,

1. if c is positive and $a < b$, then $\frac{a}{c} < \frac{b}{c}$, and
if c is positive and $a > b$, then $\frac{a}{c} > \frac{b}{c}$.

2. if c is negative and $a < b$, then $\frac{a}{c} > \frac{b}{c}$, and
if c is negative and $a > b$, then $\frac{a}{c} < \frac{b}{c}$.

Numbers: If $9 > 6$, then $\frac{9}{3} > \frac{6}{3}$ or $3 > 2$.
If $9 > 6$, then $\frac{9}{-3} < \frac{6}{-3}$ or $-3 < -2$.

Examples
Sports Link

1 **Carmen runs at least 15 miles in the park every day. If she runs 5 miles per hour, how long does she run every day? Recall that rate times time equals distance, or $rt = d$.**

$5t \geq 15$

$\frac{5t}{5} \geq \frac{15}{5}$ *Divide each side by 5.*

$t \geq 3$ *Keep the symbol facing the same direction.*

Carmen runs at least 3 hours every day.

Prerequisite Skills Review
Operations with Decimals, p. 684

2 **Solve $-10x \leq 25.6$. Check your solution.**

$-10x \leq 25.6$

$\frac{-10x}{-10} \geq \frac{25.6}{-10}$ *Divide each side by –10 and reverse the symbol.*

$x \geq -2.56$

Check: Substitute -2.56 and a number greater than -2.56, such as 0, into the inequality.

Let $x = -2.56$.	Let $x = 0$.
$-10x \leq 25.6$	$-10x \leq 25.6$
$-10(-2.56) \stackrel{?}{\leq} 25.6$	$-10(0) \stackrel{?}{\leq} 25.6$
$25.6 \leq 25.6$ true	$0 \leq 25.6$ true

The solution set is $\{x \mid x \geq -2.56\}$.

Your Turn

Solve each inequality. Check your solution.

a. $8x > 40$ **b.** $-x \leq 4.7$

We can also solve inequalities by multiplying. Use the Multiplication Property for Inequalities.

This property is also true for $\leq$ and $\geq$.

Multiplication Property for Inequalities

Words: If you multiply each side of an inequality by a positive number, the inequality remains true. If you multiply each side of an inequality by a negative number, the inequality symbol must be reversed for the inequality to remain true.

Symbols: For all numbers a, b, and c,

1. if c is positive and $a < b$, then $ac < bc$, and if c is positive and $a > b$, then $ac > bc$.
2. if c is negative and $a < b$, then $ac > bc$, and if c is negative and $a > b$, then $ac < bc$.

Numbers: If $3 > -7$, then $3(2) > -7(2)$ or $6 > -14$.
If $3 > -7$, then $3(-2) < -7(-2)$ or $-6 < 14$.

Example 3 Solve $-\frac{x}{3} < -6$. Check your solution.

$$-\frac{x}{3} < -6$$

$$-3\left(-\frac{x}{3}\right) > -3(-6) \quad \text{Multiply each side by } -3 \text{ and reverse the symbol.}$$

$$x > 18$$

Check: Substitute 18 and a number greater than 18, such as 21, into the inequality.

Let $x = 18$.	Let $x = 21$.
$-\frac{x}{3} < -6$	$-\frac{x}{3} < -6$
$-\frac{18}{3} \stackrel{?}{<} -6$	$-\frac{21}{3} \stackrel{?}{<} -6$
$-6 < -6$ false	$-7 < -6$ true

The solution is $\{x \mid x > 18\}$.

Your Turn

Solve each inequality. Check your solution.

c. $\frac{x}{-4} \leq 8$

d. $\frac{2}{3}x > 22$

Check for Understanding

Communicating Mathematics

1. **Compare and contrast** the Division Property for Inequalities and the Multiplication Property for Inequalities.
2. **Explain** why 5 is *not* a solution of the inequality $4x > 20$.
3. **Graph** the solution of $-2x \geq -8$ on a number line.

Guided Practice

Solve each inequality. Check your solution.

4. $2z > 12$
5. $-3x \leq 27$
6. $-7b < 14$ *(Examples 1 & 2)*
7. $\frac{x}{4} \geq 5$
8. $-\frac{r}{6} < 6$
9. $\frac{2}{5}a \leq -12$ *(Example 3)*

10. **Time** Cherise drives to the restaurant where she works by different routes, but the trip is at most 10 miles. Considering the stops at traffic lights, she thinks her average driving speed is about 40 miles per hour. How much time does it take Cherise to get to work? *(Example 1)*

Exercises

Practice

Homework Help	
For Exercises	See Examples
11–13, 17–19, 23–25, 29–31, 35–38	1, 2
14–16, 20–22, 26–29, 32–34	3
Extra Practice	
See page 716.	

Solve each inequality. Check your solution.

11. $-6h \leq 12$
12. $-9n < 18$
13. $8d \leq -24$
14. $\frac{a}{6} > 5$
15. $\frac{g}{3} > -12$
16. $-\frac{b}{8} \geq 5$
17. $7x \geq 49$
18. $-4y \leq -40$
19. $3z > -9$
20. $-\frac{p}{2} > -1$
21. $\frac{c}{9} > -6$
22. $-\frac{a}{8} \leq -4$
23. $3k > 5$
24. $-2t \leq 11$
25. $-3 > -6w$
26. $\frac{3}{4}x < 3$
27. $5 \leq \frac{5}{6}y$
28. $-\frac{2}{5}v \leq 20$
29. $5.3v \geq 10.6$
30. $4.1x < -6.15$
31. $-28 \leq -0.1s$
32. $-\frac{h}{3.8} \geq 2$
33. $\frac{n}{10.5} < 10$
34. $-\frac{x}{0.5} < -7$

Applications and Problem Solving

35. **Geometry** An *acute angle* has a measure less than 90°. If the measure of an acute angle is $2x$, what is the value of x?

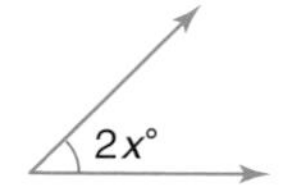

36. **Production** The ink cartridge that Bill just bought for his printer can print up to 900 pages of text. Bill is printing handbooks that are 32 pages each. How many complete handbooks can he print with this cartridge?

37. **Budgeting** Jenny mows lawns to earn money. She wants to earn at least \$200 to buy a new stereo system. If she charges \$12 a lawn, at least how many lawns does she need to mow?

38. **Critical Thinking** Use a counterexample to show that if $x < y$, then $x^2 < y^2$ is not always true.

Mixed Review

Solve each inequality. Check your solution. *(Lesson 12–2)*

39. $z + 1 \leq 5$
40. $-3 \geq b + 11$

41. **Contests** At a beach museum in San Pedro, California, more than 600 people built a life-size sand sculpture of a whale. Use an inequality to represent the number of people who built the sculpture. *(Lesson 12–1)*

Whale sculpture in San Pedro, California

42. Solve $x^2 + 5x + 4 = 0$ by factoring. *(Lesson 11–3)*

43. Find the product of $2v - 1$ and $2v + 1$. *(Lesson 9–5)*

44. **Earthquakes** In a recent year, the state of Illinois experienced an earthquake tremor. It measured 3.5 on the Richter scale, releasing about 1.6×10^7 Joules of energy. The largest earthquake ever recorded in Illinois measured 5.5 on the Richter scale, releasing about 5.7×10^{11} Joules of energy. How many times as strong was the largest earthquake as the tremor? *(Lesson 8–4)*

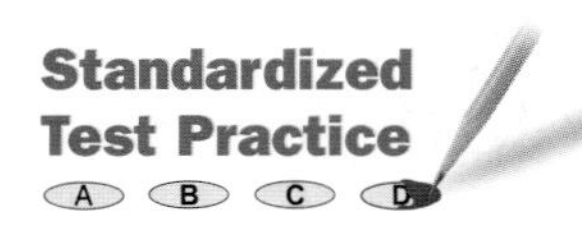

45. **Short Response** There are 20 students in a class. Each student's name is written on a separate slip of paper and placed in a box. A name is randomly drawn to determine who will read the daily announcements. The slip of paper is then returned to the box. What is the probability that the same name is drawn two days in a row? *(Lesson 5–7)*

46. **Multiple Choice** If the figure at the right is a square, then what is the value of x? *(Lesson 4–4)*

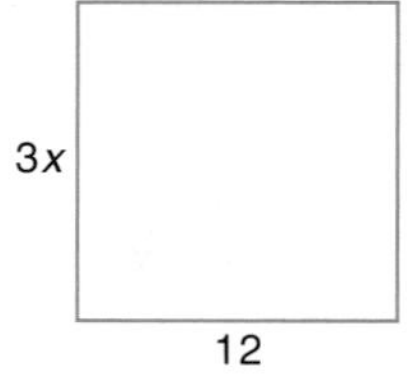

A 3 B 4
C 12 D 36

Quiz 1 Lessons 12–1 through 12–3

1. Write an inequality for the graph. *(Lesson 12–1)*

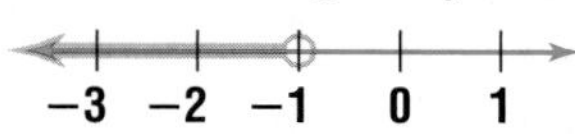

Solve each inequality. Graph the solution. *(Lesson 12–2)*

2. $a + 2 < 10$

3. $-4 + x > 1$

4. $-5 \leq s - (-3)$

5. $-3 + 13z > 14z$

Solve each inequality. Check your solution. *(Lesson 12–3)*

6. $3b \geq 30$

7. $-5x < 25$

8. $-\frac{k}{2} > -8$

9. $\frac{4}{5}v \leq 6$

10. **Weather** The record high temperature in Jackson, Mississippi, is 100° F. Suppose a recent temperature in Jackson is 84° F. How many degrees must the temperature rise to break the record? *(Lesson 12–2)*

www.algconcepts.com/self_check_quiz

12–4 Solving Multi-Step Inequalities

What You'll Learn
You'll learn to solve inequalities involving more than one operation.

Why It's Important
School You can determine what score is needed to receive a certain class grade. *See Example 5.*

Some inequalities involve more than one operation. The best strategy to solve multi-step inequalities is to undo the operations in reverse order. In other words, work backward just as you did to solve multi-step equations. For example, $2x + 5 > 11$ is a multi-step inequality. You can solve this inequality by following these steps.

Step 1: Undo addition. *Subtract 5 from each side.*

$$2x + 5 - 5 > 11 - 5$$
$$2x > 6$$

Step 2: Undo multiplication. *Divide each side by 2.*

$$\frac{2x}{2} > \frac{6}{2}$$
$$x > 3$$

Example 1

Solve $9 + 3x < 27$. Check your solution.

$9 + 3x < 27$

$9 + 3x - 9 < 27 - 9$ *Subtract 9 from each side.*

$3x < 18$

$\frac{3x}{3} < \frac{18}{3}$ *Divide each side by 3.*

$x < 6$

Look Back
Solving Multi-Step Equations: Lesson 4–5

Check: Substitute 0 and 6 into the inequality.

Let $x = 0$.

$9 + 3x < 27$

$9 + 3(0) \stackrel{?}{<} 27$

$9 + 0 \stackrel{?}{<} 27$

$9 < 27$ true

Let $x = 6$.

$9 + 3x < 27$

$9 + 3(6) \stackrel{?}{<} 27$

$9 + 18 \stackrel{?}{<} 27$

$27 < 27$ false

The solution is $\{x \mid x < 6\}$.

Your Turn

Solve each inequality. Check your solution.

a. $4 + 2x \leq 12$

b. $8x - 5 \geq 11$

Real World

Example 2 Weather Link

During Rafael's trip to Mexico, the temperature was always warmer than 30° Celsius. Use the formula $\frac{5}{9}(F - 32) = C$, where F is degrees Fahrenheit and C is degrees Celsius, to write and solve an inequality for the temperature in degrees Farenheit.

$$\frac{5}{9}(F - 32) > 30$$

$$\frac{9}{5} \cdot \frac{5}{9}(F - 32) > \frac{9}{5} \cdot 30 \quad \textit{Multiply each side by } \tfrac{9}{5}\textit{, the reciprocal of } \tfrac{5}{9}.$$

$$F - 32 > 54$$

$$F - 32 + 32 > 54 + 32 \quad \textit{Add 32 to each side.}$$

$$F > 86 \quad \textit{Check your solution.}$$

Therefore, Rafael can expect it to be warmer than 86°F in Mexico.

Remember to *reverse the inequality symbol* if you multiply or divide by a negative number.

Example 3

Solve $-4x + 3 \geq 23 + 6x$. Check your solution.

$$-4x + 3 \geq 23 + 6x$$

$$-4x + 3 - 6x \geq 23 + 6x - 6x \quad \textit{Subtract 6x from each side.}$$

$$-10x + 3 \geq 23$$

$$-10x + 3 - 3 \geq 23 - 3 \quad \textit{Subtract 3 from each side.}$$

$$-10x \geq 20$$

$$\frac{-10x}{-10} \leq \frac{20}{-10} \quad \textit{Divide each side by –10 and reverse the symbol.}$$

$$x \leq -2$$

The solution is $\{x \mid x \leq -2\}$. *Check your solution.*

Your Turn **Solve each inequality. Check your solution.**

c. $10 - 5x < 25$

d. $-3x + 1 > -17$

To solve inequalities that contain grouping symbols, you may use the Distributive Property first.

Example 4

Solve $8 \leq -2(x - 5)$. Check your solution.

$$8 \leq -2(x - 5)$$

$$8 \leq -2x + 10 \quad \textit{Distributive Property}$$

$$8 - 10 \leq -2x + 10 - 10 \quad \textit{Subtract 10 from each side.}$$

$$-2 \leq -2x$$

$$\frac{-2}{-2} \geq \frac{-2x}{-2} \quad \textit{Divide each side by } -2 \textit{ and reverse the symbol.}$$

$$1 \geq x$$

The solution is $\{x \mid x \leq 1\}$. *Check your solution.*

Your Turn **Solve each inequality. Check your solution.**

e. $2 > -(x + 7)$ **f.** $3(x - 4) \leq x - 5$

You can use a graphing calculator to solve multi-step inequalities.

Graphing Calculator Tutorial
See pp. 724–727.

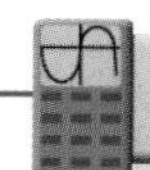

Graphing Calculator Exploration

Solve $-5 - 8x \geq 43$ by using a graphing calculator.

Step 1 Clear the [Y=] list to enter the inequality $-5 - 8x \geq 43$. (The $\geq$ symbol is item 4 on the TEST menu.)

Step 2 Press [ZOOM] 6. Use the [TRACE] and arrow keys to to move the cursor along the graph. You should see a line above the x-axis for values of x that are less than or equal to -6.

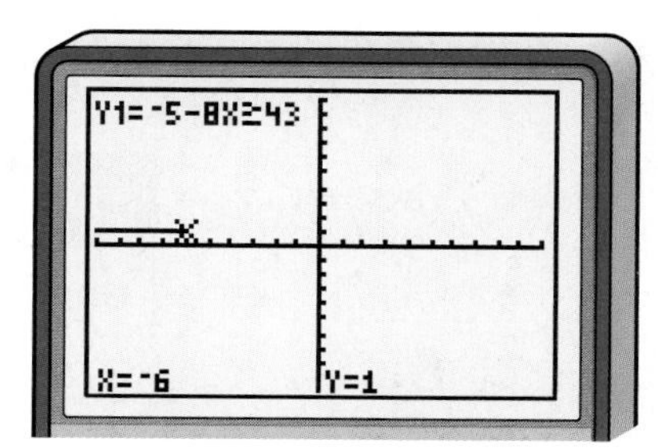

This represents the solution $\{x \mid x \leq -6\}$.

Try These

Solve each inequality. Check your solution with a graphing calculator.

1. $-9x + 2 \leq 20$ **2.** $-5(x + 4) \geq 3(x - 4)$

School Link

5 **Hannah's scores on the first three of four 100-point tests were 85, 92, and 90. What score must she receive on the fourth test to have a mean score of more than 92 for all tests?**

Explore Let s = Hannah's score on the fourth test.

Plan The sum of Hannah's four test scores, divided by 4, will give the mean score. The mean must be *more than* 92.

Solve

$$\frac{85 + 90 + 92 + s}{4} > 92$$

$$4\left(\frac{85 + 90 + 92 + s}{4}\right) > 4(92) \quad \textit{Multiply each side by 4.}$$

$$85 + 90 + 92 + s > 368$$

$$267 + s > 368$$

$$267 + s - 267 > 368 - 267 \quad \textit{Subtract 267 from each side.}$$

$$s > 101$$

(continued on the next page)

Examine Substitute a number greater than 101, such as 102, into the original problem. Hannah's average would be $\frac{85 + 90 + 92 + 102}{4}$ or 92.25. Since $92.25 > 92$ is a true statement, the solution is correct. Hannah must score more than 101 points out of a 100-point test. Without extra credit, this is not possible. So, Hannah cannot have a mean score over 92.

Check for Understanding

Communicating Mathematics

1. **Name** the operations used to solve $5 - 2x < -9$.
2. **Write** an inequality requiring more than one operation to solve.

Guided Practice

State which operation you would perform first to solve the inequality.

Sample: $12x + 4 > 20$ **Solution:** Subtract 4 from each side.

3. $-15z + 7 \geq -10$
4. $24 < 8b - 3$
5. $4.5a - (-3.1) > 8.2$
6. $\frac{n + 3}{9} \leq -11$

Solve each inequality. Check your solution. *(Examples 1–4)*

7. $2y + 4 > 12$
8. $8 - 3h \leq 20$
9. $5 - 2x > 7$
10. $7z - 4 \geq 10$
11. $10 - 5w < w + 22$
12. $3(n - 4) > 2(n + 6)$
13. **School** Kira wants her average math grade to be at least 90. Her test scores are 88, 93, and 87. What score does she need on her fourth test to earn an average score of at least 90? *(Example 5)*

Exercises

Practice

Solve each inequality. Check your solution.

Homework Help

For Exercises	See Examples
14–25, 29, 30, 32–36	1, 2, 4
26–28, 31, 37	3

Extra Practice

See page 716.

14. $4t - 8 > 16$
15. $3b + 9 < 45$
16. $2 - 3n \leq 17$
17. $-20 \geq 8 - 7x$
18. $5 \leq 7c - 2$
19. $-6g - 1 < -13$
20. $-12 + 11r \leq 54$
21. $8 - 4v \geq 6$
22. $7 + 0.1a > 9$
23. $0.3m - 2.1 \leq -3.0$
24. $\frac{s}{4} - 6 < -11$
25. $\frac{11 - 6d}{5} > 1$
26. $3h - 5 \geq 2h + 4$
27. $5x + 3 \leq 2x + 9$
28. $6j - 9 > j + 6$
29. $2(7 - 2y) > 10$
30. $-\frac{1}{6}(z + 2) \leq -1$
31. $\frac{2}{3}(b + 1) < \frac{1}{2}(b + 5)$

Write and solve an inequality for each situation.

32. The sum of twice a number and 17 is no greater than 41.
33. Five times the sum of a number and 6 is less than 35.
34. Two thirds of a number decreased by 7 is at least 9.

Applications and Problem Solving

35. **Employment** Jeremy is a receptionist at a hair salon. He earns \$7.00 an hour, plus 10% of any hair products he sells. Suppose he works 22 hours a week. How much money in hair products must he sell to earn at least \$180?

36. **Recreation** The admission fee to a state fair is \$5.00. Each ride costs an additional \$1.50.
   a. Suppose Pilar does not want to spend more than \$20. How many rides can she go on?
   b. The fair has a special admission price for \$14, which includes unlimited rides. For how many rides is this a better deal than paying for each ride separately?

37. **Finance** At a bank, an advertisement reads, "In one year, your earnings will be greater than your original investment plus 6% of the investment." Suppose Diego invests \$1400. How much money can he expect to have at the end of the year?

38. **Critical Thinking** Would the solution of $x^2 > 4$ be $x > 2$? Justify your answer.

Mixed Review

Solve each inequality. Check your solution. *(Lesson 12–3)*

39. $5p > 35$
40. $-24 \leq -8v$
41. $-\frac{x}{4} \geq 10$
42. $\frac{2}{5}z < -6$

43. **Budgeting** Haley earned \$36 babysitting. She plans to buy two books that cost \$8.25 each. With the rest of the money, she plans to go to dinner and see a movie with her friends. At most, how much money can she spend on a movie and dinner? *(Lesson 12–2)*

Standardized Test Practice

44. **Grid In** Find the value of c that makes $x^2 - 6x + c$ a perfect trinomial square. *(Lesson 11–5)*

45. **Multiple Choice** If $x \neq 5$, then $\frac{x^2 - 3x - 10}{x - 5}$ is equivalent to which of the following? *(Lesson 10–3)*

A $x + 2$
B $x - 3$
C $x - 5$
D $x + 10$

12–5 Solving Compound Inequalities

What You'll Learn
You'll learn to solve compound inequalities.

Why It's Important
Nutrition
Pharmacists use inequalities to write prescriptions.
See Example 2.

Lamar is buying vitamins for his dog. The daily dose for the vitamins is based on the dog's weight. Lamar's dog weighs 32 pounds. Since 32 is greater than 25, but less than or equal to 50, he will give his dog 2 tablets.

Daily Dose	Dog's Weight (pounds)
1 tablet	25 or less
2 tablets	greater than 25 and less than or equal to 50
3 tablets	more than 50

Another way to write this information is to use an inequality. Let w represent the weight that requires 2 tablets.

Weight is greater than 25. $\quad$ *Weight is less than or equal to 50.*

$w > 25$ *and* $w \leq 50$

These two inequalities form a **compound inequality**. The compound inequality $w > 25$ and $w \leq 50$ can be written without using the word *and*.

Method 1 $25 < w \leq 50$

This can be read as 25 is less than w, which is less than or equal to 50.

Method 2 $50 \geq w > 25$

This can be read as 50 is greater than or equal to w, which is greater than 25.

Note that in each, both inequality symbols are facing the same direction.

Example 1 **Write $x \geq 2$ and $x < 7$ as a compound inequality without using *and*.**

$x \geq 2$ and $x < 7$ can be written as $2 \leq x < 7$ or as $7 > x \geq 2$.

Your Turn

a. $x < 10$ and $x \geq -4$

b. $x \leq 6$ and $x \geq 2$

A compound inequality using *and* is true if and only if *both* inequalities are true. Thus, the graph of a compound inequality using *and* is the **intersection** of the graphs of the two inequalities.

Consider the inequality $-2 < x < 3$. It can be written using *and*: $x > -2$ and $x < 3$. To graph, follow the steps below.

Step 1 Graph $x > -2$.

Step 2 Graph $x < 3$.

Step 3 Find the intersection of the graphs.

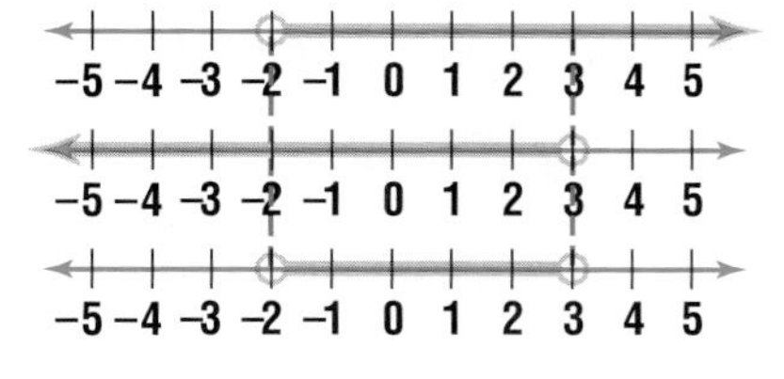

The solution, shown by the graph of the intersection, is $\{x \mid -2 < x < 3\}$.

Example 2 **Pet Care Link**

Refer to the application at the beginning of the lesson. Graph the solution of $25 < w \leq 50$.

Rewrite the compound inequality using *and*.
$25 < w \leq 50$ is the same as $w > 25$ and $w \leq 50$.

Step 1 Graph $w > 25$.

10 15 20 25 30 35 40 45 50 55 60

Step 2 Graph $w \leq 50$.

10 15 20 25 30 35 40 45 50 55 60

Step 3 Find their intersection.

10 15 20 25 30 35 40 45 50 55 60

The solution is $\{w \mid 25 < w \leq 50\}$.

Reading Algebra

Most of the time, $<$ or $\leq$ symbols are used with compound inequalities.

Your Turn

c. Graph the solution of $3 \leq x \leq 5$.

Often, you must solve a compound inequality before graphing it.

Example 3

Solve $4 < x + 3 \leq 12$. Graph the solution.

Step 1 Rewrite the compound inequality using *and*.

$$4 < x + 3 \leq 12$$

$$x + 3 > 4 \quad \text{and} \quad x + 3 \leq 12$$

Step 2 Solve each inequality.

$$x + 3 > 4 \quad \text{and} \quad x + 3 \leq 12$$
$$x + 3 - 3 > 4 - 3 \qquad x + 3 - 3 \leq 12 - 3$$
$$x > 1 \qquad x \leq 9$$

Step 3 Rewrite the inequality as $1 < x \leq 9$.

The solution is $\{x \mid 1 < x \leq 9\}$. The graph of the solution is shown at the right.

Your Turn

d. Solve $-2 < x - 4 < 2$. Graph the solution.

Another type of compound inequality uses the word *or*. This type of inequality is true if one or more of the inequalities is true. The graph of a compound inequality using *or* is the **union** of the graphs of the two inequalities.

Example 4 Graph the solution of $x > 0$ or $x \leq -1$.

Step 1 Graph $x > 0$.

−5 −4 −3 −2 −1 0 1 2 3 4 5

Step 2 Graph $x \leq -1$.

−5 −4 −3 −2 −1 0 1 2 3 4 5

Step 3 Find the union of the graphs.

−5 −4 −3 −2 −1 0 1 2 3 4 5

Your Turn

d. Graph the solution of $x > -3$ or $x < -5$.

Sometimes you must solve compound inequalities containing the word *or* before you are able to graph the solution.

Example 5 Solve $3x \geq 15$ or $-2x < 4$. Graph the solution.

$$3x \geq 15 \qquad \text{or} \qquad -2x < 4$$

$$\frac{3x}{3} \geq \frac{15}{3} \qquad\qquad \frac{-2x}{-2} > \frac{4}{-2}$$

$$x \geq 5 \qquad\qquad x > -2$$

Now graph the solution.

Step 1 Graph $x \geq 5$.

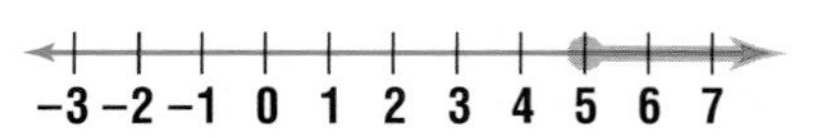

Step 2 Graph $x > -2$

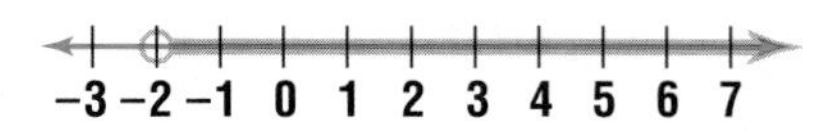

Step 3 Find the union of the graphs.

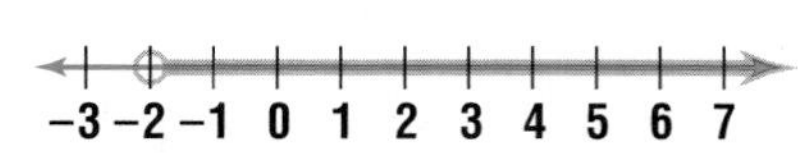

The last graph shows the solution $\{x \mid x > -2\}$.

Your Turn

e. Solve $-6x > 18$ or $x - 2 < 1$. Graph the solution.

Check for Understanding

Communicating Mathematics

1. **Define** *compound inequality* in your own words.
2. **Write** a compound inequality for *x is greater than 3 and less than or equal to 5*.

Vocabulary
compound inequality
intersection
union

Guided Practice

Getting Ready **State whether to find the intersection or union of the two inequalities to graph the solution.**

Sample 1: $-8 \leq x + 4 < 12$ **Solution:** Find the intersection since the inequalities can be joined by the word *and*.

Sample 2: $x > -9$ or $x < 5$ **Solution:** Find the union since the inequalities are joined by the word *or*.

3. $x > -10$ and $x < 3$

4. $x < 7$ or $x < 4$

5. $x - 5 < 2$ or $x \leq 15$

6. $-13 < 9x \leq -11$

Write each compound inequality without using *and*. *(Example 1)*

7. $\ell > 5$ and $\ell < 10$

8. $b < 3$ and $b \geq -2$

Graph the solution of each compound inequality. *(Examples 2 & 4)*

9. $y < 5$ and $y \geq 0$

10. $x > 3$ or $x < -7$

Solve each compound inequality. Graph the solution. *(Examples 3 & 5)*

11. $10 > c + 2 > 5$

12. $2 > s + 4 \geq -3$

13. $0 \leq 2v \leq 8$

14. $-5x > 10$ or $7x < 28$

15. $j + 6 > 6$ or $-4j \geq 4$

16. $-1 + r < -4$ or $-1 + r > 3$

17. Construction Odyssey of the Mind competitions encourage students to use creativity in solving difficult problems. One year, students had to construct a balsa-wood structure. The structure needed to be at least 9 inches tall and no more than 11.5 inches tall. Write a compound inequality describing the height of the structure. Graph the solution. *(Example 2)*

Exercises

Practice

Homework Help	
For Exercises	See Examples
18–23	1
56–59	2, 3
24–29	2, 4
30–35	3
36–55	3, 4, 5

Extra Practice

See page 717.

Write each compound inequality without using *and*.

18. $b > 0$ and $b < 5$

19. $h > -8$ and $h < 8$

20. $y \leq 4$ and $y \geq -1$

21. $g \geq 2$ and $g \leq 5$

22. $-2 < r$ and $1 > r$

23. $6 < x$ and $x \leq 8$

Graph the solution of each compound inequality.

24. $x > 0$ or $x \leq -4$

25. $z \leq 2$ and $z > -2$

26. $k < 7$ and $k > 5$

27. $y \geq 16$ and $y \leq 21$

28. $b > 4$ or $b < 0$

29. $m \leq 10$ or $m < 6$

Solve each compound inequality. Graph the solution.

30. $2 \leq a + 3 < 7$

31. $9 \geq x + 1 \geq 5$

32. $-16 < 8s < 16$

33. $9 \geq 3w > 0$

34. $-6 > r - 2 > -10$

35. $2 \leq h + 5 \leq 8$

Solve each compound inequality. Graph the solution.

36. $-5 < y - 1 < -4$
37. $16 < -4c < 20$
38. $6 \geq x - (-5) \geq 0$
39. $z + 3 > 7$ or $z - 5 \leq -12$
40. $v - 8 > 12$ or $v + 4 < 20$
41. $r \leq -1$ or $-8r < 0$
42. $3j > 18$ or $j - 3 \geq 5$
43. $p - 5 > -3$ or $-2p \leq 2$
44. $c - 2.4 \geq 7.6$ or $c - 8.8 \leq 0$
45. $d + 2.1 > 3.6$ or $d - 4 > 0.5$
46. $8x \geq 4$ or $x + \frac{3}{4} < 1$
47. $\frac{w}{3} \leq -2$ or $-w < 2$

Write a compound inequality for each solution shown below.

48. −5 −4 −3 −2 −1 0 1 2 3 4 5

49. 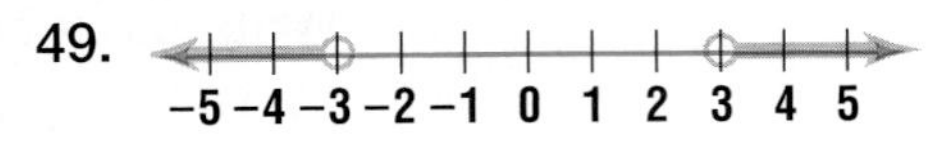

50. −8 −7 −6 −5 −4 −3 −2 −1 0 1 2

51. −4 −3 −2 −1 0 1 2 3 4 5 6

Solve each compound inequality.

52. $2 > 3y + 2 > -13$
53. $4x - 7 \leq 5$ or $-(x - 8) < -1$
54. $2x - 1 > -5$ or $-3(x + 1) > 6$
55. $10 \leq 2(k - 6) \leq 14$

Applications and Problem Solving

56. **Taxes** Matthew Brooks is single and has a part-time job while attending college. Last year, he paid \$649 in federal income tax. Write an inequality for his taxable income. Use the table that describes the different tax brackets.

If Form 1040A, line 24, is —		And you are —			
At least	But less than	Single	Married filing jointly	Married filing separately	Head of a household
		Your tax is —			
4,200	4,250	634	634	634	634
4,250	4,300	641	641	641	641
4,300	4,350	649	649	649	649
4,350	4,400	656	656	656	656
4,400	4,450	664	664	664	664
4,450	4,500	671	671	671	671
4,500	4,550	679	679	679	679
4,550	4,600	686	686	686	686

Source: Ohio Department of Taxation

57. **Cooking** A box of macaroni and cheese lists two sets of directions for cooking. It says to heat the macaroni for 11 to 13 minutes on the stove or 12 to 14 minutes in the microwave.
    a. Write an inequality that represents possible heating times.
    b. Graph the solution.

58. **Geometry** To construct any triangle, the sum of the lengths of two sides must be greater than the length of the third. Suppose that two sides of a triangle have lengths of 4 inches and 12 inches. What are the possible values for the length of the third side? Express your answer as a compound inequality.

59. **Chemistry** Soil pH is measured on a scale from 0 to 14. The pH level describes the soil as shown below.

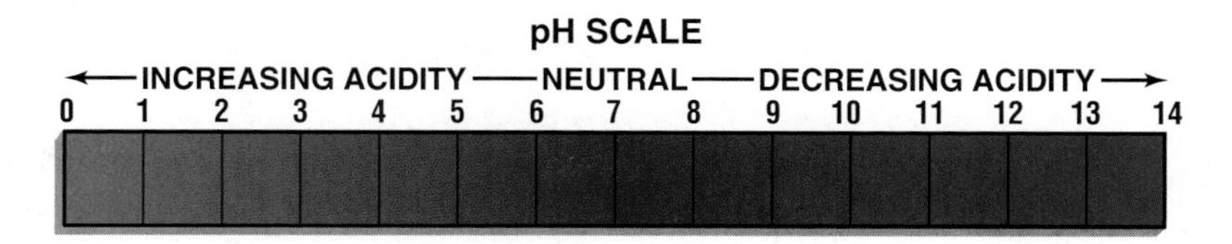

Most plants grow best if the soil pH is between 6.5 and 7.2. Mr. Cohen took samples of soil from his garden and found pH values of 6.3, 6.4, and 6.7. What range of values must the fourth sample have if the soil pH is the best for growing plants? (*Hint:* Find the mean soil pH.)

60. **Critical Thinking** Graph each compound inequality. Then state the solution.

a. $x > 4$ or $x \leq 4$

b. $2 > 3z + 2 > 14$

Mixed Review

Solve each inequality. Check your solution. *(Lessons 12–3 & 12–4)*

61. $2y + 4 < 4$

62. $-3n - 8 > 22$

63. $-2 \leq 0.6x - 5$

64. $\frac{2}{3}p - 3 \leq 7$

65. $-10b < -60$

66. $3r \geq -12$

67. **Welding** Maxwell is welding two pieces of iron together. During this process, the iron melts and begins to boil. Small droplets erupt and follow the paths of parabolic arcs. The paths can be modeled by the quadratic function $h(d) = -d^2 + 4d + 30$, where $h(d)$ represents the height of the arc above the ground at any horizontal distance d from the two pieces of iron. All measures are in inches. *(Lesson 11–1)*

a. Graph the function.

b. How high above the pieces of iron do the iron droplets jump?

68. Find the GCF of $8x^2$, $2x$, and $4xy$. *(Lesson 10–1)*

69. Simplify $3(b - 6)$. *(Lesson 9–3)*

Standardized Test Practice

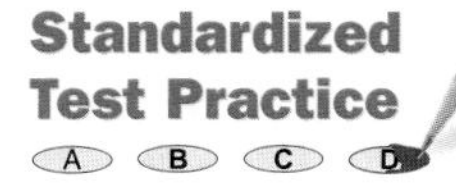

70. **Short Response** Simplify $(c^2 + 5) - (c^2 - 8c + 1)$. *(Lesson 9–2)*

71. **Multiple Choice** Emily drove 8 miles in 12 minutes. At this rate, how many miles will she drive in 1 hour? *(Lesson 5–1)*

A 4 mi
B 20 mi
C 40 mi
D 56 mi
E 96 mi

12–6 Solving Inequalities Involving Absolute Value

What You'll Learn

You'll learn to solve inequalities involving absolute value.

Why It's Important

Manufacturing Employees use inequalities involving absolute value to determine tolerances. *See Example 3.*

There are three types of open sentences that can involve absolute value. They are listed below. Note that in each case, n is nonnegative since the absolute value of a number can only equal 0 or a positive number.

$$|x| = n \qquad |x| < n \qquad |x| > n$$

You have already studied equations involving absolute value. Inequalities involving absolute value are similar. Consider the graphs and solutions of the three open sentences below.

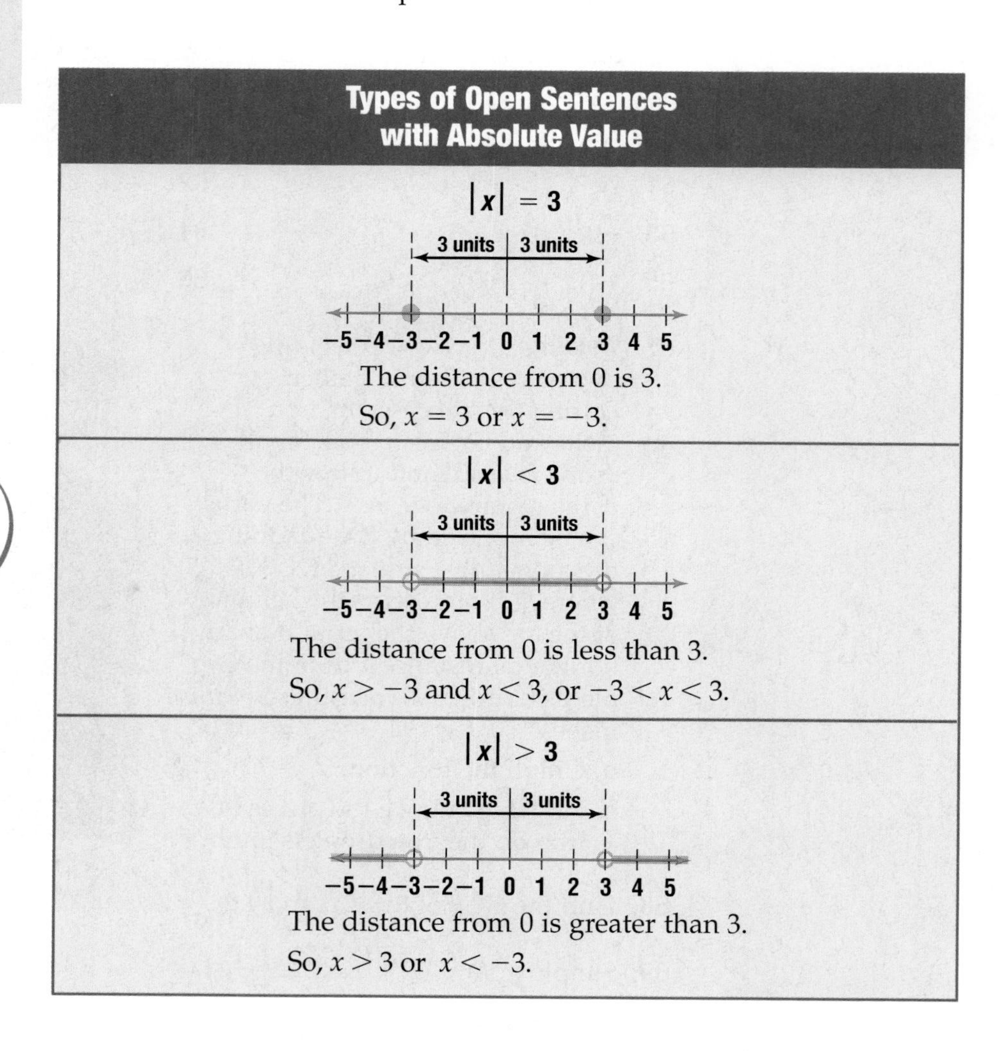

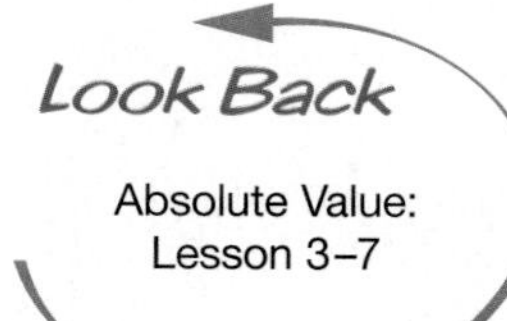

Absolute Value: Lesson 3–7

For both equations and inequalities involving absolute value, there are two cases to consider.

Case 1 The value within the absolute value symbols is positive.

Case 2 The value within the absolute value symbols is negative.

Example 1 **Solve $|x - 5| \leq 2$. Graph the solution.**

Case 1 $x - 5$ is positive.

$x - 5 \leq 2$

$x - 5 + 5 \leq 2 + 5$ *Add 5.*

$x \leq 7$

Case 2 $x - 5$ is negative.

$-(x - 5) \leq 2$

$-(x - 5)(-1) \geq 2(-1)$ *Multiply by –1 and reverse the symbol.*

$x - 5 \geq -2$

$5 + 5 \geq -2 + 5$ *Add 5.*

$x \geq 3$

So, the solution is $\{x \mid 3 \leq x \leq 7\}$.

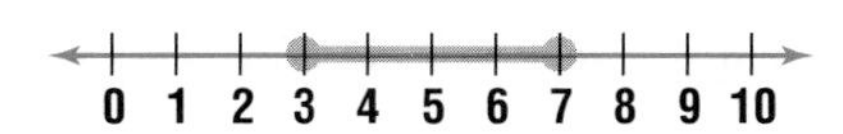

The solution makes sense since 3 and 7 are at most 2 units from 5.

Your Turn

a. Solve $|x - 7| < 4$. Graph the solution.

As in Example 1, when solving an inequality involving absolute value and the symbols $<$ or $\leq$, the solution can be written as an inequality using *and*. However, when solving an inequality involving absolute value and the symbols $>$ or $\geq$, the solution can be written as an inequality using *or*.

Example  **Solve $|6x| > 18$. Graph the solution.**

Case 1 $6x$ is positive.

$6x > 18$

$\frac{6x}{6} > \frac{18}{6}$ *Divide by 6.*

$x > 3$

Case 2 $6x$ is negative.

$-6x > 18$

$\frac{-6x}{-6} < \frac{18}{-6}$ *Divide by -6 and reverse the symbol.*

$x < -3$

So, the solution is $\{x \mid x < -3 \text{ or } x > 3\}$.

−5 −4 −3 −2 −1 0 1 2 3 4 5

Your Turn

b. Solve $|x - 2| \geq 3$. Graph the solution.

Inequalities involving absolute value are often used to indicate *tolerance*. Tolerance is the amount of error or uncertainty that is allowed when taking measurements.

Example 3 Measurement Link

When producing $\frac{1}{2}$-inch bolts for bicycle parts, the tolerance is 0.005 inch. What is the range of acceptable bolt measures?

Explore The difference in the actual size of the bolt and its expected size has to be less than or equal to 0.005 inch.

Plan Let m = the actual measure of the bolt.

Then, $\left|m - \frac{1}{2}\right| \leq 0.005$.

Solve $|m - 0.5| \leq 0.005$ *Write $\frac{1}{2}$ as 0.5.*

Case 1 $m - 0.5$ is positive.

$m - 0.5 \leq 0.005$

$m - 0.5 + 0.5 \leq 0.005 + 0.5$

$m \leq 0.505$

Case 2 $m - 0.5$ is negative.

$-(m - 0.5) \leq 0.005$

$-(m - 0.5)(-1) \geq 0.005(-1)$

$m - 0.5 \geq -0.005$

$m - 0.5 + 0.5 \geq -0.005 + 0.5$

$m \geq 0.495$

The solution is $\{m \mid 0.495 \leq m \leq 0.505\}$.

Examine An acceptable bolt must measure from 0.495 inch to 0.505 inch, inclusive. To check the solution, choose a value for m within this range and one outside of this range. Substitute them into the original problem. Which value results in a true inequality?

Check for Understanding

Communicating Mathematics

1. **Compare and contrast** the graphs of the solutions for $|x| < 7$ and $|x| > 7$.

2. **Graph** the solutions for $|x - 2| > 1$ and $|x - 2| < 1$. Which inequality is the *intersection* of the graphs of two inequalities? Which inequality is the *union* of the graphs of two inequalities?

3. **You Decide?** Madison says that the solution for $|x| \leq 0$ is the same as the solution for $|x| \geq 0$. Mia says it is not. Who is correct? Explain.

Guided Practice

Getting Ready **Write two inequalities to describe the solution.**

Sample 1: $|x| \leq 5$
Solution: $x \leq 5$ and $x \geq -5$

Sample 2: $|x| > 1$
Solution: $x > 1$ or $x < -1$

4. $|x| < 10$
5. $|x| \leq 3$
6. $|x| \geq 2$
7. $|x| > 8$

Solve each inequality. Graph the solution. *(Examples 1 & 2)*

8. $|n - 4| < 5$
9. $|x - 2| < 6$
10. $|3j| \leq 12$
11. $|t - 5| \geq 3$
12. $|2y| > 2$
13. $|s + 4| \geq 3$

14. **Measurement** Refer to Example 3. What are the possible measures for the bolt if the tolerance is 0.05 inch? Does a lesser or greater tolerance ensure more accurate measurements? Explain. *(Example 3)*

Exercises

Practice

Solve each inequality. Graph the solution.

15. $|m + 1| < 5$
16. $|3v| < 15$
17. $|z + 7| \leq 2$
18. $|x + 3| \leq 8$
19. $|p - 1| < 2$
20. $|r + 4| < 4$
21. $|7t| \leq 14$
22. $|a - 3| \leq 4$
23. $|k + 2| < 3.5$
24. $|5n| \geq 30$
25. $|y + 3| \geq 6$
26. $|z - 1| \geq 1$
27. $|9x| > 18$
28. $|w + 2| > 5$
29. $|r - 4| \geq 1$
30. $|a - 8| \geq 3$
31. $|h - 3| > 9$
32. $|d + 9| > 0.2$

Homework Help

For Exercises	See Examples
15, 17, 18–20 22, 23, 25, 26, 28–35, 38–43	1, 3
18, 21, 24, 27	2

Extra Practice

See page 717.

Write an inequality involving absolute value for each statement. Do not solve.

33. Quincy's golf score s was within 4 strokes of his average score of 90.
34. The measure m of a board used to build a cabinet must be within $\frac{1}{4}$ inch of 46 inches, inclusive, to fit properly.
35. The cruise control of a car set at 65 mph should keep the speed s within 3 mph, inclusive, of 65 mph.

For each graph, write an inequality involving absolute value.

36. −5 −4 −3 −2 −1 0 1 2 3 4 5
37. −5 −4 −3 −2 −1 0 1 2 3 4 5

Solve each inequality. Graph the solution.

38. $|2x - 11| \geq 7$
39. $|3x - 12| < 12$
40. $4|x + 3| \leq 8$
41. $10|x + 1| > 90$

Applications and Problem Solving

For Exercises 42–43, write and solve an absolute value inequality.

42. **Finance** Ms. Gibson is a bank teller. She must balance her drawer and be within \$2 of the expected balance. If Ms. Gibson's expected balance is \$8758.20, what are acceptable balances for her drawer?
43. **Chemistry** For a chemistry project, Marvin must pour 3.25 milliliters of solution into a beaker. If he does not pour within 0.05, inclusive, of 3.25 milliters, the results will be inaccurate. How many milliliters of solution can Marvin use?

44. Critical Thinking The *percent of error* is the ratio of the greatest possible error (tolerance) to a measurement.
You can find the percent of error using this formula.

$$\text{percent of error} = \left|\frac{\text{greatest possible error}}{\text{measurement}}\right| \cdot 100$$

One rating system for in-line skate bearings is based on tolerances. The table shows tolerances for the outside diameter of bearings measuring 22 millimeters.

Rating	Tolerance (mm)
1	0.010
3	0.008
5	0.005

a. Find the percent of error for each rating to the nearest hundredth.

b. What can you conclude about the rating system?

Mixed Review

Graph the solution of each compound inequality. *(Lesson 12–5)*

45. $m < -7$ or $m \geq 0$

46. $x \geq -2$ and $x \leq 5$

47. $y > -5$ and $y < 0$

48. $r > 2$ or $r < -2$

49. Solve $1 \geq 4y + 5$. *(Lesson 12–4)*

50. **Sales** Grant bought a sweater for $44.52. This cost included 6% sales tax. What was the cost of the sweater before tax? *(Lesson 5–5)*

51. Solve $|x - 2| = 4$. *(Lesson 3–7)*

Standardized Test Practice

52. **Short Response** Write three fractions whose sum is $1\frac{5}{8}$. *(Lesson 3–2)*

Quiz 2 Lessons 12–4 through 12–6

Solve each inequality. *(Lesson 12–4)*

1. $3x + 8 < 11$

2. $5 - 6n > -19$

3. $9d - 4 \geq 8 - d$

Solve each compound inequality. Graph the solution. *(Lesson 12–5)*

4. $1 + x < -4$ or $1 + x < 4$

5. $-2 \leq n + 3 < 4$

6. $-6 < -2f < 10$

Solve each inequality. Graph the solution. *(Lesson 12–6)*

7. $|y - 7| < 2$

8. $|a + 8| \geq 1$

9. $|5x| < 20$

10. **Travel** Before the meter begins ticking, the charge for a taxi is $1.60. For each mile driven, there is an additional charge of 80 cents. What is the greatest distance you can travel in this taxi if you do not want to pay more than $10.00? *(Lesson 12–4)*

www.algconcepts.com/self_check_quiz

12–7 Graphing Inequalities in Two Variables

What You'll Learn

You'll learn to graph inequalities on the coordinate plane.

Why It's Important

Budgeting By graphing inequalities, you can solve problems where there are many solutions. *See Example 3.*

Mr. Wheat is planning to take the Jazz Club to a music festival. Lawn tickets cost \$20, and pavilion tickets cost \$30. If he plans to spend at most \$300, how many of each ticket can Mr. Wheat purchase?

Let x represent the number of lawn tickets. Let y represent the number of pavilion tickets. Then the inequality below represents the solution.

The cost of lawn tickets	*plus*	*the cost of pavilion tickets*	*is at most*	*\$300.*
$20x$	$+$	$30y$	$\leq$	300

The inequality is written in two variables. It is similar to an equation written in two variables. An easy way to show the solution of an inequality is to graph it in the coordinate plane. *This problem will be solved in Example 3.*

The solution set of an inequality in two variables contains many ordered pairs. The graph of these ordered pairs fills an area on the coordinate plane called a **half-plane**. The graph of an equation defines the **boundary** or edge for each half-plane. Use these steps to graph $y > 3$.

Reading Algebra

The related equation for $y > 3$ is $y = 3$.

Step 1 Determine the boundary by graphing the related equation, $y = 3$.

Step 2 Draw a *dashed* line since the boundary is *not* part of the graph.

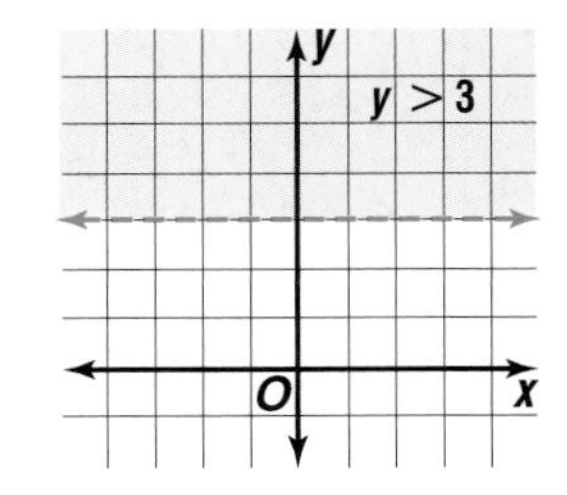

Step 3 Determine which half-plane is the solution. To do this, substitute a point from each half-plane into the inequality. Find which point results in a true statement.

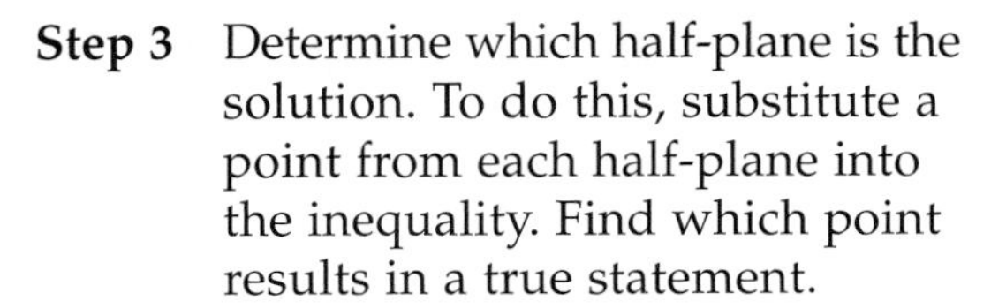

Test the point at (5, 8).
$y > 3$
$8 > 3$ *Replace y with 8.*
true

Test the point at (−3, 1).
$y > 3$
$1 > 3$ *Replace y with 1.*
false

The half-plane that contains (5, 8) is the solution. Shade that half-plane. Any point in the shaded region is a solution of the inequality $y > 3$.

Example 1 **Graph $y > -x + 3$.**

Step 1 Determine the boundary by graphing the related equation, $y = -x + 3$.

x	$-x + 3$	y
-2	$-(-2) + 3$	5
-1	$-(-1) + 3$	4
0	$-(0) + 3$	3
1	$-1 + 3$	2
2	$-2 + 3$	1

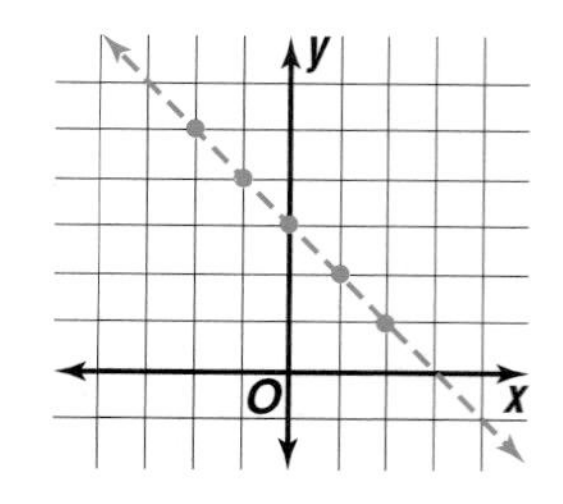

Step 2 Draw a *dashed* line since the boundary is not included.

Step 3 Test any point to find which half-plane is the solution. Use (0, 0) since it is the easiest point to use in calculations.

$y > -x + 3$

$0 > -(0) + 3$ *$x = 0, y = 0$*

$0 > 3$ false

Since (0, 0) does *not* result in a true inequality, the half-plane containing (0, 0) is *not* the solution. Thus, shade the other half-plane.

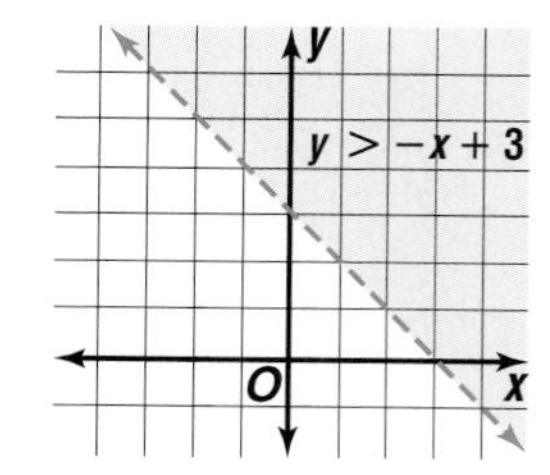

Your Turn

a. Graph $y < x - 7$.

When graphing inequalities, the boundary line is not always dashed. Consider the graph of $y \geq 3$. Since the inequality means $y > 3$ or $y = 3$, the boundary is part of the solution. This is indicated by graphing a solid line.

Example 2 **Graph $4x + y \leq 12$.**

To make a table or graph for the boundary line, solve the inequality for y in terms of x.

$$4x + y \leq 12$$

$$4x + y - 4x \leq 12 - 4x \quad \textit{Subtract } 4x \textit{ from each side.}$$

$$y \leq -4x + 12 \quad \textit{Rewrite } 12 - 4x \textit{ as } -4x + 12.$$

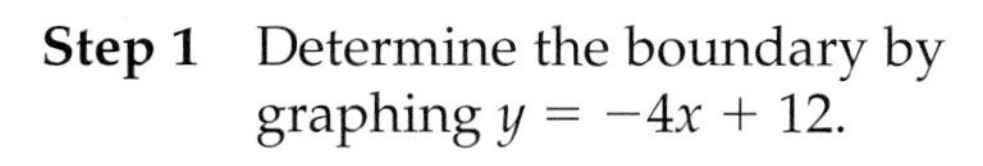

Step 1 Determine the boundary by graphing $y = -4x + 12$.

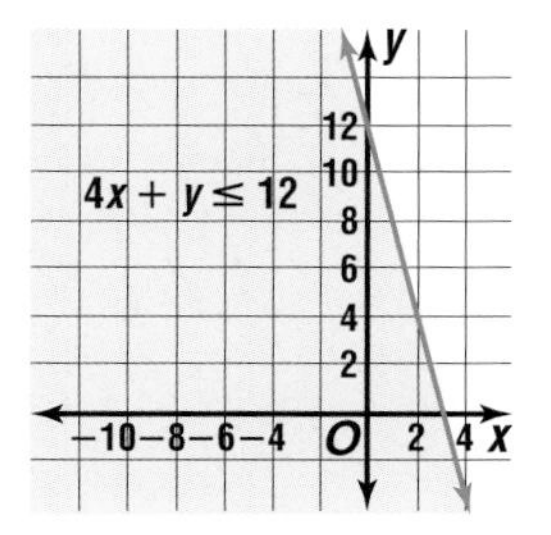

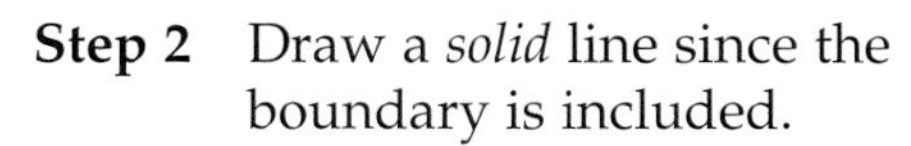

Step 2 Draw a *solid* line since the boundary is included.

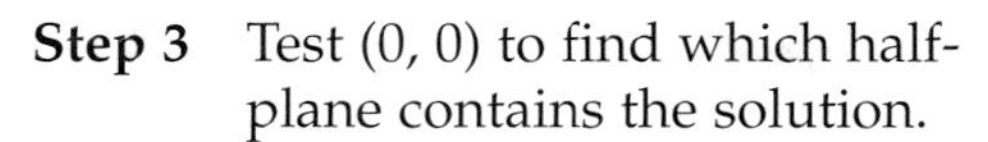

Step 3 Test (0, 0) to find which half-plane contains the solution.

$$\begin{aligned} 4x + y &\leq 12 \\ 4(0) + 0 &\leq 12 \quad x = 0, y = 0 \\ 0 &\leq 12 \quad \text{true} \end{aligned}$$

The half-plane that contains (0, 0) should be in the solution, which is indicated by the shaded region.

Your Turn

b. Graph $-2x + y \leq 10$.

When solving real-life inequalities, the domain and range of the inequality are often restricted to nonnegative numbers or whole numbers.

Real World

Example 3
Budgeting Link

Refer to the application at the beginning of the lesson. How many lawn and pavilion tickets can Mr. Wheat purchase?

First, solve for y in terms of x.

$$\begin{aligned} 20x + 30y &\leq 300 \\ 20x + 30y - 20x &\leq 300 - 20x \quad \textit{Subtract } 20x \textit{ from each side.} \\ 30y &\leq -20x + 300 \\ \frac{30y}{30} &\leq \frac{-20x + 300}{30} \quad \textit{Divide each side by 30.} \\ y &\leq \frac{-20x}{30} + \frac{300}{30} \\ y &\leq -\frac{2}{3}x + 10 \end{aligned}$$

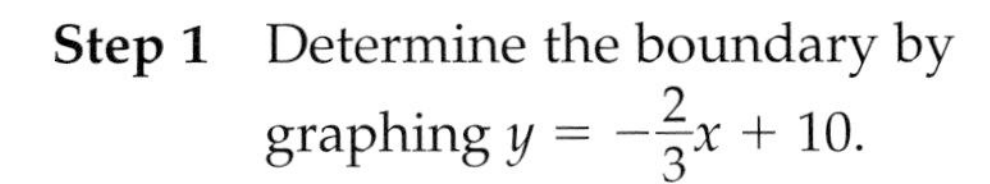

Step 1 Determine the boundary by graphing $y = -\frac{2}{3}x + 10$.

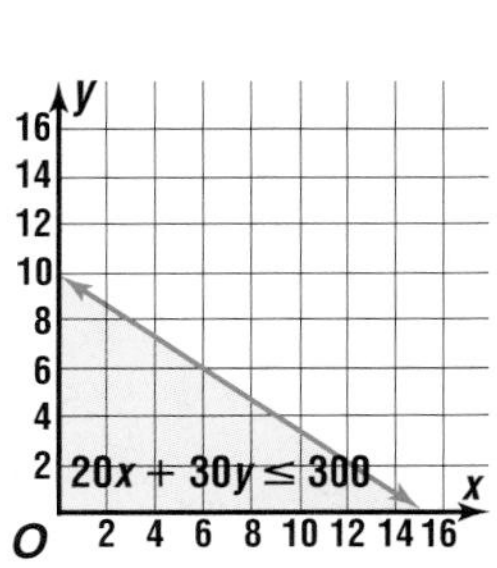

Step 2 Draw a *solid* line since the boundary is included.

Step 3 Test (0, 0) to find which half-plane contains the solution.

(continued on the next page)

Mr. Wheat cannot buy a negative number of tickets, nor can he buy portions of tickets. The solution is positive ordered pairs that are whole numbers beneath or on the graph of the line $y = -\frac{2}{3}x + 10$.

One solution is (12, 2). This represents 12 lawn tickets and 2 pavilion tickets costing $300.

Check for Understanding

Communicating Mathematics

1. **Explain** how to determine whether the boundary is a solid line or dashed line when graphing inequalities in two variables.

Math Journal

2. **Describe** how you could check whether a point is part of the solution of an inequality.

Vocabulary
half-plane
boundary

Guided Practice

Getting Ready **Test (0, 0) to find which half-plane is the solution of each inequality.**

Sample: $3x + y > 4$

Solution: $3x + y > 4$

$3(0) + 0 > 4$

$0 > 4$ false

Shade the half-plane *not* containing the point at (0, 0).

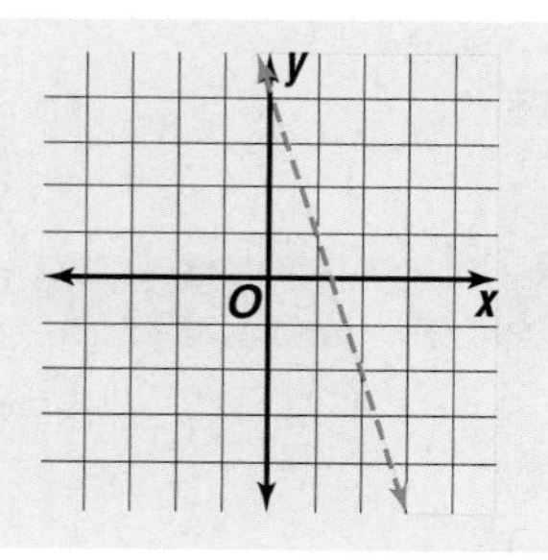

3. $x < -2$

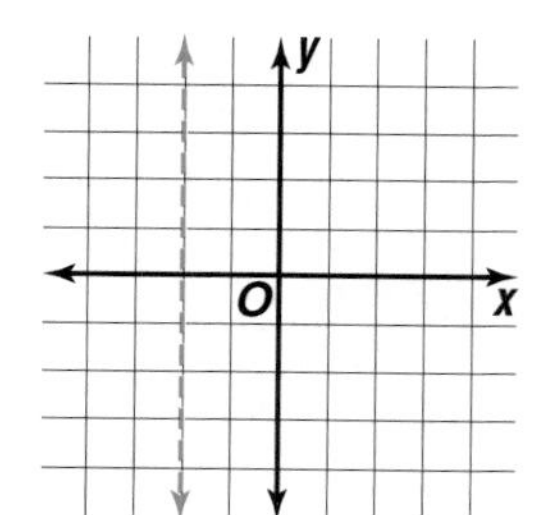

4. $y \leq 1$

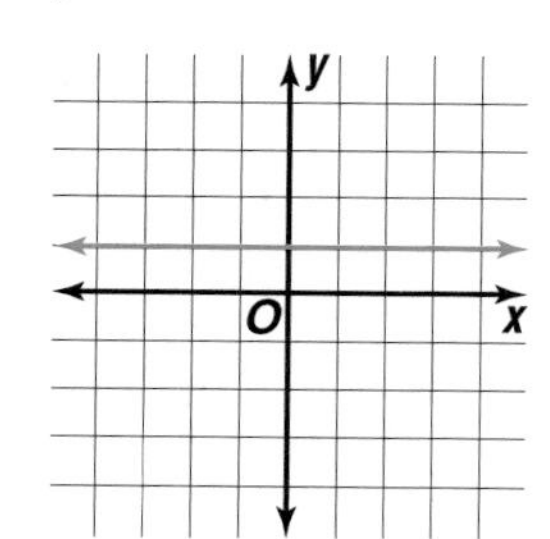

5. $5x - y > 5$

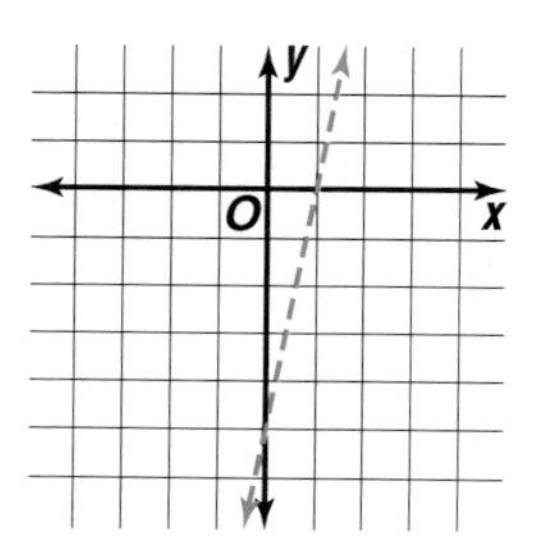

Graph each inequality. *(Examples 1 & 2)*

6. $y \leq -1$
7. $y \geq x - 7$
8. $y < 3x + 1$
9. $2x + y < 0$
10. $-4x + y > -8$
11. $x + 3y \geq 9$

12. **Sales** Tickets for the winter dance are $5 for singles and $8 for couples. To cover the deejay, photographer, and decoration expenses, a minimum of $1000 must be made from ticket sales. Write and graph an inequality that describes the number of singles' and couples' tickets that must be sold. Name at least one solution. *(Example 3)*

Exercises

Practice

Graph each inequality.

13. $y \leq 7$
14. $x \geq 5$
15. $y < x - 6$
16. $y < -x + 3$
17. $x + y > 8$
18. $y > x$
19. $-y \leq x$
20. $y \leq 2x$
21. $y < -3x + 4$
22. $-x + y < -5$
23. $x + 2y < 10$
24. $2x + y \leq 6$
25. $-3x + y \geq -1$
26. $3x - 2y > -12$
27. $y \leq 2(2x + 1)$
28. $3y > 3(4x - 3)$
29. $2(x + y) \leq 14$
30. $-3(5x - y) > 0$

Homework Help	
For Exercises	**See Examples**
13–21, 27, 28, 32	1
22–26, 29–31, 33–34	2, 3

Extra Practice
See page 717.

For Exercises 31–32, write an inequality and graph the solution.

31. The sum of two numbers is greater than four.
32. Twice a number is less than or equal to another number.

Applications and Problem Solving

Real World

33. **Animals** Amara and Toshi have set a goal to find homes for more than twelve pets through the Humane Society.
    a. Write and graph an inequality to determine how many homes each girl must find to reach the goal.
    b. List three of the solutions.
    c. Describe the limitations on the solution set.

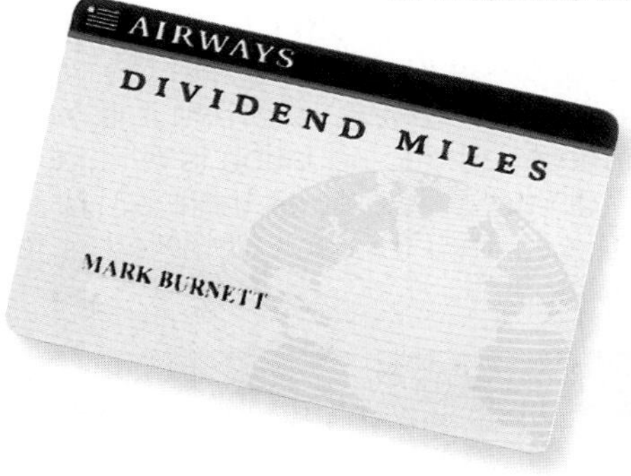

34. **Airlines** For each mile that Mr. Burnett flies, he earns one point toward a free flight. His credit card company is associated with the airline, and he earns one-half point for each dollar charged. Suppose he needs at least 20,000 points for a free flight. Show with a graph the miles that Mr. Burnett must fly and the money he must charge to get a free flight.

35. **Critical Thinking** Graph the intersection of the solutions of $4x + 2y \geq 8$ and $y < x$.

Mixed Review

Solve each inequality. Graph the solution. *(Lessons 12–5 & 12–6)*

36. $|t + 4| \geq 3$
37. $|h - 7| < 2$
38. $-1 < p + 1 \leq 6$
39. $-5b > 10$ or $b + 4 > 5$

40. **Travel** Ben and Pam left the park at the same time. Ben traveled north at 45 miles per hour. Pam traveled east at 60 miles per hour. After 1 hour, how far apart are Ben and Pam? *(Lesson 8–7)*

Standardized Test Practice

41. **Short Response** 848.3 in scientific notation. *(Lesson 8–4)*

42. **Multiple Choice** Suppose you toss 2 coins at the same time. What is the probability that both land heads up? *(Lesson 5–6)*

A 0 B 0.25 C 0.50 D 0.75 E 1

Chapter 12 Investigation

Parabolas and Pavilions

Materials

grid paper

ruler

yellow and blue colored pencils

Quadratic Inequalities

In this investigation, you will learn how to solve and graph **quadratic inequalities**.

Investigate

1. Graph the quadratic equation $y = x^2 + 2x - 3$ on a piece of graph paper.
   a. With a yellow colored pencil, shade the region inside the parabola.
   b. Make a table like the one below. Fill in column 1 with five points that appear in the yellow region, such as (0, 0). Compare the value of the y-coordinate with the value of $y = x^2 + 2x - 3$ when it is evaluated at the x-coordinate. When the quadratic equation is evaluated at 0, the result is -3. The y-coordinate, 0, is greater than -3. Place the correct inequality symbol in column 3 for each of the other four points that fall in the yellow region.

Point	y-coordinate	$<$ or $>$	$y = x^2 + 2x - 3$
(0, 0)	0	$>$	$y = (0)^2 + 2(0) - 3$ or -3

   c. Write a quadratic inequality that compares the points of the yellow region with the quadratic equation $y = x^2 + 2x - 3$.
   d. With a blue colored pencil, shade the region outside the parabola. Repeat parts b–c for the blue region.
2. Graph the quadratic equation $y = -x^2 - 2x + 3$ on a separate piece of graph paper.
   a. Shade the region inside the parabola yellow. Then shade the region outside the parabola blue. Make tables similar to those in Step 1. Write inequalities describing the yellow and blue regions.
   b. Suppose you wanted to include the values on the boundary line of a quadratic inequality. Explain how you would write the inequality to include the boundary line.
   c. Now suppose you did *not* want to include the boundary line of a quadratic inequality. Explain how you would draw the graph to show that the boundary line was *not* included in the solution.

3. Spring Town is building a community center, the Spring Town Pavilion. The building is to be 20 feet longer than it is wide and to have an area greater than or equal to 1500 square feet.

   a. Let w represent the width of the building. Then $w + 20$ represents the length. The area of the building, $w(w + 20)$, must be greater than or equal to 1500 square feet. That is, $w^2 + 20w - 1500 \geq 0$. Graph $f(w) = w^2 + 20w - 1500$.

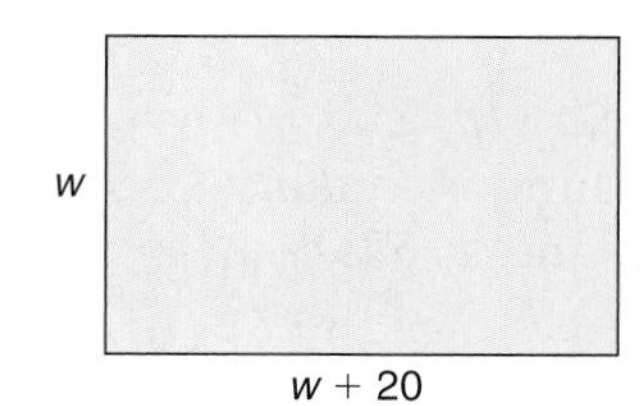

   b. Find the values for w that satisfy the inequality in part a. To do this, choose values for w both inside and outside of your parabola.

   c. What restrictions are placed on your solution for w since it represents width?

   d. Use your graph to find two possible dimensions for the Spring Town Pavilion.

Extending the Investigation

In this extension, you will continue to investigate quadratic inequalities.

Graph each quadratic inequality.

1. $y < x^2 + 4x - 8$
2. $y \geq -x^2 + 2x + 15$
3. $y \leq -3x^2 + 3$
4. The Spring Town Pavilion will have a deck for small outdoor gatherings. The length of the deck is to be 6 feet more than the width. The area of the deck is at least one-fifth of the area of the community center building. Write and graph a quadratic inequality that describes this situation. Then give three possible dimensions for the new deck.

Presenting Your Investigation

Here are some ideas to help you present your conclusions to the class.

- Make a poster that describes how to solve quadratic inequalities. Include at least two different inequalities. Show how the graphs are shaded to represent the solutions.
- Write and solve a problem similar to the one in Step 3. Make a brochure that describes the problem, the graph of its solution, and a list of possible solutions.

*inter***NET** CONNECTION **Investigation** For more information on graphing quadratic inequalities, visit: www.algconcepts.com

CHAPTER 12 Study Guide and Assessment

Understanding and Using the Vocabulary

After completing this chapter, you should be able to define each term, property, or phrase and give an example or two of each.

interNET CONNECTION **Review Activities**
For more review activities, visit: www.algconcepts.com

boundary *(p. 535)*
compound inequality *(p. 524)*
half-plane *(p. 535)*
intersection *(p. 524)*
quadratic inequalities *(p. 540)*
set-builder notation *(p. 510)*
union *(p. 525)*

Complete each sentence using a term from the vocabulary list.

1. The solution of a compound inequality using *or* can be found by the ___?___ of the graphs of the two inequalities.
2. Graph the related equation of an inequality to find the ___?___ of the half-plane.
3. Use a test point from each ___?___ to find the solution of an inequality in two variables.
4. An inequality of the form $x < y < z$ is called a(n) ___?___.
5. The ___?___ for the graph of an inequality in two variables will either be a dashed or solid line.
6. The ___?___ of two graphs is the area where they overlap.
7. A(n) ___?___ is an area on the coordinate plane representing the solution for an inequality in two variables.
8. $2x + 3 < 5$ or $x \geq -1$ is an example of a(n) ___?___.
9. A solution written in the form $\{x \mid x \leq 3\}$ is written in ___?___.
10. The solution of a compound inequality using *and* can be found by the ___?___ of the graphs of the two inequalities.

Skills and Concepts

Objectives and Examples

- **Lesson 12–1** Graph inequalities on a number line.

Graph $x < 5$ on a number line.

The graph begins at 5, but 5 is not included. The arrow is to the left. The graph describes values that are less than 5.

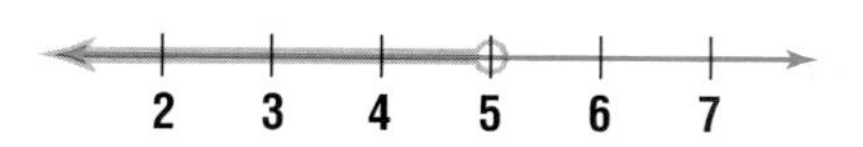

Review Exercises

Graph each inequality on a number line.

11. $x > -3$
12. $z \leq 2$
13. $3.5 > x$
14. $a \geq -1\frac{1}{2}$

Write an inequality for each graph.

15. −7 −6 −5 −4 −3

16. 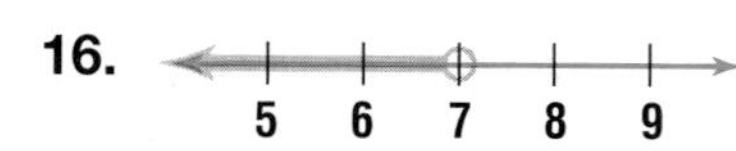

Objectives and Examples

• **Lesson 12–2** Solve inequalities involving addition and subtraction.

Solve $x - 6 > 2$.

$x - 6 > 2$
$x - 6 + 6 > 2 + 6$ *Add 6 to each side.*
$x > 8$

The solution is $\{x \mid x > 8\}$.

Review Exercises

Solve each inequality. Check your solution.

17. $x + 3 > 7$

18. $a - 4 \geq 2$

19. $\frac{2}{3} \leq y - \frac{1}{2}$

20. $12x - 11x + 5 < 3$

21. $3(x + 1) \geq 4x$

• **Lesson 12–3** Solve inequalities involving multiplication and division.

Solve $-4x < -8$.

$-4x < -8$
$\frac{-4x}{-4} > \frac{-8}{-4}$ *Divide each side by −4 and reverse the symbol.*
$x > 2$

The solution is $\{x \mid x > 2\}$.

Solve each inequality. Check your solution.

22. $3y \geq 12$

23. $-6n > 30$

24. $0.2w \leq -1.8$

25. $\frac{x}{5} > 3$

26. $-\frac{t}{2} \geq -14$

27. $\frac{3}{4}w \leq 6$

28. $\frac{h}{2.3} < -7$

• **Lesson 12–4** Solve inequalities involving more than one operation.

Solve $2(x + 1) < 4 - 3x$.

$2(x + 1) < 4 - 3x$
$2x + 2 < 4 - 3x$ *Distributive Property*
$5x + 2 < 4$ *Add 3x to each side.*
$5x < 2$ *Subtract 2 from each side.*
$x < \frac{2}{5}$ *Divide each side by 5.*

The solution is $\left\{x \mid x < \frac{2}{5}\right\}$.

Solve each inequality. Check your solution.

29. $2x + 6 \geq 14$

30. $9 < -0.2y - 1$

31. $\frac{2}{3}t - 4 \leq 2$

32. $3(4 + n) < 21$

Write and solve an inequality.

33. Four times a number decreased by 3 is greater than 25.

34. Seven minus two times a number is no less than nine.

• **Lesson 12–5** Solve compound inequalities and graph the solution.

Solve $3 < x + 2 \leq 8$.

$x + 2 > 3$ and $x + 2 \leq 8$
$x + 2 - 2 > 3 - 2$ and $x + 2 - 2 \leq 8 - 2$
$x > 1$ and $x \leq 6$

The solution is $\{x \mid 1 < x \leq 6\}$.

Solve each compound inequality. Graph the solution.

35. $-4 \leq y + 3 < 2$

36. $2 \leq 3 - t$ or $3 - t > 5$

37. $3a \geq 6$ or $-5 - a \geq -6$

38. $9 < 2x + 1 < 13$

Chapter 12 Study Guide and Assessment

• Extra Practice
See pages 715–717.

Objectives and Examples

• **Lesson 12–6** Solve inequalities involving absolute value and graph the solution.

Solve $|x + 1| < 1$.

Case 1 $x + 1$ is positive.

$x + 1 < 1$

$x < 0$ *Subtract 1 from each side.*

Case 2 $x + 1$ is negative.

$-(x + 1) < 1$

$-(x + 1)(-1) > 1(-1)$ *Multiply by −1 and reverse the symbol.*

$x + 1 > -1$

$x > -2$ *Subtract 1 from each side.*

The solution is $\{x \mid -2 < x < 0\}$.

Review Exercises

Solve each inequality. Graph the solution.

39. $|t - 1| \leq 5$

40. $|a + 7| < -2$

41. $|x + 3| < 1$

42. $|s + 4| \geq 3$

43. $|y - 2| > 0$

Write an inequality involving absolute value for each statement. Do not solve.

44. Bianca's guess g was within \$6 of the actual value of \$25.

45. The difference between Greg's score s on his final exam and his mean grade of 80 is more than 5 points.

• **Lesson 12–7** Graph inequalities in the coordinate plane.

Graph $2x + y < 5$.

First, solve for y.

$2x + y < 5$

$2x + y - 2x < 5 - 2x$

$y < -2x + 5$

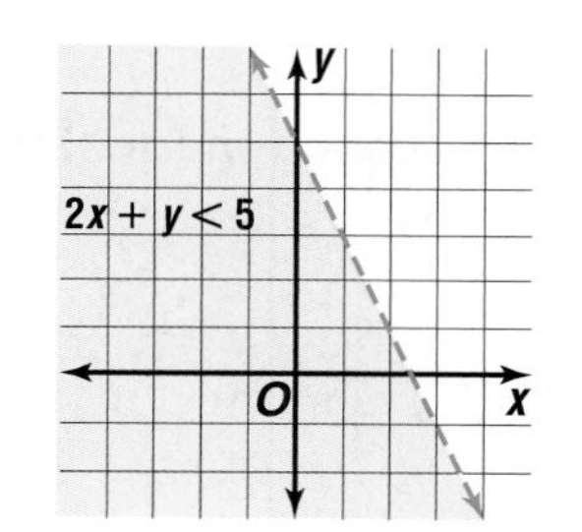

Graph the related function $y = -2x + 5$ as a dashed line and shade.

Graph each inequality.

46. $y \geq x - 1$

47. $y < -2x + 4$

48. $\frac{y}{2} > x - 2$

49. $x - y \geq 3$

Applications and Problem Solving

50. Shipping An empty book crate weighs 30 pounds. The weight of a book is 1.5 pounds. For shipping, the crate can weigh no more than 60 pounds. What is the acceptable number of books that can be packed in a crate? *(Lesson 12–4)*

51. Geometry An obtuse angle measures more than 90° but less than 180°. If the measure of an obtuse angle is $3x$, what are the possible values for x? *(Lesson 12–5)*

$3x°$

CHAPTER 12 Test

1. **List** at least three verbal phrases that are used to describe inequalities.
2. If you multiply or divide each side of an inequality by a negative number, what must happen to the symbol for the inequality to remain true?
3. Before solving an inequality involving absolute value, which two cases must you consider?

Write an inequality for each graph.

4. 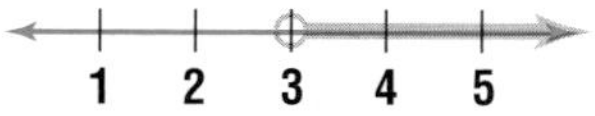

5. 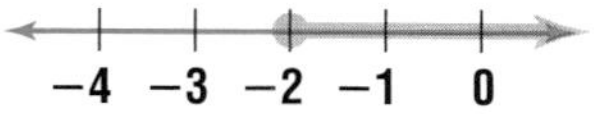

Solve each inequality. Check your solution.

6. $2 + x \geq 12$
7. $5t + 6 \leq 4t - 3$
8. $8 < -4t$
9. $-0.2x > -6$
10. $\frac{t}{4} > 1$
11. $-\frac{2}{5}m \leq 10$
12. $-3r - 1 \geq -16$
13. $7x - 12 < 30$
14. $2(h - 3) > 6$
15. $8(1 - 2z) \leq 25 + z$

Solve each inequality. Graph the solution.

16. $x + 1 > -2$ and $x + 1 < 6$
17. $4 < 3j - 2 \leq 7$
18. $2n + 5 \geq 15$ or $2n + 5 \leq 3$
19. $-6c > -24$ or $c + 0.25 < 1.3$
20. $|x + 3| \geq 4$
21. $|4b| \leq 16$

Graph each inequality.

22. $y \geq 5x - 6$
23. $4x - 2y > -6$

24. **Car Rental** Justine is renting a car that costs \$32 a day with free unlimited mileage. Since she is under the age of 25, it costs her an additional \$10 per day. Justine does not want to pay any more than \$200 on car rental costs. For what number of days can she rent a car?

25. **Manufacturing** Ball bearings are used to connect moving parts and minimize friction. Ball bearings for an automobile will work properly only if their diameter is within 0.01 inch, inclusive, of 5 inches. Write and solve an inequality to represent the range of acceptable diameters for these ball bearings.

CHAPTER 12

Preparing for Standardized Tests

Angle, Triangle, and Quadrilateral Problems

If there is no figure or diagram, draw one yourself.

You are already aware of some geometry concepts that you need to know for standardized tests. Be sure that you also know the meanings and definitions of the following terms.

right	obtuse	acute	triangle	quadrilateral
equilateral	isosceles	similar ($\approx$)	congruent ($\cong$)	collinear
reflection	translation	rotation	dilation	

State Test Example

The triangles below are similar. Find the length of side *KL*.

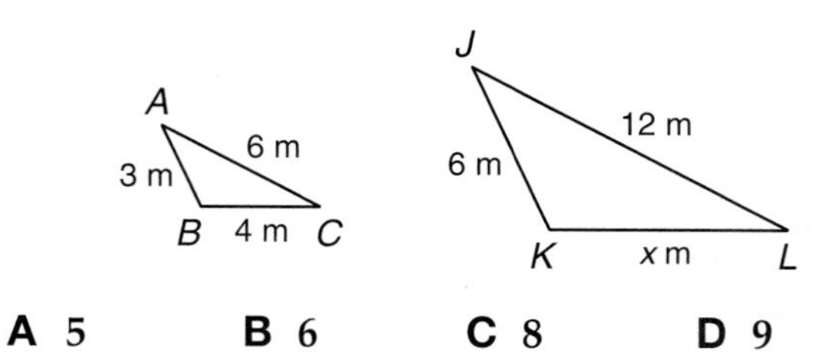

A 5 **B** 6 **C** 8 **D** 9

Hint The measures of corresponding sides of similar polygons are proportional.

Solution Find the corresponding sides of the two triangles. Side *AB* of $\triangle ABC$ corresponds to side *JK* of $\triangle JKL$. Side *BC* of $\triangle ABC$ corresponds to side *KL* of $\triangle JKL$. Using these two pairs of corresponding sides, write a proportion.

$\frac{AB}{JK} = \frac{BC}{KL}$

$\frac{3}{6} = \frac{4}{x}$ *Substitute side measures.*

$3x = 4(6)$ *Cross multiply.*

$3x = 24$

$x = 8$ *Divide each side by 3.*

Side *KL* measures 8 meters. The answer is C.

ACT Example

In the figure below, $\overline{ON}$ is congruent to $\overline{LN}$, $m\angle LON = 30$, and $m\angle LMN = 40$. What is $m\angle NLM$?

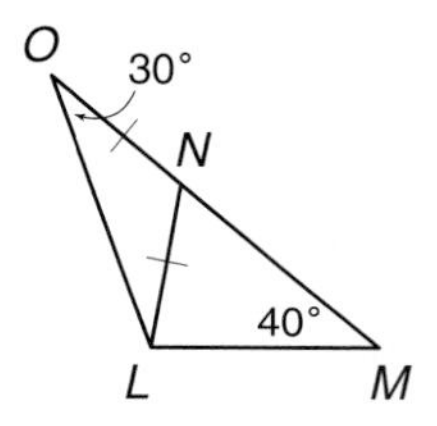

A 40 **B** 80 **C** 90 **D** 120

Hint Examine *all* of the triangles in the figure.

Solution Label the unknown angles as 1 and 2. Find $m\angle 2$.

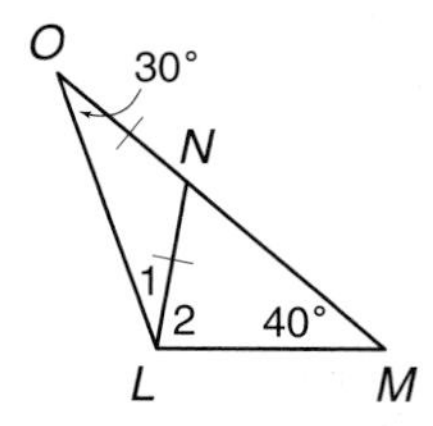

Since $\overline{ON} \cong \overline{LN}$, $\triangle ONL$ is isosceles and the base angles are equal. So, $m\angle 1 = 30$. Since the sum of the angle measures in any triangle is 180, find the sum of the angle measures in $\triangle OML$.

$180 = 30 + 40 + (30 + m\angle 2)$

$180 = 100 + m\angle 2$

$80 = m\angle 2$ *Subtract 100 from each side.*

The answer is B.

After you work each problem, record your answer on the answer sheet provided or on a sheet of paper.

Multiple Choice

1. The rectangles are similar. Find the value of x.

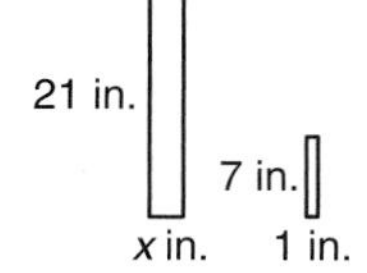

A 3 **B** 7
C 14 **D** 15

2. $\triangle ABC$ is isosceles. What is $m\angle C$?

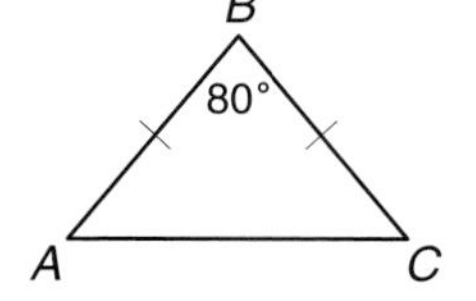

A 45 **B** 50
C 80 **D** 100

3. What is the sum of a, b, and c?

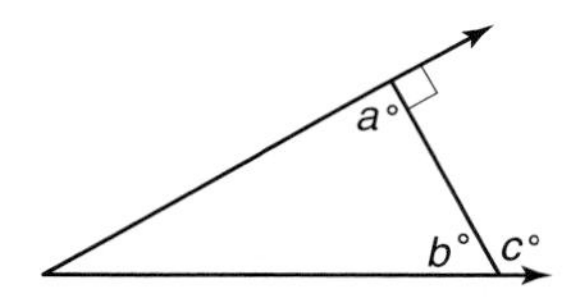

A 180 **B** 240
C 270 **D** 360
E It cannot be determined from the information given.

4. The formula for the area of a trapezoid is $A = \frac{1}{2}h(b_1 + b_2)$, where b_1 and b_2 are the lengths of the bases and h is the height. If the area of trapezoid $QRST$ is 72 square centimeters, then what is the height?

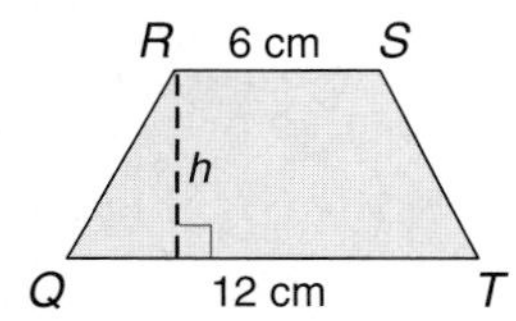

A 1 cm **B** 2 cm **C** 4 cm **D** 8 cm

5. Find the x-intercept of $y = -\frac{2}{3}x + 4$.

A 6 **B** −6 **C** 4 **D** 0

6. Simplify the expression $\frac{(-5)(4)|-6|}{-3}$.

A −120 **B** −40
C 40 **D** 120

7. The square of a number is 255 greater than twice the number. What is the number?

A 15 or −17 **B** 17 or −15
C 31 or −33 **D** 33 or −31

8. What is the value of $b + c$?

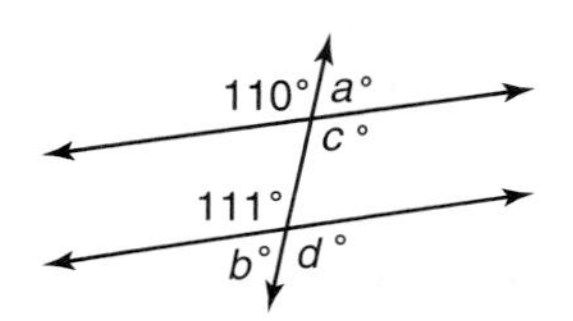

A 179 **B** 180
C 181 **D** 221

Grid In

9. In the figure, what is the sum of a, b, and c?

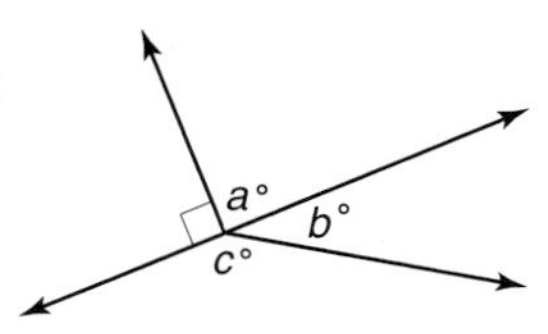

Extended Response

10. Draw a triangle that fits each description. If a drawing is not possible, explain why.

Part A

i. an obtuse equilateral triangle

ii. an acute equilateral triangle

Part B

i. an obtuse isosceles triangle

ii. an acute isosceles triangle

CHAPTER 13

Systems of Equations and Inequalities

FOLDABLES™ Study Organizer

Make this Foldable to help you organize information about the material in this chapter. Begin with four sheets of grid paper.

1. **Stack** sheets of paper with edges 4 grids apart to create tabs.

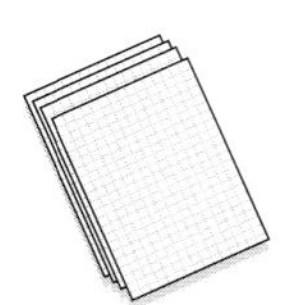

2. **Fold** up bottom edges. All tabs should be the same size.

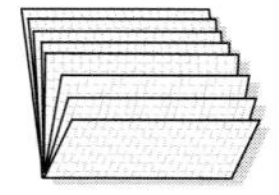

3. **Staple** along the fold.

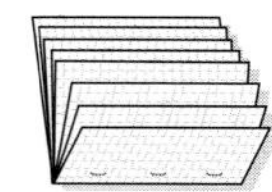

4. **Label** the tabs using lesson numbers and titles.

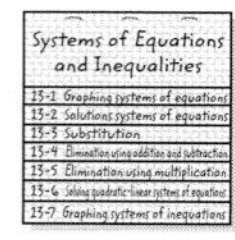

Reading and Writing As you read and study the chapter, use each page to write notes and examples.

Problem-Solving Workshop

Project

What system of inequalities can be used to describe the design at the right?

With a partner, make a design and write a system of inequalities that describes it. Exchange designs with another pair of students and write a system of inequalities that describes their design.

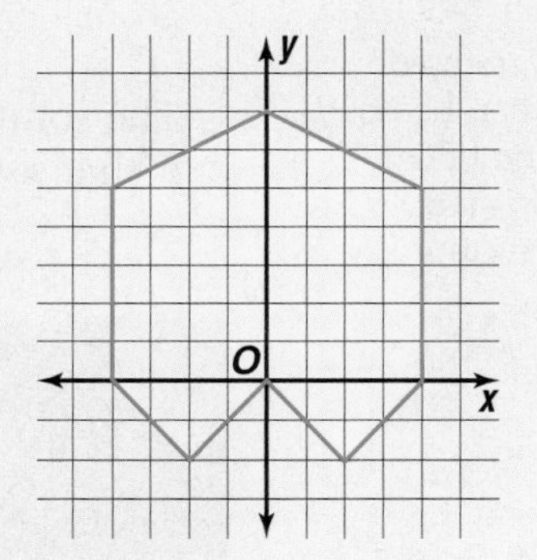

Working on the Project

Work with a partner and choose a strategy to help analyze and complete this project. Here are some suggestions to help you get started.

- Write an equation for the line that passes through points at $(-4, 0)$ and $(-4, 5)$. Write an inequality for the shaded area to the right of this line.
- Suppose your design has several colors. How could you describe the design so that another person could draw your design exactly without seeing the original?

Strategies

- Look for a pattern.
- Draw a diagram.
- Make a table.
- Work backward.
- Use an equation.
- Make a graph.
- Guess and check.

Technology Tools

- Use **drawing software** to make your design.
- Use a **graphing calculator** to find equations for the lines in the given design.

*inter*NET CONNECTION **Research** For more information about graphs of equations and inequalities, visit: www.algconcepts.com

Presenting the Project

Your portfolio should contain the following items:

- a system of inequalities that describes the design above,
- a sketch of your design on grid paper, and
- a system of inequalities that describes your design.

13–1 Graphing Systems of Equations

What You'll Learn

You'll learn to solve systems of equations by graphing.

Why It's Important

Sales Systems of equations can be used to determine how many items need to be sold in order to make a profit.
See Example 3.

A **system of equations** is a set of two or more equations with the same variables. Below are some examples of systems.

$$x + y = 5 \qquad\qquad \frac{a}{b} = 2$$
$$3x - 4y = -13 \qquad\qquad a + 2b = 8$$

The solution of a system of equations in two variables is an ordered pair that satisfies both equations.

System of Equations

Words: A system of equations is a set of two or more equations with the same variables. The solution is the ordered pair that satisfies all of the equations.

Model:

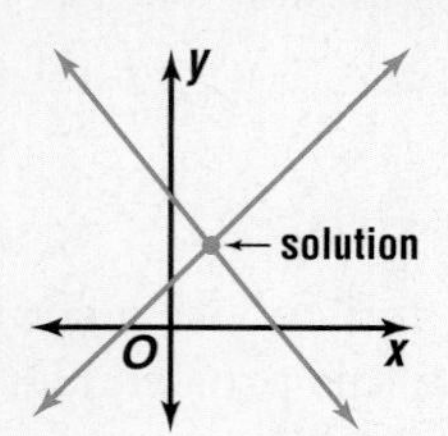

One method for solving a system of equations is to graph the equations on the same coordinate plane. The coordinates of the point of intersection are the solution.

Examples

Solve each system of equations by graphing.

1 $y = 2x$
$y = -x + 3$

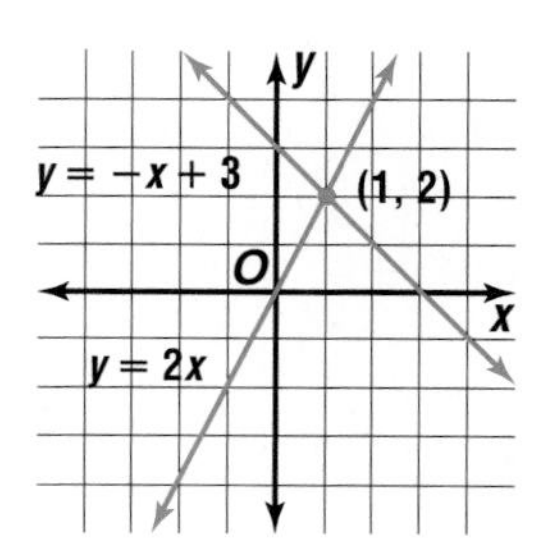

The graphs appear to intersect at (1, 2). Check this estimate by substituting the coordinates into each equation.

Look Back

Graphing Linear Equations: Lesson 7–5

Check:

$y = 2x$
$2 \stackrel{?}{=} 2(1)$ *Replace x with 1 and y with 2.*
$2 = 2 \checkmark$

$y = -x + 3$
$2 \stackrel{?}{=} -1 + 3$ *Replace x with 1 and y with 2.*
$2 = 2 \checkmark$

The solution of the system of equations is (1, 2).

$x - y = -2$
$x + y = 4$

The graphs appear to intersect at (1, 3). *Check this estimate.*

The solution of the system of equations is (1, 3).

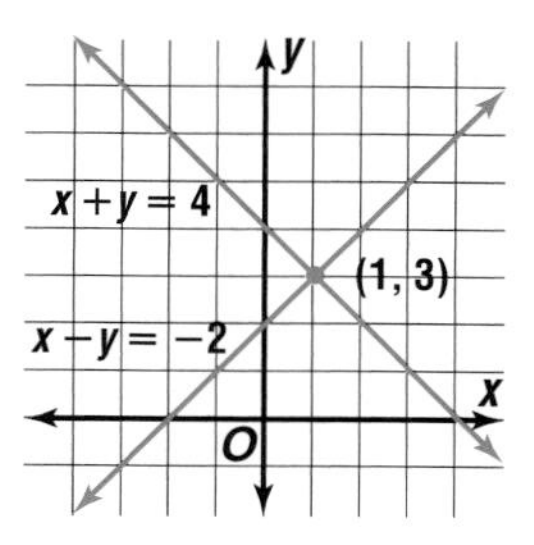

Your Turn

a. $y = 2x$
$y = -x + 6$

b. $x + y = 1$
$2x + y = 4$

A graphing calculator can be used to solve systems of equations or to check solutions.

Graphing Calculator Tutorial
See pp. 724–727.

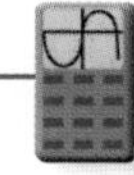

Graphing Calculator Exploration

You can use a graphing calculator to solve systems of equations.

$3x + y = 1$
$-x + 2y = 16$

Step 1 Solve each equation for y.

$$3x + y = 1 \rightarrow y = -3x + 1$$
$$-x + 2y = 16 \rightarrow y = \frac{1}{2}x + 8$$

Step 2 Graph the equations in the standard viewing window.

Press [Y=] to enter $-3x + 1$ as Y_1. Scroll down to enter $\frac{1}{2}x + 8$ as Y_2. Then press [ZOOM] 6.

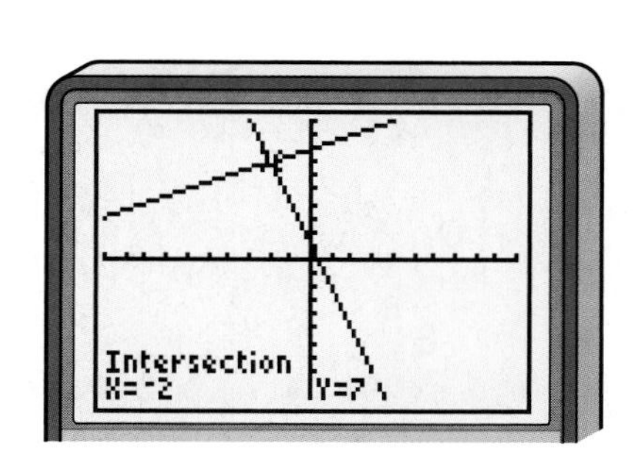

Step 3 Use the INTERSECT feature to find the intersection point.

Enter: [2nd] [CALC] 5 [ENTER] [ENTER] [ENTER]

The solution is (−2, 7). *Check by substituting the coordinates into the equations.*

Try These

Use a graphing calculator to solve each system of equations.

1. $y = x + 7$
$y = -x + 9$

2. $y = -3x$
$4x + y = 2$

3. $2x - y = 5$
$x + y = 16$

Real World

Example 3

Sales Link

The Math Club is selling T-shirts for a profit of $4 each and caps for a profit of $5 each. The club wants to sell 50 items and make a profit of $230. How many of each item should the club try to sell?

Let x = the number of T-shirts and y = the number of caps. You can write two equations to represent this situation.

$x + y = 50$ ← *the number of items*

$4x + 5y = 230$ ← *the total profit*

Graph $x + y = 50$ and $4x + 5y = 230$. The graphs appear to intersect at (20, 30). Check this estimate.

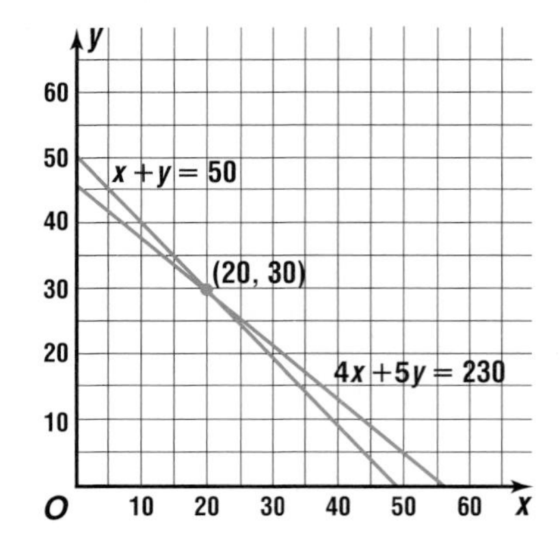

Check:

$x + y = 50$

$20 + 30 \stackrel{?}{=} 50$ $(x, y) = (20, 30)$

$50 = 50$ ✓

$4x + 5y = 230$

$4(20) + 5(30) \stackrel{?}{=} 230$ $(x, y) = (20, 30)$

$80 + 150 \stackrel{?}{=} 230$

$230 = 230$ ✓

They should try to sell 20 mascot beanbags and 30 caps.

Check for Understanding

Communicating Mathematics

1. **Describe** the solution of a system of equations.
2. **Sketch** a system of linear equations that has (2, 3) as its solution.
3. **Determine** the solution of the system of equations represented by each pair of lines.
   a. ℓ and m
   b. m and n
   c. n and ℓ

Vocabulary
system of equations

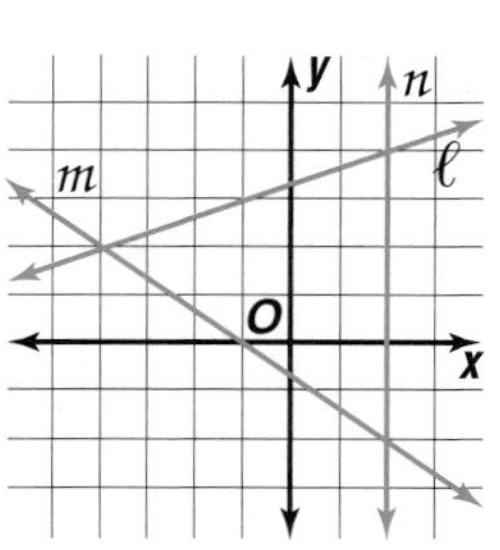

Guided Practice

Solve each system of equations by graphing. *(Examples 1–3)*

4. $y = x - 3$
   $y = -x + 3$
5. $x = -4$
   $y = \frac{1}{2}x + 3$
6. $x + y = 6$
   $y = 2$
7. $x + y = -2$
   $x - y = 4$

8. **Shopping** Randall would like to buy 10 bouquets. The standard bouquet costs $7, and the deluxe one costs $12. He can only afford to spend $100. *(Example 3)*
   a. Write a system of equations for the number of standard bouquets x and the number of deluxe bouquets y that he can buy.
   b. Use a graphing calculator to find the number of each type of bouquet he can buy.

Exercises

Practice

Homework Help

For Exercises	See Examples
9–14	1
15–24	2, 3

Extra Practice
See page 718.

Solve each system of equations by graphing.

9. $x = 5$
   $y = -4$
10. $x = 4$
    $y = x - 5$
11. $y = x + 2$
    $y = -x + 2$
12. $y = -2x + 3$
    $y = -\frac{1}{2}x$
13. $y = 6$
    $y = \frac{4}{3}x + 2$
14. $y = 2x - 7$
    $y = \frac{3}{2}x - 6$
15. $y = 2$
    $x + y = 7$
16. $x + y = 3$
    $x - y = 1$
17. $x - y = 1$
    $y = -2x - 7$
18. $2x - y = 4$
    $x + y = -4$
19. $x + y = -3$
    $\frac{1}{2}x + y = 0$
20. $5x + 4y = 10$
    $2x + y = 1$
21. Find the solution of the system $y = x + 4$ and $3x + 2y = 18$.
22. What is the solution of the system $2x + 10y = 0$ and $x + y = 4$?

Applications and Problem Solving

23. **Submarines** Two submarines began dives in the same vertical position to meet at a designated point. If one submarine was on a course approximated by the equation $x + 4y = -14$ and the other was on a course approximated by the equation $x + 3y = -8$, at what location would they meet? Write the coordinates of the point.

24. **Geometry** The graphs of the equations $y = x + 2$, $3x + y = 6$, and $y = 5x + 6$ contain the sides of a triangle.
    a. Graph the equations.
    b. Find the coordinates of the vertices of the triangle.

25. **Critical Thinking** Use your knowledge of slope and y-intercepts to write a system of equations with a solution of (0, 4).

Mixed Review

26. Which ordered pair, $(-2, 0)$, $(-1, 2)$, $(0, -1)$, or $(2, -2)$, is a solution of $3x + 4y \geq 2$? *(Lesson 12–7)*

Solve each inequality. Then graph the solution set. *(Lesson 12–6)*

27. $|x - 3| < 2$
28. $|a + 5| \geq 3$
29. $|2m + 2| > 6$

30. **School** Everyone in Mr. McClain's algebra class is at least 15 years old. Write an inequality to represent this information and then make a graph of the inequality. *(Lesson 12–1)*

Standardized Test Practice

31. **Short Response** Solve $2x^2 - 7x + 6 = 0$ by factoring. *(Lesson 11–4)*

32. **Multiple Choice** Which graph represents the function $y = x^2 + 2x - 3$? *(Lesson 11–1)*

A
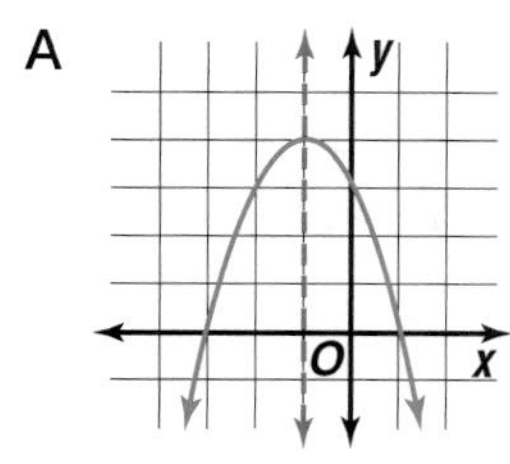

B
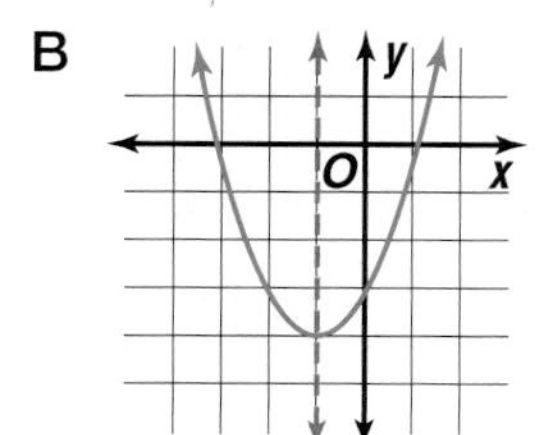

C
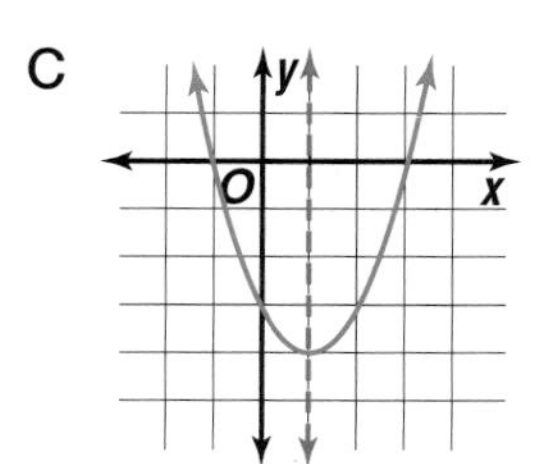

13–2 Solutions of Systems of Equations

What You'll Learn

You'll learn to determine whether a system of equations has one solution, no solution, or infinitely many solutions by graphing.

Why It's Important

Trains Train engineers must know if the tracks they are running on intersect another track. *See Example 6.*

Graphs of systems of linear equations may be intersecting lines, parallel lines, or the same line. Systems of equations can be described by the number of solutions they have.

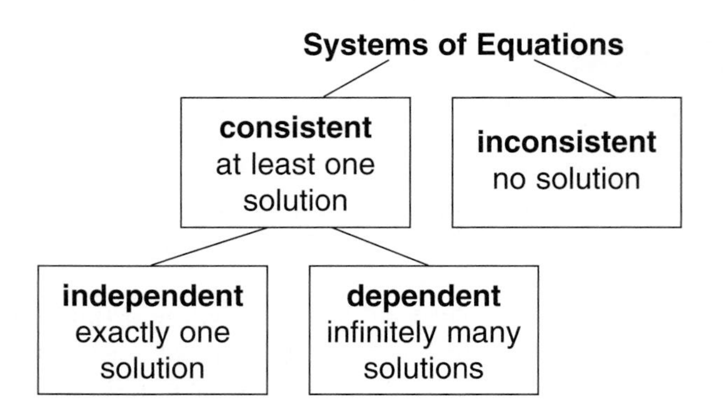

The different possibilities for the graphs of two linear equations are summarized in the following table.

Graph	Description of Graph	Slopes and Intercepts of Lines	Number of Solutions	Type of System
$y = -x + 3$, $y = x + 1$, $(1, 2)$	intersecting lines	different slopes	1	consistent and independent
$x + 2y = 2$, $2x + 4y = 4$	same line	same slope, same intercepts	infinitely many	consistent and dependent
$y = -2x + 4$, $y = -2x$	parallel lines	same slope, different intercepts	0	inconsistent

Examples

State whether each system is *consistent and independent, consistent and dependent,* or *inconsistent.*

1

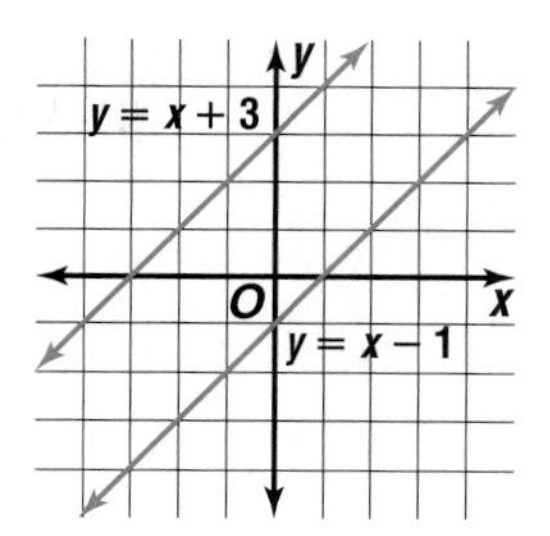

The graphs appear to be parallel lines. Since they do not intersect, there is no solution. This system is inconsistent.

2

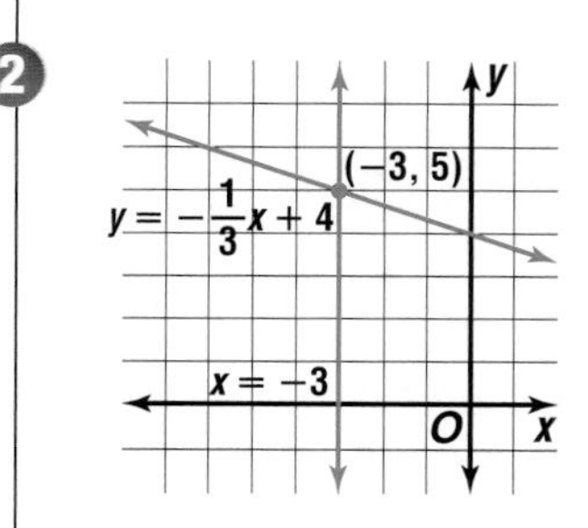

The graphs appear to intersect at the point at $(-3, 5)$. Because there is one solution, this system of equations is consistent and independent.

3

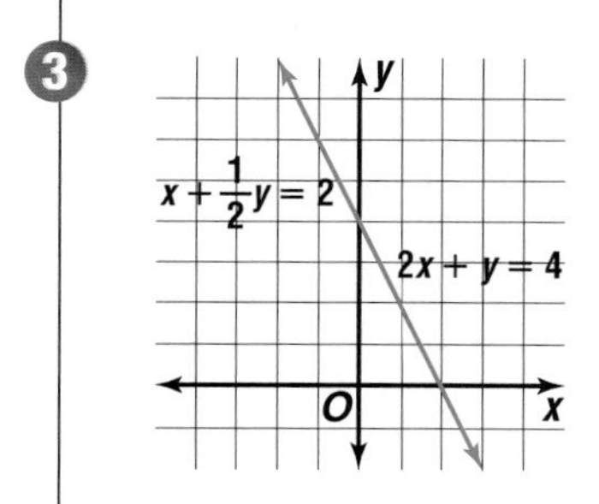

Both equations have the same graph. Because any ordered pair on the graph will satisfy both equations, there are infinitely many solutions. The system is consistent and dependent.

Your Turn

a.

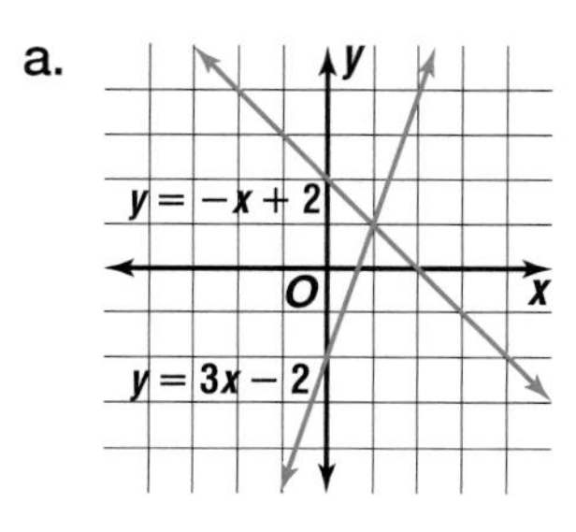

b.

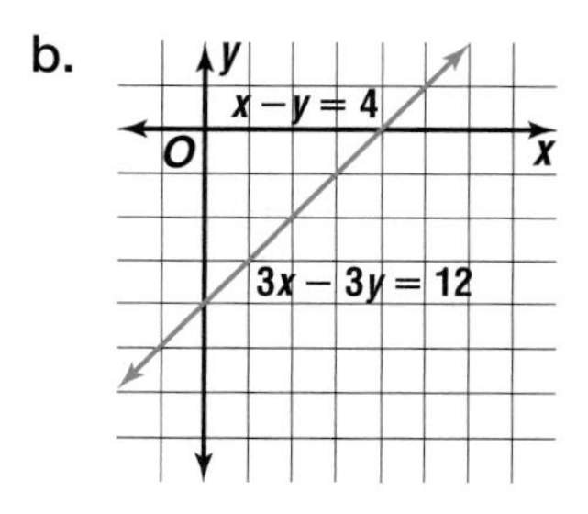

You can determine the number of solutions to a system of equations by graphing.

Examples

Determine whether each system of equations has *one* solution, *no* solution, or *infinitely many* solutions by graphing. If the system has one solution, name it.

4 $y = x + 2$
$y = -3x - 6$

The graphs appear to intersect at $(-2, 0)$. Therefore, this system of equations has one solution, $(-2, 0)$. Check that $(-2, 0)$ is a solution to each equation.

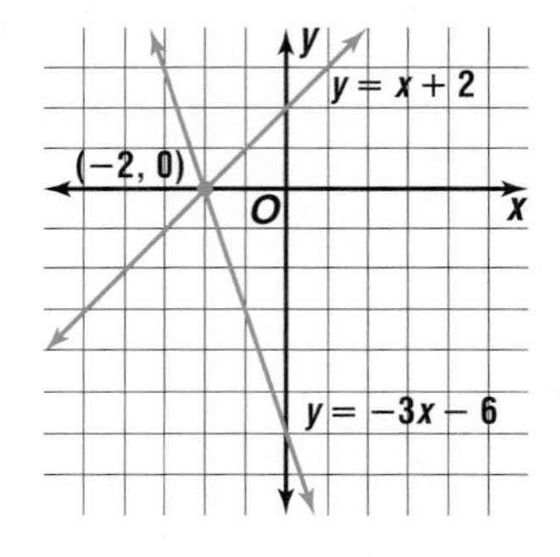

Check:

$y = x + 2$		$y = -3x - 6$	
$0 \stackrel{?}{=} -2 + 2$	*Replace x with −2*	$0 \stackrel{?}{=} -3(-2) - 6$	*Replace x with −2*
$0 = 0$ ✓	*and y with 0.*	$0 = 0$ ✓	*and y with 0.*

The solution of the system of equations is $(-2, 0)$.

5 $2x + y = 4$
$2x + y = 6$

Write each equation in slope-intercept form.

$2x + y = 4 \rightarrow y = -2x + 4$
$2x + y = 6 \rightarrow y = -2x + 6$

The graphs have the same slope and different y-intercepts. The system of equations has no solution.

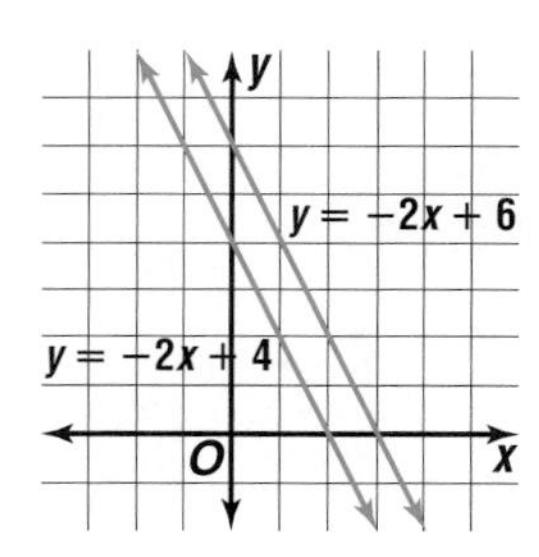

Your Turn

c. $y = x + 3$
$y = -2x + 3$

d. $2x + y = 6$
$4x + 2y = 12$

e. $3x - y = 3$
$3x - y = 0$

f. $y - 2x = 0$
$y + x = 6$

Example 6 **Transportation Link**

The system of equations below represents the tracks of two trains. Do the tracks intersect, run parallel, or are the trains running on the same track? Explain.

$$x + 2y = 4$$
$$3x + 6y = 12$$

$x + 2y = 4$

$3x + 6y = 12$

$3(x + 2y) = 3(4)$

$x + 2y = 4$ *Divide each side by 3.*

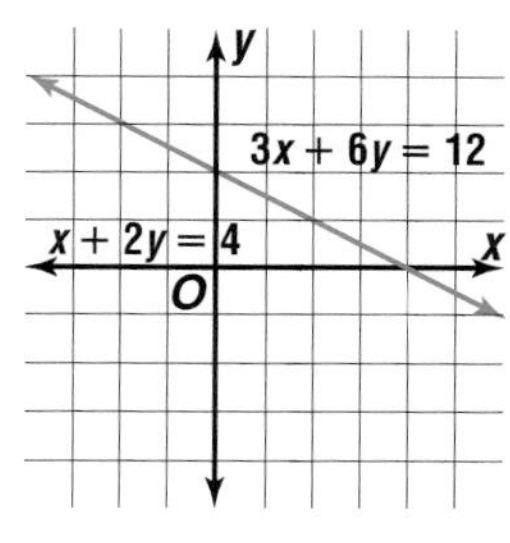

One equation is a multiple of the other. Each equation has the same graph and there are infinitely many solutions. Therefore, the trains are running on the same track.

Check for Understanding

Communicating Mathematics

1. **Describe** the possible graphs of a system of two linear equations.
2. **State** the number of solutions for each system of equations described below.
   a. One equation is a multiple of the other.
   b. The equations have the same y-intercept and different slopes.
   c. The equations have the same slope and different y-intercepts.

Vocabulary
consistent
independent
dependent
inconsistent

Math Journal

3. **Create** memory devices or other ways to remember the definitions of the terms in the Vocabulary box. Describe them in your journal.

Guided Practice

State whether each system is *consistent and independent*, *consistent and dependent*, or *inconsistent*. *(Examples 1–3)*

4.

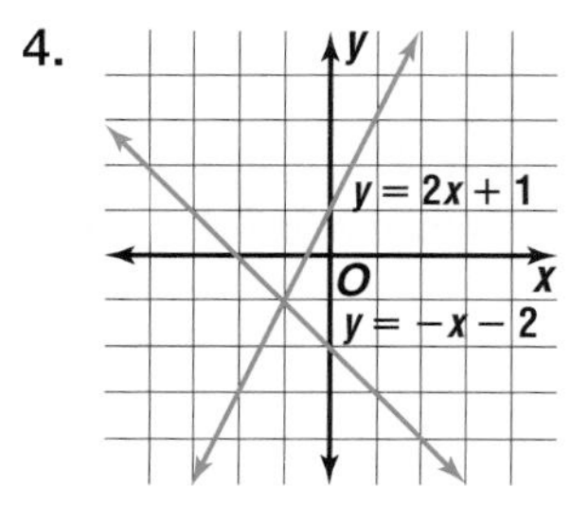

5.

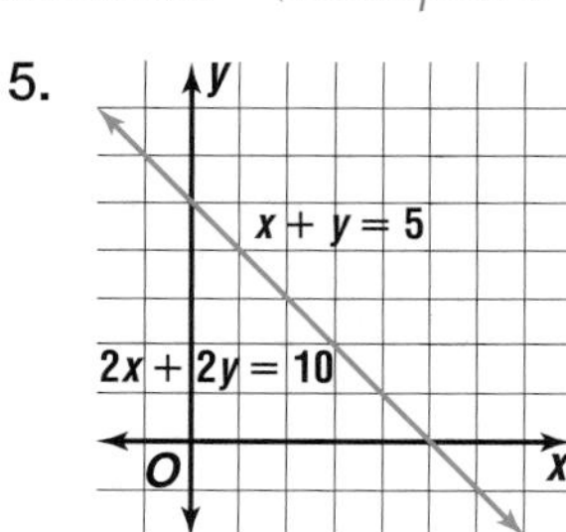

Determine whether each system of equations has *one* solution, *no* solution, or *infinitely many* solutions by graphing. If the system has one solution, name it. *(Examples 4 & 5)*

6. $y = x$
$y = x + 5$

7. $y = -x$
$y = 3x - 4$

8. $2x + y = -3$
$6x + 3y = -9$

9. **Animals** A dog sees a cat 60 feet away and starts running after it at 50 feet per second. At the same time, the cat runs away at 30 feet per second. This situation can be represented by the following system of equations and the graph at the right. *(Example 6)*

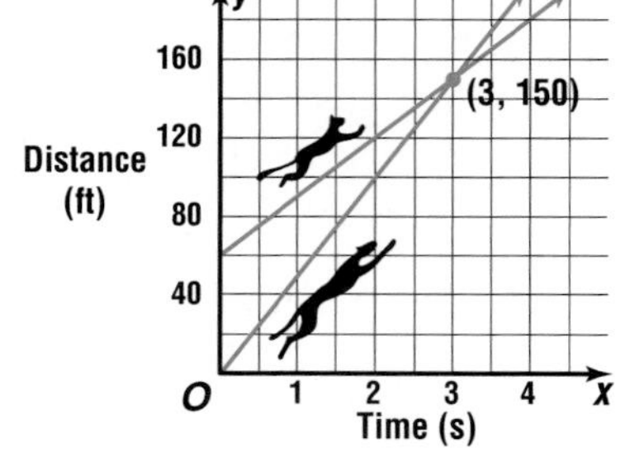

$y = 50x$ ← *dog* $\quad y = 30x + 60$ ← *cat*

a. Is this system of equations *consistent and independent*, *consistent and dependent*, or *inconsistent*? Explain.

b. Explain what the point at (3, 150) represents.

Exercises

Practice

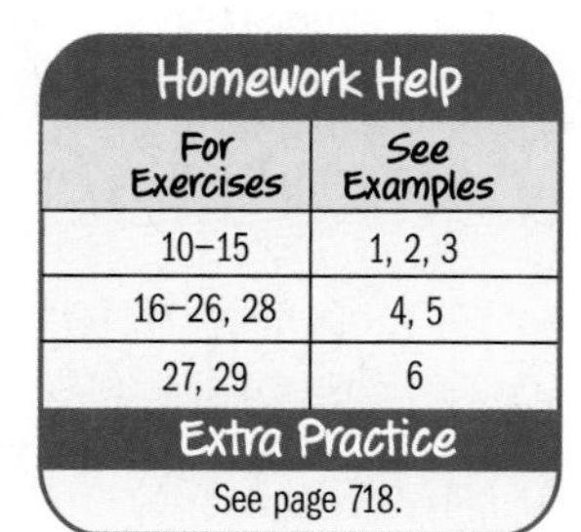

Homework Help	
For Exercises	See Examples
10–15	1, 2, 3
16–26, 28	4, 5
27, 29	6
Extra Practice	
See page 718.	

State whether each system is *consistent and independent*, *consistent and dependent*, or *inconsistent*.

10.

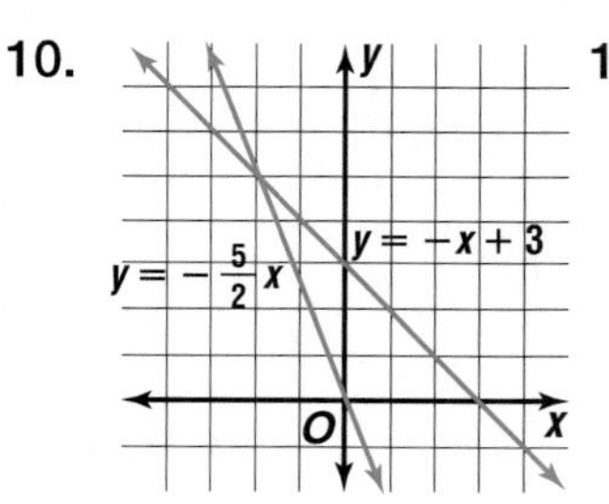

11.

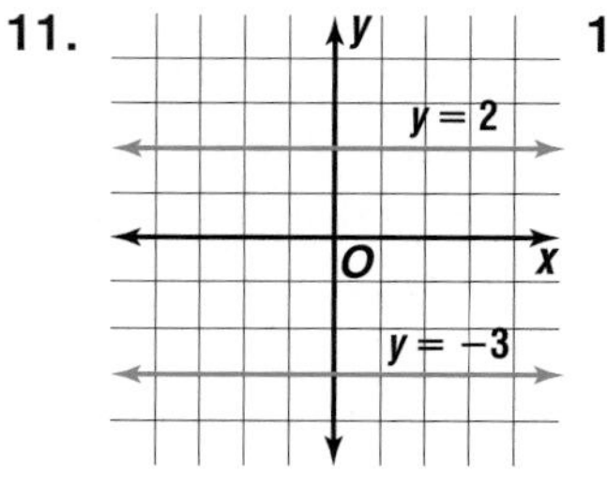

12.

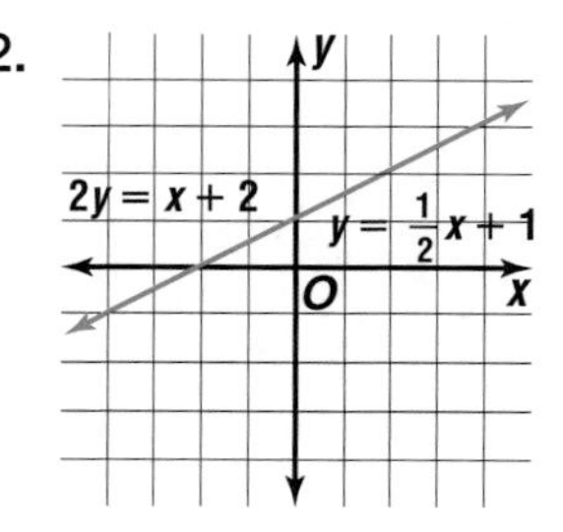

13.

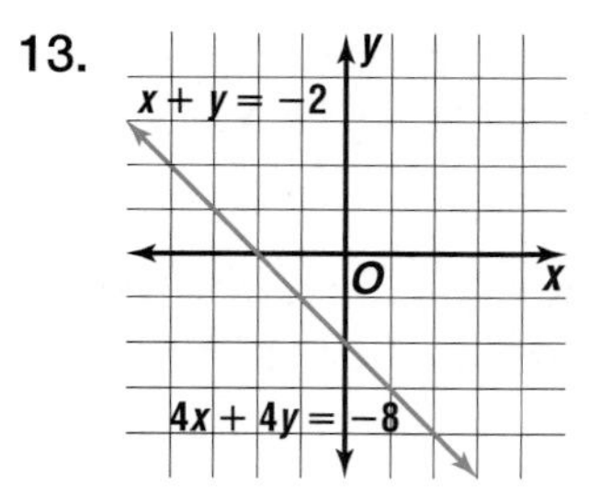

14.

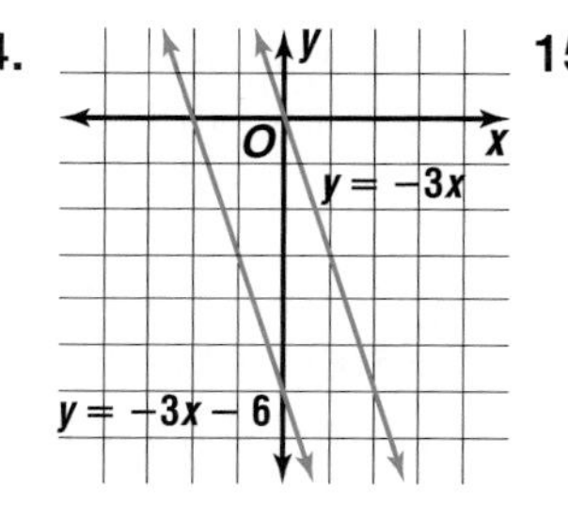

15.

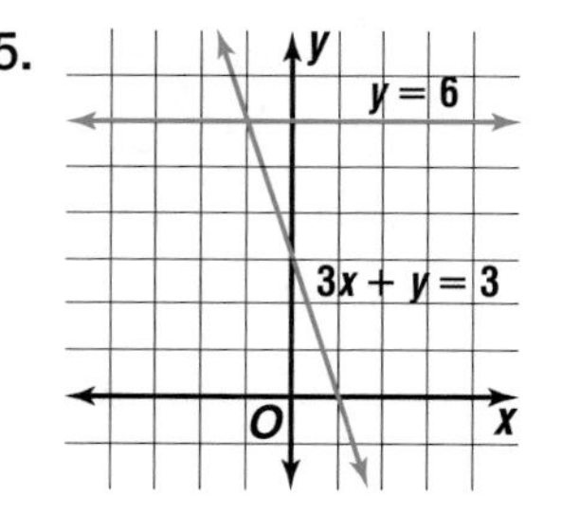

Determine whether each system of equations has *one* solution, *no* solution, or *infinitely many* solutions by graphing. If the system has one solution, name it.

16. $y = 4x - 6$
$y = 4x - 1$

17. $y = 3x - 2$
$4y = 12x - 8$

18. $y = \frac{1}{2}x$
$y = -2x + 5$

19. $y = -x + 3$
$y = \frac{1}{5}x - 3$

20. $x = 3$
$2x - 3y = 0$

21. $3x - 2y = -6$
$3x - 2y = 6$

22. $x + y = 5$
$2x + 2y = 8$

23. $6x + y = -3$
$-x + y = 4$

24. $x - 4y = -4$
$\frac{1}{2}x - 2y = -2$

25. Does the system $x - y = 4$ and $x - 3y = 2$ have *one* solution, *no* solution, or *infinitely many* solutions? If the system has one solution, name it.

26. Without graphing, determine whether the system $x - 3y = 11$ and $2x - 6y = -5$ has *one* solution, *no* solution, or *infinitely many* solutions. Explain how you know.

Applications and Problem Solving

27. **Animals** Refer to Exercise 9. Suppose the dog is chasing another dog whose distance y can be represented by the equation $y = 50x + 20$. What is the solution? Explain what it represents in terms of the dogs.

28. **Ballooning** A hot air balloon is 10 meters above the ground and rising at a rate of 15 meters per minute. Another balloon is 150 meters above the ground and descending at a rate of 20 meters per minute.
    a. Write a system of equations to represent the balloons.
    b. What is the solution of the system of equations?
    c. Explain what the solution means.

29. **Critical Thinking** Write an equation of a line in slope-intercept form that, together with the equation $x + 3y = 9$, forms a system that is inconsistent.

Mixed Review

30. What is the solution of the system $y = x + 4$ and $y = -3x - 4$? *(Lesson 13–1)*

31. Graph $y \geq 2x + 1$. *(Lesson 12–7)*

32. **Number Theory** Find two numbers if one number is 6 less than the other and whose product is 7. *(Lesson 11–5)*

Factor each polynomial. If the polynomial cannot be factored, write *prime*. *(Lesson 10–2)*

33. $4x - 8$

34. $13x + 2m$

35. $3a^2b^2 + 6ab - 9a$

Standardized Test Practice

36. **Short Response** Write a square root whose best whole number estimate is 12. *(Lesson 8–6)*

13–3 Substitution

What You'll Learn

You'll learn to solve systems of equations by the substitution method.

Why It's Important

Metallurgy Systems of equations can be used to make metal alloys.
See Example 6.

Solving Equations with Algebra Tiles: Lesson 3–5

In Lesson 13–2, you learned to solve systems of equations by graphing. But sometimes the exact coordinates of the point where lines intersect cannot be easily determined from a graph. The solution of a system can also be found by using an algebraic method called **substitution**.

Materials: 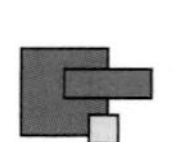 algebra tiles  equation mat

Use a model to solve the system $y = x + 1$ and $3x + y = 9$.

Step 1 Let a green tile represent x. Then, since $y = x + 1$, 1 green tile and 1 yellow tile represent y.

$x =$ [x] $\quad y =$ [x] [1]

Step 2 Represent $3x + y = 9$ on the equation mat. On one side of the mat, place three green tiles for $3x$ and 1 green tile and 1 yellow tile for y. On the other side, place 9 yellow tiles.

$3x + y = 9$

Try These

1. Use what you know about equation mats and zero pairs to solve the equation. What is the value of x?
2. Use the value of x from Exercise 1 and the equation $y = x + 1$ to find the value of y.
3. What is the solution of the system of equations?

Examples

Use substitution to solve each system of equations.

1 $y = 2x$
$3x + y = 5$

The first equation tells you that y is equal to $2x$. So, substitute $2x$ for y in the second equation. Then solve for x.

$$3x + y = 5$$
$$3x + 2x = 5 \quad \textit{Replace } y \textit{ with } 2x.$$
$$5x = 5$$
$$\frac{5x}{5} = \frac{5}{5} \quad \textit{Divide each side by 5.}$$
$$x = 1$$

Now substitute 1 for x in either equation and solve for y. *Choose the equation that is easier to solve.*

$y = 2x$

$y = 2(1)$ or 2 *Replace x with 1.*

The solution of this system of equations is (1, 2). You can see from the graph that the solution is correct. *You can also check by substituting (1, 2) into each of the original equations.*

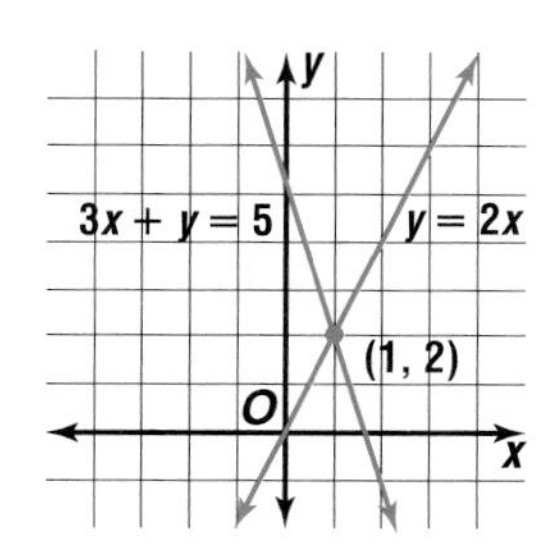

$x + y = 1$
$x = y + 6$

Substitute $y + 6$ for x in the first equation. Then solve for y.

$$x + y = 1$$
$$y + 6 + y = 1 \quad \textit{Replace } x \textit{ with } y + 6.$$
$$2y + 6 = 1$$
$$2y + 6 - 6 = 1 - 6 \quad \textit{Subtract 6 from each side.}$$
$$2y = -5$$
$$\frac{2y}{2} = \frac{-5}{2} \quad \textit{Divide each side by 2.}$$
$$y = -\frac{5}{2}$$

Now substitute $-\frac{5}{2}$ for y in either equation and solve for x.

$x = y + 6$

$x = -\frac{5}{2} + 6$ *Replace y with* $-\frac{5}{2}$.

$x = \frac{7}{2}$

The solution of this system of equations is $\left(\frac{7}{2}, -\frac{5}{2}\right)$. *Check by substituting* $\left(\frac{7}{2}, -\frac{5}{2}\right)$ *into each of the original equations.*

Example 3

$x - 3y = 3$
$2x - y = 11$

Solve the first equation for x since the coefficient of x is 1.

$x - 3y = 3 \quad \rightarrow \quad x = 3 + 3y$

Next, find the value of y by substituting $3 + 3y$ for x in the second equation.

$$2x - y = 11$$
$$2(3 + 3y) - y = 11$$
$$6 + 6y - y = 11$$
$$6 + 5y = 11$$
$$6 + 5y - 6 = 11 - 6$$
$$5y = 5$$
$$\frac{5y}{5} = \frac{5}{5}$$
$$y = 1$$

Now substitute 1 for y in either equation and solve for x.

$$x - 3y = 3$$
$$x - 3(1) = 3 \quad \textit{Replace } y \textit{ with 1.}$$
$$x - 3 = 3$$
$$x - 3 + 3 = 3 + 3$$
$$x = 6$$

The solution is (6, 1).

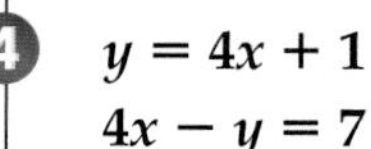

a. $x = 1 + 6y$
$x + 2y = 9$

b. $x + y = 3$
$-2x - 7y = 4$

In Lesson 13–2, you learned how to tell whether a system has one solution, no solution, or infinitely many solutions by looking at the graph. You can also determine this information algebraically.

Examples

Use substitution to solve each system of equations.

4 $y = 4x + 1$
$4x - y = 7$

Find the value of x by substituting $4x + 1$ for y in the second equation.

$$4x - y = 7$$
$$4x - (4x + 1) = 7 \quad \textit{Replace } y \textit{ with } 4x + 1.$$
$$4x - 4x - 1 = 7 \quad \textit{Distributive Property}$$
$$-1 = 7$$

The statement $-1 = 7$ is false. This means that there are no ordered pairs that are solutions to both equations. Compare the slope-intercept forms of the equations, $y = 4x + 1$ and $y = 4x - 7$. Notice that the graphs of these equations have the same slope but different y-intercepts. Thus, the lines are parallel, and the system has no solution.

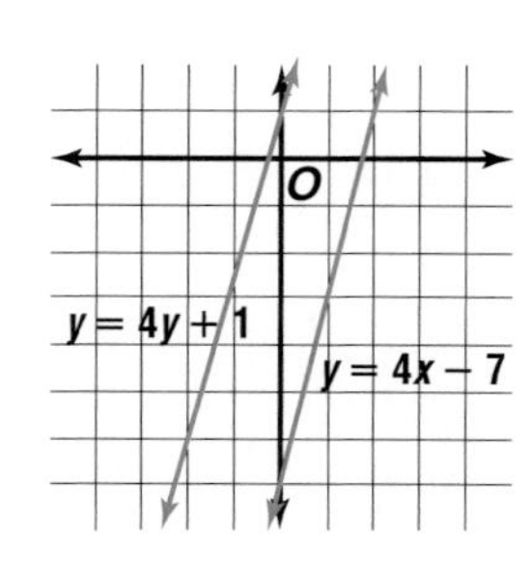

5 $x = 3 - 2y$
$2x + 4y = 6$

$$\begin{aligned} 2x + 4y &= 6 \\ 2(3 - 2y) + 4y &= 6 \quad \textit{Replace } x \textit{ with } 3 - 2y. \\ 6 - 4y + 4y &= 6 \quad \textit{Distributive Property} \\ 6 &= 6 \end{aligned}$$

The statement $6 = 6$ is true. This means that an ordered pair for any point on either line is a solution to both equations. The system has infinitely many solutions.

Your Turn

Use substitution to solve each system of inequalities.

c. $y = 2x + 1$
$4x - 2y = -9$

d. $x - 6y = 5$
$2x = 12y + 10$

Example 6
Metals Link

Prerequisite Skills Review
Decimals and Percents, p. 689

Systems of equations can be used to solve mixture problems.

A certain metal alloy is 25% copper. Another metal alloy is 50% copper. How much of each alloy should be used to make 1000 grams of a metal alloy that is 45% copper?

Explore Let a = the number of grams of the 25% copper alloy.
Let b = the number of grams of the 50% copper alloy.

	25% Copper	50% Copper	45% Copper
Total Grams	a	b	1000
Grams of Copper	$0.25a$	$0.50b$	$0.45(1000)$

Plan Write two equations to represent the information.

$a + b = 1000$ ← *total grams of alloy*
$0.25a + 0.50b = 0.45(1000)$ ← *grams of copper*

Solve Use substitution to solve this system. Since $a + b = 1000$, $a = 1000 - b$.

$$\begin{aligned} 0.25a + 0.50b &= 0.45(1000) \\ 0.25(1000 - b) + 0.50b &= 0.45(1000) \quad \textit{Replace } a \textit{ with } 1000 - b. \\ 250 - 0.25b + 0.50b &= 450 \quad \textit{Distributive Property} \\ 250 + 0.25b &= 450 \\ 250 + 0.25b - 250 &= 450 - 250 \quad \textit{Subtract 250.} \\ 0.25b &= 200 \\ \frac{0.25b}{0.25} &= \frac{200}{0.25} \quad \textit{Divide each side by 0.25.} \\ b &= 800 \end{aligned}$$

(continued on the next page)

Now substitute 800 for b in either equation and solve for a.

$$a + b = 1000$$
$$a + 800 = 1000 \quad \textit{Replace b with 800.}$$
$$a + 800 - 800 = 1000 - 800 \quad \textit{Subtract 800 from each side.}$$
$$a = 200$$

So, 200 grams of the 25% copper alloy and 800 grams of the 50% copper alloy should be used.

Examine Check by substituting (200, 800) into the original equations. The solution is correct.

Check for Understanding

Communicating Mathematics

Vocabulary
substitution

1. **Explain** when you might use substitution instead of graphing to solve a system of equations.
2. **State** what you would conclude if the solution of a system of equations yields the equation $4 = 4$.
3. **You Decide?** Faith and Todd are using substitution to solve the system $x + 3y = 8$ and $4x - y = 9$. Faith says that the first step is to solve for x in $x + 3y = 8$. Todd disagrees. He says to solve for y in $4x - y = 9$. Who is right and why?

Guided Practice

Use substitution to solve each system of equations. *(Examples 1–5)*

4. $y = x - 4$; $3x + 2y = 2$
5. $y = 3x$; $7x - y = 16$
6. $x = 2y$; $4x + 2y = 15$
7. $3x - 7y = 12$; $x - 2y = 4$
8. $x = 2y + 5$; $3x - 6y = 15$
9. $4y = -3x + 8$; $3x + 4y = 6$
10. **Mixtures** MX Labs needs to make 500 gallons of a 34% acid solution. The only solutions available are 25% acid and 50% acid. How many gallons of each should be mixed to make the 34% solution? *(Example 6)*

Exercises

Practice

Homework Help	
For Exercises	See Examples
11, 13, 15–17, 31	1, 4
12, 14, 18	2, 5
19–30	3
Extra Practice	
See page 718.	

Use substitution to solve each system of equations.

11. $y = x$; $3x + y = 4$
12. $x = 2y$; $x + y = 3$
13. $y = 3x$; $x + 2y = -21$
14. $2y = x$; $x - y = 10$
15. $y = 3x - 8$; $3x - y = 12$
16. $y = x + 7$; $x + y = 1$
17. $y = \frac{1}{2}x + 3$; $y - 5 = x$
18. $x = 7 - y$; $2x - y = 8$
19. $x - 3y = -9$; $5x - 2y = 7$
20. $x + y = 0$; $4x + 4y = 0$
21. $x - 2y = -2$; $5x - 4y = 2$
22. $2x - y = 6$; $3x - 5y = 9$

23. $x - y = 12$
$3x - y = 16$

24. $4x - 3y = -6$
$x + 5y = 10$

25. $x - 3y = 3$
$2x + 9y = 11$

26. $4x - y = -3$
$y + 2 = x$

27. $x - 6y = 5$
$2x - 12y = 10$

28. $x - 5y = 11$
$3x + y = 7$

29. Use substitution to solve $2x - y = -4$ and $-3x + y = -9$.

30. What is the solution of the system $x - 2y = 5$ and $3x - 5y = 8$?

Applications and Problem Solving

31. **Driving** Anita and Tionna were both driving the same route from college to their hometown. Anita left an hour before Tionna. Anita drove at an average speed of 55 miles per hour, and Tionna averaged 65 miles per hour.
   a. After how many hours did Tionna catch up with Anita?
   b. How many miles did she drive?

32. **Exercise** Laura and Ji-Yong were jogging on a 10-mile path. Laura had a 2-mile head start on Ji-Yong. If Laura ran at an average rate of 5 miles per hour and Ji-Yong ran at an average rate of 8 miles per hour, how long would it take for Ji-Yong to catch up with Laura?

33. **Critical Thinking** Suppose you have three equations *A*, *B*, and *C*. Suppose a system containing any two of the equations is consistent and independent. Must the three equations together be consistent? Give an example or a counterexample to support your answer.

Mixed Review

34. Determine whether the system $y = -2x$ and $2y + 4x = 0$ has *one* solution, *no* solution, or *infinitely many* solutions by graphing. If the system has one solution, name it. *(Lesson 13–2)*

35. Solve $x - y = -5$ and $x + y = 3$ by graphing. *(Lesson 13–1)*

Write a compound inequality for each solution set shown. *(Lesson 12–5)*

36. 37.

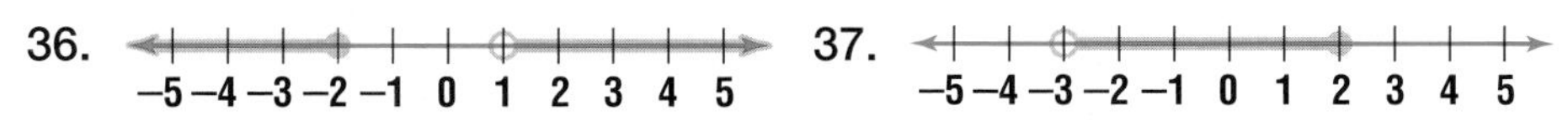

Standardized Test Practice

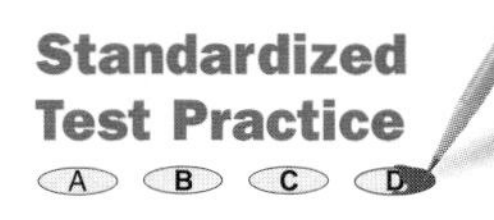

38. **Short Response** Solve $-5 + x > 12.8$. *(Lesson 12–2)*

39. **Grid In** Mr. Drew is a salesperson. The formula for his daily income is $t(s) = 0.15s + 25$, where *s* is his total sales for the day. Suppose Mr. Drew earns $94 on Monday. What were his total sales in dollars for that day? *(Lesson 6–4)*

Quiz 1 Lessons 13–1 through 13–3

Determine whether each system of equations has *one* solution, *no* solution, or *infinitely many* solutions by graphing. If the system has one solution, name it. *(Lessons 13–1 & 13–2)*

1. $y = x - 5$
$y = -\frac{1}{4}x$

2. $y = 2x + 5$
$y = x + 3$

3. $x + y = 3$
$x + y = 1$

4. $y = -2x + 3$
$2y = -4x + 6$

5. **Mixtures** Ms. Williams mixed nuts that cost $3.90 per pound with nuts that cost $4.30 per pound to make a 50-pound mixture of nuts worth $4.20 per pound. How many pounds of each type of nut did she use? *(Lesson 13–3)*

13–4 Elimination Using Addition and Subtraction

What You'll Learn

You'll learn to solve systems of equations by the elimination method using addition and subtraction.

Why It's Important

Entertainment Solving systems of equations using elimination could be used to determine entertainment costs. *See Exercise 35.*

Another algebraic method for solving systems of equations is called **elimination**. You can eliminate one of the variables by adding or subtracting the equations.

Use elimination when the coefficients of one of the variables are equal or additive inverses. For example, consider the system below.

$$5x + 11y = 12 \qquad 2x + 11y = 36$$

Notice that the coefficient of y in both equations is the same. You can solve this system of equations in three steps.

Step 1: Subtract the two equations, so that y is eliminated. The result is an equation only in x: $3x = -24$.

Step 2: The next step is to solve for x: $x = -8$.

Step 3: Substitute the value of x in any one of the two original equations and solve for y: $y = \frac{52}{11}$.

Example 1 **Use elimination to solve the system of equations.**

$\mathbf{2x + y = 3}$

$\mathbf{x + y = 1}$

$$\begin{array}{rl} 2x + y = 3 & \textit{Write the equations in column form.} \\ (-)\ x + y = 1 & \textit{Subtract the equations to eliminate the } y \textit{ terms.} \\ \hline x + 0 = 2 & \\ x = 2 & \text{The value of } x \text{ is 2.} \end{array}$$

Now substitute in either equation to find the value of y. *Choose the equation that is easier for you to solve.*

$$\begin{array}{rl} x + y = 1 & \\ 2 + y = 1 & \textit{Replace } x \textit{ with 2.} \\ 2 + y - 2 = 1 - 2 & \textit{Subtract 2 from each side.} \\ y = -1 & \text{The value of } y \text{ is } -1. \end{array}$$

The solution of the system of equations is $(2, -1)$.

Check:

$$\begin{array}{rl} 2x + y = 3 & \\ 2(2) + (-1) \stackrel{?}{=} 3 & \textit{Replace } (x, y) \textit{ with } (2, -1). \\ 4 + (-1) \stackrel{?}{=} 3 & \\ 3 = 3 & \checkmark \end{array}$$

$$\begin{array}{rl} x + y = 1 & \\ 2 + (-1) \stackrel{?}{=} 1 & \textit{Replace } (x, y) \textit{ with } (2, -1). \\ 1 = 1 & \checkmark \end{array}$$

Your Turn **Use elimination to solve each system of equations.**

a. $x + y = 5$
$2x + y = 4$

b. $3x + 5y = -2$
$3x - 2y = -16$

Example 2
Entertainment Link

A group of 3 adults and 10 students paid \$102 for a cavern tour. Another group of 3 adults and 7 students paid \$84 for the tour. Find the admission price for an adult and for a student.

Let a = the price for an adult and s = the price for a student.

Write two equations to represent this situation.

$3a + 10s = 102$ ← *total cost for the first van*
$3a + 7s = 84$ ← *total cost for the second van*

$$\begin{array}{rl} 3a + 10s = 102 & \text{Write the equations in column form.} \\ (-)\ 3a + 7s = 84 & \text{Subtract the equations to eliminate the } a \text{ terms.} \\ \hline 0 + 3s = 18 & \\ 3s = 18 & \\ \frac{3s}{3} = \frac{18}{3} & \text{Divide each side by 3.} \\ s = 6 & \text{The value of } s \text{ is 6.} \end{array}$$

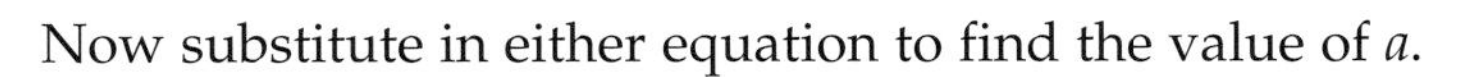

Now substitute in either equation to find the value of a.

$$\begin{array}{rl} 3a + 7s = 84 & \\ 3a + 7(6) = 84 & \text{Replace } s \text{ with 6.} \\ 3a + 42 = 84 & \\ 3a + 42 - 42 = 84 - 42 & \text{Subtract 42 from each side.} \\ 3a = 42 & \\ \frac{3a}{3} = \frac{42}{3} & \text{Divide each side by 3.} \\ a = 14 & \text{The value of } a \text{ is 14.} \end{array}$$

The solution of the system of equations is (14, 6). This means that the cost for adults was \$14 and the cost for students was \$6.

Check:

$$\begin{array}{rl} 3a + 10s = 102 & \\ 3(14) + 10(6) \stackrel{?}{=} 102 & \text{Replace } (a, s) \text{ with } (14, 6). \\ 42 + 60 \stackrel{?}{=} 102 & \\ 102 = 102 & \checkmark \end{array}$$

$$\begin{array}{rl} 3a + 7s = 84 & \\ 3(14) + 7(6) \stackrel{?}{=} 84 & \text{Replace } (a, s) \text{ with } (14, 6). \\ 42 + 42 \stackrel{?}{=} 84 & \\ 84 = 84 & \checkmark \end{array}$$

In some systems of equations, the coefficients of one of the variables are additive inverses. For these systems, apply the elimination method by adding the equations.

Example 3

Use elimination to solve the system of equations.

$4x - 6y = 10$

$3x + 6y = 4$

$$\begin{array}{rl} 4x - 6y = 10 & \textit{−6y and +6y are additive inverses.} \\ \underline{(+)\ 3x + 6y = \ \ 4} & \textit{Add the equations to eliminate the y terms.} \\ 7x + 0 = 14 & \\ 7x = 14 & \\ \frac{7x}{7} = \frac{14}{7} & \textit{Divide each side by 7.} \\ x = 2 & \text{The value of } x \text{ is 2.} \end{array}$$

Now substitute x into either equation to find the value of y.

$$\begin{array}{rl} 4x - 6y = 10 & \\ 4(2) - 6y = 10 & \textit{Replace x with 2.} \\ 8 - 6y = 10 & \\ 8 - 6y - 8 = 10 - 8 & \textit{Subtract 8 from each side.} \\ -6y = 2 & \\ \frac{-6y}{-6} = \frac{2}{-6} & \textit{Divide each side by −6.} \\ y = -\frac{2}{6} & \\ y = -\frac{1}{3} & \text{The value of } y \text{ is } -\frac{1}{3}. \end{array}$$

Prerequisite Skills Review
Simplifying Fractions, p. 685

The solution of the system is $\left(2, -\frac{1}{3}\right)$. *Check this result.*

Your Turn

Use elimination to solve each system of equations.

c. $x + y = 8$
$x - y = -2$

d. $-4x + y = 15$
$3y = 5 - 4x$

Systems of equations can be used to solve **digit problems**. Digit problems explore the relationships between digits of a number.

Example 4
Number Theory Link

The sum of the digits of a two-digit number is 8. If the tens digit is 4 more than the units digit, what is the number?

Let t represent the tens digit and let u represent the units digit.

$t + u = 8$ ← *the sum of the digits*

$t = u + 4$ ← *the relationship between the digits*

Rewrite the second equation so that the t and u are on the same side of the equation.

$$t = u + 4 \rightarrow t - u = 4$$

Then use elimination to solve.

$$\begin{aligned} t + u &= 8 \quad && \textit{Write the equations in column form.} \\ (+)\ t - u &= 4 && \textit{Add the equations to eliminate the u terms.} \\ \hline 2t + 0 &= 12 \\ 2t &= 12 \\ \frac{2t}{2} &= \frac{12}{2} && \textit{Divide each side by 2.} \\ t &= 6 && \text{The tens digit is 6.} \end{aligned}$$

Now substitute to find the units digit.

$$\begin{aligned} t + u &= 8 \\ 6 + u &= 8 \quad && \textit{Replace t with 6.} \\ 6 + u - 6 &= 8 - 6 && \textit{Subtract 6 from each side.} \\ u &= 2 && \text{The units digit is 2.} \end{aligned}$$

Since t is 6 and u is 2, the number is 62. *Check this solution.*

Check for Understanding

Communicating Mathematics

1. **Explain** when you would use elimination to solve a system of equations.
2. **Describe** the result when you add each pair of equations. What does the result tell you about the system of equations?
   a. $2x - y = 12$
      $-2x + y = 14$
   b. $x + 5y = 3$
      $-x - 5y = -3$

Vocabulary
elimination
digit problems

Guided Practice

Getting Ready **State whether *addition*, *subtraction*, *either*, or *neither* should be used to solve each system.**

Sample: $3x + y = 6$
$4x + y = 7$

Solution:
$$\begin{aligned} 3x + y &= 6 \\ (-)\ 4x + y &= 7 \\ \hline -x \quad &= -1 \end{aligned}$$
Subtraction should be used because it eliminates the y terms.

3. $x - 2y = 3$
   $3x + 2y = 8$
4. $x + y = 5$
   $x - y = 9$
5. $4x - 5y = 18$
   $2x - 5y = 1$
6. $x + 7y = 3$
   $2x + y = 4$

Use elimination to solve each system of equations. *(Examples 1–3)*

7. $x + y = 7$
   $x - y = 9$

8. $x + y = 5$
   $2x + y = 4$

9. $x + 2y = 6$
   $3x - 2y = 2$

10. $-7x + 4y = 6$
    $7x + y = 19$

11. $3x - 2y = 10$
    $3x - 5y = 1$

12. $4x = 5y - 9$
    $4x + 3y = -1$

13. **Number Theory** The sum of two numbers is 42. The greater number is three more than twice the lesser number. *(Example 4)*
    a. Write a system of equations to represent the problem.
    b. What are the numbers?

Exercises

Practice

For Exercises	See Examples
16, 17, 34, 35	1, 2
14, 15, 18–25, 32	3
26–31, 33, 36, 37	4

Homework Help

Extra Practice

See page 719.

Use elimination to solve each system of equations.

14. $x + y = 3$
    $x - y = 3$

15. $x - y = 9$
    $x + y = 19$

16. $x + 4y = 6$
    $x + 3y = 5$

17. $x + y = 10$
    $3x + y = 0$

18. $3x + y = 13$
    $2x - y = 2$

19. $9x + 2y = 12$
    $7x - 2y = -12$

20. $-3x + 4y = 5$
    $3x - 2y = -7$

21. $2x - 3y = 19$
    $2x + 3y = 13$

22. $11x - 3y = 10$
    $-2x + 3y = 8$

23. $2x - y = -8$
    $x - y = 7$

24. $2x - 5y = 9$
    $2x - 3y = 11$

25. $3x - 2y = 10$
    $2x - 2y = 5$

26. $x + 2y = 8$
    $3x = 6 - 2y$

27. $x + y = 7$
    $21 - 3x = 3y$

28. $y = 3x - 5$
    $x - y = 13$

29. $x = y - 7$
    $2x - 5y = -2$

30. $5x - y = -3$
    $2y = 10x - 7$

31. $x = 10 - y$
    $2x - y = -4$

32. What is the solution of the system $3x + 5y = -16$ and $3x - 2y = -2$?

33. Find the solution of the system $5s + 4t = 12$ and $3s = 4 + 4t$.

Applications and Problem Solving

34. **Spas** The Feel Better Spa has two specials for new members. They can receive 3 facials and 5 manicures for \$114 or 3 facials and 2 manicures for \$78. What are the prices for facials and manicures?

35. **Entertainment** The cost of admission to Water World was \$137.50 for 13 children and 2 adults in one party. The admission was \$103.50 for 9 children and 2 adults in another party.
    a. Write a system of equations to represent this problem.
    b. How much is admission for children and adults?

36. **Number Theory** The difference between two numbers is 15. The greater number is two less than twice the lesser number.

a. Write a system of equations to represent this situation.

b. Find the numbers.

37. **Number Theory** The sum of a number and twice a second number is 29. The second number is ten less than three times the first number.

a. Write a system of equations to represent this situation.

b. Find both numbers.

38. **Critical Thinking** The solution of a system of equations is $(-2, 5)$, and the first equation is $3x + 4y = 14$.

a. Write a second equation for this system.

b. Is this the only equation that could be in the system? Explain.

Mixed Review

39. Use substitution to solve the system of equations. *(Lesson 13–3)*

$y = -3x - 5$
$4x + y = 6$

State whether each system is *consistent and independent*, *consistent and dependent*, or *inconsistent*. *(Lesson 13–2)*

40.

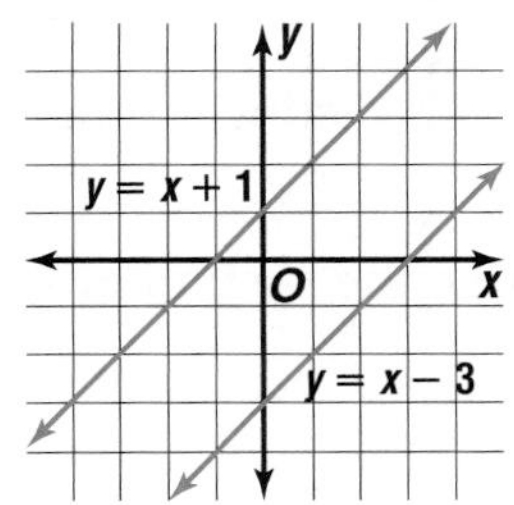

41.

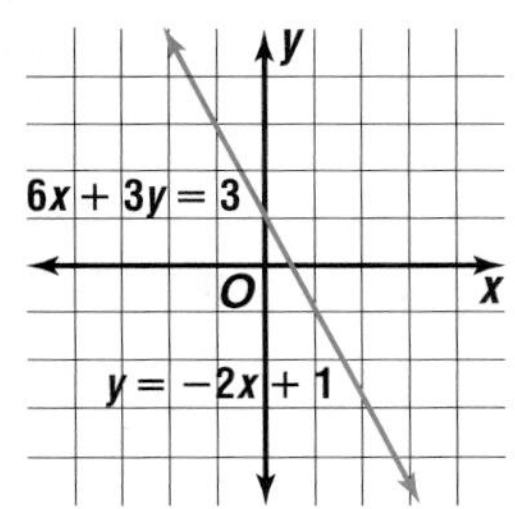

42.

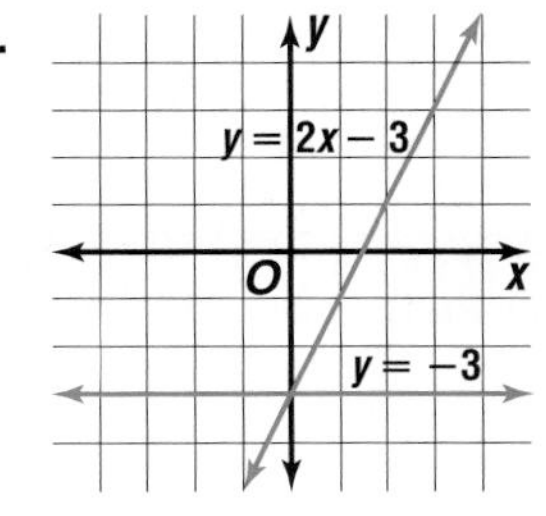

Solve each inequality. *(Lesson 12–4)*

43. $3y < 12$

44. $6 - 4m \leq 22$

45. $9 - 3x > 15$

Find each product. *(Lesson 9–4)*

46. $(m + 4)(m + 7)$

47. $(a - 3)(a - 3)$

48. $(2x - 2)(3x + 6)$

Standardized Test Practice

A B C D

49. **Short Response** Write the point-slope form of an equation for a line passing through points $(6, 1)$ and $(7, -4)$. *(Lesson 7–2)*

50. **Multiple Choice** In a scale model of a classic car, the front bumper measures 1.5 inches. If the actual bumper measures 3 feet, what is the scale of the model? *(Lesson 5–2)*

A 1 in. = 16 in.　B 1 in. = 18 in.

C 1 in. = 24 in.　D 1 in. = 43 in.

13–5 Elimination Using Multiplication

What You'll Learn

You'll learn to solve systems of equations by the elimination method using multiplication and addition.

Why It's Important

Transportation Systems of equations can be used to determine the rate of a river's current. *See Example 5.*

You have learned when and how to solve systems of equations by graphing, substitution, and elimination using addition or subtraction. The best times to use these methods are summarized in the table below.

Method	The Best Time to Use	Example
graphing	if the equations are easy to graph, or if you want an estimate because graphing usually does not give an exact solution	$y = x + 4$ $y = -x + 1$
substitution	if one of the variables in either equation has a coefficient of 1 or −1	$y = 2x + 3$ $3x - 5y = 8$
elimination using addition	if the coefficients of one variable are additive inverses	$4x - y = 9$ $-4x + 3y = 6$
elimination using subtraction	if the coefficients of one variable are the same	$2x + y = 4$ $3x + y = 1$

Sometimes neither of the variables in a system of equations can be eliminated by simply adding or subtracting the equations. In this case, another method is to multiply one or both of the equations by some number so that adding or subtracting eliminates one of the variables.

Examples

1 **Use elimination to solve the system of equations.**

$$2x + 3y = 9$$
$$8x - 5y = 19$$

Multiply the first equation by −4 so that the x terms are additive inverses.

$2x + 3y = 9$ **Multiply by −4.** → $-8x - 12y = -36$

$8x - 5y = 19$ → $(+)\ 8x - 5y = 19$

$$0 - 17y = -17$$
$$\frac{-17y}{-17} = \frac{-17}{-17}$$
$$y = 1$$

Now find the value of x by replacing y with 1 in either equation.

$2x + 3y = 9$

$2x + 3(1) = 9$ *Replace y with 1.*

$2x + 3 = 9$

$2x + 3 - 3 = 9 - 3$ *Subtract 3 from each side.*

$2x = 6$

$\frac{2x}{2} = \frac{6}{2}$ *Divide each side by 2.*

$x = 3$

The solution of this system of equations is (3, 1). *Check this solution.*

2 **Use elimination to solve the system of equations.**

$5x - 6y = 25$

$4x + 2y = 3$

Multiply the second equation by 3 so that the y terms are additive inverses.

$$5x - 6y = 25 \qquad 4x + 2y = 3 \quad \xrightarrow{\text{Multiply by 3.}}$$

$$\begin{aligned} 5x - 6y &= 25 \\ (+)\ 12x + 6y &= 9 \\ \hline 17x + 0 &= 34 \\ \frac{17x}{17} &= \frac{34}{17} \\ x &= 2 \end{aligned}$$

Now find the value of y by replacing x with 2 in either equation.

$4x + 2y = 3$

$4(2) + 2y = 3$ *Replace x with 2.*

$8 + 2y = 3$

$8 + 2y - 8 = 3 - 8$ *Subtract 8 from each side.*

$2y = -5$

$\frac{2y}{2} = \frac{-5}{2}$ *Divide each side by 2.*

$y = -\frac{5}{2}$

The solution of the system of equations is $\left(2, -\frac{5}{2}\right)$. *Check this solution.*

Your Turn

a. $2x - y = 16$
$5x + 3y = -4$

b. $4x - 3y = -2$
$2x + 7y = 16$

c. $2x - 6y = -8$
$4x - 3y = 11$

Example
Transportation Link

3 **Chris and Alana both take the Metro train to work. In May, Chris took the train 15 times during rush hour and 29 times during non-rush hour for $64.80. Alana took the train 30 times during rush hour and 14 times during non-rush hour for $76.80. What are the rush hour and non-rush hour fares?**

Prerequisite Skills Review
Operations with Decimals, p. 684

r = the rush hour fare $\qquad$ n = the non-rush hour fare

$15r + 29n = 64.80$ ← *Chris' expenses*

$30r + 14n = 76.80$ ← *Alana's expenses*

Multiply the first equation by −2 to eliminate the r terms.

$$15r + 29n = 64.80 \qquad 30r + 14n = 76.80 \quad \xrightarrow{\text{Multiply by } -2.}$$

$$\begin{aligned} -30r - 58n &= -129.60 \\ (+)\ 30r + 14n &= 76.80 \\ \hline 0 - 44n &= -52.80 \\ \frac{-44n}{-44} &= \frac{-52.80}{-44} \\ n &= 1.20 \end{aligned}$$

(continued on the next page)

Now find the value of r by replacing n with 1.20 in either equation.

$$15r + 29n = 64.80$$
$$15r + 29(1.20) = 64.80 \quad \textit{Replace n with 1.20.}$$
$$15r + 34.80 = 64.80$$
$$15r + 34.80 - 34.80 = 64.80 - 34.80 \quad \textit{Subtract 34.80 from each side.}$$
$$15r = 30$$
$$\frac{15r}{15} = \frac{30}{15} \quad \textit{Divide each side by 15.}$$
$$r = 2$$

The solution is (2, 1.20). This means that the rush hour fare is \$2.00 and the non-rush hour fare is \$1.20. Do these fares seem reasonable? *Check this solution by substituting (2, 120) for (r, n) in the original equations.*

Sometimes it is necessary to multiply each equation by a different number and then add in order to eliminate one of the variables.

Example 4 **Use elimination to solve the system of equations.**

$\mathbf{3x + 4y = -25}$

$\mathbf{2x - 3y = 6}$

Multiply the first equation by 2 and the second equation by -3 so that the x terms are additive inverses.

$3x + 4y = -25$ **Multiply by 2.** $\rightarrow$ $6x + 8y = -50$

$2x - 3y = 6$ **Multiply by −3.** $\rightarrow$ $(+)\,-6x + 9y = -18$

$$0 + 17y = -68$$
$$\frac{17y}{17} = \frac{-68}{17}$$
$$y = -4$$

Now find the value of x by replacing y with -4 in either equation.

$$3x + 4y = -25$$
$$3x + 4(-4) = -25 \quad \textit{Replace y with −4.}$$
$$3x - 16 = -25$$
$$3x - 16 + 16 = -25 + 16 \quad \textit{Add 16 to each side.}$$
$$3x = -9$$
$$\frac{3x}{3} = \frac{-9}{3} \quad \textit{Divide each side by 3.}$$
$$x = -3$$

The solution of the system of equations is $(-3, -4)$. You can also solve this system of equations by multiplying the first equation by 3 and the second equation by 4. *Why?*

Your Turn

Use elimination to solve each system of equations.

d. $5x + 3y = 12$
$4x - 5y = 17$

e. $4x + 3y = 19$
$3x - 4y = 8$

Systems of equations can be used to solve rate problems.

Example 5
Transportation Link

A barge on the Mississippi River travels 36 miles upstream in 6 hours. The return trip takes the barge only 4 hours. Find the rate of the current.

Explore Let r represent the rate of the barge in still water. Let c represent the rate of the current.

The rate of the barge traveling downstream *with* the current is $r + c$, and the rate of the barge traveling upstream *against* the current is $r - c$.

Plan Use the formula distance = rate × time, or $d = rt$, to write a system of equations. Then solve the system to find c.

	d	r	t	$d = rt$
Downstream	36	$r + c$	4	$36 = 4r + 4c$
Upstream	36	$r - c$	6	$36 = 6r - 6c$

$4(r + c) = 4r + 4c$
$6(r - c) = 6r - 6c$

Solve

$4r + 4c = 36$ → Multiply by 3. → $12r + 12c = 108$

$6r - 6c = 36$ → Multiply by −2. → $(+)\ -12r + 12c = -72$

$$0 + 24c = 36$$
$$24c = 36$$
$$\frac{24c}{24} = \frac{36}{24}$$
$$c = 1.5$$

The rate of the current is 1.5 miles per hour.

Examine Find the value of r for this system and then check the solution.

Barge on the Mississippi

Check for Understanding

Communicating Mathematics

1. **Explain** when you would use elimination with multiplication to solve a system of equations.
2. **Write** a system of equations in which you can eliminate the variable x by multiplying one equation by 3 and then adding the equations.

3. **Match** each system of equations with the method listed below that would best solve the system.

a. $y = 3x$
$6x - 2y = 9$

b. $2x + 4y = 5$
$3x - 2y = -3$

c. $10x - 2y = 12$
$-10x + 3y = -13$

d. $x + 7y = 5$
$3x + 7y = 1$

substitution
elimination (+)
elimination (−)
elimination (×)

Guided Practice

Getting Ready **Explain the steps you would take to eliminate the variable *y* in each system of equations.**

Sample: $4x + 3y = -7$
$3x + y = 1$

Solution: Multiply second equation by −3. Then add.

4. $2x + 6y = 10$
$5x + 3y = 1$

5. $3x - 2y = 3$
$8x + y = 27$

6. $2x + y = 4$
$7x - 4y = 29$

Use elimination to solve each system of equations. *(Examples 1–4)*

7. $x + 2y = 6$
$-4x + 5y = 2$

8. $x - 5y = 0$
$2x - 3y = 7$

9. $5x + 8y = 1$
$2x - 4y = -14$

10. $x + 2y = 5$
$3x + y = 7$

11. $3x - y = -5$
$6x - 2y = 8$

12. $2x - 5y = -10$
$7x - 3y = -6$

13. **Recreation** A fishing boat traveled 48 miles upstream in 4 hours. Returning at the same rate, it took 3 hours. *(Example 5)*

a. Find the rate of the current.

b. Find the rate of the boat in still water.

Exercises

Practice

Use elimination to solve each system of equations.

Homework Help	
For Exercises	See Examples
14, 15, 17, 20, 29	1
16, 18, 30	2
35	3
23, 24, 26–28	4
34	5
Extra Practice	
See page 719.	

14. $x + 8y = 3$
$2x + 3y = -7$

15. $4x + y = 8$
$x - 7y = 2$

16. $2x + y = 6$
$3x - 7y = 9$

17. $2x + 5y = 13$
$4x - 3y = -13$

18. $-3x + 2y = 10$
$-2x - y = -5$

19. $3x - 2y = 0$
$x + 6y = 5$

20. $-5x + 8y = 21$
$10x + 3y = 15$

21. $6x - 4y = 11$
$2x + 2y = 7$

22. $6x - 5y = -6$
$3x + 10y = 72$

23. $2x - 7y = 9$
$-3x + 4y = 6$

24. $5x + 3y = 4$
$-4x + 5y = -18$

25. $6x - 3y = 7$
$18x - 9y = 21$

26. $7x - 4y = 16$
$2x + 3y = 17$

27. $-5x - 2y = 12$
$11x - 5y = -17$

28. $7x - 3y = -9$
$5x - 4y = -25$

29. What is the solution of the system $9x + 8y = 7$ and $18x - 15y = 14$?

30. Use elimination to find the solution of the system $\frac{1}{3}x - y = -1$ and $\frac{1}{5}x - \frac{2}{5}y = -1$.

Determine the best method to solve each system of equations. Then solve.

31. $x + 2y = 6$
$2x - 4y = -20$

32. $y = 3x - 2$
$x - 5y = -4$

33. $y = \frac{1}{2}x + 6$
$2x - 4y = 8$

Applications and Problem Solving

34. **Travel** A riverboat on the Mississippi River travels 30 miles upstream in 2 hours and 30 minutes. The return trip downstream takes only 2 hours.

a. Find the rate of the current.

b. Find the rate of the riverboat in still water.

35. **Entertainment** The science club purchased tickets for a magic show. They paid $108 for 6 tickets in section A and 10 tickets in section B. The following week, they paid $104 for 4 tickets in section A and 12 tickets in section B.

a. Write a system of equations to represent the problem.

b. What are the prices of the tickets in section A and B?

36. **Critical Thinking** The solution of the system $5x + 6y = -9$ and $10x + 8y = c$ is $(9, b)$. Find values for c and b.

Mixed Review

Use elimination to solve each system of equations. *(Lesson 13–4)*

37. $x - y = 8$
$x + y = 6$

38. $x + y = -4$
$4x + y = 2$

39. $2x + 4y = 7$
$2x - y = 2$

40. Use substitution to solve $x = 5 + 2y$ and $3x - 4y = 3$. *(Lesson 13–3)*

Solve each inequality. *(Lesson 12–3)*

41. $-4m > 16$

42. $\frac{x}{2} < 6$

43. $\frac{3a}{4} \geq -3$

44. Use the Quadratic Formula to solve $n^2 - 5n + 12 = 0$. *(Lesson 11–6)*

Standardized Test Practice

45. **Grid In** Suppose a baseball player hits a fly ball into right field above first base. The player hits the ball with his bat at a height of 2 feet above the ground and sends the ball upward at a velocity of 25 meters per second. The height of the ball t seconds after the hit can be approximated by the formula $h = -5t^2 + 25t + 2$. If the ball is not caught, how many seconds will it take to hit the ground? Round to the nearest second. *(Lesson 11–3)*

46. **Multiple Choice** The graph shows the top turkey producing states. What percent of turkeys does the state of North Carolina produce? Round to the nearest percent. *(Lesson 5–3)*

A 5% B 19%
C 33% D 45%

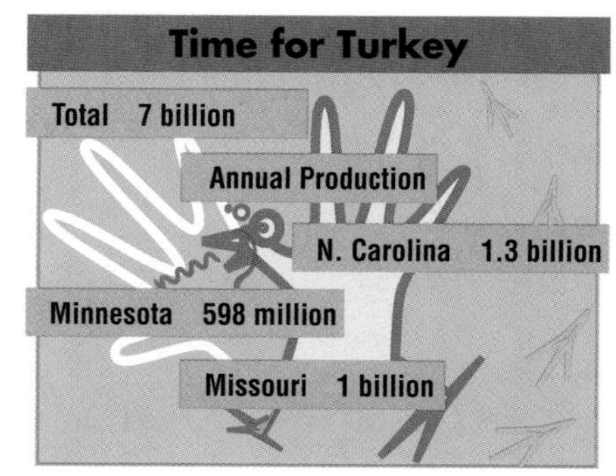

Source: U.S. Census Bureau

Chapter 13 Investigation

Crazy Computers

Matrices

In this investigation, you will use matrices to solve systems of equations. Consider the following problem.

Look Back

Matrices: Pages 80–81

Kristin and Scott are sales associates at Crazy Computers. On Monday, Kristin sold 4 Model A computers and 1 Model B computer for a total of \$6395. Scott sold 2 Model A computers and 3 Model B computers for a total of \$7195. What was the price of each computer?

Investigate

You can write a system of equations to solve this problem. You may want to use matrices to solve this system of equations.

1. Write each equation in standard form. Then write the coefficients and constants of the system in a special matrix called an **augmented matrix**. Write the coefficients in the matrix to the left of the dashed line. Write the constants to the right of the dashed line.

$$\begin{aligned} 3x + 5y &= 7 \\ 6x - 1y &= -8 \end{aligned} \quad \Rightarrow \quad \left[\begin{array}{cc|c} 3 & 5 & 7 \\ 6 & -1 & -8 \end{array}\right]$$

   a. Write $2x - 3y = 5$ and $x + 4y = -7$ as an augmented matrix.

   b. Write $x + 6y = -1$ and $y = 5$ as an augmented matrix.

2. You can use **row operations** to simplify an augmented matrix.

   - Switch any two rows.
   - Replace any row with a nonzero multiple of that row.
   - Replace any row with the sum or difference of that row and a multiple of another row.

   a. Switch rows 1 and 2 of $\left[\begin{array}{cc|c} 0 & -5 & 2 \\ 1 & 3 & 4 \end{array}\right]$.

   b. Multiply row 2 of $\left[\begin{array}{cc|c} 1 & -3 & 5 \\ 0 & 2 & -8 \end{array}\right]$ by 0.5.

   c. Replace row 2 of $\left[\begin{array}{cc|c} 1 & -2 & -5 \\ -1 & 3 & 4 \end{array}\right]$ with the sum of rows 1 and 2.

   d. In $\left[\begin{array}{cc|c} 1 & -3 & 2 \\ 0 & 3 & -5 \end{array}\right]$, what row operation would you use to get a 0 in the second column of row 1?

3. The goal of using row operations to solve a system of equations is to get to the **identity matrix**, which is $\begin{bmatrix} 1 & 0 \\ 0 & 1 \end{bmatrix}$. Suppose the solution of a system of equations is $\left[\begin{array}{cc|c} 1 & 0 & -4 \\ 0 & 1 & 2 \end{array}\right]$. What does this matrix represent?

4. Solve the problem. Let x = the cost of the Model A computer and let y = the cost of the Model B computer.

$$4x + 1y = 6395 \quad \leftarrow \textit{Kristin's sales}$$
$$2x + 3y = 7195 \quad \leftarrow \textit{Scott's sales}$$

Step 1 Write the system as an augmented matrix. $\left[\begin{array}{cc|c} 4 & 1 & 6395 \\ 2 & 3 & 7195 \end{array}\right]$

Step 2 Multiply row 2 by 2.
$2 \times 2 = 4$; $2 \times 3 = 6$; $2 \times 7195 = 14{,}390$ $\left[\begin{array}{cc|c} 4 & 1 & 6395 \\ 4 & 6 & 14{,}390 \end{array}\right]$

Step 3 Replace row 2 with the difference of row 1 and row 2. *$4 - 4 = 0$; $1 - 6 = -5$; $6395 - 14{,}390 = -7995$* $\left[\begin{array}{cc|c} 4 & 1 & 6395 \\ 0 & -5 & -7995 \end{array}\right]$

Step 4 Divide row 2 by -5. *$0 \div (-5) = 0$; $-5 \div (-5) = 1$; $-7995 \div (-5) = 1599$* $\left[\begin{array}{cc|c} 4 & 1 & 6395 \\ 0 & 1 & 1599 \end{array}\right]$

Step 5 Replace row 1 with the difference of row 1 and row 2. *$4 - 0 = 4$; $1 - 1 = 0$; $6395 - 1599 = 4796$* $\left[\begin{array}{cc|c} 4 & 0 & 4796 \\ 0 & 1 & 1599 \end{array}\right]$

Step 6 Divide row 1 by 4.
$4 \div 4 = 1$; $0 \div 4 = 0$; $4796 \div 4 = 1199$ $\left[\begin{array}{cc|c} 1 & 0 & 1199 \\ 0 & 1 & 1599 \end{array}\right]$

This means that Model A computer is \$1199 and Model B computer is \$1599.

Extending the Investigation

In this extension, you will use matrices as a problem-solving tool.

1. Suppose Scott sells 5 Model TX laser printers and 6 Model DM ink jet printers for a total of \$4185. Kristin sells 6 Model TX laser printers and 5 Model DM ink jet printers for a total of \$4549. What is the cost of each printer?
2. At Crazy Computers, there are two prices for computer software, Price A and Price B. Suppose Aislyn purchases 3 Price A software and 1 Price B software for a total of \$250 and Devin purchases 5 Price A software and 2 Price B software for a total of \$450. What are the two prices for the software?

Presenting Your Investigation

Write a real-world problem that can be solved by using matrices and a system of equations. Present your problem and solution on a poster board.

*inter***NET** CONNECTION **Investigation** For more information on matrices, visit: www.algconcepts.com

13–6 Solving Quadratic-Linear Systems of Equations

What You'll Learn
You'll learn to solve systems of quadratic and linear equations.

Why It's Important
Meteorology Quadratic-linear systems of equations can be used to determine when airplanes will cross jet streams and experience turbulence. *See Exercise 24.*

In late November, the jet stream moving across North America could be described by the quadratic equation $y = \frac{1}{4}x^2 - 12$, where Chicago was at the origin. Suppose a plane's route is described by the linear equation $y = -\frac{1}{2}x$. What are the coordinates of the point at which turbulence will occur? *This problem will be solved in Example 7.*

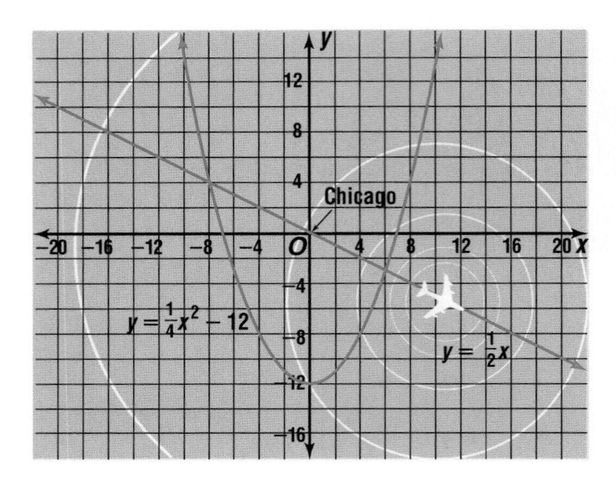

Like a linear system of equations, the solution of a **quadratic-linear** system of equations is the ordered pair that satisfies both equations. A quadratic-linear system can have 0, 1, or 2 solutions, as shown below.

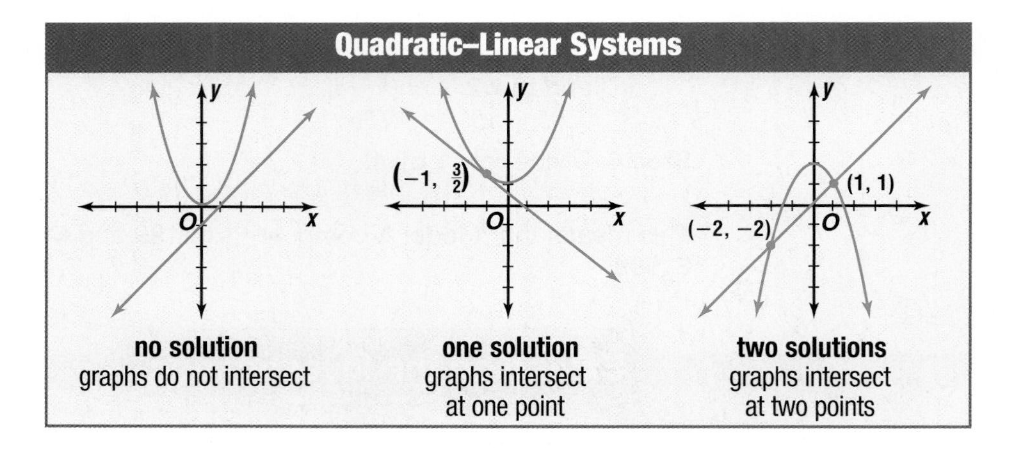

You can solve quadratic-linear systems of equations by using some of the methods you used for solving systems of linear equations. One method is graphing.

Examples

Determine whether each system of equations has *one* solution, *two* solutions, or *no* solution by graphing. If the system has one solution or two solutions, name them.

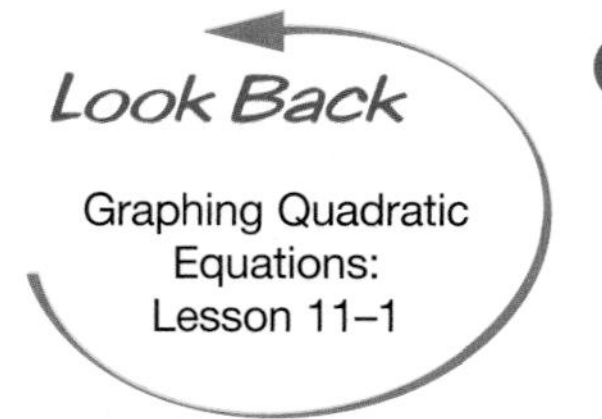

1 $y = x^2$
$y = x + 2$

The graphs appear to intersect at $(-1, 1)$ and $(2, 4)$. Check this estimate by substituting the coordinates into each equation.

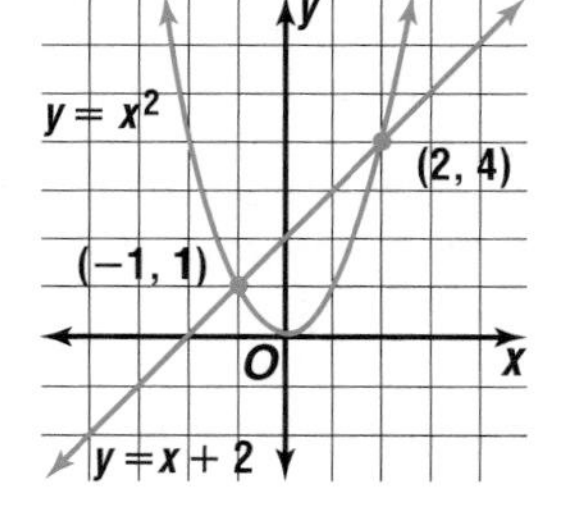

Check: $y = x^2$ | $y = x + 2$

$1 \stackrel{?}{=} (-1)^2$ *$(x, y) = (-1, 1)$* | $1 \stackrel{?}{=} -1 + 2$ *$(x, y) = (-1, 1)$*

$1 = 1$ ✓ | $1 = 1$ ✓

Check that the ordered pair (2, 4) satisfies both equations.

The solutions of the system of equations are $(-1, 1)$ and $(2, 4)$.

2 $y = 2x^2 + 5$
$y = -x + 3$

Because the graphs do not intersect, there is no solution to this system of equations.

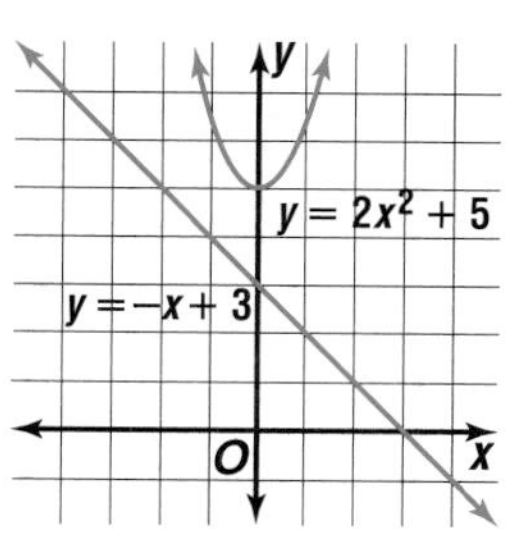

3 $y = -x^2 + 3$
$y = 2x + 4$

The graphs appear to intersect at $(-1, 2)$.

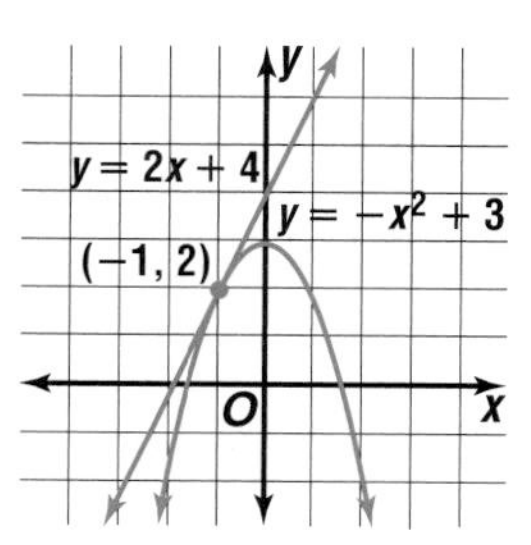

Check: $y = -x^2 + 3$ | $y = 2x + 4$

$2 \stackrel{?}{=} -(-1)^2 + 3$ *$(x, y) = (-1, 2)$* | $2 \stackrel{?}{=} 2(-1) + 4$ *$(x, y) = (-1, 2)$*

$2 \stackrel{?}{=} -(1) + 3$ | $2 \stackrel{?}{=} -2 + 4$

$2 = 2$ ✓ | $2 = 2$ ✓

The solution of the system of equations is $(-1, 2)$.

Your Turn

a. $y = x^2$
$y = x - 2$

b. $y = -2x^2 + 1$
$y = 1$

c. $y = -x^2 + 3$
$y = -2x$

You can also solve quadratic-linear systems of equations by using the substitution method.

Examples

4 **Use substitution to solve each system of equations.**

$y = -4$
$y = x^2 - 4$

Substitute -4 for y in the second equation. Then solve for x.

$y = x^2 - 4$

$-4 = x^2 - 4$ *Replace y with −4.*

$-4 + 4 = x^2 - 4 + 4$ *Add 4 to each side.*

$0 = x^2$

$0 = x$ *Take the square root of each side.*

The solution of the system of equations is $(0, -4)$.

(continued on the next page)

Check:

Sketch the graphs of the equations. The parabola and line appear to intersect at $(0, -4)$. The solution is correct.

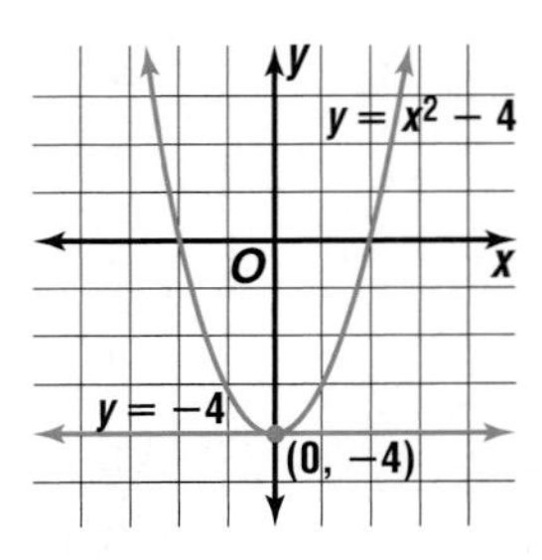

5 $y = x^2$

$y = -3$

Substitute -3 for y in the first equation.

$y = x^2$

$-3 = x^2$ *Replace y with −3.*

$\sqrt{-3} = x$ *Take the square root of each side.*

There is no real solution because the square root of a negative number is not a real number. *Check by graphing.*

6 $y = x^2 - 4x + 6$

$y = -x + 4$

Substitute $-x + 4$ for y in the first equation.

$y = x^2 - 4x + 6$

$-x + 4 = x^2 - 4x + 6$ *Replace y with −x + 4.*

$-x + 4 - 4 = x^2 - 4x + 6 - 4$ *Subtract 4 from each side.*

$-x = x^2 - 4x + 2$

$-x + x = x^2 - 4x + 2 + x$ *Add x to each side.*

$0 = x^2 - 3x + 2$

$0 = (x - 2)(x - 1)$ *Factor.*

$0 = x - 2$ or $0 = x - 1$ *Zero Product Property*

$x = 2$ $\quad$ $x = 1$

Substitute the values of x in either equation to find the corresponding values of y. *Choose the equation that is easier for you to solve.*

$y = -x + 4$

$y = -2 + 4$ *Replace x with 2.*

$y = 2$

$y = -x + 4$

$y = -1 + 4$ *Replace x with 1.*

$y = 3$

The solutions of the system of equations are $(2, 2)$ and $(1, 3)$. The graph shows that the solutions are probably correct. *You can also check by substituting the ordered pairs into the original equations.*

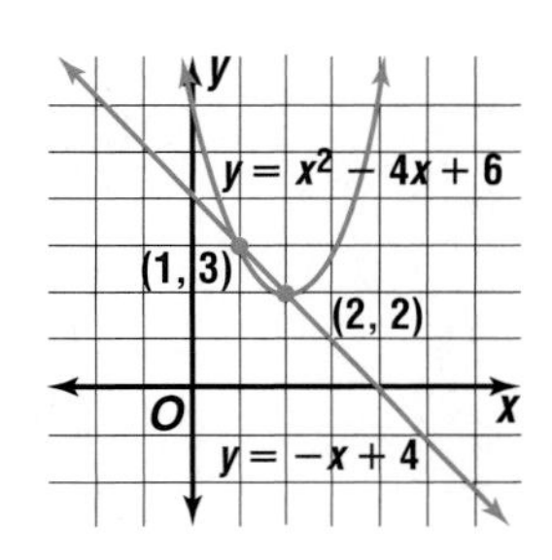

Your Turn

Use substitution to solve each system of equations.

d. $y = x^2 - 12$
$x = 3$

e. $y = -x^2$
$y = x + 2$

f. $y = x^2 - 3x$
$y = 2x$

Example 7 Meteorology Link

Refer to the application at the beginning of the lesson. What are the coordinates of the point at which turbulence will occur?

Use substitution to solve the system $y = \frac{1}{4}x^2 - 12$ and $y = -\frac{1}{2}x$.
Substitute $-\frac{1}{2}x$ for y in the first equation.

$$y = \frac{1}{4}x^2 - 12$$

$$-\frac{1}{2}x = \frac{1}{4}x^2 - 12 \quad \text{Replace } y \text{ with } -\tfrac{1}{2}x.$$

$$4\left(-\frac{1}{2}x\right) = 4\left(\frac{1}{4}x^2 - 12\right) \quad \text{Multiply each side by 4.}$$

$$-2x = x^2 - 48$$

$$-2x + 2x = x^2 - 48 + 2x \quad \text{Add } 2x \text{ to each side.}$$

$$0 = x^2 + 2x - 48$$

$$0 = (x + 8)(x - 6) \quad \text{Factor.}$$

$$0 = x + 8 \quad \text{or} \quad 0 = x - 6 \quad \text{Zero Product Property}$$

$$x = -8 \qquad\qquad x = 6$$

Substitute the values of x to find the corresponding values of y.

$$y = -\frac{1}{2}x \qquad\qquad y = -\frac{1}{2}x$$

$$y = -\frac{1}{2}(-8) \text{ or } 4 \quad \text{Replace } x \text{ with } -8. \qquad y = -\frac{1}{2}(6) \text{ or } -3 \quad \text{Replace } x \text{ with } 6.$$

The solutions of the system are $(-8, 4)$ and $(6, -3)$. This means that turbulence will occur as the plane passes through points having these coordinates. *Check by substituting the coordinates into each equation and by looking at the graph at the beginning of the lesson.*

***inter*NET CONNECTION**

Data Update For the latest information about the jet stream, visit: www.algconcepts.com

Check for Understanding

Communicating Mathematics

1. **State** the solution of the system of equations shown at the right.
2. **Sketch** a system of quadratic-linear equations that has solutions $(-2, 1)$ and $(2, 1)$.
3. **List** the different methods of solving linear and quadratic-linear systems of equations. Describe the situation in which each method is most useful.

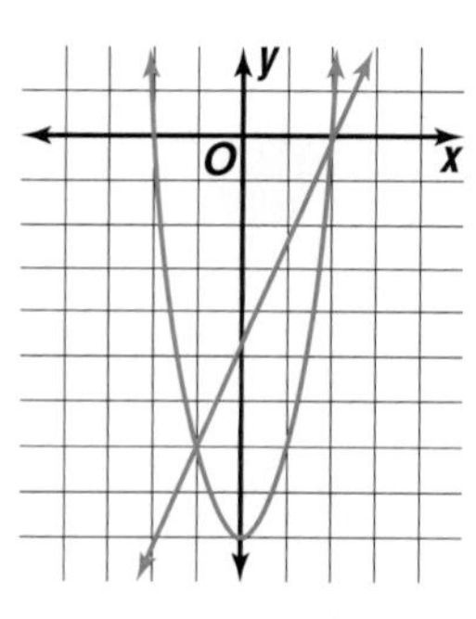

Exercise 1

Guided Practice

Solve each system of equations by graphing. *(Examples 1–3)*

4. $y = x^2$
   $y = -2x - 1$

5. $y = x^2$
   $y = 2x$

Use substitution to solve each system of equations. *(Examples 4–6)*

6. $y = x^2 - 6$
   $y = -5x$

7. $y = x^2 + 5$
   $y = -3$

8. **Business** Students in an Algebra I class at Banneker High School are simulating the start-up of a company. The income y can be described by the equation $y = \frac{1}{8}x^2$, where x represents the time in months. The expenses have been growing at a constant rate and can be defined by the equation $y = x$. When will income equal expenses? *(Example 7)*

Exercises

Practice

Homework Help	
For Exercises	See Examples
9–14	1, 2, 3
15–22, 24	4, 5, 6
23–25	7
Extra Practice	
See page 719.	

Solve each system of equations by graphing.

9. $y = x^2 - 4$
   $y = 2x - 4$

10. $x = 2$
    $y = x^2 + 1$

11. $y = 2x + 1$
    $y = x^2 + 4$

12. $y = x^2 - 6$
    $y = x - 4$

13. $y = -x^2 + 5$
    $y = -x + 3$

14. $y = -x^2 + 4$
    $y = \frac{1}{2}x + 5$

Use substitution to solve each system of equations.

15. $y = \frac{1}{8}x^2 + 5$
    $x = -4$

16. $y = -3x^2$
    $y = 2$

17. $y = \frac{1}{2}x^2$
    $y = 3x + 8$

18. $y = \frac{1}{2}x^2 - 4$
    $y = 3x + 4$

19. $y = x^2 + 3$
    $y = -\frac{1}{2}x - 5$

20. $y = x^2 - x - 3$
    $y = -1$

21. What is the solution of the system $y = x^2 - 9$ and $y = 3x - 9$?

22. Find the solution of the system $y = -x^2 + 2x - 3$ and $y = -2x + 1$.

Applications and Problem Solving

23. **Geometry** Four corners are cut from a rectangular piece of cardboard that is 10 feet by 4 feet. The cuts are x feet from the corners, as shown in the figure at the right. After the cuts are made, the sides of the rectangle are folded to form an open box. The area of the bottom of the box is 16 square feet.

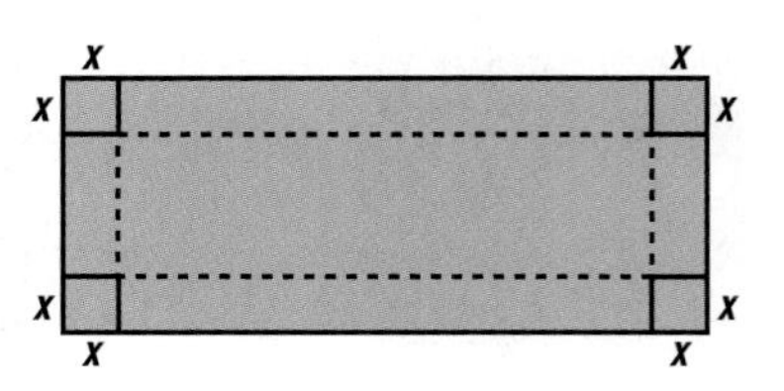

a. Write two equations to represent the area A of the bottom of the box.

b. What are the dimensions of the box?

c. What is the volume of the box?

24. **Meteorology** Refer to the application at the beginning of the lesson. If a plane's route is described by the linear equation $y = x - 14$, will it experience turbulence by the jet stream that is described by the quadratic equation $y = \frac{1}{4}x^2 - 12$? Explain.

25. **Critical Thinking** Suppose the perimeter and area of a square have the same measure. Find the length of the sides of the square. Explain how you can use a system of quadratic-linear equations to find the answer.

Mixed Review

26. **Sports** Diego kayaks 16 miles downstream in 2 hours. It takes him 8 hours to make the return trip. What is the rate of the current? *(Lesson 13–5)*

27. Solve the system $4x + 3y = 5$ and $-3x - 2y = -4$ by using elimination. *(Lesson 13–4)*

28. Describe how the graph of $y = -(x - 3)^2$ changes from its parent graph of $y = x^2$. *(Lesson 11–2)*

Solve each equation. *(Lesson 3–7)*

29. $2 + |x| = 9$ 30. $|y| + 4 = 3$ 31. $12 = 6 + |w + 2|$

Standardized Test Practice

32. **Multiple Choice** Which expression shows *3 less than the product of 5 and d*? *(Lesson 1–1)*

A $3 - 5d$ B $5d + 3$ C $5d - 3$ D $5 + d - 3$

Quiz 2 Lessons 13–4 through 13–6

Use elimination to solve each system of equations. *(Lessons 13–4 & 13–5)*

1. $x + y = 6$
 $2x - y = 6$

2. $2x - 3y = -9$
 $x - 2y = -5$

3. $3x - 5y = 8$
 $4x - 7y = 10$

4. Use substitution to find the solution of the system $y = x^2 + 5$ and $y = 6x - 3$. *(Lesson 13–6)*

5. **Number Theory** The difference between two numbers is 38. The greater number is three times the lesser number minus two. *(Lesson 13–4)*
 a. Write a system of equations to represent the situation.
 b. What are the numbers?

13–7 Graphing Systems of Inequalities

What You'll Learn
You'll learn to solve systems of inequalities by graphing.

Why It's Important
Money Systems of inequalities can be useful in helping you get the most for your money.
See Example 3.

Thomas Edison was a newsboy during the Civil War. One day, he persuaded the editor to give him 300 copies of the paper instead of the usual 100. He went to the train station where people were eager for news of the war. He was able to sell the papers for 10¢ and 25¢ instead of the usual 5¢. If he wanted to earn at least $52, how many papers could he have sold for 10¢ and 25¢?

x = the number of 10¢ papers $\qquad$ y = the number of 25¢ papers

The following **system of inequalities** can be used to represent the conditions of this problem.

$x + y \leq 300$ *He can sell as many as 300 papers.*
$0.10x + 0.25y \geq 52$ *He wants to earn at least $52.*

Since both x and y represent the number of papers, neither can be a negative number. Thus, $x \geq 0$ and $y \geq 0$. The solution of the system is the set of all ordered pairs that satisfy both inequalities and lie in the first quadrant. The solution can be determined by graphing each inequality in the same coordinate plane as shown below.

Look Back
Graphing Linear Inequalities in Two Variables: Lesson 12–7

Recall that the graph of each inequality is called a *half-plane.*

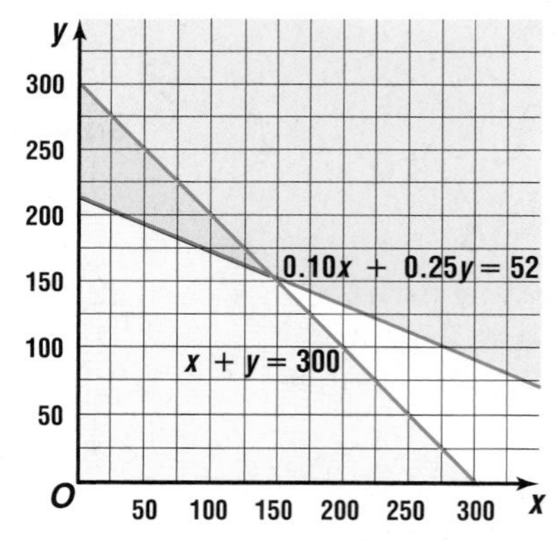

- Any point in the yellow region satisfies $x + y \leq 300$.
- Any point in the blue region satisfies $0.10x + 0.25y \geq 52$.
- Any point in the green region satisfies both inequalities. The intersection of the two half-planes represents the solution to the system of inequalities.
- The graphs of $x + y = 300$ and $0.10x + 0.25y = 52$ are the boundaries of the region and are included in the graph of the system.

This solution is a region that contains the graphs of an infinite number of ordered pairs. An example is (50, 225). This means that Edison could have sold 50 papers at 10¢ and 225 papers at 25¢ to earn at least $52.

Check:

$$x + y \leq 300$$
$$50 + 225 \overset{?}{\leq} 300 \quad (x, y) = (50, 225)$$
$$275 \leq 300 \quad \checkmark$$

$$0.10x + 0.25y \geq 52$$
$$0.10(50) + 0.25(225) \overset{?}{\geq} 52 \quad (x, y) = (50, 225)$$
$$61.25 \geq 52 \quad \checkmark$$

Remember that the boundary lines of inequalities are only included in the solution if the inequality symbol is greater than or equal to, $\geq$, or less than or equal to, $\leq$.

Examples

Solve each system of inequalities by graphing. If the system does not have a solution, write *no solution*.

1 $y \leq -1$
$y > x - 3$

The solution is the ordered pairs in the intersection of the graphs of $y \leq -1$ and $y > x - 3$. The region is shaded in green at the right. The graphs of $y = -1$ and $y = x - 3$ are the boundaries of this region. The graph of $y = x - 3$ is a dashed line and is not included in the solution of the system. *Choose a point and check the solution.*

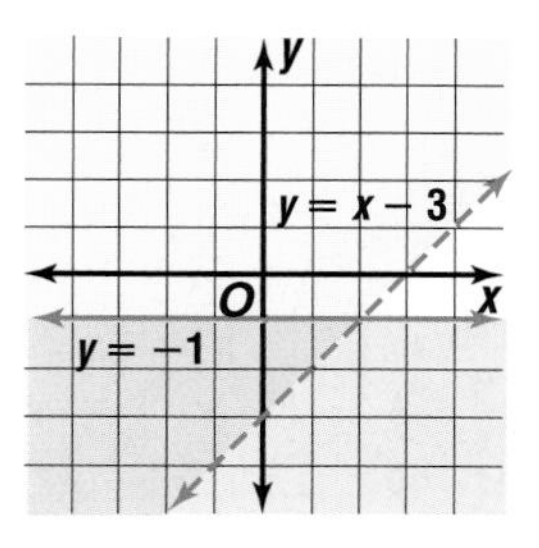

2 $2y < x - 2$
$-3x + 6y \geq 12$

The graphs of $2y = x - 2$ and $-3x + 6y = 12$ are parallel lines. *Check this by graphing or by comparing the slopes.*

Because the regions in the solution of $2y < x - 2$ and $-3x + 6y \geq 12$ have no points in common, the system of inequalities has no solution.

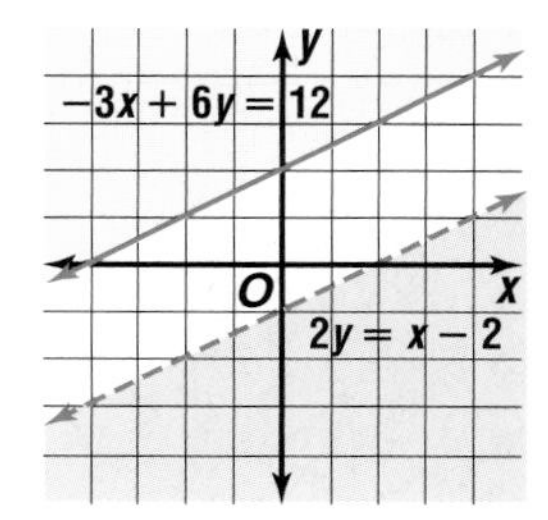

Your Turn

a. $x \geq -1$
$y < x - 2$

b. $y \geq 4x + 5$
$x + y \geq 3$

Real World

Example

Spending Link

3 **Luisa has \$96 to spend on gifts for the holidays. She must buy at least 9 gifts. She plans to buy puzzles that cost \$8 or \$12. How many of each puzzle can she buy?**

x = the number of \$8 puzzles $\quad$ y = the number of \$12 puzzles

The following system of inequalities can be used to represent the conditions of this problem.

$x + y \geq 9$ *She wants to buy at least 9 puzzles.*
$8x + 12y \leq 96$ *The total cost must be no more than \$96.*

Because the number of puzzles she can buy cannot be negative, both $x \geq 0$ and $y \geq 0$.

(continued on the next page)

The solutions are all of the ordered pairs in the intersection of the graphs of these inequalities. Only the first quadrant is used because $x \geq 0$ and $y \geq 0$.

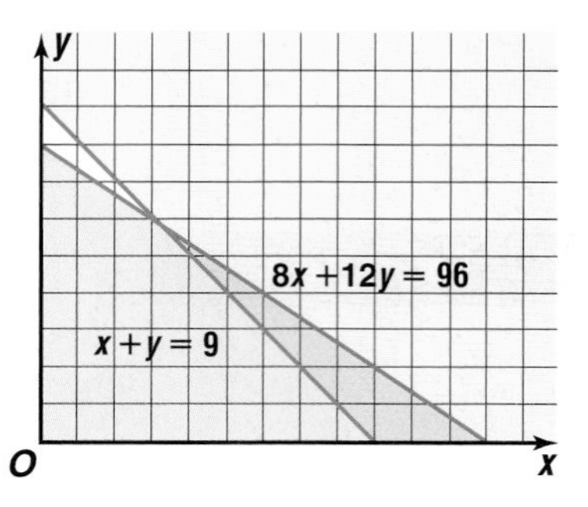

Any point with whole-number coordinates in the region is a possible solution. For example, since the point at (9, 1) is in the region, Luisa could buy 9 puzzles for \$8 and 1 for \$12. The greatest number of gifts that she could buy would be 12 puzzles for \$8 and no puzzles for \$12.

A graphing calculator can be helpful in solving systems of inequalities or in checking solutions.

Graphing Calculator Tutorial
See pp. 724–727.

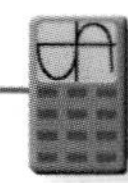

Graphing Calculator Exploration

Use a graphing calculator to solve the system of inequalities.

$y \geq 4x - 3$
$y \leq -2x + 1$

The graphing calculator graphs equations and shades above the first equation entered and below the second equation entered. (Note that inequalities that have $>$ or $\geq$ are lower boundaries and inequalities that have $<$ or $\leq$ are upper boundaries.)

Step 1 Press [2nd] [DRAW] 7 to choose the SHADE feature.

Step 2 Enter the equation that is the lower boundary of the region to be shaded, 4 [X,T,θ,n] [−] 3.

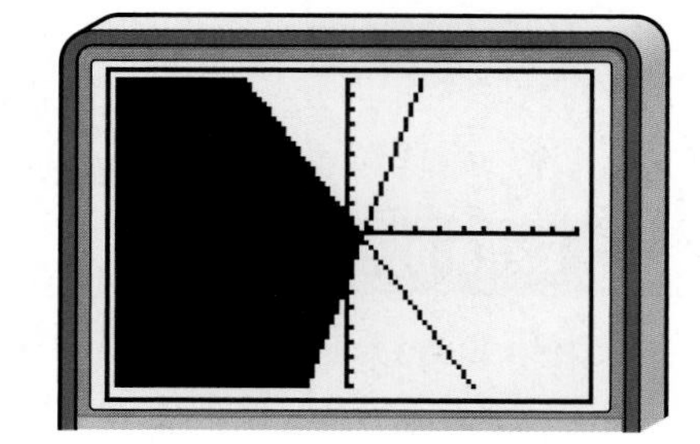

Step 3 Press [,] and enter the equation that is the upper boundary of the region to be shaded, [(−)] 2

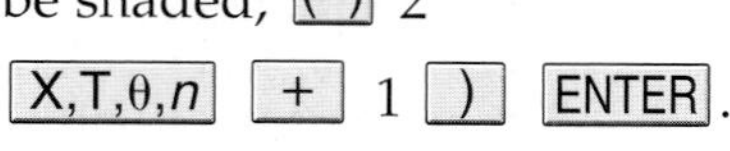

[X,T,θ,n] [+] 1 [)] [ENTER].

Try These

Use a graphing calculator to graph each system of inequalities. State one possible solution.

1. $y \geq x - 1$
   $y < x + 1$
2. $y \geq x - 2$
   $y \leq 3$
3. $y > 2x$
   $y \leq -x + 1$

Check for Understanding

Communicating Mathematics

1. **Compare and contrast** the solution of a system of equations and the solution of a system of inequalities.
2. **Sketch** a system of linear equations that has no solution.
3.  Kyle says that the solution of $y \geq 2x + 2$ and $y \leq -x - 1$ is all of the points in region B. Tarika says that the solution is all of the points in region D. Who is correct? Explain.

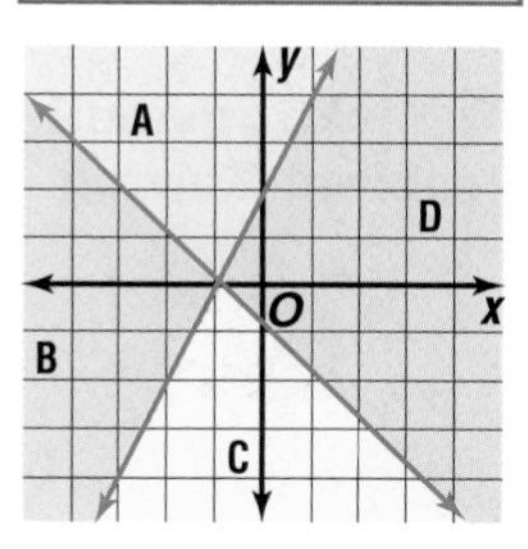

Exercise 3

Guided Practice

Getting Ready **State whether each ordered pair is a solution of the system of inequalities $x \geq -2$ and $y \leq 5$.**

Sample: $(-3, 5)$ **Solution:** $x \geq -2$
$-3 \overset{?}{\geq} -2$ *Replace x with −3.*
$(-3, 5)$ is not a solution.

4. $(6, -1)$
5. $(8, 7)$
6. $(0, 0)$

Solve each system of inequalities by graphing. If the system does not have a solution, write *no solution*. *(Examples 1 & 2)*

7. $x \geq -1$
$y \geq -4$
8. $y \leq 5$
$y \geq -x + 1$
9. $x + y > 2$
$y < x + 6$
10. $y \geq \frac{1}{2}x - 2$
$x - y < 3$

11. **Shopping** Nathaniel must buy two types of cookies for a banquet dinner. He only has \$25 to spend and needs at least 6 dozen cookies. The grocery store has small sugar cookies for \$3 a dozen and chocolate chip cookies for \$4 a dozen. How many cookies of each type can he buy? List three possible solutions. *(Example 3)*

Exercises

Practice

Solve each system of inequalities by graphing. If the system does not have a solution, write *no solution*.

Homework Help	
For Exercises	See Examples
12–25	1, 2
28	2
26	3
Extra Practice	
See page 720.	

12. $x < 2$
$y > -1$
13. $y > 0$
$x \leq 0$
14. $y > 2$
$y > -x + 2$
15. $y \geq x$
$y \leq -x$
16. $y \leq 2x$
$y \geq -3x$
17. $y \geq -3$
$y < -3x + 1$
18. $x > 2$
$y > -2x + 3$
19. $y \leq 2x - 1$
$y \geq 2x + 2$
20. $x + y \geq 2$
$x + y \leq 6$
21. $y \leq 2x + 2$
$2x + y \leq 4$
22. $y + 4 < x$
$2y + 4 > -3x$
23. $2y + x < 4$
$3x - y > 1$

24. Use the graphs of $y \geq x - 3$ and $y \geq -x - 1$ to determine the solution of the system.

25. Find the solution of the system $3x + 2y \leq -6$ and $y > x + 2$ by graphing.

Applications and Problem Solving

26. **Sports** Nykia must score at least 20 points in the last basketball game of the season to tie the school record for total points in a season. Usually she has no more than 40 opportunities to shoot the ball (both field goals and free throws combined) and makes half of her shots. What combination of shots can Nykia make to score the 20 points? Field goals are worth 2 points and free throws are worth 1 point.

27. **Communication** A long-distance carrier offers three fee plans to their customers.
    - 15 cents a minute with no monthly service charge
    - 10 cents a minute and a monthly service charge of $5.25
    - 5 cents a minute and a monthly service charge of $7.50

    Depending on how much people use the phone each month, which plan should they select?

28. **Critical Thinking** Write an inequality involving absolute value that has no solution.

Mixed Review

Determine whether each system of equations has *one* solution, *two* solutions, or *no* solution by graphing. If the system has one or two solutions, name them. *(Lesson 13–6)*

29. $y = x^2$
    $y = 4$

30. $y = x^2 + 3$
    $y = 2x - 1$

31. $y = -x^2 + 2$
    $y = -x + 5$

32. Use elimination to find the solution of the system $5x + 4y = -2$ and $3x - 2y = 1$. *(Lesson 13–5)*

33. **Mixtures** The Coffee Hut mixes coffee beans that cost $5.25 per pound with coffee beans that cost $6.50 per pound. They need a mixture of 5 pounds of coffee beans that costs $5.50 per pound. How many pounds of each type of coffee bean should they use? *(Lesson 13–3)*

Solve each equation. *(Lesson 4–4)*

34. $-4w = 32$

35. $13 = 2.6a$

36. $\frac{2}{5}y = -8$

Standardized Test Practice

Ⓐ Ⓑ Ⓒ Ⓓ

37. **Multiple Choice** Find $-15 + (-3) + (-7) + 6$. *(Lesson 2–3)*

    A -19
    B 1
    C 11
    D 16

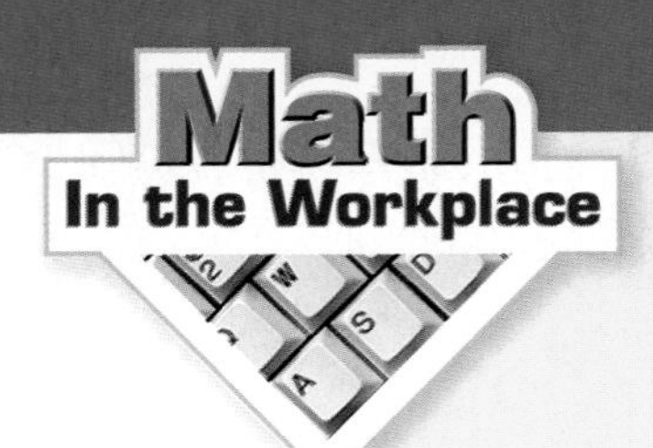

Car Dealer

Do you work well with people? Do cars interest you? If so, you may want to consider a career as a car dealer. Car dealers must keep track of the number of cars they need to sell each week to ensure maximum profit.

Suppose a car dealer receives a profit of $500 for each mid-sized car *m* sold and $750 for each sport-utility vehicle *s* sold. The dealer must sell at least two mid-sized cars for each sport-utility vehicle and must earn at least $3500 per week. The table and graph shown represent this situation.

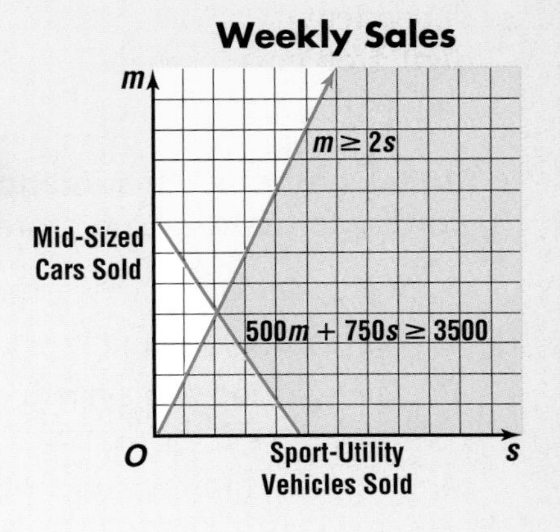

Situation	Inequality
At least 2 mid-sized cars must be sold for each sport-utility vehicle.	$m \geq 2s$
Earn a profit of at least $3500.	$500m + 750s \geq 3500$

1. Suppose a car dealer sells 2 sport-utility vehicles. How many mid-sized cars must be sold to earn at least $3500?
2. If a car dealer sells only one sport-utility vehicle, how many mid-sized cars must be sold to meet the goal?

FAST FACTS About Car Dealers

Working Conditions

- usually work in a comfortable environment
- evening, weekend, and holiday work
- work a 40-hour week, with some overtime

Education

- good communication skills and computer skills are helpful
- high school math and business classes
- college degree in business is helpful

Job Outlook

Expected Growth in Earnings

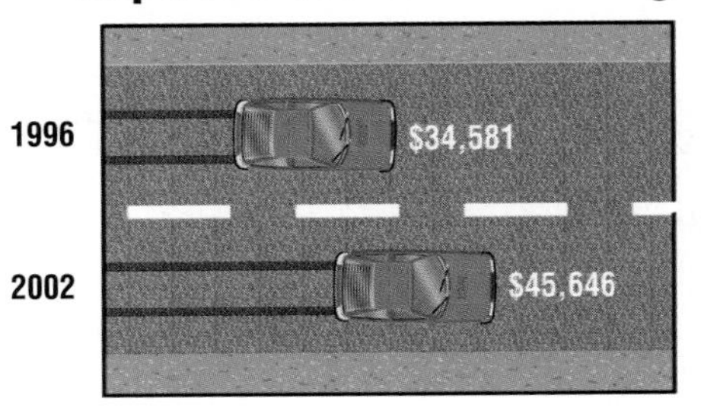

interNET CONNECTION **Career Data** For the latest information on car dealers, visit: www.algconcepts.com

CHAPTER 13

Study Guide and Assessment

Understanding and Using the Vocabulary

After completing this chapter, you should be able to define each term, property, or phrase and give an example or two of each.

interNET CONNECTION **Review Activities** For more review activities, visit: www.algconcepts.com

augmented matrix *(p. 578)*
consistent *(p. 554)*
dependent *(p. 554)*
digit problems *(p. 568)*
elimination *(p. 566)*
identity matrix *(p. 579)*
inconsistent *(p. 554)*
independent *(p. 554)*
quadratic-linear *(p. 580)*
row operations *(p. 578)*
substitution *(p. 560)*
system of equations *(p. 550)*
system of inequalities *(p. 586)*

State whether each sentence is *true* or *false*. If false, replace the underlined word(s) to make a true statement.

1. Inconsistent is the description used for a system of equations that has <u>no solution</u>.
2. A system of inequalities can be solved by <u>graphing</u> the inequalities.
3. The solution to a system of equations is the ordered pair that satisfies <u>two</u> of the equations in the system.
4. A system of linear equations that has <u>at least</u> one solution is called independent.
5. The exact coordinates of the point where two or more lines intersect can be determined by using <u>substitution</u>.
6. A set of two or more equations is called a <u>system of equations</u>.
7. <u>Dependent</u> systems of equations have infinitely many solutions.
8. The description of a system of linear equations depends on the number of <u>solutions</u>.
9. To solve a system of equations by <u>substitution</u>, one of the variables is eliminated by either adding or subtracting the equations.
10. A system of equations that has at least one solution can be described as <u>inconsistent</u>.

Skills and Concepts

Objectives and Examples

• **Lesson 13–1** Solve systems of equations by graphing.

$y = x + 1$
$x + y = 3$

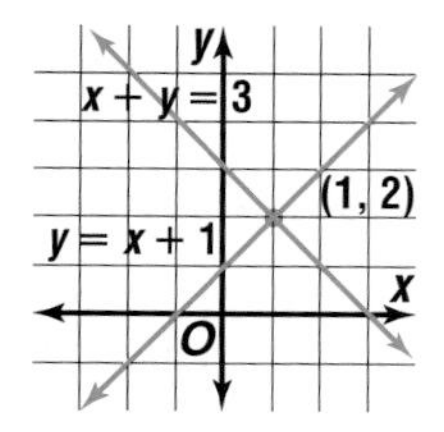

The solution is (1, 2).

Review Exercises

Solve each system of equations by graphing.

11. $y = 4x$
 $y = -x + 5$
12. $x = y$
 $y = 2x - 3$
13. $y = \frac{1}{2}x + 3$
 $y = -x$
14. $y = 5$
 $2y = x + 3$
15. $x - 3y = -6$
 $y = -\frac{1}{2}x - 3$
16. $y = 3x + 1$
 $y = -3x + 1$

Objectives and Examples

- **Lesson 13–2** Determine whether a system of equations has one solution, no solution, or infinitely many solutions by graphing.

$y = x + 5$
$y = -4x$

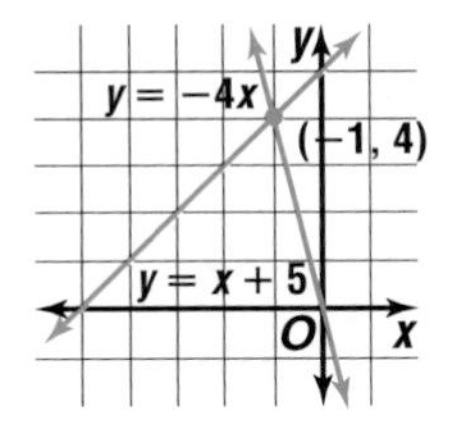

This system has one solution, $(-1, 4)$.

Review Exercises

Determine whether each system of equations has *one* solution, *no* solution, or *infinitely many* solutions by graphing. If the system has one solution, name it.

17. $y = -x + 1$; $y = x - 5$

18. $y = 2x + 3$; $y = 2x - 2$

19. $x + y = -2$; $4x + 4y = -8$

20. $y = \frac{1}{2}x - 1$; $y = x$

- **Lesson 13–3** Solve systems of equations by the substitution method.

To solve $y = 3x$ and $2x + y = 10$, substitute $3x$ for y in the second equation.

$$2x + 3x = 10$$
$$5x = 10$$
$$x = 2$$

$$y = 3x$$
$$y = 3(2) \text{ or } 6$$

The solution is $(2, 6)$.

Use substitution to solve each system of equations.

21. $y = 2x$; $4x + y = 3$

22. $3y = x$; $4x + 2y = 14$

23. $y = 3x + 1$; $3x - y = 4$

24. $y = x + 1$; $3x - 2y = -12$

25. $2x - 3y = 4$; $4x - 6y = 8$

26. $y - 3 = x$; $2x + y = 4$

- **Lesson 13–4** Solve systems of equations by the elimination method using addition and subtraction.

$$\begin{array}{r} 3x - 4y = 5 \\ (-)\ 3x + 2y = -7 \\ \hline -6y = 12 \\ y = -2 \end{array}$$

$$3x - 4(-2) = 5$$
$$3x + 8 = 5$$
$$3x = -3$$
$$x = -1$$

The solution is $(-1, -2)$.

Use elimination to solve each system of equations.

27. $2x + y = 2$; $x + y = -2$

28. $x - 2y = 4$; $2x + 2y = 2$

29. $x - y = 5$; $-x + y = 3$

30. $5x + 3y = 11$; $2x + 3y = 2$

31. $4x + 3y = 3$; $-4x - 3y = -3$

32. $y = 3 - 4x$; $x + y = 2$

- **Lesson 13–5** Solve systems of equations by the elimination method using multiplication and addition.

$3x + 2y = 5$
$x + y = 1$ **Multiply by −2.**

$$\begin{array}{r} 3x + 2y = 5 \\ (+)\ -2x - 2y = -2 \\ \hline x = 3 \end{array}$$

Substitute to find the value of y. The solution is $(3, -2)$.

Use elimination to solve each system of equations.

33. $2x + 3y = 4$; $x + 2y = 2$

34. $4x + 3y = 3$; $-2x + 4y = 4$

35. $3x + 4y = 2$; $2x - y = 1$

36. $x - 3y = 2$; $-3x + 5y = 2$

37. $3x - 2y = 4$; $4x - 3y = -5$

38. $5x + 3y = 4$; $6x + 4y = 3$

• Extra Practice
See pages 718–720.

Objectives and Examples

• **Lesson 13–6** Solve systems of quadratic and linear equations.

$y = 2x + 3$
$y = x^2 - 5$

Substitute $x^2 - 5$ for y in the first equation.

$$y = 2x + 3$$
$$x^2 - 5 = 2x + 3$$
$$x^2 - 5 - 2x - 3 = 2x + 3 - 2x - 3$$
$$x^2 - 2x - 8 = 0$$
$$(x + 2)(x - 4) = 0$$

$x + 2 = 0$ or $x - 4 = 0$
$x = -2$ $\quad x = 4$

Solve by finding the corresponding values of y. The solutions are $(-2, -1)$ and $(4, 11)$.

Review Exercises

Determine whether each system of equations has *one* solution, *two* solutions, or *no* solution by graphing. If the system has one or two solutions, name them.

39. $y = -x^2$
$y = x - 2$

40. $y = x^2 + 2$
$y = 6$

41. $y = 2x^2$
$y = -x - 4$

42. $y = x^2 - 1$
$y = -4x - 5$

Use substitution to solve each system of equations.

43. $y = x^2 + 2$
$y = 3x + 6$

44. $y = -x^2 + 4$
$y = 4x + 10$

45. Solve the system $y = x^2 - 2x$ and $y = 4x - 9$ by using substitution.

• **Lesson 13–7** Solve systems of inequalities by graphing.

Solve the system of inequalities $x \geq -2$ and $y < x - 3$ by graphing.

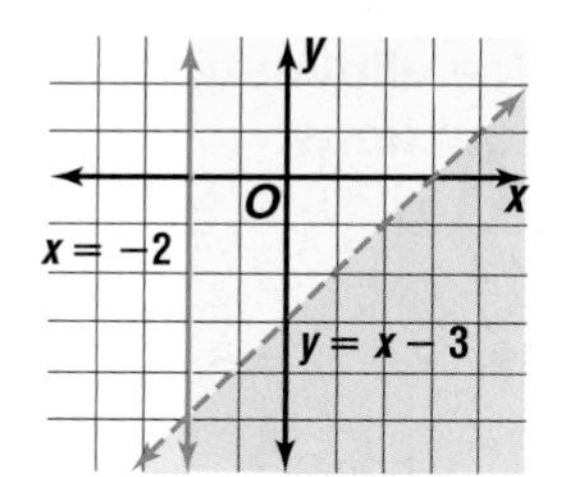

Solve each system of inequalities by graphing.

46. $x < 4$
$y \geq -2$

47. $x \geq -3$
$y \leq x + 2$

48. $x + y > -4$
$x + y \leq 2$

49. $y \geq 2x + 1$
$y > -x - 1$

50. Use the graphs of $-6x + 3y > 9$ and $y < 2x - 4$ to determine the solution of the system. If the system does not have a solution, write *no solution*.

Applications and Problem Solving

51. Mixtures Mr. Collins mixed almonds that cost $4.25 per pound with cashews that cost $6.50 per pound. He now has a mixture of 20 pounds of nuts that costs $5.60 per pound. How many pounds of each type of nut did he use? *(Lesson 13–3)*

52. Number Theory The difference between the tens digit and the units digit of a two-digit number is 3. Suppose the tens digit is one less than twice the units digit. What is the number? *(Lesson 13–4)*

53. Geometry The difference between the length and width of a rectangle is 7 feet. Find the dimensions of the rectangle if its perimeter is 50 feet. *(Lesson 13–5)*

CHAPTER 13 Test

1. **Graph** a system of equations that has infinitely many solutions.
2. **Explain** when you would use elimination with subtraction to solve a system of equations.

Solve each system of equations by graphing.

3. $y = 3$
   $y = x + 4$
4. $x + y = -2$
   $2x - y = -4$
5. $y = -x^2 - 1$
   $y = -5$

State whether each system is *consistent and independent*, *consistent and dependent*, or *inconsistent*.

6. 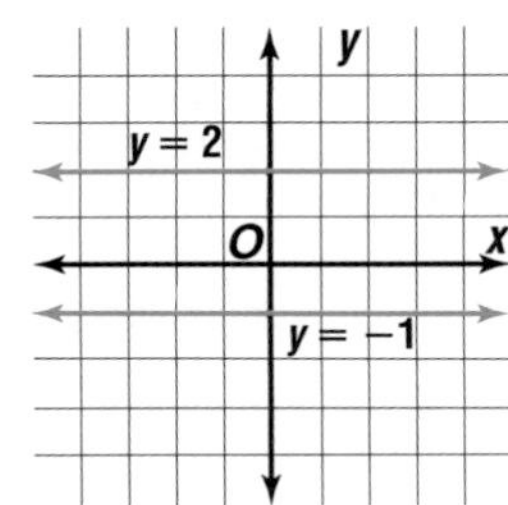

7. 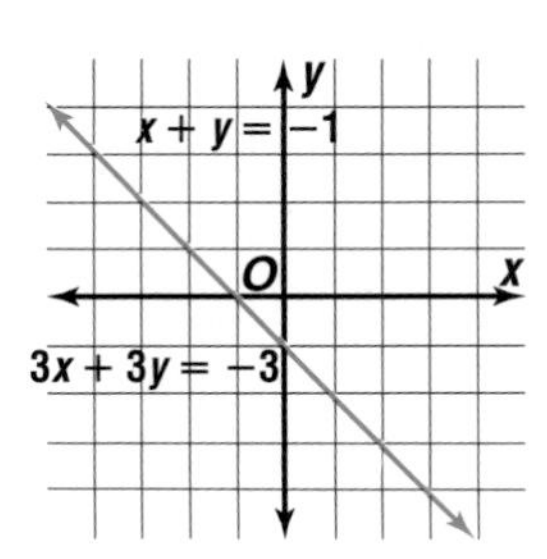

8. Use graphing to determine whether the system $y = -2x$ and $2x + y = 4$ has *one* solution, *no* solution, or *infinitely many* solutions. If the solution has one solution, name it.

Use substitution to solve each system of equations.

9. $y = 3x$
   $x + y = 4$
10. $x + y = -2$
    $x = y + 10$
11. $y = 5x - 3$
    $10x - 2y = -2$
12. $y = x^2 - 15$
    $y = 2x$
13. $y = 5x + 4$
    $y = x^2 + 5x$
14. $y = 3x + 2$
    $y = x^2 + 6$

Use elimination to solve each system of equations.

15. $x + y = 5$
    $x - y = -9$
16. $4x - 5y = 7$
    $x + 5y = 8$
17. $2x - y = 32$
    $y = 60 - 2x$
18. $x + 3y = -1$
    $2x + 4y = -2$
19. $5x - 2y = 3$
    $15x - 6y = 9$
20. $-5x + 8y = 21$
    $10x + 3y = 15$

Solve each system of inequalities by graphing.

21. $y \leq -3$
    $y > -x - 2$
22. $y < x + 4$
    $y > x - 2$
23. $x \leq 2y$
    $2x + 3y \leq 6$

24. Find two numbers whose sum is 64 and whose difference is 42.

25. **Transportation** Two trains travel toward each other on parallel tracks at the same time from towns 450 miles apart. Suppose one train travels 6 miles per hour faster than the other train. What is the rate of each train if they meet in 5 hours?

CHAPTER 13

Preparing for Standardized Tests

Perimeter, Area, and Volume Problems

Standardized tests often include questions on perimeter, area, circumference, and volume. You'll need to know and apply formulas for each of these measurements. Be sure you know these terms.

Use the information given in the figure.

area	circumference	height	perimeter	width
base	diameter	length	radius	

State Test Example

Richard plans to increase the floor area of his health club's weight room. The figure below shows the existing floor area with a solid line and the additional floor area with a dotted line. Find the length, in feet, of the new weight room.

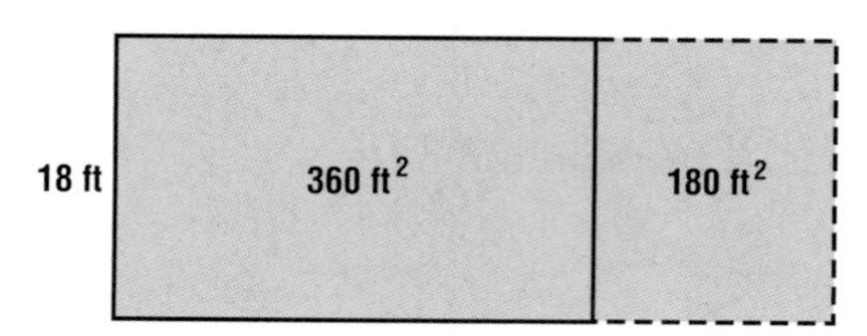

Hint Familiarize yourself with the formulas for the area and perimeter of quadrilaterals.

Solution The existing room has an area of 360 square feet and a width of 18 feet. The addition has an area of 180 square feet and a width of 18 feet.

To find the length of the new weight room, first find the area of the new room.

$$A = 360 + 180 \text{ or } 540 \text{ square feet}$$

Next, use the formula for the area of a rectangle to find the length of the room.

$A = \ell w$

$540 = \ell \cdot 18$ *Replace A with 540 and w with 18.*

$\frac{540}{18} = \frac{18\ell}{18}$ *Divide each side by 18.*

$30 = \ell$

The length of the new room is 30 feet.

SAT Example

What is the diameter of a circle with a circumference of 5 inches?

A $\frac{5}{\pi}$ in. **B** $\frac{10}{\pi}$ in. **C** 5 in.

D 5π in. **E** 10π in.

Hint If a geometry problem has no figure, sketch one.

Solution

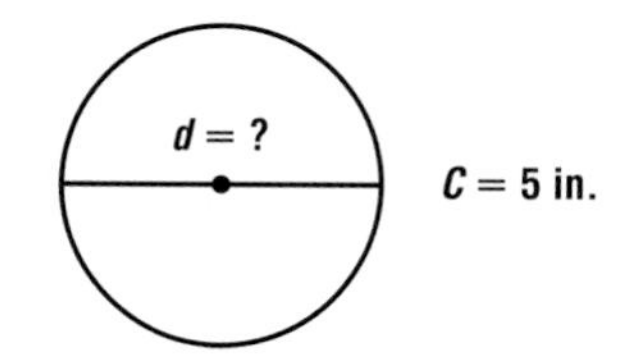

You know the circumference of the circle. You need to find the diameter of the circle.

Use the formula $C = \pi d$, where C is the circumference and d is the diameter to find the circumference of the circle.

First, replace C with 5. Then solve the equation for d.

$C = \pi d$

$5 = \pi d$ *Replace C with 5.*

$\frac{5}{\pi} = \frac{\pi d}{\pi}$ *Divide each side by π.*

$\frac{5}{\pi} = d$

The answer is A.

After you work each problem, record your answer on the answer sheet provided or on a sheet of paper.

Multiple Choice

1. Quinn's neighborhood is a rectangle 2 miles by 1.5 miles. How many miles does Quinn jog if he jogs around the boundary of his neighborhood?

A 3 **B** 3.5 **C** 6 **D** 7

2. If $AC = 4$, what is the area of $\triangle ABC$?

A $\frac{1}{2}$ **B** 2
C 4 **D** 8

3. Micela is making a poster that has a length of 36 inches. If the maximum perimeter is 96 inches, which inequality can be used to determine the width w of the poster?

A $96 \geq 2(36) + 2w$ **B** $96 \leq 2(36) + 2w$
C $96 \geq 36 + 2w$ **D** $96 \geq 36 + w$

4. If the area of a circle is 16 square meters, what is its radius in meters?

A $\frac{8}{\pi}$ **B** $\frac{16}{\pi}$ **C** $\frac{4\sqrt{\pi}}{\pi}$
D 12π **E** $144\pi^2$

5. If you double the length and the width of a rectangle, how does its perimeter change?

A It increases by $1\frac{1}{2}$.
B It doubles.
C It quadruples.
D It does not change.

6. Mr. Tremaine needs to find the area of his backyard so that he can buy the right amount of sod for it. What is the area?

A 192 ft^2
B 360 ft^2
C 456 ft^2
D 720 ft^2

7. The table shows the speed of a car and the distance needed to safely stop the car. Which equation represents the data?

A $y = x + 25$
B $y = x^2 + 20$
C $y = x^2 \times 20$
D $y = x^2 \div 20$
E $y = x^2 \div 25$

Speed (x)	Distance (y)
20	20
30	45
40	80
50	125

8. Which number could be the diameter of the circle?

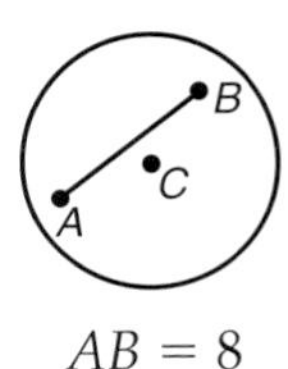

$AB = 8$

A 12 **B** 8 **C** 6 **D** 4

Grid In

9. Allie sketched her walking route. The curved part is a semicircle with a radius of 10.5 meters. If Allie walks this route 10 times, how many meters does she walk? Round to the nearest meter.

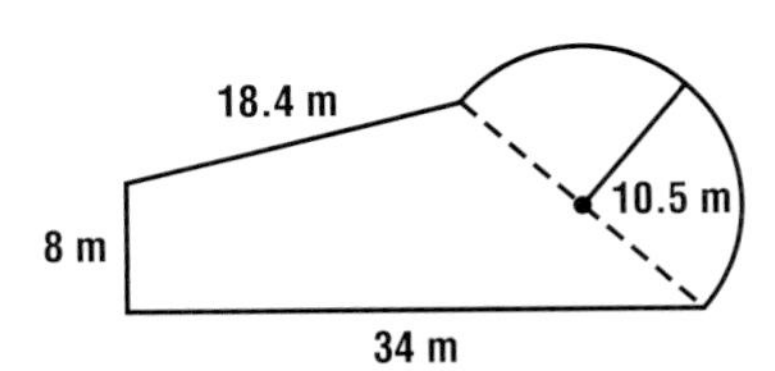

Extended Response

10. Suppose you have 120 feet of fence.

Part A What is the greatest possible rectangular area you can enclose?

Part B What are the dimensions of the rectangle that has this area?

CHAPTER 14

Radical Expressions

FOLDABLES™ Study Organizer

Make this Foldable to help you organize information about the material in this chapter. Begin with a sheet of 11" by 17" paper.

1. **Fold** the short sides to meet in the middle.

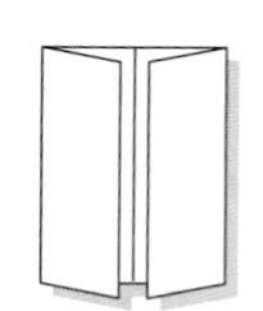

2. **Fold** the top to the bottom.

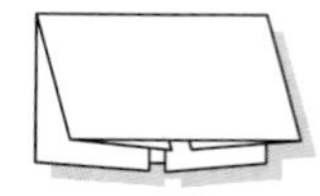

3. **Open.** Cut along second fold to make four tabs.

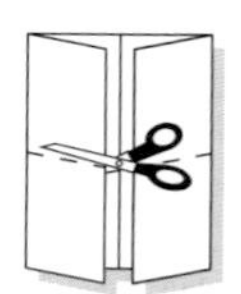

4. **Label** each tab as shown.

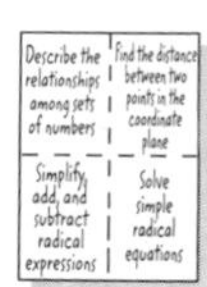

Reading and Writing As you read and study the chapter, write examples and notes under each tab.

Problem-Solving Workshop

Project

You are to design three swimming pools that hold a total of 200,000 gallons of water or less. You will need to meet the following requirements.
Pool #1: a square with semicircles on each side, at least 5 feet deep
Pool #2: a rectangle with semicircles on the sides, at least 4 feet deep
Pool #3: a circle, at least 3 feet deep

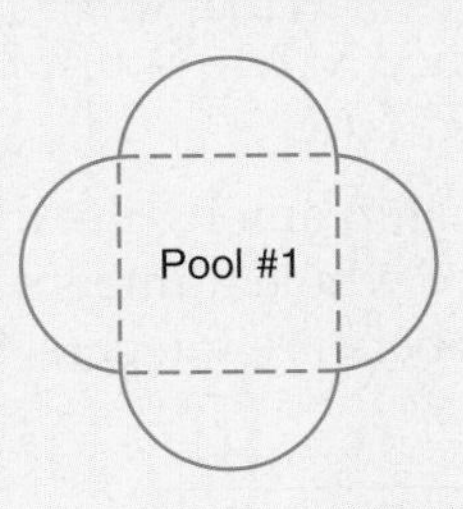

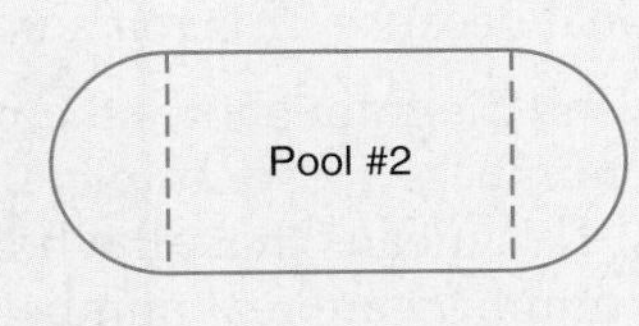

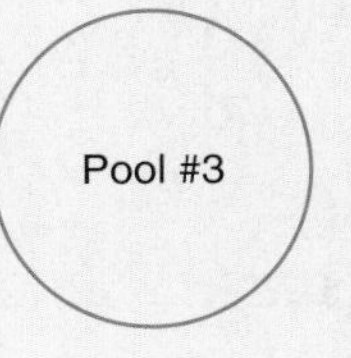

Working on the Project

Work with a partner to solve the problem.

- Suppose a square swimming pool measures 20 feet on a side. If the pool is 5 feet deep, how many cubic inches of water does it hold? How many gallons? (*Hint:* 1 gal $\approx$ 231 in^3)
- Find the diameter of a circular pool that is 3.5 feet deep and holds 16,000 gallons of water.

Technology Tools

- Use a **spreadsheet** to find the dimensions.
- Use **drawing software** to draw each pool.

Strategies

Look for a pattern.
Draw a diagram.
Make a table.
Work backward.
Use an equation.
Make a graph.
Guess and check.

interNET CONNECTION **Research** For more information about swimming pools, visit: www.algconcepts.com

Presenting the Project

Prepare a brochure of your designs. Include the following information:

- diagrams of the three pools with dimensions labeled,
- the volume of each pool and the total volume of the water, and
- a paragraph describing how you found the dimensions of each pool.

14–1 The Real Numbers

What You'll Learn
You'll learn to describe the relationships among sets of numbers.

Why It's Important
Meteorology Meteorologists use real numbers when determining the duration of a thunderstorm. *See Exercise 16.*

In Lesson 3–1, you learned about *rational numbers.* Natural numbers, whole numbers, and integers are all rational numbers. These sets are listed below.

Natural Numbers:	$\{1, 2, 3, 4, \ldots\}$
Whole Numbers:	$\{0, 1, 2, 3, \ldots\}$
Integers:	$\{\ldots, -2, -1, 0, 1, 2, \ldots\}$
Rational Numbers:	{all numbers that can be expressed in the form $\frac{a}{b}$, where a and b are integers and $b \neq 0$}

Recall that repeating or terminating decimals are also rational numbers because they can be expressed as $\frac{a}{b}$, where a and b are integers and $b \neq 0$. The square roots of perfect squares are also rational numbers. For example, $\sqrt{0.16}$ is a rational number since $\sqrt{0.16} = 0.4$. However, $\sqrt{21}$ is irrational because 21 is not a perfect square.

The Venn diagram shows the relationship among the different types of rational numbers. For example, the set of whole numbers is a subset of the integers. This means that all whole numbers are integers. Similarly, all rational numbers are real numbers.

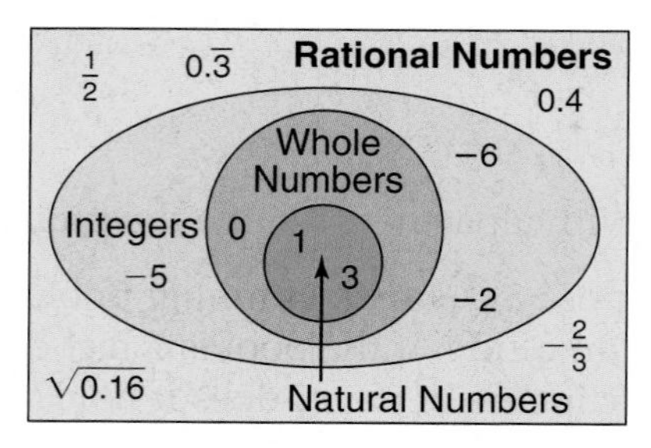

In Lesson 8–6, you learned about *irrational numbers.* A few examples of irrational numbers are shown below.

$0.1010010001\ldots$ π $0.153768\ldots$ $-\sqrt{23}$

The set of rational numbers and the set of irrational numbers together form the set of real numbers. *Numbers such as $\sqrt{-1}$ and $4 + \sqrt{-9}$ are called **complex numbers**. The set of complex numbers includes all of the real numbers as well as numbers involving square roots of negative numbers. We will not deal with complex numbers in this text.*

Reading Algebra

Natural numbers are also called *counting numbers*.

Real Numbers

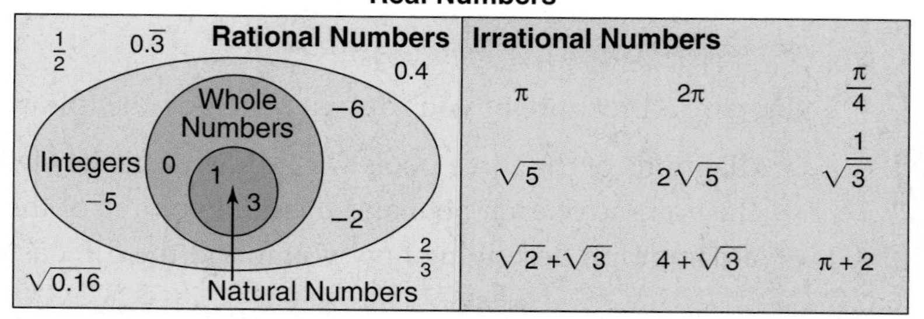

Examples

Name the set or sets of numbers to which each real number belongs.

1. -5 This number is an integer and a rational number.

2. $\frac{12}{3}$ Since $\frac{12}{3} = 4$, this number is a natural number, a whole number, an integer, and a rational number.

3. $\sqrt{95}$ $\sqrt{95} = 9.746794345\ldots$ It is not the square root of a perfect square. So, it is irrational.

4. $0.\overline{7}$ This repeating decimal is a rational number since it is equivalent to $\frac{7}{9}$. *$7 \div 9 = 0.7777777\ldots$*

5. $-\sqrt{4}$ Since $-\sqrt{4} = -2$, this number is an integer and a rational number.

Your Turn **Name the set or sets of numbers to which each real number belongs. Let N = natural numbers, W = whole numbers, Z = integers, Q = rational numbers, and I = irrational numbers.**

a. $3.141592\ldots$ b. $-\sqrt{9}$ c. $0.1666666\ldots$ d. $\frac{20}{4}$ e. 0

If you graph all of the rational numbers, you will still have some "holes" in the number line. The irrational numbers "fill in" the number line. The graph of all real numbers is the entire number line without any "holes."

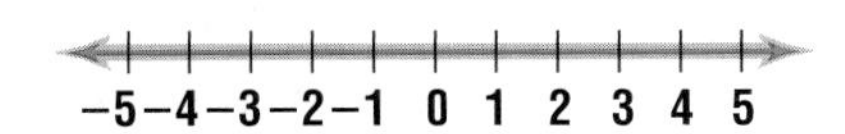

This property of real numbers is called the Completeness Property.

Completeness Property for Points on the Number Line	Each real number corresponds to exactly one point on the number line. Each point on the number line corresponds to exactly one real number.

You have learned how to graph rational numbers. Irrational numbers can also be graphed. Therefore, every real number can be graphed. Use a calculator or a table of squares and square roots to find approximate values of square roots that are irrational. These values can be used to approximate the graphs of square roots.

Examples

Find an approximation, to the nearest tenth, for each square root. Then graph the square root on a number line.

6 $\sqrt{3}$

Enter: [2nd] [$\sqrt{\ }$] 3 [ENTER] 1.732050808

An approximate value for $\sqrt{3}$ is 1.7.

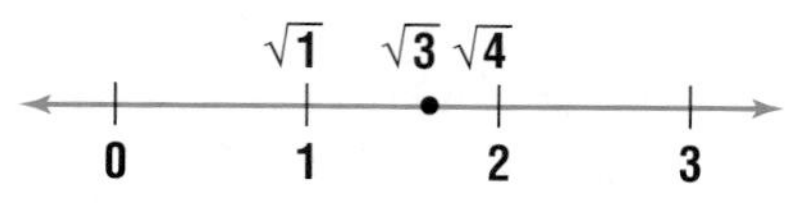

7 $-\sqrt{12}$

Enter: [(–)] [2nd] [$\sqrt{\ }$] 12 [ENTER] −3.464101615

An approximate value for $-\sqrt{12}$ is -3.5.

$-\sqrt{16}$ $-\sqrt{12}$ $-\sqrt{9}$

-5 -4 -3 -2

Your Turn

f. $\sqrt{6}$

g. $-\sqrt{23}$

Determine whether each number is rational or irrational. If it is irrational, find two consecutive integers between which its graph lies on the number line.

8 $\sqrt{47}$

$$\sqrt{36} < \sqrt{47} < \sqrt{49}$$
$$6 < \sqrt{47} < 7$$

47 is not a perfect square. So, its square root is irrational. The graph of $\sqrt{47}$ lies between 6 and 7.

9 $-\sqrt{26}$

$$-\sqrt{36} < -\sqrt{26} < -\sqrt{25}$$
$$-6 < -\sqrt{26} < -5$$

26 is not a perfect square. So, its square root is irrational. The graph of $-\sqrt{26}$ lies between -6 and -5.

Your Turn

h. $\sqrt{64}$

i. $-\sqrt{28}$

You have solved equations with rational number solutions. Some equations have solutions that are irrational numbers.

Example 10

Science Link

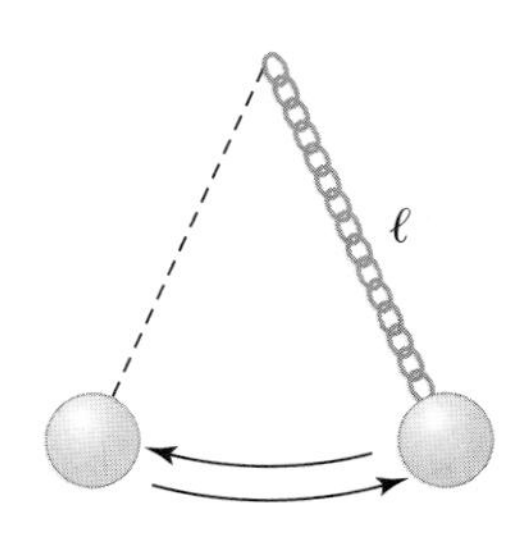

The time t in seconds it takes for a pendulum to complete one full swing (back and forth) is given by the equation $t = 2\pi\sqrt{\frac{\ell}{9.8}}$, where ℓ is the length of the pendulum in meters. Suppose a pendulum has a length of 4.9 meters. How long does it take the pendulum to complete one full swing?

$t = 2\pi\sqrt{\frac{\ell}{9.8}}$

$= 2\pi\sqrt{\frac{4.9}{9.8}}$ *Replace ℓ with 4.9.*

$= 2\pi\sqrt{0.5}$

Enter: 2 [2nd] [π] [2nd] [√] 0.5 [ENTER] 4.442882938

The pendulum will complete one full swing in about 4.4 seconds.

Check for Understanding

Communicating Mathematics

1. **Give an example** of a number that is an integer and a rational number.
2. **Give** two counterexamples for the statement *all square roots are irrational numbers.*
3. **Write** a square root that is an irrational number.

Guided Practice

Name the set or sets of numbers to which each real number belongs. Let N = natural numbers, W = whole numbers, Z = integers, Q = rational numbers, and I = irrational numbers. *(Examples 1–5)*

4. $-\sqrt{36}$
5. 0.3131131113 . . .
6. 4
7. $0.\overline{2}$

Find an approximation, to the nearest tenth, for each square root. Then graph the square root on a number line. *(Examples 6 & 7)*

8. $\sqrt{7}$
9. $-\sqrt{20}$
10. $\sqrt{32}$
11. $\sqrt{54}$

Determine whether each number is *rational* or *irrational*. If it is irrational, find two consecutive integers between which its graph lies on the number line. *(Examples 8 & 9)*

12. $\sqrt{63}$
13. $-\sqrt{25}$
14. $-\sqrt{147}$
15. $\sqrt{49}$

16. **Meteorology** To estimate the amount of time t in hours that a thunderstorm will last, meteorologists use the formula $t = \sqrt{\frac{d^3}{216}}$, where d is the diameter of the storm in miles. *(Example 10)*

a. Estimate how long a thunderstorm will last if its diameter is 8.4 miles.

b. Find the diameter of a storm that will last for 1 hour.

Exercises

Practice

Name the set or sets of numbers to which each real number belongs. Let N = natural numbers, W = whole numbers, Z = integers, Q = rational numbers, and I = irrational numbers.

17. $-\frac{1}{2}$
18. $\sqrt{17}$
19. 0
20. $\frac{15}{3}$
21. -5
22. 1.202002 . . .
23. $\sqrt{81}$
24. 0.125
25. $-\sqrt{121}$
26. $\frac{6}{18}$
27. 0.834834 . . .
28. $-\frac{3}{1}$

Find an approximation, to the nearest tenth, for each square root. Then graph the square root on a number line.

29. $\sqrt{3}$
30. $\sqrt{8}$
31. $-\sqrt{10}$
32. $\sqrt{17}$
33. $-\sqrt{40}$
34. $\sqrt{37}$
35. $\sqrt{52}$
36. $-\sqrt{99}$
37. $\sqrt{108}$
38. $-\sqrt{112}$
39. $\sqrt{250}$
40. $\sqrt{300}$

Homework Help	
For Exercises	See Examples
17–28	1–5
29–40, 53, 54, 57, 58	6, 7
41–52, 56	8, 9
55	10

Extra Practice

See page 720.

Determine whether each number is *rational* or *irrational*. If it is irrational, find two consecutive integers between which its graph lies on the number line.

41. $\sqrt{16}$
42. $-\sqrt{12}$
43. $\sqrt{41}$
44. $\sqrt{24}$
45. $-\sqrt{81}$
46. $\sqrt{66}$
47. $-\sqrt{7}$
48. $\sqrt{98}$
49. $\sqrt{125}$
50. $\sqrt{144}$
51. $-\sqrt{169}$
52. $\sqrt{220}$
53. Graph $\sqrt{2}$, π, and $-\sqrt{5}$ on a number line.
54. Write $2.\overline{4}$, 2.41, and $\sqrt{6}$ in order from least to greatest.

Applications and Problem Solving

55. **Science** Refer to Example 10. Suppose a pendulum has a length of 18 meters. How long does it take the pendulum to complete one full swing?

56. **Geometry** The radius of a circle is given by $r = \sqrt{\frac{A}{\pi}}$, where A is the area of the circle. Suppose a circle has an area of 25 square centimeters. What is the measure of the radius?

57. **Electricity** The voltage V in a circuit is given by $V = \sqrt{PR}$, where P is the power in watts and R is the resistance in ohms. A circuit is designed with two resistance settings, 4.6 ohms and 5.2 ohms, and two power settings, 1200 watts and 1500 watts. Which settings are required so the voltage is between 75 and 85 volts?

58. **Geometry** Find the area of each rectangle. Round answers to the nearest tenth. (*Hint:* Use the Pythagorean Theorem.)

a. 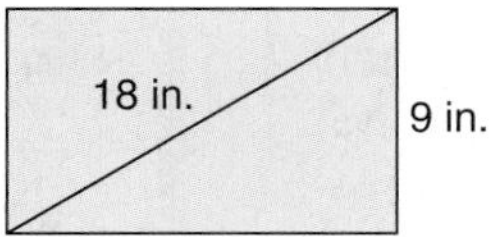

b. 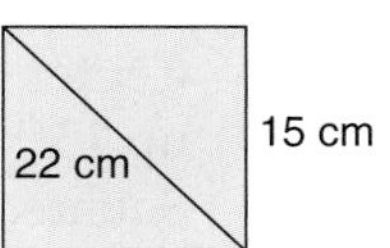

59. **Critical Thinking** Find all numbers of the form $\sqrt{n}$ such that n is a natural number and the graph of $\sqrt{n}$ lies between each pair of numbers on the number line.

a. 4 and 5 b. 4.25 and 4.5

Mixed Review

60. Find the solution of the system of inequalities $y \leq x + 3$ and $y > -x - 4$. *(Lesson 13–7)*

61. What is the solution of the system of equations $y = x^2 + 2$ and $y = 4x - 1$? *(Lesson 13–6)*

62. Graph the solution of $|4n| > 16$ on a number line. *(Lesson 12–6)*

Solve each inequality. *(Lesson 12–2)*

63. $x + 3 < -4$ 64. $w - 14 \geq -8$ 65. $3.4 + m \leq 1.6$

66. Write an inequality for the graph shown at the right. *(Lesson 12–1)*

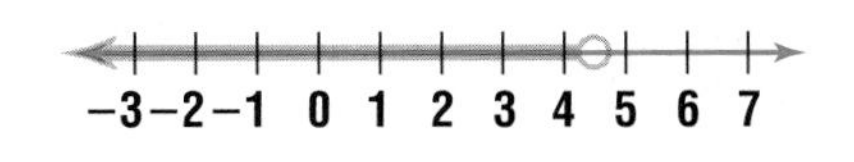

67. Graph $y = 2^x - 1$. Then state the y-intercept. *(Lesson 11–7)*

Factor each polynomial. If the polynomial cannot be factored, write *prime*. *(Lesson 10–5)*

68. $x^2 - 25y^2$ 69. $4a^2 - 36$ 70. $m^2 + 2m + 4$

71. Find the degree of $5x^3 - 3x^3y - 6xy$. *(Lesson 9–4)*

Standardized Test Practice

72. **Short Response** Cleavon has $35 to spend on CDs. Suppose CDs cost $13.85 each, including tax. Write an equation that can be used to determine how many CDs Cleavon can buy. *(Lesson 4–4)*

73. **Multiple Choice** What is the value of a if $6 - (-8) = a$? *(Lesson 2–3)*

A 14 B 6 C −14 D −6

14–2 The Distance Formula

What You'll Learn
You'll learn to find the distance between two points in the coordinate plane.

Why It's Important
Engineering
Engineers can use the distance formula to determine the amount of cable needed to install a cable system. *See Exercise 9.*

A coordinate system is superimposed over a map of Washington, D.C. Jessica and Omar walk from the Metro Center to 15th Street and then up 15th Street to McPherson Square. How far is the Metro center from McPherson Square? *This problem will be solved in Example 2.*

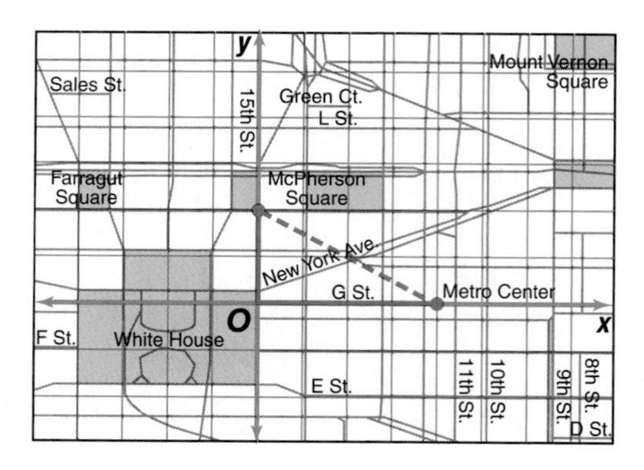

Recall that subtraction is used to find the distance between two points that lie on a vertical or horizontal line.

In the following activity, you will find the distance between two points that do not lie on a horizontal or vertical line.

Look Back

Graphing Ordered Pairs: Lesson 2–2

Hands-On Algebra

Materials: 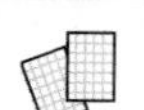 grid paper  straightedge

Step 1 Graph $M(-4, 3)$ and $N(2, -5)$ on a coordinate plane.

Step 2 Draw a vertical segment from M and a horizontal segment from N. Label the point of intersection P.

Step 3 Find the coordinates of P.

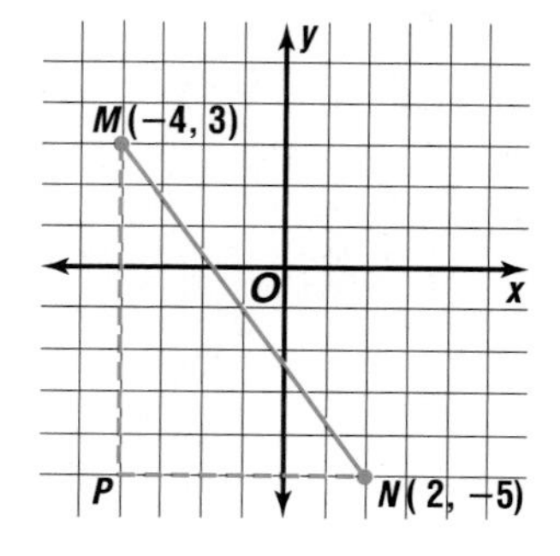

Try These

1. Find the distance between M and P.
2. Find the distance between N and P.
3. What kind of triangle is $\triangle MNP$?
4. What theorem can be used to find MN if MP and NP are known?
5. Find the distance between M and N.

In the activity, you discovered that $(MN)^2 = (MP)^2 + (NP)^2$. You can solve this equation for MN to find the distance between points M and N.

$$(MN)^2 = (MP)^2 + (NP)^2$$
$$\sqrt{(MN)^2} = \sqrt{(MP)^2 + (NP)^2} \quad \textit{Take the positive square root of each side.}$$
$$MN = \sqrt{(MP)^2 + (NP)^2}$$

The **Distance Formula** can be used to find the distance between any two points (x_1, y_1) and (x_2, y_2).

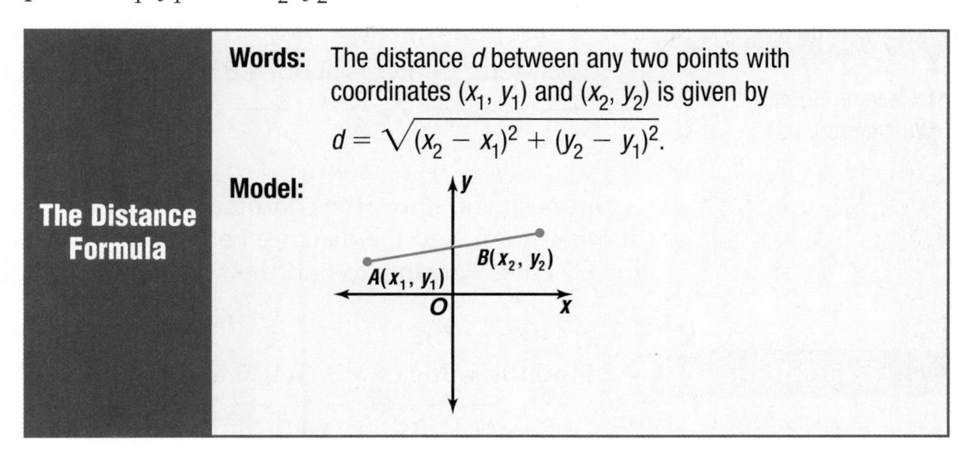

Example 1

Find the distance between points $A(1, 2)$ and $B(-3, 7)$.

$$d = \sqrt{(x_2 - x_1)^2 + (y_2 - y_1)^2} \quad \textit{Distance Formula}$$
$$= \sqrt{(-3 - 1)^2 + (7 - 2)^2} \quad (x_1, y_1) = (1, 2), \text{ and } (x_2, y_2) = (-3, 7)$$
$$= \sqrt{(-4)^2 + 5^2}$$
$$= \sqrt{16 + 25}$$
$$= \sqrt{41} \text{ or about 6.4 units}$$

Your Turn

a. Find the distance between points $C(3, 5)$ and $D(6, 4)$.

Example 2

Map Link

Refer to the application at the beginning of the lesson. The Metro Center is located 4 blocks east of the intersection of 15th Street and G Street. McPherson Square is located 2 blocks north of the intersection. How far is the Metro Center from McPherson Square?

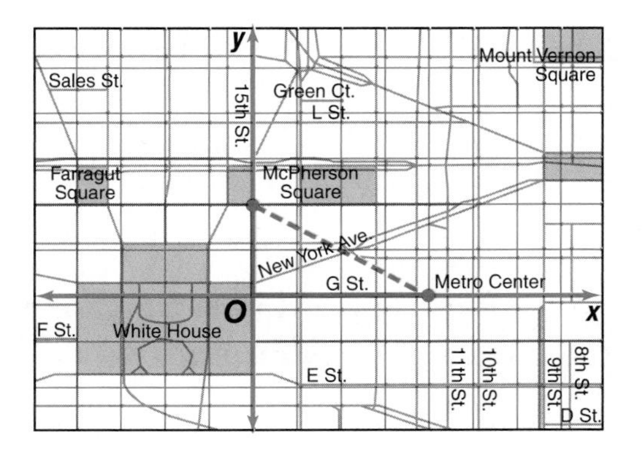

(continued on the next page)

Let the Metro Center be represented by (4, 0) and McPherson Square by (0, 2). So, $x_1 = 4$, $y_1 = 0$, $x_2 = 0$, and $y_2 = 2$.

$$\begin{aligned} d &= \sqrt{(x_2 - x_1)^2 + (y_2 - y_1)^2} && \textit{Distance Formula} \\ &= \sqrt{(0 - 4)^2 + (2 - 0)^2} && (x_1, y_1) = (4, 0), \textit{ and } (x_2, y_2) = (0, 2) \\ &= \sqrt{(-4)^2 + (2)^2} \\ &= \sqrt{16 + 4} \\ &= \sqrt{20} \text{ or about 4.5 blocks} \end{aligned}$$

The Metro Center is about 4.5 blocks from McPherson Square.

McPherson Square, Washington, D.C.

Suppose you know the coordinates of a point, one coordinate of another point, and the distance between the two points. You can use the Distance Formula to find the missing coordinate.

Example 3 **Find the value of a if $G(4, 7)$ and $H(a, 3)$ are 5 units apart.**

$$\begin{aligned} d &= \sqrt{(x_2 - x_1)^2 + (y_2 - y_1)^2} && \textit{Distance Formula} \\ 5 &= \sqrt{(a - 4)^2 + (3 - 7)^2} && (x_1, y_1) = (4, 7), \textit{ and } (x_2, y_2) = (a, 3) \\ 5 &= \sqrt{(a - 4)^2 + (-4)^2} \\ 5 &= \sqrt{a^2 - 8a + 16 + 16} && (a - 4)^2 = a^2 - 8a + 16 \textit{ and } (-4)^2 = 16 \\ 5 &= \sqrt{a^2 - 8a + 32} \\ 5^2 &= (\sqrt{a^2 - 8a + 32})^2 && \textit{Square each side.} \\ 25 &= a^2 - 8a + 32 \\ 0 &= a^2 - 8a + 7 && \textit{Subtract 25 from each side.} \\ 0 &= (a - 7)(a - 1) && \textit{Factor.} \end{aligned}$$

$a - 7 = 0$ or $a - 1 = 0$ *Zero Product Property.*

$a = 7$ $\quad$ $a = 1$

The value of a is 7 or 1.

Look Back

Zero Product Property: Lesson 11–4

Your Turn

b. Find the value of a if $J(a, 5)$ and $K(-7, 3)$ are $\sqrt{29}$ units apart.

Check for Understanding

Communicating Mathematics

1. **Name** the theorem that is used to derive the Distance Formula.

Vocabulary

Distance Formula

2. **Tell** how to find the distance between the points $A(10, 3)$ and $B(2, 3)$ without using the Distance Formula.

3. **You Decide** Anna says that to find the distance between $A(5, -8)$ and $B(7, -6)$, you should take the square root of $(7 - 5)^2 + [-6 - (-8)]^2$. Nick disagrees. He says to take the square root of $(5 - 7)^2 + [-8 - (-6)]^2$. Who is correct? Explain your reasoning.

Guided Practice

Find the distance between each pair of points. Round to the nearest tenth, if necessary. *(Example 1)*

4. $C(5, -1), D(2, 2)$
5. $M(6, 8), N(3, 4)$
6. $R(-3, 8), S(5, 4)$

Find the value of *a* if the points are the indicated distance apart. *(Example 3)*

7. $A(3, -1), B(a, 7); d = 10$
8. $J(5, a), K(6, 1); d = \sqrt{10}$

9. **Engineering** The MAXTEC Company is installing a fiber optic cable system between two of their office buildings. One of the buildings, MAXTEC West, is located 5 miles west and 2 miles north of the corporate office. The other building, MAXTEC East, is located 4 miles east and 5 miles north of the corporate office. How much cable will be needed to connect MAXTEC West and MAXTEC East? Round to the nearest tenth. *(Example 2)*

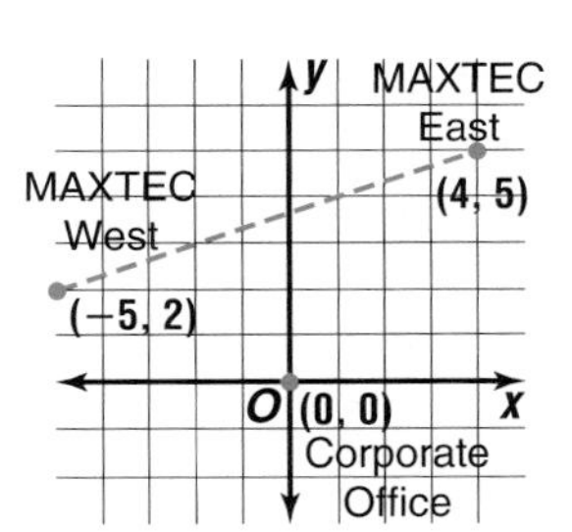

Exercises

Practice

Find the distance between each pair of points. Round to the nearest tenth, if necessary.

10. $X(5, 0), Y(12, 0)$
11. $A(-6, -4), B(-6, 8)$
12. $M(-2, -5), N(3, 7)$
13. $P(-4, 0), Q(3, -3)$
14. $V(-3, -4), W(-1, -2)$
15. $C(7, 2), D(-4, 10)$
16. $E(3, -6), F(9, -2)$
17. $G(-4, -6), H(-7, -3)$

Homework Help

For Exercises	See Examples
10–17, 24, 25	1, 2
18–23, 26, 27	3

Extra Practice

See page 720.

Find the value of *a* if the points are the indicated distance apart.

18. $A(a, -5), B(-3, -2); d = 5$
19. $D(-3, a), E(5, 2); d = 17$
20. $Q(7, 2), R(-1, a); d = 10$
21. $G(7, -3), H(5, a); d = \sqrt{85}$
22. $T(6, -3), U(-3, a); d = \sqrt{130}$
23. $U(1, -6), V(10, a); d = \sqrt{145}$

24. Find the distance between $J(-9, 5)$ and $K(-4, -2)$.

25. What is the distance between $C(-8, 1)$ and $D(5, 6)$?

26. What is the value of c if $W(1, c)$ and $V(-4, 9)$ are 13 units apart?

27. Suppose $M(b, 9)$ and $N(20, -5)$ are $\sqrt{340}$ units apart. What is the value of b?

Applications and Problem Solving

28. **Geometry** Triangle MNP has vertices $M(-6, 14)$, $N(2, -1)$, and $P(-2, 2)$. Find the perimeter of $\triangle MNP$. Round to the nearest tenth.

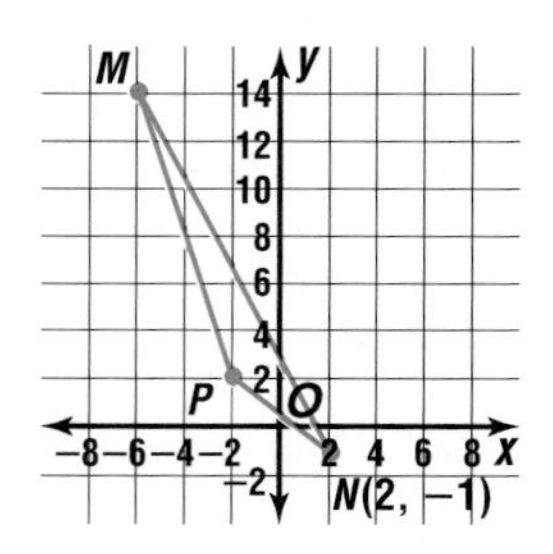

29. **Geometry** An isosceles triangle is a triangle with at least two congruent sides. Determine whether $\triangle CDE$ with vertices $C(1, 6)$, $D(2, 1)$, and $E(4, 1)$ is isosceles. Explain your reasoning.

30. **School** At Dannis State University, Webber Hall is located 35 meters west and 55 meters south of the student union. Packard Hall is located 42 meters east and 30 meters south of the student union.

    a. Draw a diagram on a coordinate grid to represent this situation.

    b. How far is Webber Hall from Packard Hall? Round to the nearest tenth.

31. **Engineering** Refer to Exercise 9. Suppose the MAXTEC Company builds another office building, MAXTEC North, 4 miles east and 2 miles north of the corporate office. How much cable will be needed to connect each of the buildings to the corporate office? Round to the nearest tenth.

y
MAXTEC East
(4, 5)
MAXTEC West
(−5, 2)
O (0, 0)
x
Corporate Office

32. **Critical Thinking** If the diagonals of a trapezoid have the same length, then the trapezoid is isosceles. The vertices of trapezoid $ABCD$ are $A(-2, 2)$, $B(10, 6)$, $C(9, 8)$, and $D(0, 5)$. Is the trapezoid isosceles? Explain.

Mixed Review

Find an approximation, to the nearest tenth, for each square root. Then graph the square root on a number line. *(Lesson 14–1)*

33. $\sqrt{12}$ **34.** $-\sqrt{27}$ **35.** $\sqrt{135}$

36. Solve $y \geq -4$ and $y < -2x + 1$ by graphing. *(Lesson 13–7)*

Solve each inequality. Check your solution. *(Lesson 12–3)*

37. $\frac{k}{3} \geq 2$ **38.** $\frac{-n}{4} > 3$ **39.** $\frac{3}{4}b \leq -6$

40. Factor $12m^2 + 7m - 10$. *(Lesson 10–4)*

Standardized Test Practice

41. Short Response Find the solution of $|3 + w| + 4 = 2$. *(Lesson 3–7)*

42. Multiple Choice The table shows the number of pets that have lived at the White House. Which two pets account for exactly half of the pets that have lived at the White House? *(Lesson 1–7)*

A dog and cat **B** horse and bird
C bird and dog **D** dog and horse

White House Pets	
dog	23
bird	16
horse	11
cat	10
cow	4
goat	4

Source: USA TODAY

Quiz 1 Lessons 14–1 and 14–2

Name the set or sets of numbers to which each real number belongs. Let N = natural numbers, W = whole numbers, Z = integers, Q = rational numbers, and I = irrational numbers. *(Lesson 14–1)*

1. 0 **2.** $\sqrt{36}$ **3.** 0.121231234 . . .

4. Find an approximation, to the nearest tenth, for $-\sqrt{35}$. *(Lesson 14–1)*

5. Write $\sqrt{2}$, 1.22, and $1.\overline{2}$ in order from least to greatest. *(Lesson 14–1)*

Find the distance between each pair of points. Round to the nearest tenth, if necessary. *(Lesson 14–2)*

6. $A(2, 0)$, $B(-1, 3)$ **7.** $C(4, 5)$, $D(-3, -2)$ **8.** $E(0, -4)$, $F(8, 7)$

9. What is the value of m if $A(m, 8)$ and $B(3, 4)$ are 5 units apart? *(Lesson 14–2)*

10. Geometry A scalene triangle is a triangle with no congruent sides. Determine whether $\triangle XYZ$ with vertices $X(5, 4)$, $Y(1, 5)$, and $Z(-1, 1)$ is scalene. Explain. *(Lesson 14–2)*

Chapter 14 Investigation

Materials

 grid paper

 ruler

 scissors

Midpoints

In Lesson 14–2, you learned how to find the distance between two points in a coordinate plane. In this Investigation, you will learn how to find the **midpoint** of a segment. The midpoint of a line segment is the point on the segment that separates it into two segments of equal length.

Investigate

1. Graph $A(-2, -1)$, $B(6, -1)$, $C(10, 3)$, and $D(2, 3)$ on a coordinate plane.
   a. Connect the points. What type of figure is formed?
   b. Make a table and record the coordinates as shown.

Coordinates of First Point	Coordinates of Second Point	Coordinates of Midpoint
$A(-2, -1)$	$B(6, -1)$	$E(,)$
$B(6, -1)$	$C(10, 3)$	$F(,)$
$C(10, 3)$	$D(2, 3)$	$G(,)$
$D(2, 3)$	$A(-2, -1)$	$H(,)$

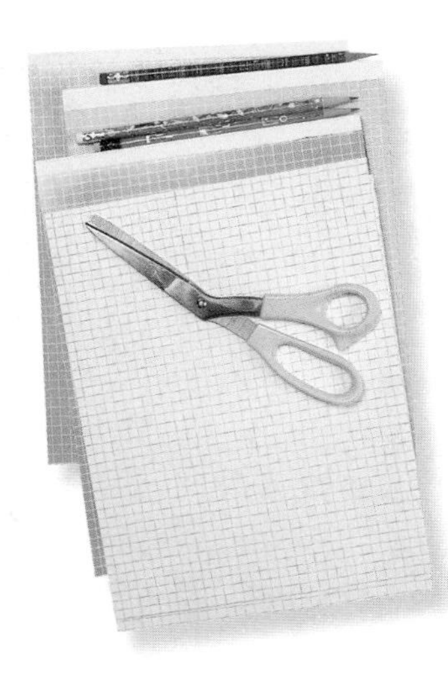

   c. Fold the paper so that points A and B coincide. Crease the paper to mark the midpoint of $\overline{AB}$. Label this point E. Find the coordinates of point E. Why is point E the midpoint of $\overline{AB}$?
   d. Fold the paper so that points B and C coincide. Crease the paper to mark the midpoint of $\overline{BC}$. Label this point F. Find the coordinates of point F. Continue this process with points C and D and points D and A. Label the midpoints G and H, respectively. Record the coordinates of these midpoints in your table.
   e. Draw $\overline{AC}$ and $\overline{BD}$. These are the diagonals of the figure. Find the midpoints of $\overline{AC}$ and $\overline{DB}$. Label the midpoints M and N, respectively. What are the coordinates of M and N? What appears to be true about the midpoints of the diagonals?
   f. Draw $\overline{EG}$ and $\overline{FH}$. What are the coordinates of their midpoints?
   g. Notice the triangles formed on the inside of the figure. Explain any relationships that exist between the triangles. You may cut out the triangles to help see the relationships.

2. Graph $X(-3, -3)$, $Y(3, -3)$, and $Z(3, 5)$ on a coordinate plane.

   a. Connect the points. What type of figure is formed?

   b. Make a table and record the coordinates as shown.

Coordinates of First Point	Coordinates of Second Point	Coordinates of Midpoint
$X(-3, -3)$	$Y(3, -3)$	$U(\ , \ )$
$Y(3, -3)$	$Z(3, 5)$	$V(\ , \ )$
$Z(3, 5)$	$X(-3, -3)$	$W(\ , \ )$

   c. Use paper folding to find the midpoints of the segments. Label the midpoints U, V, and W as shown in the table. What are the coordinates of the midpoints? Record the results.

   d. Draw $\overline{YW}$. Compare the measures of $\overline{YW}$, $\overline{XW}$, and $\overline{ZW}$. What appears to be true? Use the Distance Formula to verify your results.

   e. Draw $\overline{WU}$ and $\overline{WV}$. Four triangles are formed. Explain any relationships that exist between the triangles. You may cut out the triangles to help see the relationships.

3. Refer to the tables. Study each pair of first point coordinates, second point coordinates, and the corresponding midpoint coordinates.

   a. Explain how the concept of *mean*, or average, can be used to find the midpoint of a segment.

   b. Write a formula to find the midpoint of two points $X(a, b)$ and $Y(c, d)$.

Extending the Investigation

In this extension, you will continue to investigate midpoints.

- Graph points $A(3, 2)$, $B(-1, -6)$, and $C(-5, -4)$ on a coordinate plane. Connect the points to form a figure. Using the Midpoint Formula you discovered in Exercise 3b, determine the coordinates of the midpoints of each side of the figure. Label the points D, E, and F. Use the Distance Formula to verify that the midpoints separate each segment into two equal lengths.
- Use the Midpoint Formula to determine the midpoint of each pair of points.
  a. $M(4, 3)$, $N(2, 5)$ **b.** $R(1, 0)$, $S(-3, 2)$ **c.** $C(3, -6)$, $D(1, -4)$

Presenting Your Investigation

Here are some ideas to help you present your conclusions to the class.

- Make a poster showing the results of your research. Include the tables, graphs, and formula for finding the midpoint of a segment. List any relationships you discovered.
- Discuss any similarities in the triangles formed inside of the figures.

Investigation For more information on midpoints, visit: www.algconcepts.com

14–3 Simplifying Radical Expressions

What You'll Learn

You'll learn to simplify radical expressions.

Why It's Important

Science Scientists can use radical expressions to find the speed of a river. *See Exercise 41.*

Why does the character from the comic yell *"the square root of sixteen"* when hitting the golf ball?

FoxTrot

The expression $\sqrt{16}$ is a **radical expression**. Since the radicand, 16, is a perfect square, $\sqrt{16} = 4$. In Lesson 8–5, you learned to simplify radical expressions using the Product Property of Square Roots and prime factorization. You can simplify radical expressions in which the radicand is not a perfect square in a similar manner. *Recall that the radicand is the number or expression under the square root symbol.*

Examples

1 **Simplify $\sqrt{75}$.**

$\sqrt{75} = \sqrt{3 \cdot 5 \cdot 5}$ *Prime factorization*

$= \sqrt{3 \cdot 25}$ *$5 \times 5 = 25$*

$= \sqrt{3} \cdot \sqrt{25}$ *Product Property of Square Roots*

$= \sqrt{3} \cdot 5$ or $5\sqrt{3}$ *Simplify $\sqrt{25}$.*

Your Turn **Simplify each square root. Leave in radical form.**

a. $\sqrt{68}$

b. $\sqrt{375}$

Sports Link

2 **On a softball field, the distance from second base to home plate is $\sqrt{800}$ meters. Express $\sqrt{800}$ in simplest radical form.**

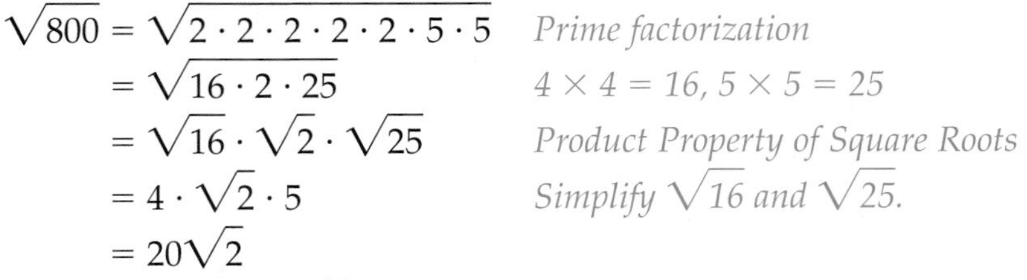

$\sqrt{800} = \sqrt{2 \cdot 2 \cdot 2 \cdot 2 \cdot 2 \cdot 5 \cdot 5}$ *Prime factorization*

$= \sqrt{16 \cdot 2 \cdot 25}$ *$4 \times 4 = 16$, $5 \times 5 = 25$*

$= \sqrt{16} \cdot \sqrt{2} \cdot \sqrt{25}$ *Product Property of Square Roots*

$= 4 \cdot \sqrt{2} \cdot 5$ *Simplify $\sqrt{16}$ and $\sqrt{25}$.*

$= 20\sqrt{2}$

The distance is $20\sqrt{2}$ meters.

interNET CONNECTION

Data Update For the latest information on softball, visit: www.algconcepts.com

The Product Property can also be used to multiply square roots.

Example 3 **Simplify $\sqrt{5} \cdot \sqrt{35}$.**

$$\begin{aligned} \sqrt{5} \cdot \sqrt{35} &= \sqrt{5} \cdot \sqrt{5 \cdot 7} && \textit{Prime factorization} \\ &= \sqrt{5} \cdot \sqrt{5} \cdot \sqrt{7} && \textit{Product Property of Square Roots} \\ &= \sqrt{5^2} \cdot \sqrt{7} && \sqrt{5} \times \sqrt{5} = \sqrt{5^2} \\ &= 5 \cdot \sqrt{7} \text{ or } 5\sqrt{7} && \textit{Simplify.} \end{aligned}$$

Your Turn **Simplify each expression. Leave in radical form.**

c. $\sqrt{3} \cdot \sqrt{15}$

d. $\sqrt{10} \cdot \sqrt{30}$

Look Back

Quotient Property of Square Roots: Lesson 8–3

To divide square roots and simplify radical expressions that involve division, use the Quotient Property of Square Roots. A fraction containing radicals is in simplest form if no radicals are left in the denominator.

Example 4 **Simplify $\frac{\sqrt{32}}{\sqrt{4}}$.**

$$\begin{aligned} \frac{\sqrt{32}}{\sqrt{4}} &= \sqrt{\frac{32}{4}} && \textit{Quotient Property of Square Roots} \\ &= \sqrt{8} && 32 \div 4 = 8 \\ &= \sqrt{2 \cdot 2 \cdot 2} && \textit{Prime factorization} \\ &= \sqrt{2^3} && 2 \cdot 2 \cdot 2 = 2^3 \\ &= \sqrt{2^2} \cdot \sqrt{2} && \\ &= 2\sqrt{2} && \textit{Simplify.} \end{aligned}$$

Your Turn **Simplify each expression. Leave in radical form.**

e. $\frac{\sqrt{72}}{\sqrt{6}}$

f. $\frac{\sqrt{160}}{\sqrt{2}}$

To eliminate radicals from the denominator of a fraction, you can use a method for simplifying radical expressions called **rationalizing the denominator**.

Example 5 **Simplify $\frac{\sqrt{5}}{\sqrt{10}}$.**

$$\begin{aligned} \frac{\sqrt{5}}{\sqrt{10}} &= \frac{\sqrt{5}}{\sqrt{10}} \cdot \frac{\sqrt{10}}{\sqrt{10}} && \frac{\sqrt{10}}{\sqrt{10}} = 1 \\ &= \frac{\sqrt{5 \cdot 10}}{\sqrt{10 \cdot 10}} && \end{aligned}$$

(continued on the next page)

$$= \frac{\sqrt{50}}{\sqrt{100}} \quad \textit{Simplify.}$$

$$= \frac{\sqrt{25 \cdot 2}}{10}$$

$$= \frac{\sqrt{25} \cdot \sqrt{2}}{10} \quad \textit{Product Property of Square Roots}$$

$$= \frac{5 \cdot \sqrt{2}}{10} \text{ or } \frac{\sqrt{2}}{2} \quad \textit{Simplify.}$$

Your Turn **Simplify each expression. Leave in radical form.**

g. $\frac{\sqrt{6}}{\sqrt{8}}$

h. $\frac{\sqrt{32}}{\sqrt{3}}$

Binomials of the form $a\sqrt{b} + c\sqrt{d}$ and $a\sqrt{b} - c\sqrt{d}$ are **conjugates** of each other because their product is a rational number.

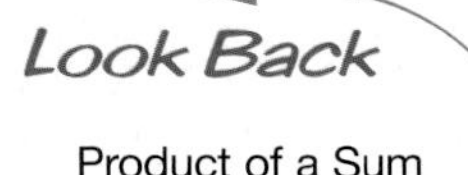

Product of a Sum and a Difference: Lesson 9–5

$$(6 + \sqrt{3})(6 - \sqrt{3}) = 6^2 - (\sqrt{3})^2 \quad \textit{Use the pattern } (a + b)(a - b) = a^2 - b^2 \textit{ to simplify the product.}$$

$$= 36 - 3$$

$$= 33$$

Conjugates are useful for simplifying radical expressions because their product is always a rational number.

Example 6 **Simplify** $\frac{6}{3 - \sqrt{2}}$.

To rationalize the denominator, multiply both the numerator and denominator by $3 + \sqrt{2}$, which is the conjugate of $3 - \sqrt{2}$.

$$\frac{6}{3 - \sqrt{2}} = \frac{6}{3 - \sqrt{2}} \cdot \frac{3 + \sqrt{2}}{3 + \sqrt{2}} \quad \textit{Notice that } \frac{3 + \sqrt{2}}{3 + \sqrt{2}} = 1.$$

$$= \frac{6(3) + 6\sqrt{2}}{3^2 - (\sqrt{2})^2} \quad \textit{Distributive Property } (a - b)(a + b) = a^2 - b^2$$

$$= \frac{18 + 6\sqrt{2}}{9 - 2} \quad \textit{Simplify.}$$

$$= \frac{18 + 6\sqrt{2}}{7}$$

Your Turn **Simplify each expression. Leave in radical form.**

i. $\frac{3}{3 - \sqrt{5}}$

j. $\frac{4}{5 + \sqrt{6}}$

Radical expressions are in simplest form if the following conditions are met.

Simplified Form for Radicals	A radical expression is in simplest form when the following three conditions have been met. **1.** No radicands have perfect square factors other than 1. **2.** No radicands contain fractions. **3.** No radicals appear in the denominator of a fraction.

Consider the expression $\sqrt{x^2}$. It appears that $\sqrt{x^2} = x$. However, if $x = -3$, then $\sqrt{(-3)^2}$ is 3, not -3. For radical expressions like $\sqrt{x^2}$, use absolute value to ensure nonnegative results. The results of simplifying a few radical expressions are listed below.

$\sqrt{x^2} = |x|$ $\quad \sqrt{x^3} = x\sqrt{x}$ $\quad \sqrt{x^4} = x^2$ $\quad \sqrt{x^5} = x^2\sqrt{x}$ $\quad \sqrt{x^6} = |x^3|$

For $\sqrt{x^3}$, absolute value is not necessary. If x were negative, then x^3 would be negative, and $\sqrt{x^3}$ is not a real number. *Why is absolute value not used for $\sqrt{x^4}$?*

Example 7 **Simplify $\sqrt{98ab^2c^4}$. Use absolute value symbols if necessary.**

$\sqrt{98ab^2c^4}$

$= \sqrt{2 \cdot 7 \cdot 7 \cdot a \cdot b^2 \cdot c^4}$ — *Prime factorization*

$= \sqrt{2 \cdot 49 \cdot a \cdot b^2 \cdot c^4}$ — *$7 \times 7 = 49$*

$= \sqrt{2} \cdot \sqrt{49} \cdot \sqrt{a} \cdot \sqrt{b^2} \cdot \sqrt{c^4}$ — *Product Property of Square Roots*

$= \sqrt{2} \cdot 7 \cdot \sqrt{a} \cdot |b| \cdot c^2$ — *Simplify.*

$= 7|b|c^2\sqrt{2a}$ — *The absolute value of b ensures a nonnegative result.*

Your Turn

Simplify each expression. Use absolute value symbols if necessary.

k. $\sqrt{63ab^2}$ **l.** $\sqrt{200x^2y^3}$

Check for Understanding

Communicating Mathematics

1. **Explain** why absolute values are sometimes needed when simplifying radical expressions containing variables.
2. **Explain** how you can show whether $8 + \sqrt{2}$ and $8 - \sqrt{2}$ are conjugates.

Vocabulary
radical expression
rationalizing the denominator
conjugates

3. **You Decide?** LaToya says that, in simplest form, the expression $\frac{2}{3+\sqrt{5}}$ is written as $\frac{6-2\sqrt{5}}{4}$. Greg disagrees. He says it should be written as $\frac{3-\sqrt{5}}{2}$. Who is correct? Explain your reasoning.

Guided Practice

Getting Ready **State the conjugate of each expression. Then multiply the expression by its conjugate.**

Sample: $1+\sqrt{7}$

Solution: The conjugate is $1-\sqrt{7}$.

$$(1+\sqrt{7})(1-\sqrt{7}) = 1^2 - (\sqrt{7})^2$$
$$= 1 - 7 \text{ or } -6$$

4. $5-\sqrt{6}$
5. $2+\sqrt{8}$
6. $2\sqrt{7}-3\sqrt{2}$

Simplify each expression. Leave in radical form. *(Examples 1–6)*

7. $\sqrt{75}$
8. $\sqrt{96}$
9. $\sqrt{6}\cdot\sqrt{15}$
10. $\frac{\sqrt{36}}{\sqrt{3}}$
11. $\frac{\sqrt{3}}{\sqrt{7}}$
12. $\frac{1}{6-\sqrt{3}}$

Simplify each expression. Use absolute value symbols if necessary. *(Example 7)*

13. $\sqrt{36x^2y}$
14. $\sqrt{50m^4n^5}$

15. **Geometry** The radius of the circle is $\sqrt{32}$ units. Express the radius in simplest form. *(Example 2)*

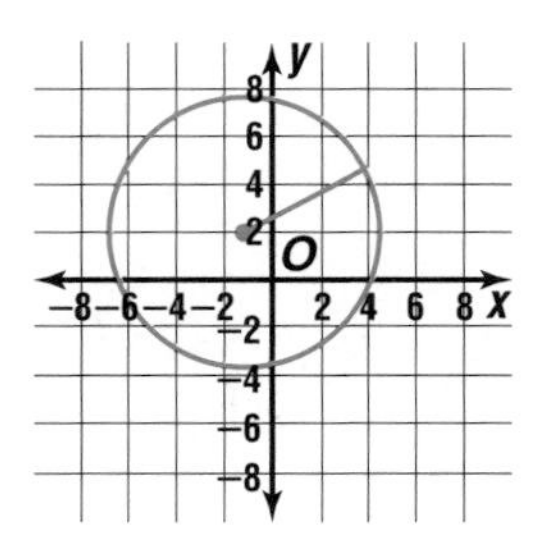

Exercises

Practice

Homework Help	
For Exercises	See Examples
16–20, 40, 41	1, 2
21–24	3
25–30	4, 5
31–33	6
34–39	7

Extra Practice
See page 721.

Simplify each expression. Leave in radical form.

16. $\sqrt{45}$
17. $\sqrt{98}$
18. $\sqrt{280}$
19. $\sqrt{500}$
20. $\sqrt{1000}$
21. $\sqrt{2}\cdot\sqrt{8}$
22. $\sqrt{6}\cdot\sqrt{18}$
23. $3\sqrt{5}\cdot\sqrt{5}$
24. $\sqrt{8}\cdot\sqrt{12}$
25. $\frac{\sqrt{48}}{\sqrt{3}}$
26. $\frac{\sqrt{52}}{\sqrt{4}}$
27. $\frac{\sqrt{80}}{\sqrt{2}}$
28. $\frac{\sqrt{3}}{\sqrt{8}}$
29. $\frac{\sqrt{4}}{\sqrt{5}}$
30. $\frac{\sqrt{5}}{\sqrt{12}}$
31. $\frac{2}{4-\sqrt{3}}$
32. $\frac{5}{3+\sqrt{2}}$
33. $\frac{4}{6-\sqrt{7}}$

Simplify each expression. Use absolute value symbols if necessary.

34. $\sqrt{16gh^4}$
35. $\sqrt{40m^2}$
36. $\sqrt{47c^6d}$
37. $\sqrt{54a^2b^3}$
38. $\sqrt{125rst}$
39. $\sqrt{36x^3y^4z^5}$

Applications and Problem Solving

40. **Quilting** The quilt pattern *Sky Rocket* is shown at the right. Suppose the legs of the indicated triangle each measure 8 centimeters. Use the Pythagorean Theorem to find the measure of the hypotenuse. Express the answer as a radical in simplest form.

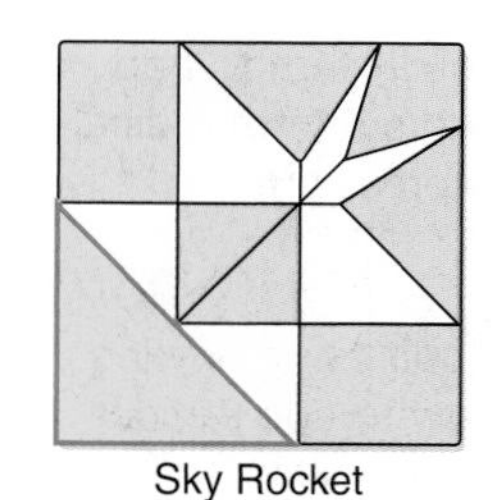
Sky Rocket

41. **Science** An L-shaped tube like the one shown can be used to measure the speed V in miles per hour of water in a river. By using the formula $V = \sqrt{2.5h}$, where h is the height in inches of the column of water above the surface, the speed of the water can be found.

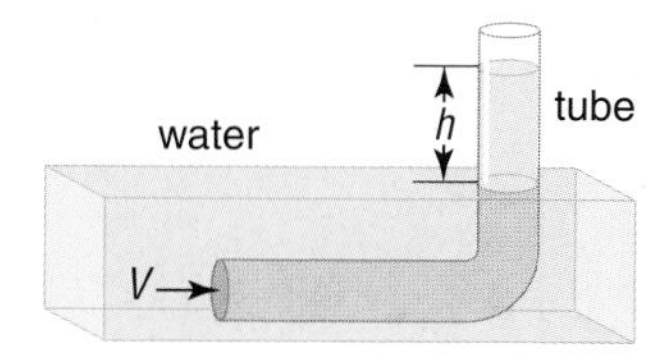

   a. Suppose the tube is placed in a river and the height of the column of water is 4.8 inches. What is the speed of the water?
   b. What would h be if the speed is exactly 5 miles per hour?

42. **Critical Thinking** Is the sentence $\sqrt{a \cdot b} = \sqrt{a} \cdot \sqrt{b}$ true for negative values of a and b? Explain your reasoning.

Mixed Review

43. Point J is located at $(-1, 5)$, and point K is located at $(m, 2)$. What is the value of m if the points are $\sqrt{18}$ units apart? *(Lesson 14–2)*

Name the set or sets of numbers to which each real number belongs. Let N = natural numbers, W = whole numbers, Z = integers, Q = rational numbers and I = irrational numbers. *(Lesson 14–1)*

44. 0.75
45. $-\sqrt{144}$
46. $\frac{20}{4}$

47. Use substitution to solve $y = 2x + 3$ and $y - 2x = -5$. *(Lesson 13–3)*

48. Solve $15 > a - 3 > 10$. Graph the solution. *(Lesson 12–5)*

Standardized Test Practice

49. **Short Response** Tell whether 9 is closer to $\sqrt{79}$ or $\sqrt{89}$. *(Lesson 8–6)*

50. **Multiple Choice** Two dice are rolled. Find the probability that a number less than 5 is rolled on one die and an odd number is rolled on the other die. *(Lesson 5–7)*

A $\frac{1}{2}$ B $\frac{2}{3}$ C $\frac{1}{3}$ D $\frac{3}{4}$

14–4 Adding and Subtracting Radical Expressions

What You'll Learn

You'll learn to add and subtract radical expressions.

Why It's Important

Hobbies Knowing how to add radical expressions can help you find the perimeter of a sail on a sailboat. *See Exercise 37.*

To find the exact perimeter of quadrilateral *ABCD*, you will need to add radical expressions.

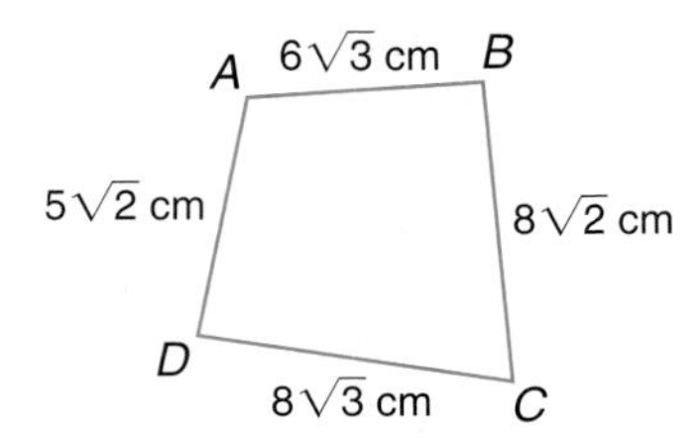

Radical expressions with the same radicands can be added or subtracted in the same way that monomials are added or subtracted.

Look Back

Adding and Subtracting Monomials: Lesson 9–2

Monomials	**Radical Expressions**
$5x + 3x = (5 + 3)x$	$5\sqrt{2} + 3\sqrt{2} = (5 + 3)\sqrt{2}$
$= 8x$	$= 8\sqrt{2}$
$8y - 2y = (8 - 2)y$	$8\sqrt{3} - 2\sqrt{3} = (8 - 2)\sqrt{3}$
$= 6y$	$= 6\sqrt{3}$

Notice that the Distributive Property was used to simplify each radical expression.

Examples

Simplify each expression.

1 $6\sqrt{7} - 2\sqrt{7}$

$6\sqrt{7} - 2\sqrt{7} = (6 - 2)\sqrt{7}$ *Distributive Property*

$= 4\sqrt{7}$

2 $5\sqrt{5} + 3\sqrt{5} - 18\sqrt{5}$

$5\sqrt{5} + 3\sqrt{5} - 18\sqrt{5} = (5 + 3 - 18)\sqrt{5}$ *Distributive Property*

$= -10\sqrt{5}$

Your Turn

a. $8\sqrt{6} + 3\sqrt{6}$

b. $5\sqrt{2} - 12\sqrt{2}$

c. $4\sqrt{3} + 7\sqrt{3} - 2\sqrt{3}$

d. $3\sqrt{13} - 2\sqrt{13} - 6\sqrt{13}$

Example 3 Geometry Link

Refer to the beginning of the lesson. Find the exact perimeter of quadrilateral *ABCD*.

$P = 6\sqrt{3} + 8\sqrt{2} + 8\sqrt{3} + 5\sqrt{2}$ *Like terms: $6\sqrt{3}$ and $8\sqrt{3}$; $8\sqrt{2}$ and $5\sqrt{2}$*

$= 6\sqrt{3} + 8\sqrt{3} + 8\sqrt{2} + 5\sqrt{2}$ *Commutative Property*

$= (6 + 8)\sqrt{3} + (8 + 5)\sqrt{2}$ *Distributive Property*

$= 14\sqrt{3} + 13\sqrt{2}$

The exact perimeter of quadrilateral *ABCD* is $14\sqrt{3} + 13\sqrt{2}$ centimeters.

In Example 3, the expression $14\sqrt{3} + 13\sqrt{2}$ cannot be simplified further for the following reasons.

- The radicands are different.
- There are no common factors.
- Each radicand is in simplest form.

If the radicals in a radical expression are not in simplest form, simplify them first. Then use the Distributive Property wherever possible to further simplify the expression.

Example 4

Simplify $2\sqrt{20} - \sqrt{45}$.

$2\sqrt{20} - \sqrt{45} = 2\sqrt{2^2 \cdot 5} - \sqrt{3^2 \cdot 5}$ *Prime factorization*

$= 2(2\sqrt{5}) - (3\sqrt{5})$ *Simplify.*

$= 4\sqrt{5} - 3\sqrt{5}$

$= (4 - 3)\sqrt{5}$ *Distributive Property*

$= 1\sqrt{5}$ or $\sqrt{5}$

Your Turn **Simplify each expression.**

e. $4\sqrt{27} - 5\sqrt{12}$

f. $7\sqrt{18} + 3\sqrt{50}$

Check for Understanding

Communicating Mathematics

Math Journal

1. **Describe** in your own words how to add radical expressions.
2. **Explain** why you should simplify each radical in a radical expression before adding or subtracting.
3. **Explain** how you use the Distributive Property to simplify the sum or difference of like radicals.

Guided Practice

Getting Ready **Express each radical in simplest form. Then determine the like radicals.**

Sample: $5\sqrt{6}, 3\sqrt{5}, 3\sqrt{20}$

Solution: $5\sqrt{6} = 5\sqrt{6}$
$3\sqrt{5} = 3\sqrt{5}$ ✓
$3\sqrt{20} = 6\sqrt{5}$ ✓

4. $4\sqrt{2}, -3\sqrt{18}, 2\sqrt{6}$
5. $-3\sqrt{10}, 4\sqrt{5}, -5\sqrt{3}$
6. $-2\sqrt{7}, 6\sqrt{14}, 5\sqrt{28}$
7. $4\sqrt{12}, -7\sqrt{6}, -6\sqrt{27}$

Simplify each expression. *(Examples 1, 2, & 4)*

8. $7\sqrt{6} + 4\sqrt{6}$
9. $4\sqrt{5} - 2\sqrt{5}$
10. $8\sqrt{3} - 3\sqrt{3} + 7\sqrt{3}$
11. $3\sqrt{7} - 6\sqrt{7} - 2\sqrt{7}$
12. $5\sqrt{27} + 2\sqrt{48}$
13. $2\sqrt{50} - 4\sqrt{32}$
14. **Geometry** Find the exact perimeter of triangle ABC. *(Example 3)*

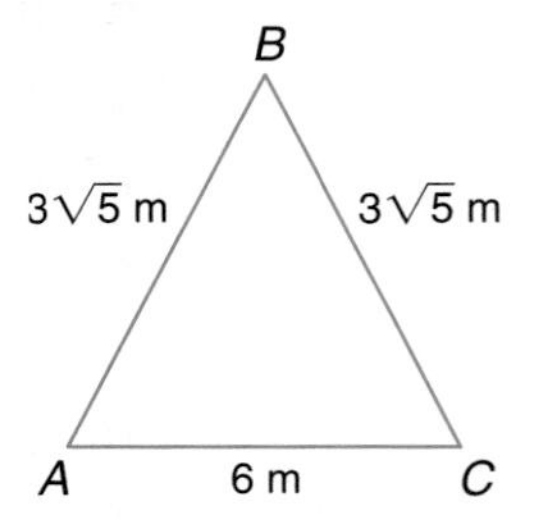

Exercises

Practice

Simplify each expression.

15. $4\sqrt{3} + 2\sqrt{3}$
16. $12\sqrt{2} + 3\sqrt{2}$
17. $15\sqrt{5} - \sqrt{5}$
18. $5\sqrt{2} - 8\sqrt{3}$
19. $-6\sqrt{7} + 10\sqrt{7}$
20. $-7\sqrt{3} - 2\sqrt{3}$
21. $5\sqrt{5} + 3\sqrt{5} - 10\sqrt{5}$
22. $6\sqrt{7} - 5\sqrt{7} + 8\sqrt{7}$
23. $7\sqrt{2} - 16\sqrt{2} + 8\sqrt{5}$
24. $5\sqrt{11} + 2\sqrt{11} - 4\sqrt{11}$
25. $-\sqrt{6} - 4\sqrt{6} + 12\sqrt{6}$
26. $-9\sqrt{14} + 2\sqrt{14} - 6\sqrt{14}$
27. $4\sqrt{3} + 5\sqrt{12}$
28. $-2\sqrt{28} + 3\sqrt{7}$
29. $2\sqrt{18} - 5\sqrt{8}$
30. $3\sqrt{63} - \sqrt{112}$
31. $\sqrt{50} + \sqrt{108} + \sqrt{18}$
32. $-4\sqrt{27} - 5\sqrt{32} + 2\sqrt{75}$

33. Simplify $2\sqrt{20} - \sqrt{180} - 3\sqrt{24}$.
34. Write $-2\sqrt{50} + 5\sqrt{48} + 7\sqrt{98}$ in simplest form.

Homework Help

For Exercises	See Examples
15–17, 19–22, 24–26, 37	1, 2
35	3
18, 23, 27–34	4

Extra Practice

See page 721.

Applications and Problem Solving

35. **Geometry** The perimeter of quadrilateral $DEFG$ is $14\sqrt{7}$ feet. What is the missing measure?

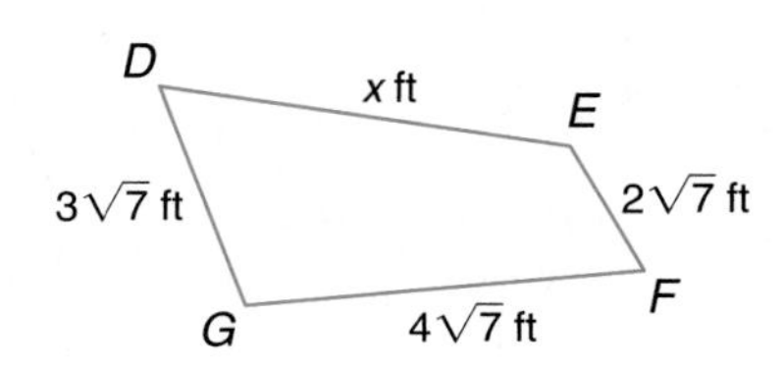

Close to Home

36. **Media** Refer to the comic. Suppose the first team member says "$\sqrt{7}$," the second says "$2\sqrt{7}$," and so on.
   a. Write an expression that could be used to determine the response of the fourth student in line.
   b. Determine the response of the fourth person in line.
   c. Suppose the coach asks the students to count off by $\sqrt{5}$. Find the response of the last person in line.

37. **Hobbies** Ling is making a model of a sailboat. The dimensions of one of the sails are shown in the diagram.
   a. Find the missing measure. (*Hint:* Use the Pythagorean Theorem.)
   b. Determine the exact perimeter of the sail.

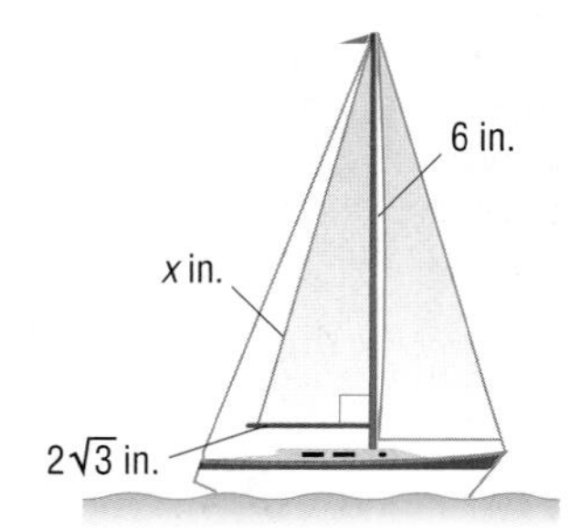

38. **Critical Thinking** Is the set of irrational numbers closed under addition? Explain your reasoning.

Mixed Review

39. Simplify $\sqrt{45mn^3p^2}$. *(Lesson 14–3)*

Find the distance between each pair of points. Round to the nearest tenth, if necessary. *(Lesson 14–2)*

40. $C(-3, 5), D(0, 8)$

41. $J(4, -2), K(-6, -1)$

42. **Geometry** The base of a triangle measures 3 feet more than its height. The area of the triangle is 20 square feet. Write and solve a quadratic equation to find the height and base of the triangle. *(Lesson 11–4)*

Factor each monomial. *(Lesson 10–1)*

43. $-26a^3b^2$

44. $36xy^4z$

Standardized Test Practice

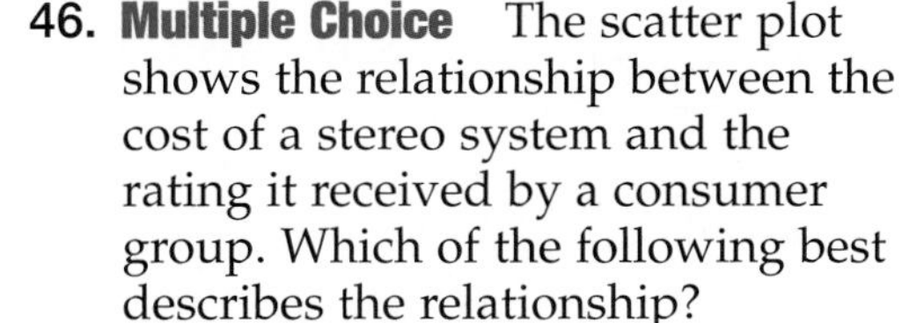

45. **Short Response** Find the product of $a + 2c$ and $a - 2c$. *(Lesson 9–5)*

46. **Multiple Choice** The scatter plot shows the relationship between the cost of a stereo system and the rating it received by a consumer group. Which of the following best describes the relationship? *(Lesson 7–4)*

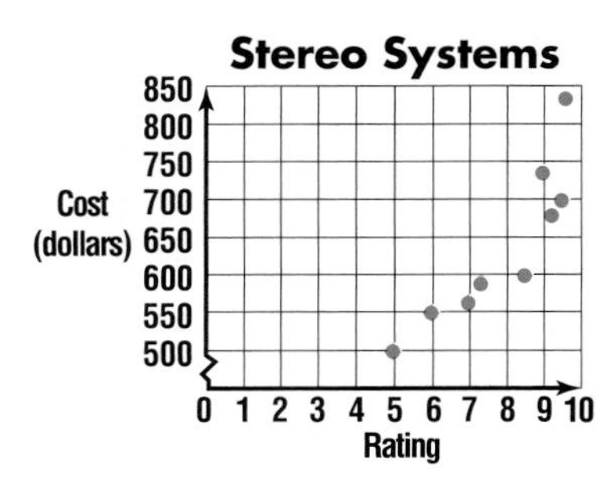

   A positive   B dependent
   C no pattern   D negative

14–5 Solving Radical Equations

What You'll Learn

You'll learn to solve simple radical equations in which only one radical contains a variable.

Why It's Important

Engineering

Engineers use radical equations to determine the velocity of roller coasters.
See Example 5.

The speed of a roller coaster as it travels through a loop depends on the height of the hill from which the coaster has just descended. The equation $s = 8\sqrt{h - 2r}$ gives the speed s in feet per second where h is the height of the hill and r is the radius of the loop.

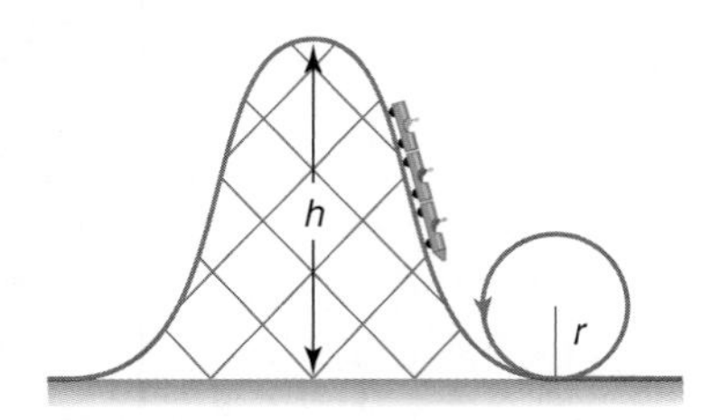

Suppose the owner of an amusement park wants to design a roller coaster that will travel at a speed of 40 feet per second as it goes through a loop with a radius of 30 feet. How high should the hill be? *This problem will be solved in Example 5.*

Equations like $s = 8\sqrt{h - 2r}$ that contain radicals with variables in the radicand are called **radical equations**. To solve these equations, first isolate the radical on one side of the equation. Then square each side of the equation to eliminate the radical.

Examples

Solve each equation. Check your solution.

1 $\sqrt{x} + 5 = 8$

$$\sqrt{x} + 5 = 8$$

$$\sqrt{x} + 5 - 5 = 8 - 5 \quad \textit{Subtract 5 from each side.}$$

$$\sqrt{x} = 3$$

$$(\sqrt{x})^2 = 3^2 \quad \textit{Square each side.}$$

$$x = 9$$

Check: $\sqrt{x} + 5 = 8$

$$\sqrt{9} + 5 \stackrel{?}{=} 8 \quad \textit{Replace x with 9.}$$

$$3 + 5 \stackrel{?}{=} 8$$

$$8 = 8 \quad \checkmark$$

The solution is 9.

2 $\sqrt{m + 3} + 2 = 6$

$$\sqrt{m + 3} + 2 = 6$$

$$\sqrt{m + 3} + 2 - 2 = 6 - 2 \quad \textit{Subtract 2 from each side.}$$

$$\sqrt{m + 3} = 4$$

$$(\sqrt{m + 3})^2 = 4^2 \quad \textit{Square each side.}$$

$$m + 3 = 16$$

$$m + 3 - 3 = 16 - 3 \quad \textit{Subtract 3 from each side.}$$

$$m = 13 \quad \textit{Check this result.}$$

The solution is 13.

Your Turn **Solve each equation. Check your solution.**

a. $\sqrt{y} - 6 = 4$

b. $\sqrt{a - 1} + 5 = 7$

You can use a graphing calculator to solve radical equations.

Graphing Calculator Tutorial

See pp. 724–727.

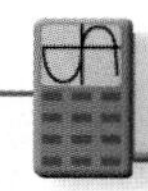

Graphing Calculator Exploration

Find the solution of $\sqrt{x} + 5 = 8$.

Step 1 Set the viewing window for x: [0, 15] by 1 and y: [0, 15] by 1.

Step 2 Press Y=. Enter the left side of the equation, $\sqrt{x} + 5$, as Y_1. Enter the right side of the equation, 8, as Y_2.

Step 3 Press GRAPH to graph the equations.

Step 4 The solution of the equation is the intersection point of the two lines. Use the INTERSECT feature to find the intersection point.

Enter: 2nd [CALC] 5

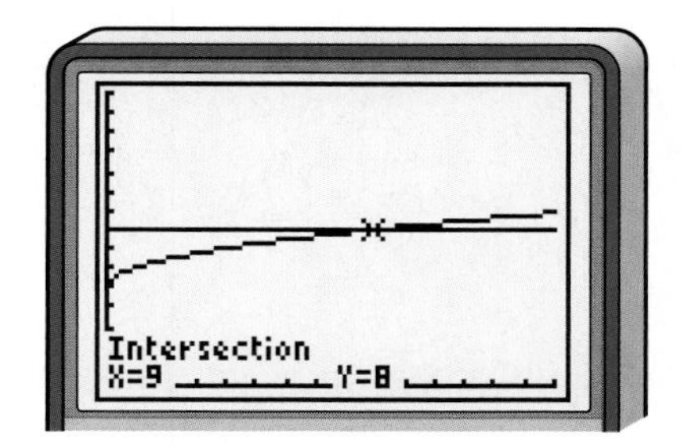

The graph intersects at the point (9, 8). The solution is the x-coordinate, 9.

Try These

1. Solve the equation in Example 2 by using a graphing calculator.
2. Use a graphing calculator to find the solution of Your Turn Exercise b.
3. What changes must be made to the graphing calculator before finding the solution of Your Turn Exercise a?

Squaring each side of an equation may produce results that do not satisfy the *original* equation. So, you must check all solutions when you solve radical equations.

Examples

Solve each equation. Check your solution.

3 $\sqrt{n + 2} = n - 4$

$$\sqrt{n + 2} = n - 4$$

$$(\sqrt{n + 2})^2 = (n - 4)^2 \quad \textit{Square each side.}$$

$$n + 2 = n^2 - 8n + 16$$

$$n - n + 2 - 2 = n^2 - 8n + 16 - n - 2 \quad \textit{Subtract n and 2 from each side.}$$

$$0 = n^2 - 9n + 14$$

$$0 = (n - 7)(n - 2) \quad \textit{Factor.}$$

(continued on the next page)

$$n - 7 = 0 \quad \text{or} \quad n - 2 = 0 \qquad \textit{Use the Zero Product Property.}$$
$$n = 7 \qquad\qquad n = 2$$

Check:

$$\sqrt{n + 2} = n - 4 \qquad\qquad \sqrt{n + 2} = n - 4$$
$$\sqrt{7 + 2} \stackrel{?}{=} 7 - 4 \qquad\qquad \sqrt{2 + 2} \stackrel{?}{=} 2 - 4$$
$$\sqrt{9} \stackrel{?}{=} 3 \qquad\qquad \sqrt{4} \stackrel{?}{=} -2$$
$$3 = 3 \ \checkmark \qquad\qquad 2 \neq -2$$

Since 2 does not satisfy the original equation, 7 is the only solution.

4 $\sqrt{3h - 5} + 5 = h$

$$\sqrt{3h - 5} + 5 = h$$
$$\sqrt{3h - 5} + 5 - 5 = h - 5 \qquad \textit{Subtract 5 from each side.}$$
$$\sqrt{3h - 5} = h - 5$$
$$(\sqrt{3h - 5})^2 = (h - 5)^2 \qquad \textit{Square each side.}$$
$$3h - 5 = h^2 - 10h + 25$$
$$3h - 3h - 5 + 5 = h^2 - 10h + 25 - 3h + 5 \qquad \textit{Add } -3h \textit{ and 5 to each side.}$$
$$0 = h^2 - 13h + 30$$
$$0 = (h - 10)(h - 3) \qquad \textit{Factor.}$$
$$h - 10 = 0 \quad \text{or} \quad h - 3 = 0 \qquad \textit{Use the Zero Product Property.}$$
$$h = 10 \qquad\qquad h = 3$$

Check:

$$\sqrt{3h - 5} + 5 = h \qquad\qquad \sqrt{3h - 5} + 5 = h$$
$$\sqrt{3(10) - 5} + 5 \stackrel{?}{=} 10 \qquad\qquad \sqrt{3(3) - 5} + 5 \stackrel{?}{=} 3$$
$$\sqrt{30 - 5} + 5 \stackrel{?}{=} 10 \qquad\qquad \sqrt{9 - 5} + 5 \stackrel{?}{=} 3$$
$$\sqrt{25} + 5 \stackrel{?}{=} 10 \qquad\qquad \sqrt{4} + 5 \stackrel{?}{=} 3$$
$$5 + 5 \stackrel{?}{=} 10 \qquad\qquad 2 + 5 \stackrel{?}{=} 3$$
$$10 = 10 \ \checkmark \qquad\qquad 7 \neq 3$$

Since 3 does not satisfy the original equation, 10 is the only solution.

Your Turn

c. $\sqrt{d + 1} = d - 1$

d. $\sqrt{3x - 14} + x = 6$

Radical equations are used in many real-life situations.

5 **Refer to the application at the beginning of the lesson. How high should the hill be?**

Explore You know the speed of the coaster and the radius of the loop. You need to know the height of the hill.

Data Update For the latest information on roller coasters, visit: www.algconcepts.com

Plan Use the equation $s = 8\sqrt{h - 2r}$ to find the height of the hill.

Solve

$$s = 8\sqrt{h - 2r}$$
$$40 = 8\sqrt{h - 2(30)} \quad \textit{Replace s with 40 and r with 30.}$$
$$40 = 8\sqrt{h - 60}$$
$$\frac{40}{8} = \frac{8\sqrt{h - 60}}{8} \quad \textit{Divide each side by 8.}$$
$$5 = \sqrt{h - 60}$$
$$5^2 = (\sqrt{h - 60})^2 \quad \textit{Square each side.}$$
$$25 = h - 60$$
$$25 + 60 = h - 60 + 60 \quad \textit{Add 60 to each side.}$$
$$85 = h$$

The height of the hill is 85 feet.

Examine Check the result.

$$s = 8\sqrt{h - 2r}$$
$$40 \stackrel{?}{=} 8\sqrt{85 - 2(30)} \quad \textit{Replace s with 40, h with 85, and r with 30.}$$
$$40 \stackrel{?}{=} 8\sqrt{85 - 60}$$
$$40 \stackrel{?}{=} 8\sqrt{25}$$
$$40 \stackrel{?}{=} 8 \cdot 5$$
$$40 = 40 \quad \checkmark$$

The answer is correct.

Check for Understanding

Vocabulary
radical equation

Communicating Mathematics

1. **Tell** the first step you should take when solving a radical equation.
2. **Explain** why it is important to check radical equations after finding possible solutions.

3. **Write** an example of a radical equation. Then write the steps for solving the equation.

Guided Practice

Getting Ready **Square each side of the following equations.**

Sample: $\sqrt{y - 1} = 4$

Solution:
$$\sqrt{y - 1} = 4$$
$$(\sqrt{y - 1})^2 = 4^2$$
$$y - 1 = 16$$

4. $\sqrt{x} = 7$
5. $\sqrt{a + 5} = 2$
6. $9 = \sqrt{2c - 3}$

Solve each equation. Check your solution. *(Examples 1–4)*

7. $\sqrt{x} - 4 = 7$
8. $\sqrt{a - 2} - 5 = 3$
9. $\sqrt{m + 5} + 1 = m$
10. $\sqrt{1 + 2x} = x + 1$

11. **Engineering** Refer to the application at the beginning of the lesson. Suppose the owner of an amusement park wants to design a roller coaster that will travel through a loop at a speed of 56 feet per second after descending a 120-foot hill. What should the radius of the loop be? *(Example 5)*

Exercises

Practice

Solve each equation. Check your solution.

12. $\sqrt{x} = 3$
13. $\sqrt{a} = -4$
14. $7 = \sqrt{7m}$
15. $\sqrt{-3d} = 6$
16. $\sqrt{y} + 5 = 0$
17. $0 = \sqrt{2c} - 2$
18. $\sqrt{m - 4} = 6$
19. $\sqrt{w + 6} = 9$
20. $3 = \sqrt{4n + 1}$
21. $11 = \sqrt{2z - 5}$
22. $\sqrt{8h + 1} - 5 = 0$
23. $2 = \sqrt{3b - 5} + 6$
24. $\sqrt{x + 6} = x$
25. $p = \sqrt{5p - 6}$
26. $\sqrt{8 - b} = b - 2$
27. $\sqrt{k - 2} + 4 = k$
28. $t - 1 = \sqrt{2t + 6}$
29. $5 + \sqrt{3j - 5} = j$

Homework Help

For Exercises	See Examples
12–23, 30–33	1, 2
24–29	3, 4

Extra Practice

See page 721.

30. Solve $\sqrt{\frac{a}{4}} = 6$.

31. Find the solution of $\sqrt{\frac{5x}{7}} - 8 = 2$.

Applications and Problem Solving

32. **Science** The speed of sound S, in meters per second, near Earth's surface can be determined using the formula $S = 20\sqrt{t + 273}$, where t is the surface temperature in degrees Celsius. Suppose a racing team has designed a car that can travel 340 meters per second, in hopes of breaking the sound barrier. At what temperature will the speed of sound be 340 meters per second?

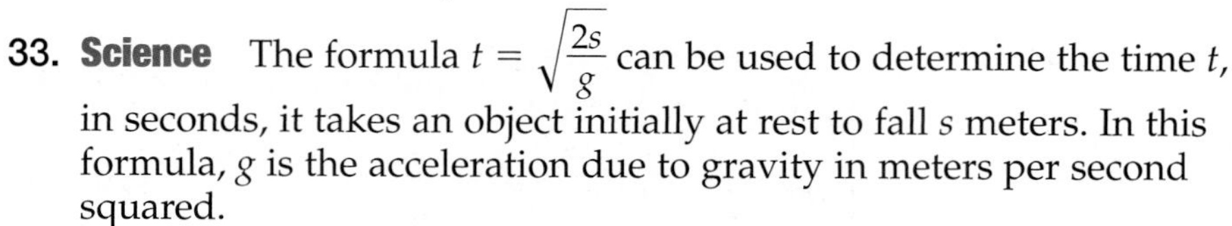

33. **Science** The formula $t = \sqrt{\frac{2s}{g}}$ can be used to determine the time t, in seconds, it takes an object initially at rest to fall s meters. In this formula, g is the acceleration due to gravity in meters per second squared.
    a. Suppose a rock falls 7.2 meters in 3 seconds on the moon. What is the acceleration due to gravity on the moon?
    b. Suppose a rock falls 78.4 meters in 4 seconds on Earth. What is the acceleration due to gravity on Earth?

34. **Critical Thinking** Find two numbers such that the square root of their sum is 5 and the square root of their product is 12.

Mixed Review

Simplify each expression. *(Lesson 14–4)*

35. $-7\sqrt{5} + 2\sqrt{5}$

36. $4\sqrt{7} - 3\sqrt{28}$

37. $-\sqrt{54} - \sqrt{18} + \sqrt{24}$

Simplify each expression. Leave in radical form. *(Lesson 14–3)*

38. $\sqrt{5} \cdot \sqrt{12}$

39. $\dfrac{\sqrt{15}}{\sqrt{3}}$

40. $\dfrac{5}{4 - \sqrt{7}}$

41. Find the solution of $y = \frac{1}{2}x - 4$ and $y = -3$ by graphing. *(Lesson 13–1)*

Write an inequality for each graph. *(Lesson 12–1)*

42. −4 −3 −2 −1 0 1 2

43. −3 −2 −1 0 1 2 3

Standardized Test Practice

A B C D

44. **Short Response** There are two field mice living in a barn. Suppose the number of mice triples every 4 months. How many mice will there be after 3 years if none of them die? *(Lesson 11–7)*

45. **Short Response** Suppose y varies inversely as x and $y = 18$ when $x = 15$. Find y when $x = 12$. *(Lesson 6–6)*

Quiz 2 Lessons 14–3 through 14–5

Simplify each expression. Leave in radical form. *(Lesson 14–3)*

1. $\sqrt{90}$

2. $\sqrt{5} \cdot \sqrt{25}$

3. $\dfrac{1}{7 + \sqrt{5}}$

4. Write $\sqrt{16x^2y^3}$ in simplest form. Use absolute value symbols if necessary. *(Lesson 14–3)*

Simplify each expression. *(Lesson 14–4)*

5. $3\sqrt{5} + 9\sqrt{5}$

6. $2\sqrt{40} - 7\sqrt{2} + 6\sqrt{10}$

7. **Geometry** The perimeter of $\triangle ABC$ is $16\sqrt{2}$ inches. Find the value of x. *(Lesson 14–4)*

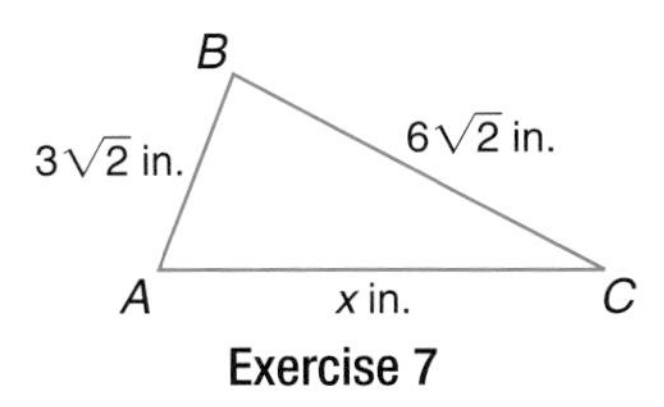

Exercise 7

Solve each equation. Check your solution. *(Lesson 14–5)*

8. $\sqrt{x} = 4$

9. $\sqrt{a} + 8 = 0$

10. $\sqrt{m + 3} = 5$

CHAPTER 14

Study Guide and Assessment

Understanding and Using the Vocabulary

After completing this chapter, you should be able to define each term, property, or phrase and give an example or two of each.

interNET CONNECTION **Review Activities**
For more review activities, visit: www.algconcepts.com

conjugates *(p. 616)*
Distance Formula *(p. 607)*
radical equations *(p. 624)*
real numbers *(p. 600)*
rationalizing the denominator *(p. 615)*

Choose the correct term to complete each sentence.

1. To find the distance d between any two points (x_1, y_1) and (x_2, y_2), use the formula $(\sqrt{(x_2 - x_1)^2 - (y_2 - y_1)^2}, \sqrt{(x_2 - x_1)^2 + (y_2 - y_1)^2})$.
2. The product of conjugates is (sometimes, always) a rational number.
3. Rationalizing the denominator is a way of eliminating (radicals, perfect squares) from the denominator of a fraction.
4. Natural numbers, whole numbers, and integers are all (irrational, rational) numbers.
5. The expression $(\sqrt{7ab}, \sqrt{8mn})$ is in simplest form.
6. Radical equations (sometimes, always) have more than one solution.
7. The number 0 is a (whole, natural) number.
8. Binomials of the form $a\sqrt{b} + c\sqrt{d}$ and $a\sqrt{b} - c\sqrt{d}$ are (conjugates, radical equations).
9. The equation $\sqrt{x} + 4 = 0$ has (one, no) real solution.
10. The set of (integers, real numbers) contains the sets of rational and irrational numbers.

Skills and Concepts

Objectives

- **Lesson 14–1** Describe the relationships among sets of numbers.

$\sqrt{16} = 4$; so $\sqrt{16}$ is a natural number, a whole number, an integer, and a rational number.

The number -3 is an integer and a rational number.

2.151151115 . . . is not the square root of a perfect square. So, it is an irrational number.

Examples

Name the set or sets of numbers to which each real number belongs. Let N = natural numbers, W = whole numbers, Z = integers, Q = rational numbers, and I = irrational numbers.

11. $-\frac{1}{4}$
12. π
13. $-\sqrt{25}$
14. $\frac{24}{6}$

Find an approximation, to the nearest tenth, for each square root. Then graph the square root on a number line.

15. $\sqrt{5}$
16. $-\sqrt{15}$
17. $\sqrt{111}$
18. $-\sqrt{260}$

Objectives and Examples

- **Lesson 14–2** Find the distance between two points in the coordinate plane.

Find the distance between points $A(3, -2)$ and $B(5, 8)$.

$$\begin{aligned} d &= \sqrt{(x_2 - x_1)^2 + (y_2 - y_1)^2} \\ &= \sqrt{(5 - 3)^2 + (8 - (-2))^2} \\ &= \sqrt{(2)^2 + (10)^2} \\ &= \sqrt{4 + 100} \\ &= \sqrt{104} \\ &\approx 10.2 \end{aligned}$$

Review Exercises

Find the distance between each pair of points. Round to the nearest tenth, if necessary.

19. $P(1, 2)$, $Q(4, 6)$

20. $J(-6, -3)$, $K(1, 0)$

21. $X(4, -1)$, $Y(-2, 5)$

22. $R(-9, -5)$, $S(-2, 19)$

Find the value of *a* if the points are the indicated distance apart.

23. $G(a, 2)$, $H(-3, 5)$; $d = \sqrt{13}$

24. $C(4, -7)$, $D(-6, a)$; $d = \sqrt{116}$

- **Lesson 14–3** Simplify radical expressions.

$$\begin{aligned} \sqrt{288} &= \sqrt{12^2 \cdot 2} \\ &= 12\sqrt{2} \end{aligned}$$

$$\begin{aligned} \sqrt{32x^3y^2} &= \sqrt{32x^3y^2} \\ &= \sqrt{2 \cdot 2 \cdot 2 \cdot 2 \cdot 2 \cdot x^3 \cdot y^2} \\ &= \sqrt{2 \cdot 16 \cdot x^3 \cdot y^2} \\ &= \sqrt{2} \cdot \sqrt{16} \cdot \sqrt{x^3} \cdot \sqrt{y^2} \\ &= \sqrt{2} \cdot 4 \cdot x \cdot \sqrt{x} \cdot |y| \\ &= 4x|y|\sqrt{2x} \end{aligned}$$

Simplify each expression. Leave in radical form.

25. $\sqrt{50}$

26. $\sqrt{3} \cdot \sqrt{6}$

27. $\dfrac{\sqrt{54}}{\sqrt{2}}$

28. $\dfrac{\sqrt{7}}{\sqrt{13}}$

29. $\dfrac{1}{4 - \sqrt{5}}$

30. $\dfrac{3}{13 - \sqrt{2}}$

Simplify each expression. Use absolute value symbols if necessary.

31. $\sqrt{300a^2bc^4}$

32. $\sqrt{121m^5n^4p^3}$

- **Lesson 14–4** Add and subtract radical expressions.

$$\begin{aligned} 8\sqrt{7} - 6\sqrt{7} &= (8 - 6)\sqrt{7} \\ &= 2\sqrt{7} \end{aligned}$$

$$\begin{aligned} 13\sqrt{2} + 4\sqrt{2} - 7\sqrt{2} &= (13 + 4 - 7)\sqrt{2} \\ &= 10\sqrt{2} \end{aligned}$$

$$\begin{aligned} \sqrt{54} + \sqrt{96} &= \sqrt{3^2 \cdot 6} + \sqrt{4^2 \cdot 6} \\ &= 3\sqrt{6} + 4\sqrt{6} \\ &= (3 + 4)\sqrt{6} \text{ or } 7\sqrt{6} \end{aligned}$$

Simplify each expression.

33. $2\sqrt{5} + 5\sqrt{5}$

34. $3\sqrt{8} - 6\sqrt{8}$

35. $-5\sqrt{6} + 14\sqrt{6} - 9\sqrt{6}$

36. $12\sqrt{3} - 14\sqrt{3} + 6\sqrt{3}$

37. $7\sqrt{2} + 5\sqrt{18}$

38. $-5\sqrt{40} + 10\sqrt{10}$

39. Simplify $2\sqrt{45} - 5\sqrt{80}$.

40. Write $2\sqrt{45} + 3\sqrt{75} - \sqrt{50}$ in simplest form.

Chapter 14 Study Guide and Assessment

Extra Practice
See pages 720–721.

Objectives and Examples

- **Lesson 14–5** Solve simple radical equations in which only one radical contains a variable.

Solve $\sqrt{x + 4} - 8 = -5$.

$$\begin{aligned} \sqrt{x+4} - 8 &= -5 \\ \sqrt{x+4} - 8 + 8 &= -5 + 8 \\ \sqrt{x+4} &= 3 \\ (\sqrt{x+4})^2 &= 3^2 \\ x + 4 &= 9 \\ x + 4 - 4 &= 9 - 4 \\ x &= 5 \end{aligned}$$

The solution is 5.

Review Exercises

Solve each equation. Check your solution.

41. $\sqrt{y} = 5$

42. $\sqrt{m} = -3$

43. $\sqrt{9h} = 9$

44. $\sqrt{-4a} = 8$

45. $7 = \sqrt{a} + 10$

46. $\sqrt{n - 12} = 15$

47. $t - 1 = \sqrt{3t + 7}$

48. $\sqrt{4x - 3} = x$

49. Solve $\sqrt{\frac{g}{6}} = 5$.

50. Find the solution of $\sqrt{\frac{3x}{4}} + 2 = 5$.

Applications and Problem Solving

51. Weather The graph shows record low temperatures. Name the set or sets of numbers to which the temperatures belong. *(Lesson 14–1)*

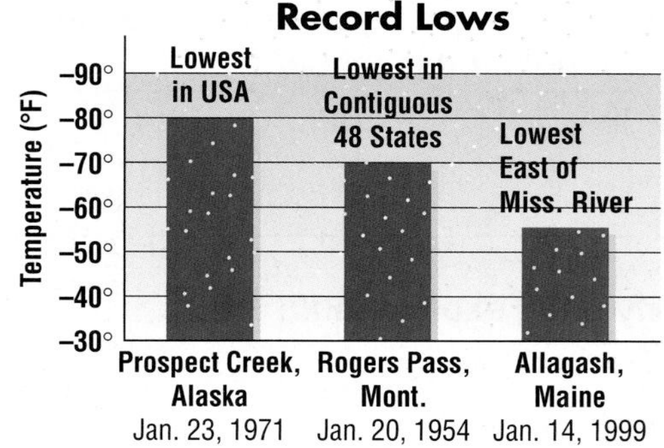

52. Geometry Quadrilateral *EFGH* is a square with vertices $E(2, 5)$, $F(6, 1)$, $G(2, -3)$, and $H(-2, 1)$. *(Lesson 14–2)*

a. Draw quadrilateral *EFGH* on a coordinate plane.

b. Determine the perimeter of the figure. Round to the nearest tenth.

53. Gems The weight of the largest pearl ever found in a giant clam is about $\sqrt{200}$ pounds. Express the weight in simplest radical form. *(Lesson 14–3)*

54. Geometry Triangle *XYZ* has a perimeter of $28\sqrt{5}$ inches. What is the measure of side *YZ*? *(Lesson 14–4)*

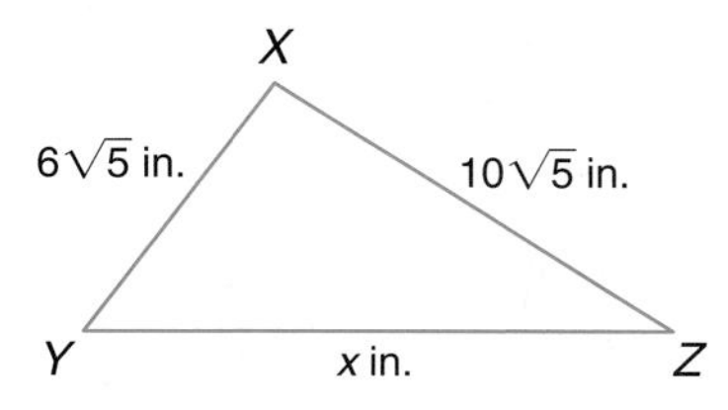

55. Nature The equation $S = 3\sqrt{d}$ gives the speed S of a tidal wave in meters per second if d is the depth of the water in meters. *(Lesson 14–5)*

a. Suppose a tidal wave is traveling 186 meters per second. What is the depth of the water?

b. What is the depth of the water if a tidal wave is traveling 153 meters per second?

CHAPTER 14 Test

1. **Name** two examples of each of the following.
   a. integer
   b. irrational number
   c. rational number
2. **Write** a pair of conjugates.

Name the set or sets of numbers to which each real number belongs. Let N = natural numbers, W = whole numbers, Z = integers, Q = rational numbers, and I = irrational numbers.

3. $\frac{5}{3}$
4. 0
5. $-0.45275563\ldots$
6. $-\sqrt{81}$

7. Graph $\sqrt{5}$, $-\sqrt{3}$, and 4.5 on a number line.

8. Write 3.22, -3.22, $\sqrt{11}$, and $3.\overline{2}$ in order from least to greatest.

Find the distance between each pair of points. Round to the nearest tenth, when necessary.

9. $G(7, 4)$, $H(-2, 4)$
10. $X(3, -3)$, $Y(-9, 2)$
11. $M(1, -1)$, $N(-5, 1)$

12. Suppose $A(5, m)$ and $B(8, 1)$ are 5 units apart. What is the value of m?

Simplify each expression. Leave in radical form.

13. $\sqrt{80}$
14. $\sqrt{3} \cdot \sqrt{27}$
15. $4\sqrt{3} \cdot \sqrt{6}$
16. $\frac{\sqrt{96}}{\sqrt{8}}$
17. $\frac{\sqrt{7}}{\sqrt{11}}$
18. $\frac{7}{7 + \sqrt{5}}$
19. $-4\sqrt{7} + 6\sqrt{7} - 12\sqrt{7}$
20. $5\sqrt{18} - 2\sqrt{50}$

Simplify each expression. Use absolute value symbols if necessary.

21. $\sqrt{25a^3b^2}$
22. $\sqrt{56m^4np}$

Solve each equation. Check your solution.

23. $\sqrt{b} + 8 = 5$
24. $\sqrt{4x - 3} = 6 - x$

25. **Science** The distance d in miles a person can see on any planet is given by the formula $d = \sqrt{\frac{rh}{2640}}$, where r is the radius of the planet in miles and h is the height of the person in feet. Suppose a person 6 feet tall is standing on Mars. If the radius of the planet is 2109 miles, how far can the person see? Round to the nearest tenth.

CHAPTER 14 Preparing for Standardized Tests

Systems of Equations Problems

On standardized tests, you will often need to translate problems into systems of equations and inequalities. You will also need to solve systems of equations using substitution or addition methods.

To solve systems of equations on standardized tests, try adding them first. This technique may work because the systems often are of the form $x + y = 6$ and $x - y = 4$

State Test Example

Hector has 20 coins. They are all quarters and nickels. The total value is \$2.20. Which system of equations can be used to determine the number of quarters q and the number of nickels n?

A $q + n = 20$
$0.30qn = 2.20$

B $q + n = 20$
$0.25q + 0.05n = 2.20$

C $q + n = 20$
$0.05q + 0.25n = 2.20$

D $q + n = 20$
$0.25q + 0.05n = 20$

Hint First write the equations in words and then translate them into symbols.

Solution The number of quarters plus the number of nickels is 20.

$$q + n = 20$$

The value of the quarters plus the value of the nickels equals the total value. The value of the quarters is 25 cents times the number of quarters or $0.25q$. The value of the nickels is 5 cents times the number of nickels or $0.05n$.

$$0.25q + 0.05n = 2.20$$

The answer is B.

SAT Example

If $2x + 3y = 20$ and $3x + 2y = 40$, what is the value of $x + y$?

Hint Look for ways to solve a problem without lengthy calculations.

Solution Read carefully. You must find the value of $x + y$, not the individual values of x and y.

Add the two equations.

$$\begin{array}{r} 2x + 3y = 20 \\ + \ 3x + 2y = 40 \\ \hline 5x + 5y = 60 \end{array}$$

Notice that the coefficients of x and y are both 5. Divide each side of the equation by 5. Simplify each term.

$$\frac{5x + 5y}{5} = \frac{60}{5}$$

$$\frac{\overset{1}{\cancel{5}}(x + y)}{\underset{1}{\cancel{5}}} = \frac{\overset{12}{\cancel{60}}}{\underset{1}{\cancel{5}}}$$

$$x + y = 12$$

The value of $x + y$ is 12. The answer is 12.

After you work each problem, record your answer on the answer sheet provided or on a sheet of paper.

Multiple Choice

1. Jared's total score on the SAT was 1340. His math score m was 400 points less than twice his verbal score v. Which system of equations will help determine his scores?

A $m + v = 1340$, $m = 2v - 400$

B $m + v = 1340$, $m = 400 - 2v$

C $m + v = 1340$, $400m = 2v$

D $m - v = 1340$, $m = 2v - 400$

2. The flag for Monroe High School has an area of 120 square feet. A new flag will have a width $1\frac{1}{2}$ times the width of the old flag. The length remains the same. What is the area, in square feet, of the new flag?

A 160 **B** 180 **C** 240 **D** 360

3. At what point do the lines with equations $y = 2x - 2$ and $7x - 3y = 11$ intersect?

A $(5, 8)$ **B** $(8, 5)$ **C** $\left(\frac{5}{8}, -1\right)$

D $\left(\frac{5}{8}, 1\right)$ **E** $\left(\frac{25}{16}, \frac{9}{8}\right)$

4. Triangle LMN is similar to triangle PQR. What is PQ?

L, M, N: 7 in., 5 in.; P, Q, R: x in., 8 in.

A 4.4 in.

B 7 in.

C 10 in.

D 11.2 in.

5. Which equation represents the line graphed?

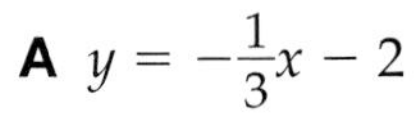

A $y = -\frac{1}{3}x - 2$

B $y = -\frac{1}{3}x + 2$

C $y = \frac{1}{3}x - 2$ **D** $y = \frac{1}{3}x + 2$

6. Gina has x marbles and Dawn has y marbles. Together they have q marbles. If Gina gives Dawn 3 of her marbles, then they will have an equal number. How many marbles does Dawn have?

A $\frac{q-6}{2}$ **B** $\frac{q-3}{2}$ **C** $\frac{q+6}{2}$

D $q - 3$ **E** $q + 3$

7. What is the slope of a line parallel to the line graphed?

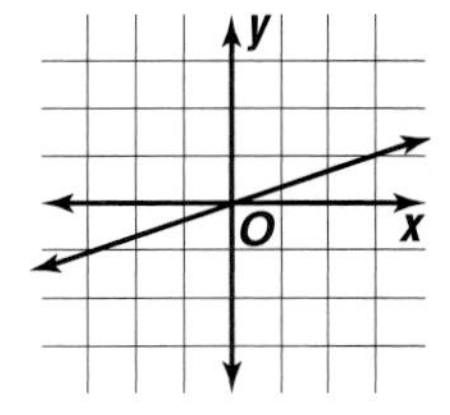

A -3 **B** $-\frac{1}{3}$

C $\frac{1}{3}$ **D** 3

8. What value of n makes the equation true?

$$9^n - 8^n = 1^n$$

A -1 **B** 0

C 1 **D** 2

Grid In

9. $\angle ABC$ is a right angle, $\overrightarrow{BC}$ bisects $\angle EBD$. If $m\angle ABD$ is 130, then what is x?

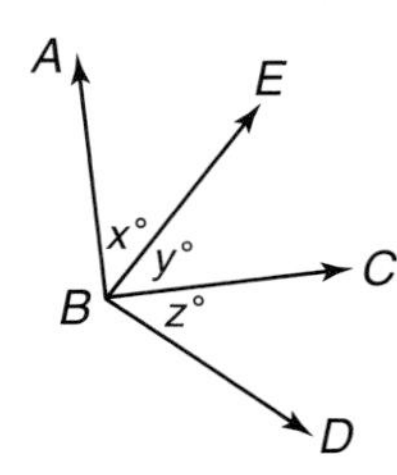

Extended Response

10. Talent show tickets are \$3 for adults and \$2 for students. Ticket sales of at least \$600 are needed to cover costs. At least 5 times as many students as adults attend.

Part A Write a system of inequalities to find the number of adult and student tickets that need to be sold. Let x represent the number of adult tickets and let y represent the number of student tickets.

Part B Graph the system of inequalities. Give one example of adult and student ticket sales that would solve the system.

CHAPTER 15

Rational Expressions and Equations

Make this Foldable to help you organize information about the material in this chapter. Begin with a sheet of notebook paper.

1. **Fold** lengthwise to the holes.

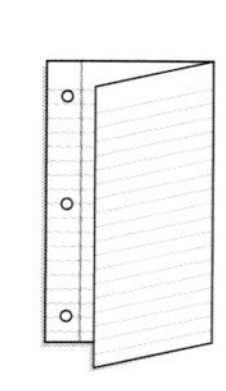

2. **Cut** along the top line and then cut ten tabs.

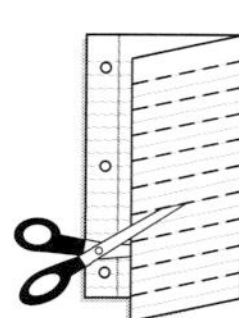

3. **Label** the tabs using the vocabulary words in the chapter.

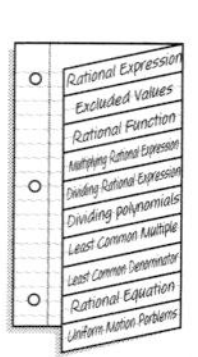

Reading and Writing Store the Foldable in a 3-ring binder. As you read and study the chapter, write notes and examples under the tabs.

Problem-Solving Workshop

Project

Although penguins waddle slowly on land, the gentoo penguin of Antarctica is thought to be the fastest swimming bird in the world. Use the formula $d = rt$ to model the distance d, rate r, and time t of the penguin. Suppose the penguin swims a distance of 20 miles. The equation that relates t and r to d is $20 = rt$.

In this project, you will graph $20 = rt$ and describe the graph. You will then choose four other values for d and graph their equations.

Working on the Project

Work with a partner and choose a strategy. Here are some suggestions to help you get started.

- Choose several values of r and solve the equation for t. Write the results as ordered pairs (r, t).
- Graph the ordered pairs on a coordinate plane.
- Research the average distance that four other animals could travel in one hour.

Strategies

Look for a pattern.
Draw a diagram.
Make a table.
Work backward.
Use an equation.
Make a graph.
Guess and check.

Technology Tools

- Use a **spreadsheet** or a **graphing calculator** to prepare your graphs.
- Use **presentation software** to prepare and give your presentation.

*inter***NET** CONNECTION **Research** For more information about animal speeds, visit: www.algconcepts.com

Presenting the Project

Prepare a presentation of your findings for your animals. Make sure that your presentation includes:

- a table of ordered pairs and a graph for each value of d,
- an explanation of your findings, including a comparison of the graphs, and
- a discussion of how these graphs differ from the graph of an equation such as $d = 20t$. Use the terms *direct variation* and *inverse variation* in your discussion.

15–1 Simplifying Rational Expressions

What You'll Learn
You'll learn to simplify rational expressions.

Why It's Important
Photography
Photographers use a rational expression to find the amount of light that falls on an object.
See Exercise 63.

A fraction denotes a quotient. In algebra, the fraction $\frac{2x}{x + 3}$ is called a **rational expression.** In a rational expression, both the numerator and denominator are polynomials.

Rational Expression	A rational expression is an algebraic fraction whose numerator and denominator are polynomials.

Every polynomial is a rational expression because it can be written as a quotient with 1 in the denominator.

Zero cannot be the denominator of a fraction because division by zero is undefined. In the expression $\frac{2x}{x + 3}$, if $x = -3$, the denominator equals zero. So, any value assigned to a variable that results in a denominator of zero must be excluded from the domain of the variable. These values are called **excluded values**.

A function that contains a rational expression is called a **rational function.** You can use the graph of the rational function of a rational expression to investigate excluded values of the variable.

Graphing Calculator Tutorial
See pp. 724–727.

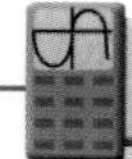

Graphing Calculator Exploration

The screen shows the graph of the rational function $y = \frac{1}{x}$. The graph is made up of two branches. One branch is in the first quadrant, and the other branch is in the third quadrant.

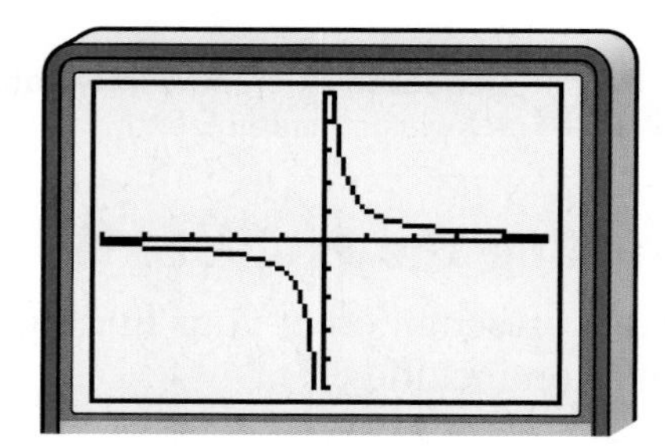

Try These

1. Explain why the graph has two branches.
2. Describe what happens to the graph as the value of x approaches 0.
3. Explain why the two branches of the graph do not touch each other. How does this relate to the excluded value of x?

4. Graph each function. State the excluded value(s) of x.

a. $y = \frac{1}{x-1}$

b. $y = \frac{1}{x-2}$

c. $y = \frac{1}{x+2}$

d. $y = \frac{1}{x(x-1)}$

e. $y = \frac{1}{(x-1)(x+2)}$

f. $y = \frac{1}{(x+3)(x+4)}$

5. **Make a conjecture** about how to find the excluded values of a rational expression without using a calculator.

Examples

Find the excluded value(s) for each rational expression.

1 $\frac{5m}{2+m}$

Exclude the values for which $2 + m = 0$.

$2 + m = 0$

$m = -2$

So, m cannot equal -2.

Zero Product Property: Lesson 11–4

2 $\frac{6}{a(a-4)}$

Exclude the values for which $a(a - 4) = 0$.

$a(a - 4) = 0$

$a = 0$ or $a - 4 = 0$ *Zero Product Property*

So, a cannot equal 0 or 4.

3 $\frac{5n}{n^2 - 25}$

Exclude the values for which $n^2 - 25 = 0$.

$n^2 - 25 = 0$

$(n + 5)(n - 5) = 0$ *Factor $n^2 - 25$.*

$n = -5$ or $n = 5$ *Zero Product Property*

So, n cannot equal -5 or 5.

Your Turn

a. $\frac{2s}{s-3}$

b. $\frac{3x}{x(x+2)}$

c. $\frac{2a-3}{a^2 - a - 6}$

Recall that you can simplify a fraction by using the following steps.

- First, factor the numerator and denominator.
- Then, divide the numerator and denominator by the greatest common factor.

Example 4

History Link

In the 1960 presidential election, more than 60% of the registered voters cast ballots. No presidential election since 1960 has had a greater voter turnout. Express 60% as a fraction in simplest form.

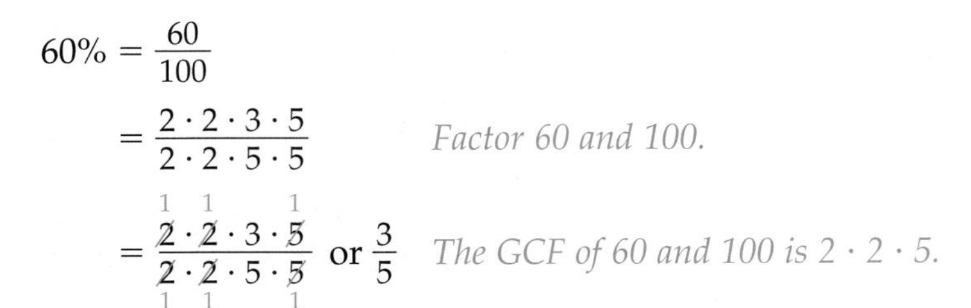

$$60\% = \frac{60}{100}$$

$$= \frac{2 \cdot 2 \cdot 3 \cdot 5}{2 \cdot 2 \cdot 5 \cdot 5} \quad \textit{Factor 60 and 100.}$$

$$= \frac{\overset{1}{\cancel{2}} \cdot \overset{1}{\cancel{2}} \cdot 3 \cdot \overset{1}{\cancel{5}}}{\underset{1}{\cancel{2}} \cdot \underset{1}{\cancel{2}} \cdot 5 \cdot \underset{1}{\cancel{5}}} \text{ or } \frac{3}{5} \quad \textit{The GCF of 60 and 100 is } 2 \cdot 2 \cdot 5.$$

Your Turn **Simplify each fraction.**

d. $\frac{9}{24}$ e. $\frac{15}{20}$ f. $\frac{12}{30}$

You can use the same procedure to simplify rational expressions that have polynomials in the numerator and denominator. To *simplify* means that the numerator and denominator have no factors in common, except 1.

Examples

From this point on, you can assume that all values that give a denominator of zero are excluded.

Simplify each rational expression.

5 $\mathbf{\frac{8a^2b}{12ab^3}}$

$$\frac{8a^2b}{12ab^3} = \frac{2 \cdot 2 \cdot 2 \cdot a \cdot a \cdot b}{2 \cdot 2 \cdot 3 \cdot a \cdot b \cdot b \cdot b} \quad \textit{Note that } a \neq 0 \textit{ and } b \neq 0.$$

$$= \frac{\overset{1}{\cancel{2}} \cdot \overset{1}{\cancel{2}} \cdot 2 \cdot \overset{1}{\cancel{a}} \cdot a \cdot \overset{1}{\cancel{b}}}{\underset{1}{\cancel{2}} \cdot \underset{1}{\cancel{2}} \cdot 3 \cdot \underset{1}{\cancel{a}} \cdot \underset{1}{\cancel{b}} \cdot b \cdot b} \text{ or } \frac{2a}{3b^2} \quad \textit{The GCF is 4ab.}$$

6 $\mathbf{\frac{2x - 2}{5x - 5}}$

$$\frac{2x-2}{5x-5} = \frac{2(x-1)}{5(x-1)} \quad \textit{Factor } 2x - 2 \textit{ and } 5x - 5.$$

$$= \frac{2\overset{1}{\cancel{(x-1)}}}{5\underset{1}{\cancel{(x-1)}}} \text{ or } \frac{2}{5} \quad \textit{The GCF is } (x - 1).$$

7 $\mathbf{\frac{a(a + 1)}{a^2 + 3a + 2}}$

$$\frac{a(a+1)}{a^2+3a+2} = \frac{a(a+1)}{(a+2)(a+1)} \quad \textit{Factor } a^2 + 3a + 2.$$

$$= \frac{a\overset{1}{\cancel{(a+1)}}}{(a+2)\underset{1}{\cancel{(a+1)}}} \text{ or } \frac{a}{a+2} \quad \textit{The GCF is } (a + 1).$$

Look Back

Factoring Trinomials: Lesson 10–3

8 $\dfrac{x^2-9}{x^2-x-6}$

$$\frac{x^2-9}{x^2-x-6} = \frac{(x-3)(x+3)}{(x+2)(x-3)}$$ *Factor $x^2 - 9$ and $x^2 - x - 6$.*

$$= \frac{\overset{1}{\cancel{(x-3)}}(x+3)}{(x+2)\underset{1}{\cancel{(x-3)}}} \text{ or } \frac{x+3}{x+2}$$ *The GCF is $(x - 3)$.*

9 $\dfrac{4-2x}{x^2-4x+4}$

$$\frac{4-2x}{x^2-4x+4} = \frac{2(2-x)}{(x-2)(x-2)}$$ *Factor $4 - 2x$ and $x^2 - 4x + 4$.*

$$= \frac{2(-1)(x-2)}{(x-2)(x-2)}$$ *Factor -1 from $(2 - x)$.*

$$= \frac{2(-1)\overset{1}{\cancel{(x-2)}}}{(x-2)\underset{1}{\cancel{(x-2)}}} \text{ or } \frac{-2}{x-2}$$ *The GCF is $(x - 2)$.*

Your Turn **Simplify each rational expression.**

g. $\dfrac{14y^3z}{8y^2z^4}$ h. $\dfrac{-12a^2b^2}{18a^5}$ i. $\dfrac{3y-9}{4y-12}$

j. $\dfrac{x(x+3)}{x^2+6x+9}$ k. $\dfrac{a^2-2a-8}{a^2+3a+2}$ l. $\dfrac{x^2-4x}{16-x^2}$

Check for Understanding

Communicating Mathematics

1. **Explain** why 2 is an excluded value for $\dfrac{x}{x-2}$.
2. **List** the steps you would use to simplify a rational expression.

Vocabulary
rational expression
excluded values
rational function

3. 

Sam and Darnell are trying to simplify $\dfrac{x+3}{x+4}$. Sam says the answer is $\dfrac{3}{4}$. Darnell says it is already in simplest form. Who is correct? Explain your reasoning.

Guided Practice

Getting Ready **Factor each expression.**

Sample 1: $5y + 15$
Solution: $5(y + 3)$

Sample 2: $x^2 + 5x + 6$
Solution: $(x + 3)(x + 2)$

4. $4x - 20$
5. $16x^2 - 20x$
6. $y^2 + 6y + 8$
7. $a^2 - a - 20$
8. $n^2 - 3n - 18$
9. $x^2 - 10x + 24$

Find the excluded value(s) for each rational expression.
(Examples 1–3)

10. $\dfrac{7}{x-2}$
11. $\dfrac{4a}{a(a+6)}$
12. $\dfrac{2x+3}{x^2-2x-15}$

Simplify each rational expression. *(Examples 4–9)*

13. $\frac{8}{20}$
14. $\frac{5x}{25x^2}$
15. $\frac{12y^2z}{18yz^4}$
16. $\frac{x(x+3)}{7(x+3)}$
17. $\frac{3y+9}{4y+12}$
18. $\frac{x^2-3x}{x^2+x-12}$
19. $\frac{x^2+2x-8}{x^2+5x+4}$
20. $\frac{x^2+6x+5}{x^2+3x-10}$
21. $\frac{25-x^2}{x^2+x-30}$

22. **Sports** The National Hockey League team that wins the first game of the best-of-seven series for the Stanley Cup has about an 80% chance of winning the series. Express 80% as a fraction in simplest form. *(Example 4)* **Source:** NHL

Exercises

Practice

Homework Help	
For Exercises	See Examples
23–31, 60, 64	1–3
61–63	4
32–44, 54	5, 6
45–58	7–9

Extra Practice
See page 722.

Find the excluded value(s) for each rational expression.

23. $\frac{x}{x+5}$
24. $\frac{3a}{2a+4}$
25. $\frac{n+6}{n-10}$
26. $\frac{5}{a(a+8)}$
27. $\frac{x+2}{(x+2)(x-3)}$
28. $\frac{5m}{(m-2)(m-3)}$
29. $\frac{9x}{x^2+5x}$
30. $\frac{x^2+6x+5}{x^2+3x-10}$
31. $\frac{y+3}{y^2-16}$

Simplify each rational expression.

32. $\frac{10}{16}$
33. $\frac{12}{36}$
34. $\frac{15}{20}$
35. $\frac{4c}{6d}$
36. $\frac{8xy}{24x^2}$
37. $\frac{-36abc^2}{9ac}$
38. $\frac{2(x+1)}{8(x+1)}$
39. $\frac{n-3}{5(n-3)}$
40. $\frac{x(x+5)}{y(x+5)}$
41. $\frac{(x+3)(x-2)}{(x-2)(x+1)}$
42. $\frac{(y+6)(y-6)}{(y+6)(y+6)}$
43. $\frac{x^2-3x}{2(x-3)}$
44. $\frac{a^2-a}{a-1}$
45. $\frac{r^2-r-6}{3r-9}$
46. $\frac{x^2+3x+2}{x^2+2x+1}$
47. $\frac{y^2+7y+12}{y^2-16}$
48. $\frac{x-3}{x^2+x-12}$
49. $\frac{y^2+4y+4}{y^2+y-2}$
50. $\frac{r^2-4r-5}{r^2-2r-15}$
51. $\frac{z^2-z-20}{z^2+7z+12}$
52. $\frac{6x^2+24x}{x^2+8x+16}$
53. $\frac{8m^2-16m}{m^2-4m+4}$
54. $\frac{r^2+6r+5}{2r^2-2}$
55. $\frac{c^2-c-20}{c^3+10c^2+24c}$
56. $\frac{9-3x}{x^2-6x+9}$
57. $\frac{x^2-4x}{16-x^2}$
58. $\frac{12-4y}{y^2+y-12}$

59. Write $\frac{4a}{3a+a^2}$ in simplest form.

60. What are the excluded values of x for $\frac{x^2-9}{x^2+5x+6}$?

Applications and Problem Solving

61. **Investing** The equation $y = \frac{72}{x}$ can be used to estimate how long it will take to double an investment. In the equation, y is the number of years, and x is the annual interest rate expressed as a percent. How long would it take to double your money if it is invested at an annual rate of 8%?

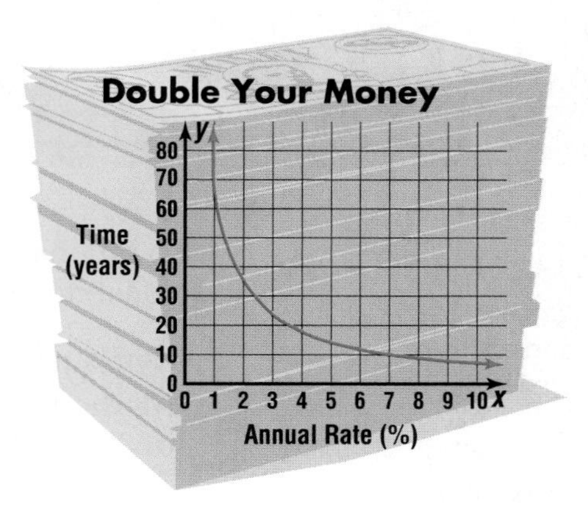

62. **Entertainment** The graph shows the number of films that adults see in a theater in a typical month.
   a. What fraction of adults see one movie?
   b. What fraction of adults see three or more movies?

63. **Photography** The intensity of light that falls on an object is given by the formula $I = \frac{P}{d^2}$. In the formula, d is the distance from the light source, P is the power in lumens, and I is the intensity of light in lumens per square meter. An object being photographed is 3 meters from a 72-lumen light source. Find the intensity.

64. **Critical Thinking** Write rational expressions that have the following as excluded values.
   a. 0 b. -3 c. -2 and 7

Mixed Review

Solve each equation. *(Lesson 14–6)*

65. $\sqrt{x} = 4$
66. $\sqrt{y} = 2\sqrt{5}$
67. $\sqrt{n - 3} = 4$
68. $\sqrt{a} + 4 = 29$

Simplify each expression. *(Lesson 14–5)*

69. $2\sqrt{3} + \sqrt{3} + 4\sqrt{3}$
70. $6\sqrt{5} - 3\sqrt{5} + 10\sqrt{5}$
71. $4\sqrt{3} + \sqrt{12}$
72. $2\sqrt{50} + 3\sqrt{5}$

73. Find the distance between $J(2, 1)$ and $K(5, 5)$. *(Lesson 14–2)*

Standardized Test Practice

74. **Short Response** The sum of two numbers is 9. Their difference is 3. Find the numbers. *(Lesson 13–4)*

75. **Multiple Choice** Which property or properties of inequalities allow(s) you to prove that if $2y + 8 > 18$, then $y > 5$? *(Lesson 12–4)*
   A Distributive Property
   B Addition Property
   C Division Property
   D Addition and Division Properties

15–2 Multiplying and Dividing Rational Expressions

What You'll Learn
You'll learn to multiply and divide rational expressions.

Why It's Important
Carpentry Carpenters divide rational expressions to find how many boards of a certain size can be cut from a piece of wood.
See Exercise 42.

Two rational expressions can be multiplied or divided just like two rational numbers.

You can use two methods to multiply rational numbers.

- **Method 1** Multiply numerators and multiply denominators. Then divide each numerator and denominator by the greatest common factor.
- **Method 2** Divide numerators and denominators by any common factors. Then multiply numerators and denominators.

Method 1
Multiply, then simplify.

$$\frac{2}{5} \cdot \frac{1}{2} = \frac{2}{10}$$

$$= \frac{\overset{1}{\cancel{2}}}{\underset{5}{\cancel{10}}} \text{ or } \frac{1}{5} \quad \textit{The GCF is 2.}$$

Method 2
Simplify, then multiply.

$$\frac{2}{5} \cdot \frac{1}{2} = \frac{\overset{1}{\cancel{2}}}{5} \cdot \frac{1}{\underset{1}{\cancel{2}}} \quad \textit{2 is a common factor.}$$

$$= \frac{1}{5}$$

Both methods have the same result.

You can use the same methods to multiply rational expressions. Multiply $\frac{2x}{(x+1)^2}$ and $\frac{(x+1)}{x}$.

Method 1
Multiply, then simplify.

$$\frac{2x}{(x+1)^2} \cdot \frac{x+1}{x} = \frac{2x(x+1)}{x(x+1)^2}$$

$$= \frac{2\cancel{x}\cancel{(x+1)}}{\cancel{x}\cancel{(x+1)}(x+1)}$$

$$= \frac{2}{(x+1)}$$

Method 2
Simplify, then multiply.

$$\frac{2x}{(x+1)^2} \cdot \frac{x+1}{x} = \frac{2\cancel{x}}{(x+1)\cancel{(x+1)}} \cdot \frac{\cancel{(x+1)}}{\cancel{x}}$$

$$= \frac{2}{(x+1)}$$

Examples

Find each product.

1 $\frac{6r^2}{5s^2} \cdot \frac{10rs}{6r^3}$

Method 1 $\frac{6r^2}{5s^2} \cdot \frac{10rs}{6r^3} = \frac{60r^3s}{30r^3s^2}$ *Multiply.*

$$= \frac{\overset{1}{\cancel{2}} \cdot 2 \cdot \overset{1}{\cancel{3}} \cdot \overset{1}{\cancel{5}} \cdot \overset{1}{\cancel{r}} \cdot \overset{1}{\cancel{r}} \cdot \overset{1}{\cancel{r}} \cdot \overset{1}{\cancel{s}}}{\underset{1}{\cancel{2}} \cdot \underset{1}{\cancel{3}} \cdot \underset{1}{\cancel{5}} \cdot \underset{1}{\cancel{r}} \cdot \underset{1}{\cancel{r}} \cdot \underset{1}{\cancel{r}} \cdot \underset{1}{\cancel{s}} \cdot s} \text{ or } \frac{2}{s} \quad \textit{Simplify.}$$

Method 2 $\frac{6r^2}{5s^2} \cdot \frac{10rs}{6r^3} = \frac{\overset{1\,1}{\cancel{6}\cancel{r^2}}}{\underset{1\,s}{\cancel{5}\cancel{s^2}}} \cdot \frac{\overset{2\,1\,1}{\cancel{10}\cancel{r}\cancel{s}}}{\underset{1\,1}{\cancel{6}\cancel{r^3}}}$ *Simplify.*

$$= \frac{2}{s} \quad \textit{Multiply.}$$

2 $\dfrac{3m}{n+2} \cdot \dfrac{n+2}{9m^2}$

$$\frac{3m}{n+2} \cdot \frac{n+2}{9m^2} = \frac{\cancel{3}\cancel{m}}{\cancel{n+2}} \cdot \frac{\cancel{n+2}}{\cancel{9m^2}} \quad \text{3, m, and n + 2 are common factors.}$$

$$= \frac{1}{3m}$$

3 $\dfrac{y-3}{y+5} \cdot \dfrac{2y^2+10y}{2y-6}$

$$\frac{y-3}{y+5} \cdot \frac{2y^2+10y}{2y-6} = \frac{y-3}{y+5} \cdot \frac{2y(y+5)}{2(y-3)} \quad \text{Factor } 2y^2+10y \text{ and } 2y-6.$$

$$= \frac{\cancel{y-3}}{\cancel{y+5}} \cdot \frac{\cancel{2}y(\cancel{y+5})}{\cancel{2}(\cancel{y-3})} \quad 2,\ y-3, \text{ and } y+5 \text{ are common factors.}$$

$$= \frac{y}{1} \text{ or } y$$

4 $\dfrac{x^2-25}{x^2-3x-10} \cdot \dfrac{x+2}{x}$

$$\frac{x^2-25}{x^2-3x-10} \cdot \frac{x+2}{x} = \frac{(x+5)(x-5)}{(x-5)(x+2)} \cdot \frac{x+2}{x} \quad \text{Factor } x^2-25 \text{ and } x^2-3x-10.$$

$$= \frac{(x+5)(\cancel{x-5})}{(\cancel{x-5})(\cancel{x+2})} \cdot \frac{\cancel{x+2}}{x} \quad x-5 \text{ and } x+2 \text{ are common factors.}$$

$$= \frac{x+5}{x}$$

Your Turn **Find each product.**

a. $\dfrac{12x}{5y} \cdot \dfrac{20y^2}{36x^2}$

b. $\dfrac{x+2}{8} \cdot \dfrac{12x}{(x+2)(x-2)}$

c. $\dfrac{n-2}{6} \cdot \dfrac{3}{n^2-2n}$

d. $\dfrac{a^2+6a+9}{a^2-4} \cdot \dfrac{a-2}{a+3}$

Look Back

Reciprocal: Lesson 4–3

To divide a rational number by any nonzero number, multiply by its reciprocal. You can use the same method to multiply rational expressions.

$$\frac{2}{3} \div \frac{1}{2} = \frac{2}{3} \cdot \frac{2}{1} = \frac{4}{3}$$

The reciprocal of $\frac{1}{2}$ is $\frac{2}{1}$.

$$\frac{5}{x} \div \frac{y}{z} = \frac{5}{x} \cdot \frac{z}{y} = \frac{5z}{xy}$$

The reciprocal of $\frac{y}{z}$ is $\frac{z}{y}$.

Example 5 **Sports Link**

The Indianapolis 500 is a 500-mile automobile race. Each lap is $2\frac{1}{2}$ miles long. How many laps does a driver complete to race the entire 500 miles?

To find the number of laps, divide 500 by $2\frac{1}{2}$.

$$500 \div 2\frac{1}{2} = \frac{500}{1} \div \frac{5}{2}$$ *Express 500 and $2\frac{1}{2}$ as improper fractions.*

$$= \frac{500}{1} \cdot \frac{2}{5}$$ *The reciprocal of $\frac{5}{2}$ is $\frac{2}{5}$.*

$$= \frac{\overset{100}{\cancel{500}}}{1} \cdot \frac{2}{\underset{1}{\cancel{5}}}$$ *5 is a common factor.*

$$= \frac{200}{1} \text{ or } 200$$

A driver completes 200 laps.

Examples

Find each quotient.

6 $\frac{6x^3}{y} \div \frac{2x}{y^2}$

$$\frac{6x^3}{y} \div \frac{2x}{y^2} = \frac{6x^3}{y} \cdot \frac{y^2}{2x}$$ *The reciprocal of $\frac{2x}{y^2}$ is $\frac{y^2}{2x}$.*

$$= \frac{\overset{3\,x^2}{\cancel{6x^3}}}{\underset{1}{\cancel{y}}} \cdot \frac{\overset{y}{\cancel{y^2}}}{\underset{1\,1}{\cancel{2x}}}$$ *2, x, and y are common factors.*

$$= \frac{3x^2y}{1} \text{ or } 3x^2y$$

7 $\frac{2m + 8}{m + 5} \div (m + 4)$

$$\frac{2m + 8}{m + 5} \div (m + 4) = \frac{2m + 8}{m + 5} \cdot \frac{1}{m + 4}$$ *The reciprocal of $(m + 4)$ is $\frac{1}{m + 4}$.*

$$= \frac{2(m + 4)}{m + 5} \cdot \frac{1}{m + 4}$$ *Factor $2m + 8$.*

$$= \frac{2\overset{1}{\cancel{(m + 4)}}}{m + 5} \cdot \frac{1}{\underset{1}{\cancel{m + 4}}}$$ *$(m + 4)$ is a common factor.*

$$= \frac{2}{m + 5}$$

Your Turn

e. $\frac{12y}{7z^2} \div \frac{3xy}{2z}$

f. $\frac{x^2 + 4x + 3}{x^2} \div (x + 3)$

Sometimes it is necessary to factor -1 from one of the terms.

Example 8 Find $\frac{x^2 - 4}{2y} \div \frac{2 - x}{6xy}$.

$$\frac{x^2 - 4}{2y} \div \frac{2 - x}{6xy} = \frac{x^2 - 4}{2y} \cdot \frac{6xy}{2 - x}$$ *The reciprocal of $\frac{2-x}{6xy}$ is $\frac{6xy}{2-x}$.*

$$= \frac{(x + 2)(x - 2)}{2y} \cdot \frac{6xy}{2 - x}$$ *Factor $x^2 - 4$.*

$$= \frac{(x + 2)(x - 2)}{2y} \cdot \frac{6xy}{-1(x - 2)}$$ *Factor -1 from $2 - x$.*

$$= \frac{(x + 2)\overset{1}{\cancel{(x - 2)}}}{\underset{1\,1}{\cancel{2y}}} \cdot \frac{\overset{3\;1}{\cancel{6xy}}}{-1\underset{1}{\cancel{(x - 2)}}}$$ *$x - 2$, y, and 2 are common factors.*

$$= \frac{3x(x + 2)}{-1} \text{ or } -3x(x + 2)$$

Reading Algebra

Instead of factoring -1 from $2 - x$, you could have factored -1 from $x - 2$ and obtained the same results.

Your Turn

g. Find $\frac{x^2 - 9}{8x^3} \div \frac{6 - 2x}{4x}$

Check for Understanding

Communicating Mathematics

1. **Identify** and correct the error that was made while finding the quotient.

$$\frac{a^2 - 25}{3a} \div \frac{a + 5}{15a^2} = \frac{3a}{(a + 5)(a - 5)} \cdot \frac{a + 5}{15a^2}$$

$$= \frac{\overset{1\,1}{\cancel{3a}}}{\underset{1}{\cancel{(a + 5)}}(a - 5)} \cdot \frac{\overset{1}{\cancel{a + 5}}}{\underset{5\,a}{\cancel{15a^2}}} \text{ or } \frac{1}{5a(a - 5)}$$

Math Journal

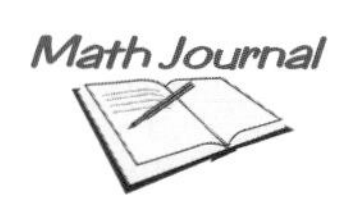

2. **Write** a short paragraph explaining which method you prefer when finding the product of rational expressions: to simplify first and then multiply or to multiply first and then simplify. Include examples to support your point of view.

Guided Practice

Getting Ready **Find the reciprocal of each expression.**

Sample 1: $\frac{m}{2}$

Solution: The reciprocal is $\frac{2}{m}$.

Sample 2: $(a + b)$

Solution: The reciprocal is $\frac{1}{a + b}$.

3. $\frac{x^2}{4}$ 4. y 5. $\frac{x - 4}{x + 5}$ 6. $y + 6$

Find each product. *(Examples 1–4)*

7. $\frac{a^2b}{b^2c} \cdot \frac{c}{d}$

8. $\frac{5(n-1)}{-3} \cdot \frac{9}{n-1}$

9. $\frac{2x-10}{x^2+x-12} \cdot \frac{x-3}{x-5}$

10. $\frac{y^2+3y-10}{2y} \cdot \frac{y^2-3y}{y^2-5y+6}$

Find each quotient. *(Examples 6–8)*

11. $\frac{12a^3}{bc} \div \frac{3a^2}{bc}$

12. $\frac{9xy}{5} \div 3x^2y^2$

13. $\frac{3x^2+6x}{x} \div \frac{2x+4}{x^2}$

14. $\frac{x^2-5x+6}{2x^2} \div \frac{3-x}{4}$

15. **Sewing** It takes $1\frac{1}{2}$ yards of fabric to make one flag for the school color guard. How many flags can be made from 18 yards of fabric? *(Example 5)*

Exercises

Practice

Homework Help	
For Exercises	See Examples
16–18	1
19–27, 40	2–4
42–43	5
28–39, 41	6–8
Extra Practice	
See page 722.	

Find each product.

16. $\frac{ab}{ac} \cdot \frac{c}{d}$

17. $\frac{3x}{2y} \cdot \frac{y^2}{6}$

18. $\frac{6a^2}{8n^2} \cdot \frac{12n}{9a}$

19. $\frac{3(a-b)}{a} \cdot \frac{a^2}{a-b}$

20. $\frac{2(a+2b)}{5} \cdot \frac{5}{3(a+2b)}$

21. $\frac{7s}{s+2} \cdot \frac{2s+4}{21}$

22. $\frac{3x+30}{2x} \cdot \frac{4x}{4x+40}$

23. $\frac{x+3}{x+4} \cdot \frac{x}{x^2+7x+12}$

24. $\frac{3a-6}{a^2-9} \cdot \frac{a+3}{a^2-2a}$

25. $\frac{x^2-9}{x+7} \cdot \frac{2x+14}{x^2+6x+9}$

26. $\frac{x}{x^2+8x+15} \cdot \frac{2x+10}{x^2}$

27. $\frac{n^2}{n^2-4} \cdot \frac{n^2-5n+6}{n^2-3n}$

Find each quotient.

28. $\frac{a^2}{b} \div \frac{a^2}{b^2}$

29. $\frac{7a^2b}{xy} \div \frac{7}{6xy}$

30. $2xz \div \frac{4xy}{z}$

31. $\frac{2a^3}{a+1} \div \frac{a^2}{a+1}$

32. $\frac{b^2-9}{4b} \div (b-3)$

33. $\frac{y^2+8y+16}{y^2} \div (y+4)$

34. $\frac{y^2}{y+2} \div \frac{y}{y+2}$

35. $\frac{m^2+2m+1}{2} \div \frac{m+1}{m-1}$

36. $\frac{x^2-4x+4}{3x} \div \frac{x^2-4}{6}$

37. $\frac{x^2+7x+10}{x-1} \div \frac{x^2+2x-15}{1-x}$

38. $\frac{x^2-16}{16-x^2} \div \frac{7}{x}$

39. $\frac{a^2-9}{9} \div \frac{6-2a}{27a^2}$

40. Find the product of $\frac{y-3}{8}$ and $\frac{12}{y-3}$.

41. What is the quotient when $\frac{x^2}{x^2-y^2}$ is divided by $\frac{x^2}{x+y}$?

Applications and Problem Solving

42. **Carpentry** How many boards, each 2 feet 8 inches long, can be cut from a board 16 feet 6 inches long?

43. **Probability** Two darts are randomly thrown one at a time. Assume that both hit the target.
   a. What is the probability that both will hit the shaded region?
   b. Find the probability if $x = 3$.

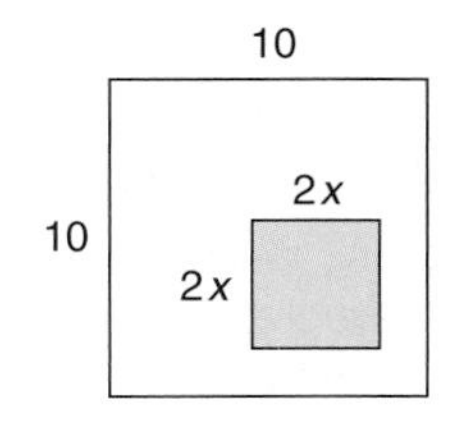

44. **Critical Thinking** Find two different pairs of rational expressions whose product is $\frac{5x^2}{8y^3}$.

Mixed Review

Simplify. *(Lesson 15–1)*

45. $\frac{14a^2b^4c}{2a^3bc}$

46. $\frac{m^2 - 16}{m^2 - 8m + 16}$

47. $\frac{8 - 4x}{x^2 - 4x + 4}$

48. **Cycling** You can use the formula $s = 4\sqrt{r}$ to find the fastest speed at which a cyclist can turn a corner safely and not tip over. In the formula, s is the speed in miles per hour, and r is the radius of the corner in feet. If $s = 8$ miles per hour, find r. *(Lesson 14–5)*

Use elimination to solve each system of equations. *(Lesson 13–5)*

49. $x + y = 4$
   $2x - 3y = -7$

50. $x + 3y = -4$
   $x - 2y = 6$

51. $3a + 4b = -25$
   $2a - 3b = 6$

52. Graph $2x + 3y \leq 12$. *(Lesson 12–7)*

Standardized Test Practice

53. **Extended Response** Paul wants to start a small lawn-care company to earn money in the summer. He estimates that his expenses will be \$200 for gasoline and equipment. He plans to charge \$25 per lawn. The equation $P = 25n - 200$ can be used to find his profits. In the equation, n represents the number of lawns he cuts. *(Lesson 6–2)*
   a. Determine the ordered pairs that satisfy the equation if the domain is $\{5, 8, 10\}$.
   b. Graph the relation.

54. **Multiple Choice** A 60-kilogram mass is 140 centimeters from the fulcrum of a lever. How far from the fulcrum must an 80-kilogram mass be to balance the lever? *(Lesson 4–4)*

   A 186.7 cm   B 34.3 cm
   C 120 cm   D 105 cm

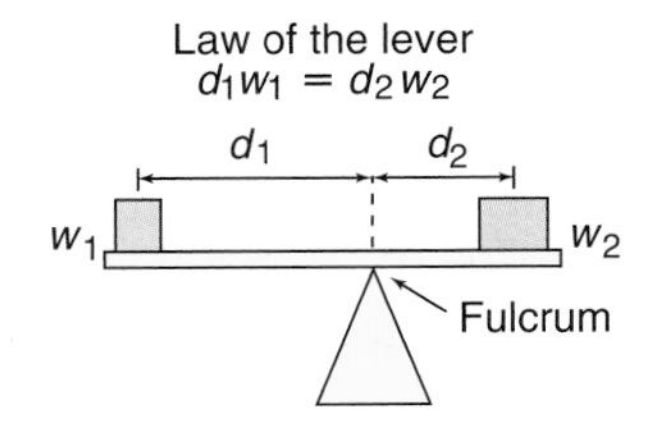

15–3 Dividing Polynomials

What You'll Learn

You'll learn to divide polynomials by binomials.

Why It's Important

Aviation You can use polynomials to determine distance, rate, and time. *See Exercise 32.*

In the previous lesson, you learned that some divisions can be performed using factoring.

$$(x^2 + 4x + 3) \div (x + 3) = \frac{x^2 + 4x + 3}{x + 3}$$

$$= \frac{\overset{1}{\cancel{(x + 3)}}(x + 1)}{\underset{1}{\cancel{(x + 3)}}} \quad \textit{Factor the dividend.}$$

$$= x + 1$$

Therefore, $(x^2 + 4x + 3) \div (x + 3) = x + 1$. Since the remainder is 0, the divisor is a factor of the dividend.

You can also use algebra tiles to model the division shown above.

Materials: 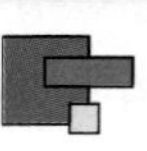 algebra tiles  product mat

Step 1 Model the polynomial $x^2 + 4x + 3$. This represents the dividend.

Step 2 Place the x^2-tile at the corner of the mat. Arrange the three 1-tiles to make a length of $x + 3$. This represents the divisor.

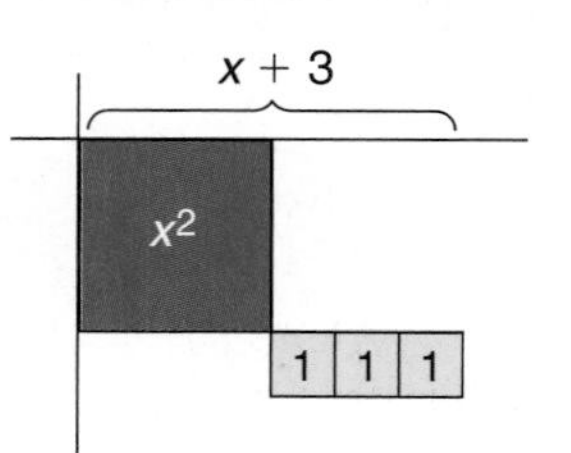

Step 3 Use the remaining tiles to make a rectangle. The width of the rectangle, $x + 1$, is the quotient.

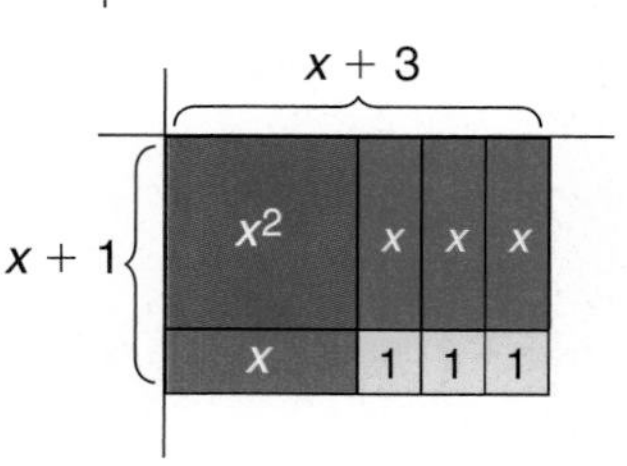

Try These

Use algebra tiles to find each quotient. Recall that you can add zero-pairs without changing the value of the polynomial.

1. $(x^2 + 6x + 9) \div (x + 3)$

2. $(x^2 - 4x + 4) \div (x - 2)$

3. $(x^2 + x - 6) \div (x + 3)$

4. $(x^2 - 9) \div (x + 3)$

5. What happens when you try to model $(x^2 + 5x + 9) \div (x + 3)$? What do you think your result means?

You can also divide polynomials using long division. Follow these steps to divide $2x^2 + 7x + 3$ by $2x + 1$.

Step 1 To find the first term of the quotient, divide the first term of the dividend, $2x^2$, by the first term of the divisor, $2x$.

$$\begin{array}{rll} & x & \\ 2x + 1 \overline{)\, 2x^2 + 7x + 3} & & 2x^2 \div 2x = x \\ \underline{(-)\ 2x^2 + 1x} & & \textit{Multiply } x \textit{ and } 2x + 1. \\ 6x & & \textit{Subtract.} \end{array}$$

Step 2 To find the next term of the quotient, divide the first term of the partial dividend, $6x$, by the first term of the divisor, $2x$.

$$\begin{array}{rl} x + 3 & \\ 2x + 1 \overline{)\, 2x^2 + 7x + 3} & \\ \underline{(-)\ 2x^2 + 1x} & \\ 6x + 3 & \textit{Bring down 3; } 6x \div 2x = 3. \\ \underline{(-)\ 6x + 3} & \textit{Multiply 3 and } 2x + 1. \\ 0 & \textit{Subtract.} \end{array}$$

Therefore, $(2x^2 + 7x + 3) \div (2x + 1) = x + 3$.

Examples

Find each quotient.

1 $(6y - 3) \div (2y - 1)$

$$\begin{array}{rl} 3 & \\ 2y - 1 \overline{)\, 6y - 3} & 6y \div 2y = 3 \\ \underline{(-)\ 6y - 3} & \textit{Multiply 3 and } 2y - 1. \\ 0 & \textit{Subtract.} \end{array}$$

Therefore, $(6y - 3) \div (2y - 1) = 3$.

2 $(x^2 - 2x - 8) \div (x + 2)$

$$\begin{array}{rl} x - 4 & \\ x + 2 \overline{)\, x^2 - 2x - 8} & \\ \underline{(-)\ x^2 + 2x} & \textit{Multiply } x \textit{ and } x + 2. \\ -4x - 8 & \textit{Subtract: } -2x - 2x = -4x\textit{; bring down } -8. \\ \underline{(-)\ -4x - 8} & \textit{Multiply } -4 \textit{ and } x + 2. \\ 0 & \textit{Subtract.} \end{array}$$

Therefore, $(x^2 - 2x - 8) \div (x + 2) = x - 4$.

Since $x - 4$ is a factor of $x^2 - 2x - 8$, you can check the quotient by simplifying $\frac{x^2 - 2x - 8}{x + 2}$ or by multiplying $(x + 2)$ by $(x - 4)$.

Your Turn

a. $(15a - 6) \div (5a - 2)$

b. $(x^2 - 7x + 10) \div (x - 2)$

If the divisor is *not* a factor of the dividend, the remainder will not be 0. The quotient can be expressed as follows.

$$\text{quotient} = \text{partial quotient} + \frac{\text{remainder}}{\text{divisor}}$$

Example 3 **Find $(8y^2 - 2y + 1) \div (2y - 1)$.**

$$\begin{array}{rl}
4y + 1 & \\
2y - 1\overline{)8y^2 - 2y + 1} & \\
(-)\ 8y^2 - 4y & \textit{Multiply } 4y \textit{ and } 2y - 1. \\
2y + 1 & \textit{Subtract. Then bring down 1.} \\
(-)\ 2y - 1 & \textit{Multiply 1 and } 2y - 1. \\
2 & \textit{Subtract. The remainder is 2.}
\end{array}$$

The quotient is $4y + 1$ with remainder 2.

So, $(8y^2 - 2y + 1) \div (2y - 1) = 4y + 1 + \frac{2}{2y - 1}$.

Your Turn

c. Find $(2a^2 + 7a + 3) \div (a + 2)$.

In an expression like $x^2 - 4$ there is no x term. In such situations, rename the dividend using zero as the coefficient of the missing term.

Example 4 **Find $(x^2 - 4) \div (x + 1)$.**

$$\begin{array}{rl}
x - 1 & \\
x + 1\overline{)x^2 + 0x - 4} & \textit{Rename } x^2 - 4 \textit{ as } x^2 + 0x - 4. \\
(-)\ x^2 + 1x & \textit{Multiply } x \textit{ and } x + 1. \\
-1x - 4 & \textit{Subtract. Then bring down } -4. \\
-1x - 1 & \textit{Multiply } -1 \textit{ and } x + 1. \\
-3 & \textit{Subtract. The remainder is } -3.
\end{array}$$

Therefore, $(x^2 - 4) \div (x + 1) = x - 1 + \frac{-3}{x + 1}$.

Your Turn

d. Find $(a^3 + 8a - 20) \div (a - 2)$.

If you know the area of a rectangle and the length of one side, you can find the width by dividing polynomials.

Example 5
Geometry Link

Find the width of a rectangle if its area is $10x^2 + 29x + 21$ square units and its length is $2x + 3$ units.

$2x + 3$

$10x^2 + 29x + 21$

To find the width, divide the area $10x^2 + 29x + 21$ by the length $2x + 3$.

$$\begin{array}{r l}
5x + 7 & \\
2x + 3 \overline{)10x^2 + 29x + 21} & \\
(-)\ 10x^2 + 15x & \textit{Multiply 5x and 2x + 3.} \\
14x + 21 & \textit{Subtract. Then bring down 21.} \\
(-)\ 14x + 21 & \textit{Multiply 7 and 2x + 3.} \\
0 & \textit{The remainder is 0.}
\end{array}$$

Therefore, the width of the rectangle is $5x + 7$ units.
You can check your answer by multiplying $(2x + 3)$ and $(5x + 7)$.

Check for Understanding

Communicating Mathematics

1. **Identify** the dividend, divisor, quotient, and remainder.
   $(3k^2 - 7k - 5) \div (3k + 2) = k - 3 + \frac{1}{3k + 2}$
2. **Write** a division problem represented by the model.
3. **Explain** how you know whether the quotient is a factor of the dividend.

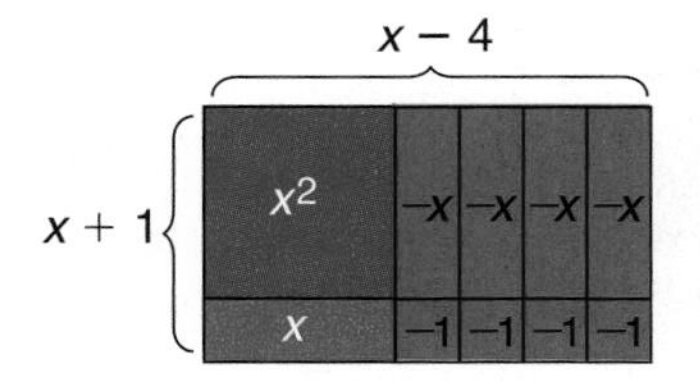

Guided Practice

Getting Ready **Find each quotient.**

Sample 1: $x^2 \div x$
Solution: x

Sample 2: $4x^2 \div 2x$
Solution: $2x$

4. $3y^2 \div y$
5. $6a^2 \div 2a$
6. $10x^3 \div 5x$

Find each quotient. *(Examples 1–4)*

7. $(x^2 + 4x) \div (x + 4)$
8. $(a^2 + 3a + 2) \div (a + 1)$
9. $(2x^2 + 3x - 3) \div (2x - 1)$
10. $(s^3 + 9) \div (s - 3)$

11. **Geometry** Find the length of a rectangle if its area is $2x^2 - 5x - 12$ square inches and its width is $x - 4$ inches. *(Example 5)*

$x - 4$

$2x^2 - 5x - 12$

Exercises

Practice

Find each quotient.

Homework Help	
For Exercises	See Examples
12–15	1
16–29, 32, 33	2–4
30	5
Extra Practice	
See page 722.	

12. $(12y - 4) \div (3y - 1)$
13. $(x^2 - 3x) \div (x - 3)$
14. $(8a^2 + 6a) \div (4a + 3)$
15. $(6r^3 - 15r^2) \div (2r - 5)$
16. $(a^2 + 6a + 5) \div (a + 5)$
17. $(x^2 + x - 12) \div (x - 3)$
18. $(s^2 + 11s + 18) \div (s + 2)$
19. $(a^2 - 2a - 35) \div (a - 7)$
20. $(c^2 + 12c + 36) \div (c + 9)$
21. $(3t^2 - 10t - 24) \div (3t - 4)$
22. $(2m^2 + 7m + 3) \div (m + 2)$
23. $(2b^2 + 3b - 6) \div (2b - 1)$
24. $(a^3 + 8a - 21) \div (a - 2)$
25. $(x^3 + 27) \div (x + 3)$
26. $(x^3 - 8) \div (x - 2)$
27. $(4x^4 - 2x^2 + x + 1) \div (x - 1)$

28. Find the quotient when $x^2 + 9x + 20$ is divided by $x + 4$.

29. What is the quotient when $2x^2 - 9x + 9$ is divided by $2x - 3$?

Applications and Problem Solving

30. Geometry The volume of a rectangular prism is $x^3 + 6x^2 + 8x$ cubic feet. If the height of the prism is $x + 4$ feet and the length is $x + 2$ feet, find the width of the prism.

31. Transportation The distance from San Francisco, CA, to New York, NY, is 2807 miles. To the nearest hour, find the number of hours it would take to travel this distance using each method of transportation at the given average speed. Use the formula $d = rt$.

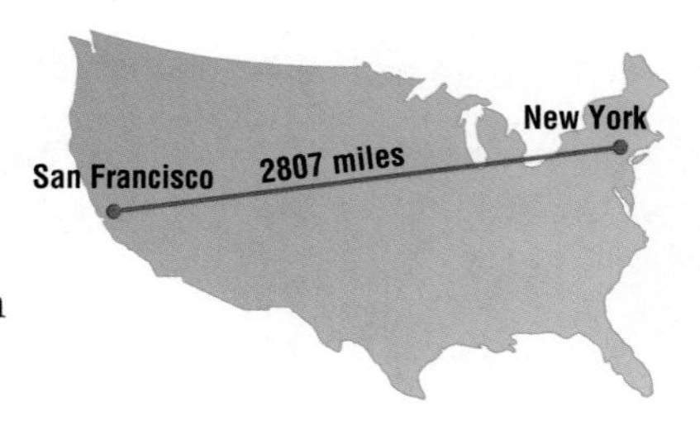

a. walking, 3 mph
b. stagecoach, 20 mph
c. automobile, 55 mph
d. jumbo jet, 608 mph

32. Aviation The distance d in miles flown by an airplane is given by the polynomial $x^2 + 501x + 500$. Suppose that $x + 500$ represents the speed r in miles per hour of the airplane. Use the formula $d = rt$ to find the following.

a. Find the polynomial that represents the time t in hours.

b. If $x = 5$, find the distance, rate, and time for the airplane.

33. Critical Thinking Find the value of k if the remainder is 15 when $x^3 - 7x^2 + 4x + k$ is divided by $x - 2$.

Mixed Review

Find each product or quotient. *(Lesson 15–2)*

34. $\frac{3x + 9}{x} \cdot \frac{x^2}{x^2 - 9}$

35. $\frac{a^2}{b^2} \div \frac{a^2}{b^2}$

36. Sports When a professional football team from the west coast plays a professional football team from the east coast on television on Monday nights, the west coast team wins 64% of the time. Express 64% as a fraction in simplest form. **Source:** Stanford University Sleep Disorders Clinic *(Lesson 15–1)*

Determine whether each system of equations has *one* solution, *no* solution, or *infinitely many* solutions. *(Lesson 13–2)*

37. 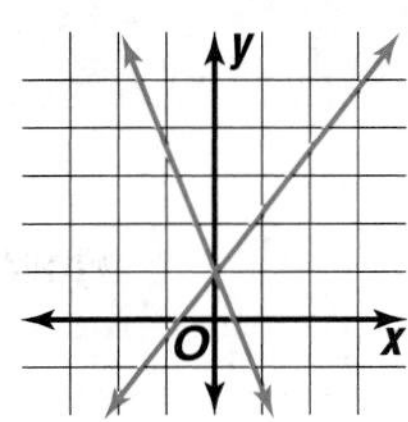

38. 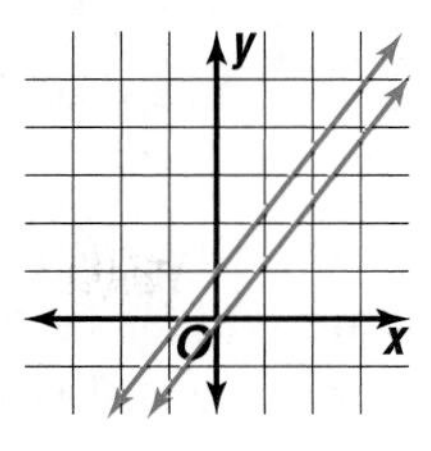

39. 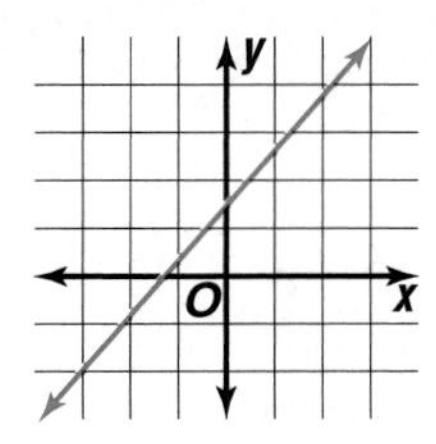

Solve each inequality. *(Lesson 12–4)*

40. $3x - 1 > 14$

41. $-7y + 6 \leq 48$

Standardized Test Practice

42. **Multiple Choice** Which is the graph of $y = x^2 + 2$? *(Lesson 11–1)*

A 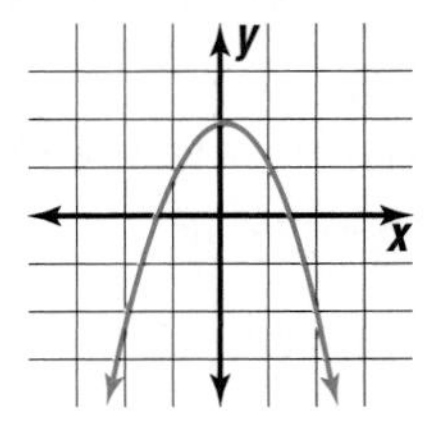

B 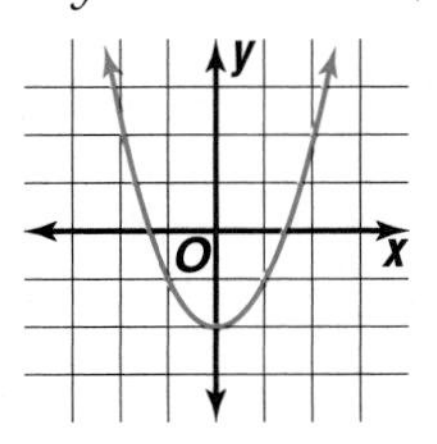

C 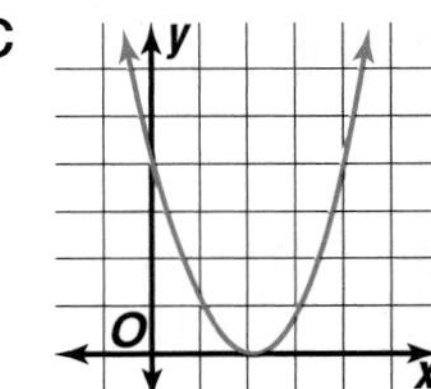

D 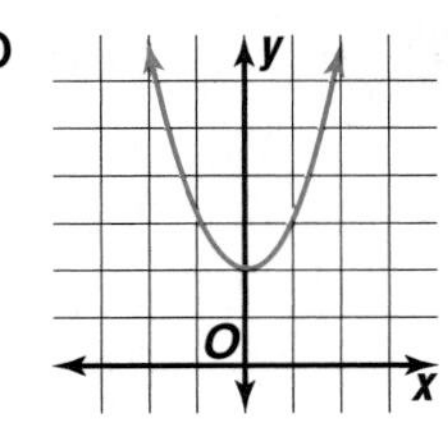

Quiz 1 Lessons 15–1 through 15–3

Find the excluded value(s) for each rational expression. Then simplify. *(Lesson 15–1)*

1. $\dfrac{m^2 - 2m}{m - 2}$

2. $\dfrac{x^2 - 4x + 4}{x^2 + 4x - 12}$

Find each product or quotient. *(Lesson 15–2)*

3. $\dfrac{y + 3}{2y + 8} \cdot \dfrac{y^2 + 5y + 4}{y^2 + 4y + 3}$

4. $\dfrac{a^2b^2}{6a + 18} \div \dfrac{ab}{a^2 - 9}$

5. $\dfrac{6 - 2a}{a^2 + a - 2} \cdot \dfrac{a^2 + 3a + 2}{a - 3}$

6. $\dfrac{x^2}{x^2 - 16} \div \dfrac{x}{x^2 - 16}$

Find each quotient. *(Lesson 15–3)*

7. $(y^2 - 5y) \div (y - 5)$

8. $(n^2 + 2n - 10) \div (n - 2)$

9. $(x^3 - 1) \div (x - 1)$

10. **Fast Food** About 90% of Americans eat fast food in a given month. Of these, 60% use the drive-through window at least once. What percent of Americans use a fast-food drive-through at least once a month?
Source: Maritz Marketing Research Inc. *(Lesson 15–2)*

15–4 Combining Rational Expressions with Like Denominators

What You'll Learn
You'll learn to add and subtract rational expressions with like denominators.

Why It's Important
Entertainment
Rational expressions can be combined to determine concert profits.
See Exercise 35.

Rational expressions with like denominators are added or subtracted in the same way as number fractions with like denominators. In the following example, two steps are used to subtract $\frac{1}{10}$ from $\frac{4}{10}$.

Step 1 Add or subtract the numerators. $\frac{4}{10} - \frac{1}{10} = \frac{4-1}{10}$

Step 2 Write the sum or difference over the common denominator. $= \frac{3}{10}$

You can use this same method to add or subtract rational expressions with like denominators. Recall that in rational expressions, both the numerator and denominator can have variables.

Examples

Find each sum or difference.

1 $\frac{2}{a} + \frac{4}{a}$

$\frac{2}{a} + \frac{4}{a} = \frac{2+4}{a}$ *The common denominator is a. Add the numerators.*

$= \frac{6}{a}$

2 $\frac{5x}{11} - \frac{2x}{11}$

$\frac{5x}{11} - \frac{2x}{11} = \frac{5x-2x}{11}$ *The common denominator is 11. Subtract the numerators.*

$= \frac{3x}{11}$

3 $\frac{b}{b+3} - \frac{3}{b+3}$

$\frac{b}{b+3} - \frac{3}{b+3} = \frac{b-3}{b+3}$ *The common denominator is b + 3.*

$= \frac{b-3}{b+3}$

Your Turn

a. $\frac{4m}{7} + \frac{m}{7}$

b. $\frac{7}{x} - \frac{3}{x}$

c. $\frac{n}{n-2} - \frac{1}{n-2}$

When adding or subtracting fractions, the result is not always in simplest form. To find the simplest form, divide the numerator and the denominator by the greatest common factor (GCF).

$$\frac{3}{8} + \frac{1}{8} = \frac{4}{8}$$

$$= \frac{\overset{1}{\cancel{4}}}{\underset{2}{\cancel{8}}}$$ *The GCF of 4 and 8 is 4.*

$$= \frac{1}{2}$$

$$\frac{9}{16} - \frac{3}{16} = \frac{6}{16}$$

$$= \frac{\overset{3}{\cancel{6}}}{\underset{8}{\cancel{16}}}$$ *The GCF of 6 and 16 is 2.*

$$= \frac{3}{8}$$

You can use similar steps to combine rational expressions.

Examples

Find each sum or difference. Write in simplest form.

4 $\frac{7}{10m} + \frac{1}{10m}$

$$\frac{7}{10m} + \frac{1}{10m} = \frac{7+1}{10m}$$ *The common denominator is 10m. Add the numerators.*

$$= \frac{8}{10m}$$

$$= \frac{\overset{4}{\cancel{8}}}{\underset{5}{\cancel{10}}m}$$ *Divide by the GCF, 2.*

$$= \frac{4}{5m}$$

5 $\frac{3}{3t} - \frac{9}{3t}$

$$\frac{3}{3t} - \frac{9}{3t} = \frac{3-9}{3t}$$ *The common denominator is 3t. Subtract the numerators.*

$$= \frac{-6}{3t}$$

$$= \frac{\overset{-2}{\cancel{-6}}}{\underset{1}{\cancel{3}}t}$$ *Divide by the GCF, 3.*

$$= \frac{-2}{t} \text{ or } -\frac{2}{t}$$

Your Turn

d. $\frac{a}{12} + \frac{2a}{12}$

e. $\frac{8}{9y} - \frac{5}{9y}$

Sometimes, the denominators of rational expressions are binomials.

Examples

Find each sum or difference. Write in simplest form.

6 $\frac{3}{x+2} + \frac{1}{x+2}$

$\frac{3}{x+2} + \frac{1}{x+2} = \frac{3+1}{x+2}$ *The common denominator is $x + 2$. Add the numerators.*

$= \frac{4}{x+2}$

7 $\frac{y+2}{y-1} - \frac{8}{y-1}$

$\frac{y+2}{y-1} - \frac{8}{y-1} = \frac{y+2-8}{y-1}$ *The common denominator is $y - 1$. Subtract the numerators.*

$= \frac{y-6}{y-1}$

Your Turn

f. $\frac{4m}{2m+3} + \frac{5}{2m+3}$

g. $\frac{5x}{x+2} - \frac{2x}{x+2}$

Sometimes you must factor in order to simplify the sum or difference of rational expressions.

Examples

Find each sum or difference. Write in simplest form.

8 $\frac{b}{b+4} + \frac{b+8}{b+4}$

$\frac{b}{b+4} + \frac{b+8}{b+4} = \frac{b+(b+8)}{b+4}$ *The common denominator is $b + 4$. Add the numerators.*

$= \frac{2b+8}{b+4}$

$= \frac{2(b+4)}{b+4}$ *Factor the numerator.*

$= \frac{2\cancel{(b+4)}^{1}}{\cancel{b+4}_{1}}$ *Divide by the GCF, $b + 4$.*

$= 2$

9 $\frac{15x}{4x-1} - \frac{3x+3}{4x-1}$

$\frac{15x}{4x-1} - \frac{3x+3}{4x-1} = \frac{15x-(3x+3)}{4x-1}$ *The common denominator is $4x - 1$. Subtract the numerators.*

$= \frac{15x-3x-3}{4x-1}$ *Distributive Property*

$= \frac{12x-3}{4x-1}$

$$= \frac{3(4x - 1)}{4x - 1}$$ *Factor the numerator.*

$$= \frac{3\cancel{(4x - 1)}^{1}}{\cancel{4x - 1}_{1}}$$ *Divide by the GCF, 4x − 1.*

$$= 3$$

Your Turn **Find each sum or difference. Write in simplest form.**

h. $\frac{3r}{r + 5} + \frac{15}{r + 5}$ **i.** $\frac{m}{m + 3} - \frac{3m + 6}{m + 3}$

You can solve some geometry problems by adding rational expressions.

Example 10
Geometry Link

Find the perimeter of rectangle *ABCD*.

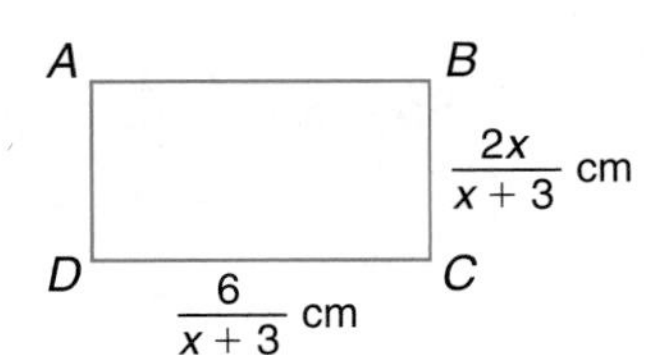

$$P = 2\ell + 2w$$

$$= 2\left(\frac{6}{x + 3}\right) + 2\left(\frac{2x}{x + 3}\right)$$ *Replace ℓ with $\frac{6}{x + 3}$ and w with $\frac{2x}{x + 3}$.*

$$= \frac{2(6)}{x + 3} + \frac{2(2x)}{x + 3}$$

$$= \frac{12}{x + 3} + \frac{4x}{x + 3}$$

$$= \frac{12 + 4x}{x + 3}$$ *Add the numerators.*

$$= \frac{4(3 + x)}{x + 3}$$ *Factor the numerator.*

$$= \frac{4\cancel{(x + 3)}^{1}}{\cancel{x + 3}_{1}}$$ *Divide by the GCF, x + 3.*

$$= 4$$

The perimeter of rectangle *ABCD* is 4 centimeters.

Check for Understanding

Communicating Mathematics

1. **Explain** how the sum of two fractions with the same denominator can have a sum of zero.
2. **Identify** the mistake in the solution to the problem below. Then write the correct sum.

$$\frac{4x}{2x + 1} + \frac{2}{2x + 1} = \frac{4x + 2}{4x + 2} \text{ or } 1$$

3. **Complete** the table.

a	b	$a + b$	$a - b$	$a \cdot b$	$a \div b$
$\frac{2}{t}$	$\frac{1}{t}$	$\frac{3}{t}$	$\frac{1}{t}$	$\frac{2}{t^2}$	2
$\frac{12}{y}$	$\frac{4}{y}$				
$\frac{x}{x-1}$	$\frac{1}{x-1}$				

Guided Practice

Getting Ready **Find each sum or difference. Write in simplest form.**

Sample 1: $\frac{4}{15} + \frac{3}{15}$

Solution: $\frac{4}{15} + \frac{3}{15} = \frac{4+3}{15}$

$= \frac{7}{15}$

Sample 2: $\frac{8}{9} - \frac{2}{9}$

Solution: $\frac{8}{9} - \frac{2}{9} = \frac{8-2}{9}$

$= \frac{6}{9}$ or $\frac{2}{3}$

4. $\frac{5}{7} + \frac{1}{7}$
5. $\frac{4}{12} + \frac{5}{12}$
6. $\frac{6}{10} - \frac{1}{10}$

Find each sum or difference. Write in simplest form. *(Examples 1–9)*

7. $\frac{5}{x} + \frac{2}{x}$
8. $\frac{2t}{3} - \frac{t}{3}$
9. $\frac{7y}{y} - \frac{8y}{y}$
10. $\frac{a}{12} + \frac{2a}{12}$
11. $\frac{2x}{x-3} - \frac{6}{x-3}$
12. $\frac{2c+3}{c-4} - \frac{c-2}{c-4}$

13. **Geometry** Find the perimeter of the rectangle. *(Example 10)*

$\frac{2y}{7x-2y}$ in.

$\frac{3y}{7x-2y}$ in.

Exercises

Practice

Homework Help	
For Exercises	See Examples
14–25, 35	1–5
26–33	6–9
34	10

Extra Practice

See page 723.

Find each sum or difference. Write in simplest form.

14. $\frac{7}{m} + \frac{4}{m}$
15. $\frac{x}{8} + \frac{6x}{8}$
16. $\frac{7}{15z} - \frac{3}{15z}$
17. $\frac{5t}{2} - \frac{4t}{2}$
18. $\frac{8p}{13} - \frac{3p}{13}$
19. $\frac{1}{6r} + \frac{6}{6r}$
20. $\frac{a}{2} + \frac{a}{2}$
21. $\frac{n}{3} + \frac{2n}{3}$
22. $\frac{5}{3y} - \frac{2}{3y}$
23. $\frac{5x}{24} - \frac{3x}{24}$
24. $\frac{8}{27m} + \frac{4}{27m}$
25. $\frac{5}{2z} + \frac{-7}{2z}$

26. $\frac{8}{y-2} - \frac{6}{y-2}$

27. $\frac{y}{a+1} - \frac{y}{a+1}$

28. $\frac{3}{x+2} - \frac{2}{x+2}$

29. $\frac{2x}{x+3} + \frac{6}{x+3}$

30. $\frac{8n+3}{3n+4} - \frac{2n-5}{3n+4}$

31. $\frac{10m-1}{4m-3} - \frac{8-2m}{4m-3}$

32. What is $\frac{3c-7}{2c-3}$ minus $\frac{c-4}{2c-3}$?

33. Find the sum of $\frac{x+y}{y-2}$ and $\frac{y-x}{y-2}$.

Applications and Problem Solving

34. **Geometry** Find the perimeter of the isosceles triangle.

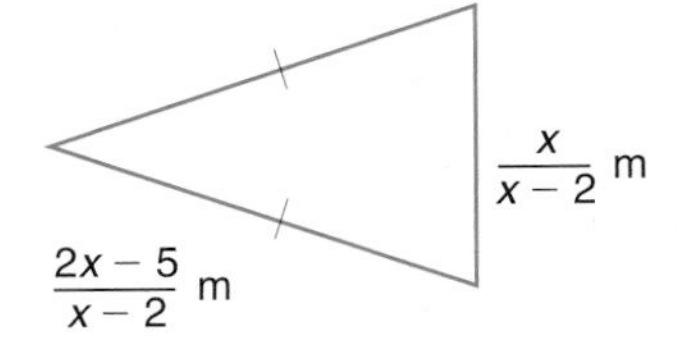

35. **Entertainment** A musical group receives $15,000 for each concert. The five members receive equal shares of the money. They must pay 0.3 of their earnings for other salaries and expenses.
    a. Write a rational expression that represents how much money each member of the music group receives for a concert after expenses.
    b. Simplify the expression you wrote in part a to find how much each member of the group receives for a concert.

36. **Critical Thinking** Which rational expression is not equivalent to the others?

    a. $\frac{4}{x-5}$ b. $-\frac{4}{5-x}$ c. $\frac{-4}{x-5}$ d. $\frac{-4}{5-x}$

Mixed Review

Find each quotient. *(Lesson 15–3)*

37. $(a^2 + 9a + 20) \div (a + 5)$

38. $(x^2 - 2x - 35) \div (x - 7)$

39. $(2y^2 - 3y - 35) \div (2y + 7)$

40. $(t^2 + 12t + 36) \div (t + 9)$

41. Find the product of $\frac{5x^2y}{8ab}$ and $\frac{12a^2b}{25x}$. *(Lesson 15–2)*

42. Solve the system $y = x^2 - 2$ and $y = -x$ by graphing. *(Lesson 13–6)*

Standardized Test Practice

43. **Short Response** Find $(3x + y)^2$. *(Lesson 9–5)*

44. **Multiple Choice** About 25% of all identical twins are "mirror" twins—a reflection of each other. In a group of identical twins, 35 pairs were found to be "mirror" twins. About how many pairs of identical twins would you expect to be in the group? *(Lesson 5–4)*

    A 9 C 70
    B 140 D 100

15–5 Combining Rational Expressions with Unlike Denominators

What You'll Learn
You'll learn to add and subtract rational expressions with unlike denominators.

Why It's Important
Teaching Teachers can use LCM when dividing their classes into groups.
See Example 3.

The **least common multiple (LCM)** is the least number that is a common multiple of two or more numbers. Suppose you want to find the LCM of 4, 6, and 9. Write multiples of each number until you find a common multiple.

multiples of 4: 0, 4, 8, 12, 16, 20, 24, 28, 32, **36**, . . .
multiples of 6: 0, 6, 12, 18, 24, 30, **36**, . . .
multiples of 9: 0, 9, 18, 27, **36**, . . .

Zero is a multiple of every number, but it cannot be the LCM. The LCM of 4, 6, and 9 is 36.

You can also use prime factorization to find the LCM.

$$4 = 2 \cdot 2 \qquad 6 = 2 \cdot 3 \qquad 9 = 3 \cdot 3$$

The prime factors are 2 and 3. The greatest number of times 2 appears is twice (in 4). The greatest number of times 3 appears is twice (in 9). So, the LCM of 4, 6, and 9 is $2 \cdot 2 \cdot 3 \cdot 3$ or 36. This is the same answer you got by listing the multiples.

Examples

Find the LCM for each pair of expressions.

1 $\mathbf{15a^2b, 18a^3}$

$15a^2b = 3 \cdot 5 \cdot a \cdot a \cdot b$ *Factor each expression.*
$18a^3 = 2 \cdot 3 \cdot 3 \cdot a \cdot a \cdot a$

$\text{LCM} = 2 \cdot 3 \cdot 3 \cdot 5 \cdot a \cdot a \cdot a \cdot b$ *Use each factor the greatest number of times it appears in either factorization.*
$= 90a^3b$

2 $\mathbf{x^2 + x - 6, 2x^2 - 3x - 2}$

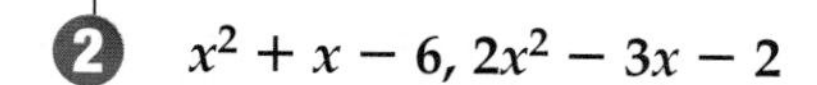

$x^2 + x - 6 = (x + 3)(x - 2)$ *Factor each expression.*
$2x^2 - 3x - 2 = (2x + 1)(x - 2)$

$\text{LCM} = (x + 3)(x - 2)(2x + 1)$ *Use each factor the greatest number of times it appears in either factorization.*

Your Turn

a. $12a, 8ab$

b. $x^2 - 9, x^2 - 2x - 3$

Real World

Example
Teaching Link

3 **Mrs. Reimer wants student groups of 4, 6, or 8. What is the minimum number of desks needed?**

Find the LCM of 4, 6, and 8.

$4 = 2 \cdot 2$

$6 = 2 \cdot 3$

$8 = 2 \cdot 2 \cdot 2$

$\text{LCM} = 2 \cdot 2 \cdot 2 \cdot 3$ or 24 *Use each factor the greatest number of times it appears in any of the factorizations.*

Mrs. Reimer must have at least 24 desks in her classroom.

To add or subtract fractions with unlike denominators, first rename the fractions so the denominators are alike. Any common denominator could be used. However, the computation is usually easier if you use the **least common denominator (LCD)**. Recall that the least common denominator is the LCM of the denominators.

Examples

Write each pair of rational expressions with the same LCD.

4 $\frac{9}{2m^2}, \frac{5}{6m}$

First find the LCD.

$2m^2 = 2 \cdot m \cdot m$

$6m = 2 \cdot 3 \cdot m$

$\text{LCD} = 2 \cdot 3 \cdot m \cdot m$ or $6m^2$

Then write each fraction with the same LCD.

$\frac{9}{2m^2} \cdot \frac{3}{3} = \frac{27}{6m^2}$ $\frac{5}{6m} \cdot \frac{m}{m} = \frac{5m}{6m^2}$

$\frac{1}{x+2}, \frac{3x}{4x+8}$

First find the LCD.

$x + 2 = x + 2$

$4x + 8 = 4(x + 2)$

$\text{LCD} = 4(x + 2)$

Then write each fraction with the same LCD.

$\frac{1}{x+2} \cdot \frac{4}{4} = \frac{4}{4(x+2)}$ $\frac{3x}{4x+8} = \frac{3x}{4(x+2)}$

Your Turn

c. $\frac{6}{b^3}, \frac{7}{ab}$

d. $\frac{x}{x-6}, \frac{x-3}{x-2}$

Use the following steps to add or subtract rational expressions with unlike denominators.

Step 1 Find the LCD.

Step 2 Change each rational expression into an equivalent expression using the LCD.

Step 3 Add or subtract as with rational expressions with like denominators.

Step 4 Simplify if necessary.

Examples

Find each sum or difference. Write in simplest form.

6 $\frac{5}{6x} + \frac{7}{12x^2}$

Step 1 Find the LCD.

$$6x = 2 \cdot 3$$
$$12x^2 = 2 \cdot 2 \cdot 3 \cdot x \cdot x$$
$$\text{LCM} = 2 \cdot 2 \cdot 3 \cdot x \cdot x \text{ or } 12x^2$$
$$\text{LCD} = 12x^2$$

Step 2 Rename each expression with the LCD as the denominator. The denominator of $\frac{7}{12x^2}$ is already $12x^2$, so only $\frac{5}{6x}$ needs to be renamed.

$$\frac{5}{6x} \cdot \frac{2x}{2x} = \frac{10x}{12x^2}$$ *Why do you multiply $\frac{5}{6x}$ by $\frac{2x}{2x}$?*

Step 3 Add.

$$\frac{5}{6x} + \frac{7}{12x^2} = \frac{10x}{12x^2} + \frac{7}{12x^2}$$

$$= \frac{10x + 7}{12x^2}$$ *This expression is in simplest form.*

7 $\frac{m}{m^2 - 9} - \frac{3}{m - 3}$

$$m^2 - 9 = (m - 3)(m + 3)$$
$$m - 3 = m - 3$$
$$\text{LCM} = (m - 3)(m + 3)$$
$$\text{LCD} = (m - 3)(m + 3)$$

$$\frac{m}{m^2 - 9} - \frac{3}{m - 3}$$

$$= \frac{m}{(m - 3)(m + 3)} - \frac{3}{m - 3} \cdot \frac{m + 3}{m + 3}$$ *Multiply by $\frac{3}{m - 3}$ by $\frac{m + 3}{m + 3}$.*

$$= \frac{m}{(m - 3)(m + 3)} - \frac{3m + 9}{(m - 3)(m + 3)}$$ *The LCD is $(m - 3)(m + 3)$.*

$$= \frac{m - (3m + 9)}{(m - 3)(m + 3)}$$ *Subtract the numerators.*

$$= \frac{m - 3m - 9}{(m - 3)(m + 3)}$$ *Distributive Property*

$$= \frac{-2m - 9}{(m - 3)(m + 3)}$$ *Simplify.*

8 $\dfrac{5}{x-3} + \dfrac{7}{2x-6}$

$$\begin{aligned}\frac{5}{x-3} + \frac{7}{2x-6} &= \frac{5}{x-3} + \frac{7}{2(x-3)} && \textit{The LCD is } 2(x-3).\\ &= \frac{5}{x-3}\cdot\frac{2}{2} + \frac{7}{2(x-3)}\\ &= \frac{10}{2(x-3)} + \frac{7}{2(x-3)}\\ &= \frac{10+7}{2(x-3)} && \textit{Add the numerators.}\\ &= \frac{17}{2(x-3)}\end{aligned}$$

Your Turn **Find each sum or difference. Write in simplest form.**

e. $\dfrac{3}{xy} + \dfrac{2}{y^2}$

f. $\dfrac{5a}{2a+6} - \dfrac{a}{a+3}$

Check for Understanding

Communicating Mathematics

Vocabulary
least common multiple
least common denominator

1. **Write** two rational expressions with unlike denominators in which one of the denominators is the LCD of the rational expressions.
2. **Explain** the steps you would take to find the LCD for the expression $\dfrac{3}{8x^2} - \dfrac{5}{12xy} + \dfrac{7}{6y^2}$. Then simplify the expression.
3. **You Decide** Ashley found the sum of $\dfrac{4}{x-1}$ and $\dfrac{5}{x}$ to be $\dfrac{9}{2x-1}$. Malik found the sum to be $\dfrac{9x-5}{x^2-x}$. Who is correct? Explain why.

Guided Practice

Getting Ready **Find the LCM for each pair of numbers.**

Sample: 10, 12 **Solution:** $10 = 2 \cdot 5$ $12 = 2 \cdot 2 \cdot 3$
LCM $= 2 \cdot 2 \cdot 3 \cdot 5$ or 60

4. 6, 10
5. 8, 18
6. 12, 15

Find the LCM for each pair of expressions. *(Examples 1 & 2)*

7. $6xy$, $15y^2$
8. $x + 1$, $x^2 + 5x + 4$

Write each pair of rational expressions with the same LCD. *(Examples 4 & 5)*

9. $\dfrac{4}{a^2}$, $\dfrac{5}{a}$
10. $\dfrac{7}{t+3}$, $\dfrac{13}{2t+6}$

Find each sum or difference. Write in simplest form. *(Examples 6–8)*

11. $\frac{t}{6} + \frac{3t}{12}$
12. $\frac{2}{3a} - \frac{4}{9a}$
13. $\frac{2}{ab^2} + \frac{3}{ab}$
14. $\frac{6}{m} + \frac{5}{n}$
15. $\frac{x}{x^2 - 4} - \frac{4}{x + 2}$
16. $\frac{2}{3a - 1} - \frac{7}{15a - 5}$

17. **Astronomy** Earth, Jupiter, and Saturn revolve around the Sun about once every 1, 12, and 30 years, respectively. The last time Earth, Jupiter, and Saturn were lined up with the sun was in 1982, and Jupiter and Saturn could be seen close together, high in the sky at midnight. In what year will this happen again? *(Example 3)*

Exercises

Practice

Homework Help	
For Exercises	See Examples
18–23	1, 2
52–54	3
24–29	4, 5
30–51	6–8
Extra Practice	
See page 723.	

Find the LCM for each pair of expressions.

18. $3t, 9t^2$
19. $12a^2, 2ab^2$
20. $16m, 6mn$
21. $y + 2, y^2 - 4$
22. $x^2 + 9x + 14, x^2 + 3x + 2$
23. $x^2 - 9, 3x^2 - 8x - 3$

Write each pair of rational expressions with the same LCD.

24. $\frac{4}{m^3}, \frac{1}{m}$
25. $\frac{7}{2t}, \frac{8}{10t}$
26. $\frac{3}{6ab}, \frac{14}{4a^2}$
27. $\frac{5}{xy}, \frac{6}{yz}$
28. $\frac{x}{x^2 - 1}, \frac{-2}{x - 1}$
29. $\frac{10k}{3k + 1}, \frac{k}{3 + 9k}$

Find each sum or difference. Write in simplest form.

30. $\frac{a}{5} - \frac{a}{15}$
31. $\frac{d}{2} - \frac{d}{5}$
32. $\frac{t}{3} + \frac{2t}{7}$
33. $\frac{9}{4x} + \frac{3}{2x}$
34. $\frac{5}{2a} - \frac{3}{6a}$
35. $\frac{m}{3t} + \frac{1}{t}$
36. $\frac{4}{a^3} - \frac{2}{a}$
37. $\frac{7}{3x} - \frac{1}{6x^2}$
38. $\frac{2t}{5n^2} - \frac{1}{3n}$
39. $\frac{2}{x} + \frac{x + 3}{y}$
40. $\frac{6z}{7w} + \frac{2z}{w^3}$
41. $\frac{1}{4xy} + \frac{y}{10x^2}$
42. $\frac{9}{y + 1} - \frac{3}{4y + 4}$
43. $\frac{4a}{2a + 6} + \frac{3}{a + 3}$
44. $\frac{7}{x - 2} + \frac{3}{x}$
45. $\frac{y}{y^2 - 2y + 1} - \frac{1}{y - 1}$
46. $\frac{3a}{a^2 + 4a + 4} - \frac{2}{a + 2}$
47. $\frac{5x + 2}{2x - 1} + \frac{2x - 3}{8x - 4}$
48. $\frac{7}{a^2 + 2a - 3} - \frac{5}{a + 5a + 6}$
49. $\frac{x^2 + 4x - 5}{x^2 - 2x - 3} + \frac{2}{x + 1}$
50. Add $\frac{2x + 3}{x^2 - 4}$ and $\frac{6}{x + 2}$.
51. Simplify $\frac{m}{m - n} - \frac{5}{m}$.

Applications and Problem Solving

52. **Entertainment** The choreographer of a musical needs enough dancers so that they can be arranged in groups of 6, 9, and 12, with no one sitting out. What is the least number of dancers needed?

53. **Fitness** Cynthia jogs around an oval track in 150 seconds. Minya walks around the track in 210 seconds. Suppose they start at the same time. After how many minutes will they meet back at their starting point?

54. **Critical Thinking** Consider the numbers 18 and 20.
   a. Find the GCF and LCM of the numbers.
   b. What is the value of GCF · LCM?
   c. Describe the relationship between 18 · 20 and GCF · LCM.
   d. Describe how you could find the GCF of two numbers if you already know the LCM.

Mixed Review

Find each sum or difference. Write in simplest form. *(Lesson 15–4)*

55. $\frac{9}{a} + \frac{7}{a}$ 56. $\frac{2}{5t} + \frac{3}{5t}$ 57. $\frac{4n}{n+1} - \frac{2n-2}{n+1}$

58. Find $(m^2 + 6m - 7) \div (m + 7)$. *(Lesson 15–3)*

Simplify each expression. *(Lesson 14–4)*

59. $7\sqrt{19} + 2\sqrt{19}$ 60. $6\sqrt{12} + 4\sqrt{3}$ 61. $8\sqrt{5} - 3\sqrt{2}$

62. **Geometry** The length of the hypotenuse of a right triangle is $\sqrt{\frac{56}{25}}$ centimeters. Express the length as a radical in simplest form. *(Lesson 14–3)*

Standardized Test Practice

63. **Short Response** What is the solution of the system $y = 7 - x$ and $x - y = -3$? *(Lesson 13–3)*

64. **Multiple Choice** Determine which polynomial can be the measure of the area of a square. *(Lesson 10–5)*

   A $x^2 - 13x + 36$ B $n^2 + 20n - 100$
   C $4a^2 + 12a + 9$ D $4a^2 - 6a + 9$

Quiz 2 Lessons 15–3 through 15–5

Find each quotient. *(Lesson 15–3)*

1. $(x^2 + 6x - 16) \div (x - 2)$ 2. $(2x^2 - 11x - 20) \div (2x + 3)$

Find each sum or difference. Write in simplest form. *(Lesson 15–4)*

3. $\frac{a}{9} + \frac{7a}{9}$ 4. $\frac{5}{21t} + \frac{4}{21t}$

5. **Parades** At the Veteran's Day parade, members of the Veterans of Foreign Wars (VFW) found that they could arrange themselves in rows of 6, 7, or 8, with no one left over. What is the least number of VFW members in the parade? *(Lesson 15–5)*

15–6 Solving Rational Equations

What You'll Learn
You'll learn to solve rational equations.

Why It's Important
Aviation Pilots can use rational equations to find wind speed. *See Exercise 9.*

Every week there is a 10-point quiz in math class. For the first five quizzes, Julia scored a total of 36 points, which gave her an average of 7.2. She is determined to get 10 points on each of the next quizzes until she brings her average up to an 8. On how many quizzes must she score 10 points in order to have an overall quiz average of 8 points?

Let x represent the number of quizzes on which she must score 10 points.

total number of quizzes $= 5 + x$

sum of scores $= 36 + 10x$

$$\text{average} = \frac{36 + 10x}{5 + x} \quad \begin{matrix} \leftarrow \textit{sum of scores} \\ \leftarrow \textit{number of quizzes} \end{matrix}$$

$$8 = \frac{36 + 10x}{5 + x} \quad \textit{Julia wants to have an average of 8.}$$

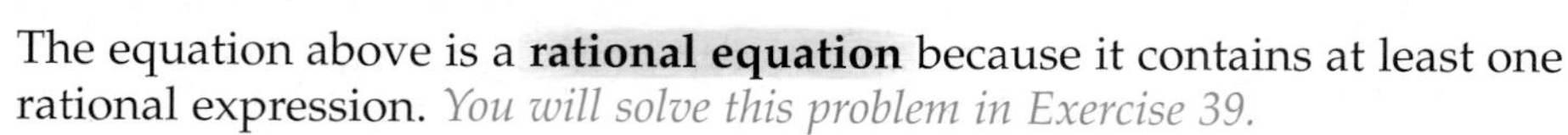

The equation above is a **rational equation** because it contains at least one rational expression. *You will solve this problem in Exercise 39.*

There are three steps in solving rational equations.

Step 1 Find the LCD of each term.
Step 2 Multiply each side of the equation by the LCD.
Step 3 Use the Distributive Property to simplify.

Examples

Solve each equation. Check your solution.

1 $\dfrac{3a}{5} + \dfrac{3}{2} = \dfrac{7a}{10}$

$$\frac{3a}{5} + \frac{3}{2} = \frac{7a}{10} \quad \textit{The LCD is 10.}$$

$$10\left(\frac{3a}{5} + \frac{3}{2}\right) = 10\left(\frac{7a}{10}\right) \quad \textit{Multiply each side by the LCD.}$$

$$10\left(\frac{3a}{5}\right) + 10\left(\frac{3}{2}\right) = 10\left(\frac{7a}{10}\right) \quad \textit{Distributive Property}$$

$$\overset{2}{\cancel{10}}\left(\frac{3a}{\underset{1}{\cancel{5}}}\right) + \overset{5}{\cancel{10}}\left(\frac{3}{\underset{1}{\cancel{2}}}\right) = \overset{1}{\cancel{10}}\left(\frac{7a}{\underset{1}{\cancel{10}}}\right)$$

$$6a + 15 = 7a$$

$$6a + 15 - 6a = 7a - 6a \quad \textit{Subtract } 6a \textit{ from each side.}$$

$$15 = a$$

Check: $\frac{3a}{5} + \frac{3}{2} = \frac{7a}{10}$

$\frac{3(15)}{5} + \frac{3}{2} \stackrel{?}{=} \frac{7(15)}{10}$ *Replace a with 15.*

$9 + \frac{3}{2} \stackrel{?}{=} \frac{21}{2}$

$\frac{21}{2} = \frac{21}{2}$ ✓

2 $\mathbf{\frac{2}{3x} - \frac{1}{2x} = \frac{1}{6}}$

$\frac{2}{3x} - \frac{1}{2x} = \frac{1}{6}$ *The LCD is 6x.*

$6x\left(\frac{2}{3x} - \frac{1}{2x}\right) = 6x\left(\frac{1}{6}\right)$ *Multiply each side by the LCD.*

$6x\left(\frac{2}{3x}\right) - 6x\left(\frac{1}{2x}\right) = 6x\left(\frac{1}{6}\right)$ *Distributive Property*

$\overset{2}{\cancel{6x}}\left(\frac{2}{\underset{1}{\cancel{3x}}}\right) - \overset{3}{\cancel{6x}}\left(\frac{1}{\underset{1}{\cancel{2x}}}\right) = \overset{1}{\cancel{6}}x\left(\frac{1}{\underset{1}{\cancel{6}}}\right)$

$4 - 3 = x$ *Simplify.*

$1 = x$

Check: $\frac{2}{3x} - \frac{1}{2x} = \frac{1}{6}$

$\frac{2}{3(1)} - \frac{1}{2(1)} \stackrel{?}{=} \frac{1}{6}$ *Replace x with 1.*

$\frac{2}{3} - \frac{1}{2} \stackrel{?}{=} \frac{1}{6}$ *The LCD is 6.*

$\frac{4}{6} - \frac{3}{6} \stackrel{?}{=} \frac{1}{6}$

$\frac{1}{6} = \frac{1}{6}$ ✓

3 $\mathbf{\frac{3x}{x+2} + \frac{1}{x+2} = 4}$

$\frac{3x}{x+2} + \frac{1}{x+2} = 4$ *The LCD is x + 2.*

$(x+2)\left(\frac{3x}{x+2} + \frac{1}{x+2}\right) = (x+2)4$ *Multiply each side by the LCD.*

$(x+2)\left(\frac{3x}{x+2}\right) + (x+2)\left(\frac{1}{x+2}\right) = 4x + 8$ *Distributive Property*

$\overset{1}{\cancel{(x+2)}}\left(\frac{3x}{\underset{1}{\cancel{x+2}}}\right) + \overset{1}{\cancel{(x+2)}}\left(\frac{1}{\underset{1}{\cancel{x+2}}}\right) = 4x + 8$

$3x + 1 = 4x + 8$ *Simplify.*

(continued on the next page)

$$3x + 1 - 3x = 4x + 8 - 3x \quad \textit{Subtract 3x from each side.}$$
$$1 = x + 8$$
$$1 - 8 = x + 8 - 8 \quad \textit{Subtract 8 from each side.}$$
$$-7 = x \quad \textit{Check the solution.}$$

4 $\frac{3}{r} - \frac{1}{r-1} = \frac{1}{r-1}$

$$\frac{3}{r} - \frac{1}{r-1} = \frac{1}{r-1} \quad \textit{The LCD is } r(r-1).$$
$$r(r-1)\left(\frac{3}{r} - \frac{1}{r-1}\right) = r(r-1)\left(\frac{1}{r-1}\right) \quad \textit{Multiply each side by the LCD.}$$
$$r(r-1)\left(\frac{3}{r}\right) - r(r-1)\left(\frac{1}{r-1}\right) = r(r-1)\left(\frac{1}{r-1}\right) \quad \textit{Distributive Property}$$
$$\overset{1}{\cancel{r}}(r-1)\left(\frac{3}{\underset{1}{\cancel{r}}}\right) - r\overset{1}{\cancel{(r-1)}}\left(\frac{1}{\underset{1}{\cancel{r-1}}}\right) = r\overset{1}{\cancel{(r-1)}}\left(\frac{1}{\underset{1}{\cancel{r-1}}}\right)$$
$$(r-1)3 - r = r$$
$$3r - 3 - r = r \quad \textit{Distributive Property}$$
$$2r - 3 = r$$
$$2r - 3 - r = r - r \quad \textit{Subtract r from each side.}$$
$$r - 3 = 0$$
$$r - 3 + 3 = 0 + 3 \quad \textit{Add 3 to each side.}$$
$$r = 3 \quad \textit{Check the solution.}$$

Your Turn **Solve each equation. Check your solution.**

a. $\frac{4x}{3} + \frac{7}{2} = \frac{9x}{12}$ b. $\frac{18}{b} = \frac{3}{b} + 3$ c. $\frac{4}{n+1} - \frac{2}{n} = \frac{5}{n}$

Recall that **uniform motion problems** can be solved by using the formula below.

$$\underbrace{\textit{distance}}_{d} = \underbrace{\textit{rate}}_{r} \cdot \underbrace{\textit{time}}_{t}$$

5 **The Milbys rented a houseboat on the Sacramento River. The maximum speed of the boat in still water is 8 miles per hour. At this rate, a 30-mile trip downstream took the same amount of time as an 18-mile trip against the current. What was the rate of the current?**

Explore Let c = the rate of the current.

Let $8 + c$ = the rate of the boat traveling downstream with the current. *When the boat is traveling with the current, you add the speed of the current to the speed of the boat.*

Let $8 - c$ = the rate of the boat traveling upstream against the current. *When the boat is traveling against the current, you subtract the speed of the current from the speed of the boat.*

interNET CONNECTION

Data Update For the latest information on various rivers, visit: www.algconcepts.com

Plan Since $d = rt$, then $t = \frac{d}{r}$.

	d	r	$t = \frac{d}{r}$
Downstream	30	$8 + c$	$\frac{30}{8 + c}$
Upstream	18	$8 - c$	$\frac{18}{8 - c}$

Solve

$$\frac{30}{8 + c} = \frac{18}{8 - c}$$ *The time downstream equals the time upstream.*

$$(8 + c)(8 - c)\left(\frac{30}{8 + c}\right) = (8 + c)(8 - c)\left(\frac{18}{8 - c}\right)$$ *The LCD is $(8 + c)(8 - c)$.*

$$\overset{1}{\cancel{(8 + c)}}(8 - c)\left(\frac{30}{\underset{1}{\cancel{8 + c}}}\right) = (8 + c)\overset{1}{\cancel{(8 - c)}}\left(\frac{18}{\underset{1}{\cancel{8 - c}}}\right)$$

$$(8 - c)(30) = (8 + c)(18)$$

$$240 - 30c = 144 + 18c$$

$$240 - 30c + 30c = 144 + 18c + 30c$$ *Add 30c to each side.*

$$240 = 144 + 48c$$

$$240 - 144 = 144 + 48c - 144$$ *Subtract 144 from each side.*

$$96 = 48c$$

$$\frac{96}{48} = \frac{48c}{48}$$ *Divide each side by 48.*

$$2 = c$$

The rate of the current was 2 miles per hour.

Examine Check the solution to see if it makes sense. The houseboat goes downstream at 8 + 2 or 10 miles per hour. A 30-mile trip would take 30 ÷ 10 or 3 hours. The houseboat goes upstream at 8 − 2 or 6 miles per hour. An 18-mile trip would take 18 ÷ 6 or 3 hours. Both trips take the same amount of time, so the solution is correct.

A houseboat on a river

Check for Understanding

Communicating Mathematics

Math Journal

1. **List** two differences between linear equations and rational equations.

2. **Describe** two different ways in which $\frac{4}{x + 1} = \frac{8}{x - 1}$ can be solved. Then determine whether there are other steps that you could use to solve the examples in this lesson.

Vocabulary

rational equation
uniform motion problems

Guided Practice

Solve each equation. Check your solution. *(Examples 1–4)*

3. $\frac{2x}{7} + \frac{1}{2} = \frac{x}{14}$
4. $\frac{6}{c} = \frac{10}{c} + 4$
5. $\frac{5}{m+3} - \frac{2}{m+3} = -9$
6. $\frac{a+1}{a} + \frac{a+4}{a} = 6$
7. $\frac{2x}{x+3} + \frac{3}{x} = 2$
8. $\frac{r-1}{r+1} - \frac{2r}{r-1} = -1$

9. **Transportation** An airplane can fly at a rate of 600 miles per hour in calm air. It can fly 2520 miles with the wind in the same time it can fly 2280 miles against the wind. Find the speed of the wind. *(Example 5)*

Exercises

Practice

For Exercises	See Examples
10–20, 22, 23	1, 2
21, 24–39, 41	3, 4
40	5

Homework Help

Extra Practice

See page 723.

Solve each equation. Check your solution.

10. $\frac{a}{6} + \frac{2a}{3} = -\frac{5}{2}$
11. $\frac{1}{4} + \frac{5r}{8} = \frac{r}{4}$
12. $\frac{b}{5} - \frac{2b}{15} = 1$
13. $\frac{x}{7} = \frac{x+3}{10}$
14. $\frac{t-1}{4} = \frac{t}{3}$
15. $\frac{2a-3}{6} = \frac{2a}{3} + \frac{1}{2}$
16. $\frac{4}{3y} + \frac{1}{y} = 7$
17. $\frac{6}{x} - \frac{3}{2x} = \frac{1}{2}$
18. $\frac{1}{2a} + \frac{3}{4a} = \frac{1}{4}$
19. $\frac{m+1}{m} + \frac{m+3}{m} = 5$
20. $\frac{t+1}{t} + \frac{t+4}{t} = 6$
21. $\frac{5}{5-p} - \frac{1}{5-p} = -2$
22. $\frac{5}{2x} - \frac{1}{6x} = 2$
23. $\frac{1}{4x} + \frac{1}{6x} = 5$
24. $\frac{3}{x} + \frac{4x}{x-3} = 4$
25. $\frac{n-3}{n} = \frac{n-3}{n-6}$
26. $\frac{3}{r+4} - \frac{1}{r} = \frac{1}{r}$
27. $\frac{5}{x+1} + \frac{1}{x} = \frac{2}{x^2+x}$
28. $\frac{5}{n-2} + \frac{2}{n} = \frac{1}{n}$
29. $\frac{6}{t+1} - \frac{3}{4t+4} = \frac{3}{4}$
30. $\frac{1}{2a+6} + \frac{3a}{a+3} = \frac{1}{2}$
31. $\frac{7}{a-1} = \frac{5}{a+3}$
32. $\frac{1}{m+1} + \frac{5}{m-1} = \frac{1}{m^2-1}$
33. $\frac{j}{j+1} + \frac{5}{j-1} = 1$
34. $\frac{6}{z+2} + \frac{3}{z^2-4} = \frac{7}{z+2}$
35. $\frac{1}{4m} - \frac{2}{m-3} = \frac{2}{m}$
36. $\frac{x+2}{2x-1} + \frac{3x-6}{8x-4} = \frac{9}{4}$

37. What is the value of w in the equation $\frac{2w-3}{w-3} - 2 = \frac{12}{w+3}$?

38. Solve $\frac{3n}{n^2-5n+4} = \frac{2}{n-4} + \frac{3}{n-1}$.

Applications and Problem Solving

39. **Quizzes** Refer to the application at the beginning of the lesson. Solve $8 = \frac{36+10x}{5+x}$ to find the number of quizzes on which Julia must score 10 points in order to have an overall quiz average of 8 points.

40. **Cycling** A long-distance cyclist pedaling at a steady rate travels 30 miles with the wind. She can travel only 18 miles against the wind in the same amount of time. If the rate of the wind is 3 miles per hour, what is the cyclist's rate without the wind? (*Hint:* Use the formula $d = rt$.)

41. **Critical Thinking** What number would you add to the numerator and denominator of $\frac{2}{11}$ to make a fraction equivalent to $\frac{1}{2}$?

Mixed Review

42. **Space** Scientists can predict when an asteroid might hit Earth or come very close to Earth by studying their orbits. Toro is an asteroid that orbits the sun every 584 days. Earth orbits the sun every 365 days. Suppose Earth and Toro are lined up with the sun today. Find the LCM of 584 and 365 to determine how long before Earth and Toro arrive together back at that same point in their orbits. *(Lesson 15–5)*

An asteroid

43. Find the sum of $\frac{11}{4n^2}$ and $\frac{7}{4n^2}$. Write your answer in simplest form. *(Lesson 15–4)*

Name the set or sets of numbers to which each real number belongs. Let N = natural numbers, W = whole numbers, Z = integers, Q = rational numbers, and I = irrational numbers. *(Lesson 14–1)*

44. $\frac{16}{2}$

45. -7

46. $\sqrt{15}$

47. **Farming** In order to have enough time in the growing season, a farmer has at most 16 days left to plant his corn and soybean crops. He can plant corn at a rate of 10 acres per day and soybeans at a rate of 15 acres per day. Suppose he has at most 200 acres available and he wants to plant both crops. *(Lesson 13–7)*

a. Let c represent the number of days that corn will be planted and let s represent the number of days that soybeans will be planted. Write a system of inequalities to represent this situation.

b. Graph the system of inequalities.

c. List two ordered pairs that are possible solutions and explain what each ordered pair represents.

Solve each system of equations by graphing. *(Lesson 13–1)*

48. $y = 2x - 1$
$y = -x + 5$

49. $x = 4$
$x + y = 3$

Standardized Test Practice

A B C D

50. **Short Response** Write an irrational square root whose graph is between points A and B on the number line below. *(Lesson 8–6)*

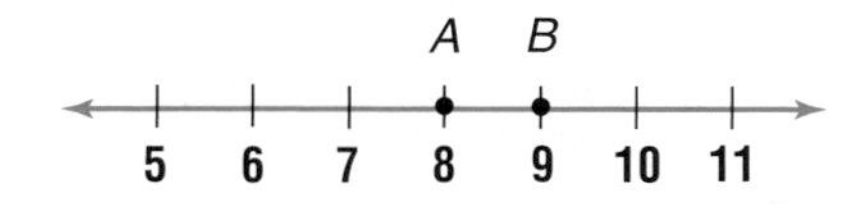

Chapter 15 Investigation

Down the Drain

Materials

 calculator

Work Problems

Suppose you have a big job to do, like painting a house. How much faster could you get the job done if you had help? In this investigation, you will use algebraic techniques to solve **work problems**.

Investigate

1. A public swimming pool holds 100,000 gallons of water and has two drain outlets. If only outlet A is open, the pool will drain in 20 hours. If only outlet B is open, the pool will drain in 16 hours.

   a. The drainage rate of outlet A is 100,000 gallons ÷ 20 hours or 5000 gallons per hour. What is the drainage rate of outlet B?

   b. Make a table like the one below and continue filling in columns 2, 3, and 4 until the total water drained in column 4 is more than 100,000 gallons.

You could also use a spreadsheet for your calculations.

Number of Hours	Water Drained by Outlet A (gal)	Water Drained by Outlet B (gal)	Total Water Drained (gal)
1	5000	6250	11,250
2	10,000	12,500	22,500

   c. Approximately how long will it take to drain the pool with both outlets open?

2. The problem above also can be solved using an algebraic equation.

   a. Since outlet A can drain the pool in 20 hours, it can drain $\frac{1}{20}$ of the pool in 1 hour. How much of the pool can outlet B drain in 1 hour?

   b. Use the table at the right and the following formula to write rational expressions for the work done w by outlets 1 and 2 in t hours.

$$\underbrace{\text{rate of work}}_{r} \cdot \underbrace{\text{time}}_{t} = \underbrace{\text{work done}}_{w}$$

	r	t	w
Outlet A	$\frac{1}{20}$	t	?
Outlet B	$\frac{1}{16}$	t	?

c. If both outlets are open, then the following equation is true.

$$\text{work of outlet A} + \text{work of outlet B} = \text{the whole job}$$

Write an equation to represent the time it takes the pool to drain if both outlets are open. Assume that 1 represents the whole job. *Why?*

d. Solve the equation in part b to find the how long it takes to drain the pool if both outlets are open. Compare this answer to the estimate you made in Exercise 1d.

3. Micheal and Gus work for a ski resort. Their job is to prepare Bear Tooth Run for skiers. It takes Micheal 4 hours to prepare the run and it takes Gus 5 hours. How long would it take them to prepare the run if they work together?

   a. In the swimming pool problem, you knew the total amount of gallons to be drained. In this problem since you do not know the size of the trail, write the work as a fraction done per hour. Micheal's rate is $\frac{1}{4}$ of the job per hour. What is Gus' rate?

   b. Write an algebraic equation to represent the work done if Micheal and Gus work together.

   c. How long would it take them both to prepare the run?

Extending the Investigation

Use a spreadsheet or write an algebraic equation to solve each problem.

- Ian and Mandy own a lawn care business. Ian can mow, trim, and fertilize one particular yard in 2 hours. Mandy can complete the same tasks for that yard in 4 hours. How long would it take them to complete this yard working together?
- In the summer, Sheila and her son, Cory, paint house exteriors. To paint a typical house, it takes Sheila about 45 hours. To paint the same size house, it takes Cory 55 hours. How long would it take them to paint one house working together?

Presenting Your Conclusions

Here are some ideas to help you present your conclusions to the class.

- Design a creative bulletin board displaying a problem from this investigation. Show two ways in which to solve the problem.
- Write a problem similar to those in this investigation. Solve the problem by using a spreadsheet and by using an equation. Write a one-page paper discussing the advantages and disadvantages of using each method to solve work problems.

Investigation For more information on work problems visit: www.algconcepts.com

CHAPTER 15

Study Guide and Assessment

Understanding and Using the Vocabulary

After completing this chapter, you should be able to define each term, property, or phrase and give an example or two of each.

interNET CONNECTION **Review Activities**
For more review activities, visit: www.algconcepts.com

excluded value *(p. 638)*
least common denominator *(p. 663)*
least common multiple *(p. 662)*
rational equation *(p. 668)*
rational expression *(p. 638)*
rational function *(p. 638)*
uniform motion problems *(p. 670)*
work problems *(p. 674)*

Choose the correct term or expression to complete each sentence.

1. An excluded value is any value assigned to a variable that results in a (denominator, numerator) of zero.
2. A rational expression is an algebraic fraction whose numerator and denominator are (rational numbers, polynomials).
3. To find $\frac{9}{x+y}$ divided by $\frac{3x+6y}{x}$, multiply $\frac{9}{x+y}$ and $\left(\frac{3x+6y}{x}, \frac{x}{3x+6y}\right)$.
4. The excluded values of $\frac{4x}{x^2-36}$ are (0 and 36, 6 and -6).
5. To simplify a rational expression, divide the numerator and denominator by their (greatest common factor, least common multiple).
6. The rational expression $\frac{x+2}{x+4}$ (is, is not) in simplest form.
7. The reciprocal of $a + b$ is $\left(-a - b, \frac{1}{a+b}\right)$.
8. To add or subtract rational expressions, first rename the expressions using the (greatest common factor, least common multiple) as the denominator.
9. The least common multiple of x^2 and $2x$ is $(2x^2, 2x^3)$.
10. The divisor in $(x^2 + 8x + 15) \div (x + 5) = (x + 3)$ is $(x^2 + 8x + 15, x + 5)$.

Skills and Concepts

Objectives and Examples

- **Lesson 15–1** Simplify rational expressions.

$$\frac{z^2 - 3z + 2}{z^2 - 2z} = \frac{(z-1)\overset{1}{\cancel{(z-2)}}}{z\underset{1}{\cancel{(z-2)}}} = \frac{z-1}{z}$$

Review Exercises

Simplify each expression.

11. $\frac{4x^2yz}{16xy^3}$
12. $\frac{a^2 - 5a}{a - 5}$
13. $\frac{n^2 - 16}{n^2 + 2n - 8}$
14. $\frac{x^3 - 2x^2 - 3x}{x^2 + x - 12}$

Objectives and Examples

- **Lesson 15–2** Multiply and divide rational expressions.

$$\frac{6x}{x^2 + x - 2} \cdot \frac{x^2 - 4}{2x^2}$$

$$= \frac{\overset{3\,1}{\cancel{6x}}}{\underset{1}{(\cancel{x+2})}(x-1)} \cdot \frac{(x-2)\overset{1}{(\cancel{x+2})}}{\underset{1\,x}{\cancel{2x^2}}}$$

$$= \frac{3(x-2)}{x(x-1)}$$

$$\frac{a^3}{2c} \div \frac{a^2}{c^3} = \frac{\overset{a}{\cancel{a^3}}}{\underset{1}{2\cancel{c}}} \cdot \frac{\overset{c^2}{\cancel{c^3}}}{\underset{1}{\cancel{a^2}}}$$ *The reciprocal of $\frac{a^2}{c^3}$ is $\frac{c^3}{a^2}$.*

$$= \frac{ac^2}{2}$$

Review Exercises

Find each product or quotient.

15. $\frac{4x^2y}{y^2z} \cdot \frac{z}{6y}$
16. $\frac{4a - 4b}{a} \cdot \frac{a^3}{8a - 8b}$
17. $\frac{x + 2}{x + 3} \cdot \frac{x}{x^2 - x - 6}$
18. $\frac{6m^2n}{10p} \div 3m$
19. $\frac{y^2 - 4y + 4}{3} \div \frac{y^2 - 4}{9}$
20. $\frac{x^2 - 9}{2x} \div \frac{6 - 2x}{8x^2}$

- **Lesson 15–3** Divide polynomials by binomials.

$$\begin{array}{r l}
2x - 3 & \\
x + 2\overline{)2x^2 + x - 6} & 2x^2 \div x = 2x \\
\underline{(-)\ 2x^2 + 4x} & \textit{Multiply } 2x \textit{ and } x + 2. \\
-3x - 6 & \textit{Subtract.} \\
\underline{(-)\ -3x - 6} & -3x \div x = -3 \\
0 & \textit{Subtract.}
\end{array}$$

Find each quotient.

21. $(a^2 + 6a) \div (a + 6)$
22. $(x^2 - 5x - 6) \div (x + 1)$
23. $(4y^2 + 8y + 5) \div (2y + 3)$
24. $(x^3 - 1) \div (x - 1)$

- **Lesson 15–4** Add and subtract rational expressions with like denominators.

$$\frac{x^2}{x+3} - \frac{9}{x+3} = \frac{x^2 - 9}{x+3}$$

$$= \frac{\overset{1}{(\cancel{x+3})}(x-3)}{\underset{1}{\cancel{x+3}}}$$

$$= x - 3$$

Find each sum or difference. Write in simplest form.

25. $\frac{2x}{8} + \frac{4x}{8}$
26. $\frac{11}{15m} - \frac{2}{15m}$
27. $\frac{3x}{x + 4} + \frac{2x}{x + 4}$
28. $\frac{8x}{2x - 3} - \frac{2x + 9}{2x - 3}$

Chapter 15 Study Guide and Assessment

• **Extra Practice**
See pages 722–723.

Objectives and Examples

• **Lesson 15–5** Add and subtract rational expressions with unlike denominators.

$$\frac{y}{y+2} - \frac{2}{y} \quad \textit{The LCD is } y(y+2).$$
$$= \frac{y}{y+2} \cdot \frac{y}{y} - \frac{2}{y} \cdot \frac{y+2}{y+2}$$
$$= \frac{y^2}{y(y+2)} - \frac{2(y+2)}{y(y+2)}$$
$$= \frac{y^2 - 2(y+2)}{y(y+2)}$$
$$= \frac{y^2 - 2y - 4}{y(y+2)}$$

Review Exercises

Find each sum or difference. Write in simplest form.

29. $\frac{3}{4x} + \frac{5}{8x^2}$

30. $\frac{5}{x} - \frac{x+1}{3x}$

31. $\frac{4}{a} + \frac{3}{a-2}$

32. $\frac{y+2}{y^2-4} - \frac{2}{y+2}$

• **Lesson 15–6** Solve rational equations.

$$\frac{x}{4} - \frac{x}{6} = \frac{1}{4}$$
$$12\left(\frac{x}{4} - \frac{x}{6}\right) = 12\left(\frac{1}{4}\right)$$
$$\overset{3}{\cancel{12}}\left(\frac{x}{\underset{1}{\cancel{4}}}\right) - \overset{2}{\cancel{12}}\left(\frac{x}{\underset{1}{\cancel{6}}}\right) = \overset{3}{\cancel{12}}\left(\frac{1}{\underset{1}{\cancel{4}}}\right)$$
$$3x - 2x = 3$$
$$x = 3$$

Solve each equation. Check your solution.

33. $\frac{x}{3} - \frac{3x}{4} = \frac{1}{12}$

34. $\frac{6}{3x} - \frac{3}{x} = 1$

35. $\frac{x+2}{x} + \frac{2}{3x} = \frac{1}{3}$

36. $\frac{m}{m+1} + \frac{5}{m-1} = 1$

Applications and Problem Solving

37. Carpentry Determine the number of pieces of $\frac{3}{8}$-inch plywood that are in a stack 30 inches high. *(Lesson 15–2)*

38. Boating The top speed of a boat in still water is 5 miles per hour. At this speed, a 21-mile trip downstream takes the same amount of time as a 9-mile trip upstream. Find the rate of the current. *(Lesson 15–6)*

39. Geometry The volume of a rectangular prism is $4x^3 + 14x^2 + 10x$ cubic inches. The height of the prism is $2x + 5$ inches, and the length of the prism is $2x$ inches. *(Lesson 15–3)*

a. Find the polynomial that represents the width.

b. If $x = 2$, find the length, width, height, and volume of the prism.

CHAPTER 15 Test

1. **Explain** why $x = -3$ is an excluded value for $\frac{2x}{x+3}$.
2. **Identify** and correct the error that was made while finding the difference.

$$\frac{4}{x+1} - \frac{x-3}{x+1} = \frac{4-x-3}{x+1}$$
$$= \frac{1-x}{x+1}$$

3. **Find** the least common denominator of $\frac{5}{6a}$ and $\frac{2}{3a^2}$.

Find the excluded value(s) for each rational expression.

4. $\frac{8}{x}$
5. $\frac{6m}{m(m-2)}$
6. $\frac{2n}{n^2-9}$

Simplify each expression.

7. $\frac{9a^3b^2}{15ab^5}$
8. $\frac{x^3-x^2}{x-1}$
9. $\frac{x-2}{x^2-5x+6}$

Find each sum, difference, product, or quotient. Write in simplest form.

10. $\frac{z^3}{8} \cdot \frac{10x^2}{z^3}$
11. $\frac{x^2-1}{3x} \div \frac{1-x}{9x}$
12. $\frac{x^2-x-2}{x^2-4} \cdot \frac{x+2}{x^2+4x+3}$
13. $\frac{4a+4b}{a} \div \frac{10a+10b}{a^2}$
14. $\frac{5}{8x} + \frac{11}{8x}$
15. $\frac{t}{t+5} - \frac{t-6}{t+5}$
16. $\frac{6}{5y} + \frac{7}{10y^2}$
17. $\frac{2}{x+4} - \frac{x}{x^2-16}$

Find each quotient.

18. $(y^2 + 10y + 16) \div (y + 2)$
19. $(x^3 + x^2 - x - 1) \div (x - 1)$
20. $(a^2 - 10) \div (a + 3)$
21. $(x^2 - 5x + 8) \div (x - 2)$

Solve each equation. Check your solution.

22. $\frac{n+2}{3} + \frac{n}{2} = \frac{1}{2}$
23. $\frac{5}{x+2} - \frac{1}{4x+8} = \frac{1}{4}$

24. **Geometry** Find the measure of the area of the rectangle in simplest form.

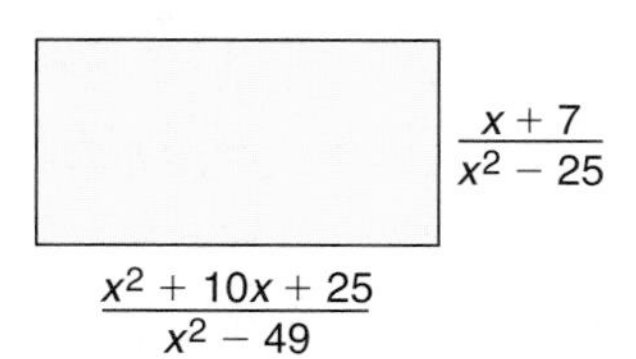

25. **Transportation** A tugboat pushing a barge up the Mississippi River takes 1 hour longer to travel 36 miles up the river than to travel the same distance down the river. If the rate of the current is 3 miles per hour, find the speed of the tugboat and barge in still water.

CHAPTER 15

Preparing for Standardized Tests

Right Triangle Problems

You'll need to know how to identify right triangles and how to apply the Pythagorean Theorem.

> If a and b are the lengths of the legs of a triangle and c is the length of the hypotenuse, then $a^2 + b^2 = c^2$.

Look for special right triangles, such as 3-4-5 triangles.

State Test Example

A 32-foot telephone pole is braced with a cable from the top of the pole to a point 7 feet from the base. What is the length of the cable, rounded to the nearest tenth?

A 31.2 ft **B** 32.8 ft
C 34.3 ft **D** 36.2 ft

Hint Since telephone poles are vertical and the ground is horizontal, the pole forms a right angle at the base.

Solution Draw a figure.

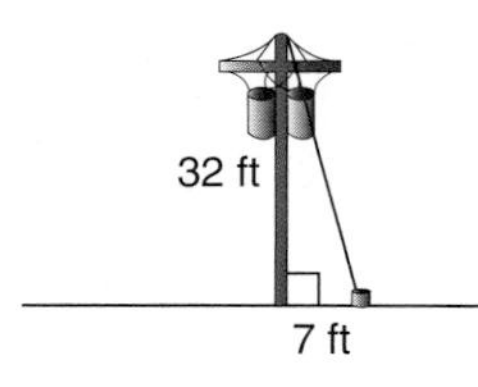

The cable forms the hypotenuse of a right triangle. Use the Pythagorean Theorem to find the length of the hypotenuse.

$a^2 + b^2 = c^2$ *Pythagorean Theorem*

$32^2 + 7^2 = c^2$ *Replace a with 32 and b with 7.*

$1024 + 49 = c^2$

$1073 = c^2$

$\sqrt{1073} = \sqrt{c^2}$ *Take the square root of each side.*

$32.8 \approx c$ *Use a calculator.*

To the nearest tenth, the length of the cable is 32.8 feet.

The answer is B.

ACT Example

In the figure, $\overline{MO}$ is perpendicular to $\overline{LN}$, LO is equal to 4, MO is equal to ON, and LM is equal to 6. What is MN?

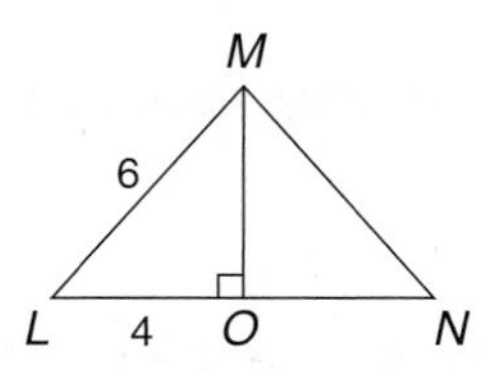

A $2\sqrt{10}$ **B** $3\sqrt{5}$ **C** $4\sqrt{5}$
D $3\sqrt{10}$ **E** $6\sqrt{4}$

Hint Identify the types of triangles shown in the figure.

Solution $\triangle LMO$ is a right triangle. Use the Pythagorean Theorem to find MO.

$a^2 + b^2 = c^2$ *Pythagorean Theorem*

$a^2 + 4^2 = 6^2$ *Replace b with 4 and c with 6.*

$a^2 + 16 = 36$

$a^2 + 16 - 16 = 36 - 16$ *Subtract 16.*

$a^2 = 20$

$\sqrt{a^2} = \sqrt{20}$ *Take the square root.*

$a = \sqrt{4(5)}$ or $2\sqrt{5}$

$\triangle MON$ is an isosceles triangle. Use the Pythagorean Theorem to find MN.

$a^2 + b^2 = c^2$ *Pythagorean Theorem*

$\left(2\sqrt{5}\right)^2 + \left(2\sqrt{5}\right)^2 = c^2$ $a = 2\sqrt{5}, b = 2\sqrt{5}$

$20 + 20 = c^2$

$\sqrt{40} = \sqrt{c^2}$

$2\sqrt{10} = c$

The answer is A.

After you work each problem, record your answer on the answer sheet provided or on a sheet of paper.

Multiple Choice

1. Quadrilateral $ABCD$ is a rectangle. $\overline{AD}$ is 5 centimeters long, and $\overline{CD}$ is 12 centimeters long. What is length of $\overline{AC}$?

A 13 cm
B 17 cm
C 30 cm
D 169 cm

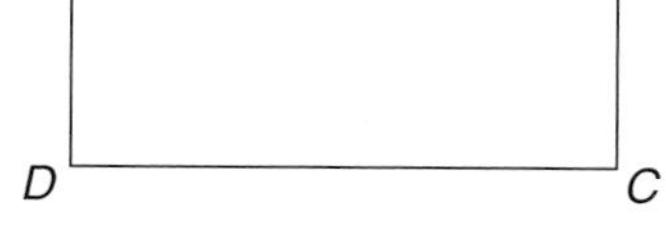

2. What is the distance between points G and H in the graph?

A 4
B $4\sqrt{2}$
C 8
D $8\sqrt{2}$

3. The hypotenuse of an isosceles right triangle has a length of 20 units. What is the length of one of the legs of the triangle?

A 10 **B** $10\sqrt{2}$ **C** $10\sqrt{3}$
D 20 **E** $20\sqrt{2}$

4. The streets on a jogging route form a right triangle. How long is the jogging route?

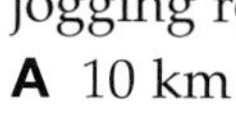

A 10 km
B 14 km
C 24 km
D 100 km

5. What is the equation of the graph?

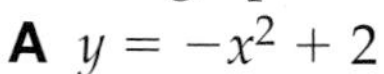

A $y = -x^2 + 2$
B $y = x^2 - 2$
C $y = (x + 2)^2$
D $y = x^2 + 2$

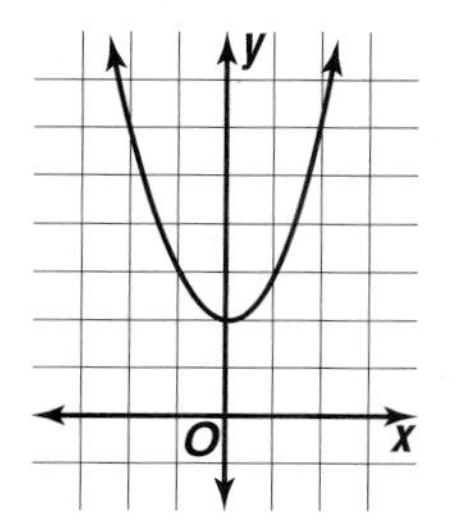

6. What is the length of $\overline{BC}$?

A 6
B $4\sqrt{3}$
C $2\sqrt{13}$
D 8
E $2\sqrt{38}$

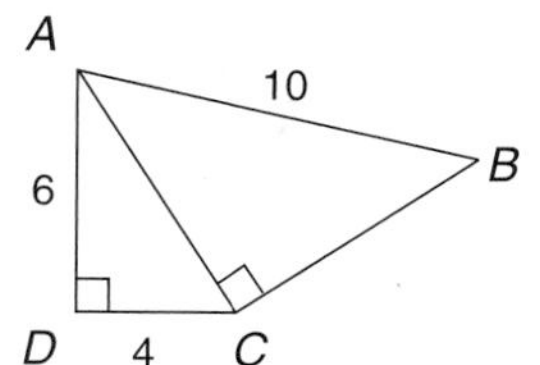

7. The base of a 25-foot ladder is placed 10 feet from the side of a house. How high does the ladder reach?

A 12.5 ft **B** 15 ft **C** 22.9 ft **D** 35 ft

8. $\sqrt{3} + \sqrt{4}$ is between which pair of numbers?

A 3 and 4
B 4 and 5
C 5 and 6
D 6 and 7

Grid In

9. Triangles ABC and XYZ are similar right triangles. If $\angle B$ measures 55°, what is the degree measure of $\angle Z$?

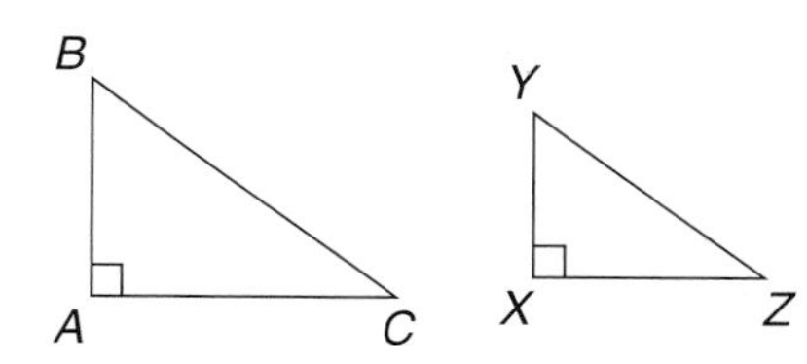

Extended Response

10. You want to fit a square cake inside a circular cake carrier that is 12 inches in diameter.

Part A Draw a diagram showing side s of the *largest* square cake that will fit inside the carrier.

Part B Use your diagram to write an equation and then find the value of s.

Student Handbook

Skills Handbook

Reference Handbook

Prerequisite Skills Review

Operations with Decimals

- To add or subtract decimals, line up the decimal points. You may want to *annex*, or place zeros at the end of the decimals, to help align the columns. Then add or subtract.

Examples

1 Find 18.39 + 6.4.

$$\begin{array}{r} 18.39 \\ +\ 6.4 \\ \hline \end{array} \quad \rightarrow \quad \begin{array}{r} 18.39 \\ +\ 6.40 \\ \hline 24.79 \end{array}$$

Annex a zero to align the columns.

2 Find 5 − 2.17.

$$\begin{array}{r} 5 \\ -\ 2.17 \\ \hline \end{array} \quad \rightarrow \quad \begin{array}{r} 5.00 \\ -\ 2.17 \\ \hline 2.83 \end{array}$$

Annex two zeros to align the columns.

- To multiply decimals, multiply as with whole numbers. Then add the total number of decimal places in the factors. Place the same number of decimal places in the product, counting from right to left.
- To divide decimals, move the decimal point in the divisor to the right, and then move the decimal point in the dividend the same number of places. Align the decimal point in the quotient with the decimal point in the dividend.

Examples

3 Find 7.3(0.61).

$$\begin{array}{r} 7.3 \\ \times\ 0.61 \\ \hline 73 \\ 438\ \\ \hline 4.453 \end{array}$$

7.3 ← *1 decimal place*
× 0.61 ← *2 decimal places*
4.453 ← *3 decimal places*

4 Find 8.12 ÷ 5.8.

$$\begin{array}{r} 1.4 \\ 5.8\overline{)8.12} \\ 5\,8\ \\ \hline 2\,32 \\ 2\,32 \\ \hline 0 \end{array}$$

Move each decimal point right 1 place.

Find each sum or difference.

1. 4.8 + 1.1
2. 10.25 + 0.16
3. 36.5 − 11.2
4. 13.6 + 20.41
5. 42.99 − 12.63
6. 5.37 − 2.47
7. 24 + 7.3
8. 7.28 − 1.1
9. 0.7 + 1.682
10. 6.2 − 4.01
11. 4.02 + 5.9 + 0.03
12. 18 − 16.39

Find each product or quotient.

13. 8(0.4)
14. 5.72 ÷ 2
15. 1.6(1.9)
16. 7.2(3.5)
17. 24(0.86)
18. 4.86 ÷ 0.3
19. 11.5 ÷ 4.6
20. 9.3(30.23)
21. 0.5(2.41)(6.7)
22. $\frac{50.4}{36}$
23. $\frac{6.46}{0.68}$
24. $\frac{9.264}{0.24}$

Simplifying Fractions

- A fraction is in simplest form when the greatest common factor (GCF) of the numerator and the denominator is 1.
- To write a fraction in simplest form, divide both the numerator and the denominator by the GCF.

Examples **Write each fraction in simplest form.**

1 $\frac{6}{9}$

The GCF of 6 and 9 is 3.

$$\frac{6}{9} = \frac{2}{3}$$ (÷ 3 above and below) *Divide 6 and 9 by 3.*

2 $\frac{4}{20}$

The GCF of 4 and 20 is 4.

$$\frac{4}{20} = \frac{1}{5}$$ (÷ 4 above and below) *Divide 4 and 20 by 4.*

- To change an improper fraction to a mixed number, divide the numerator by the denominator. Write the remainder as a fraction in simplest form.
- To change a mixed number to an improper fraction, multiply the whole number by the denominator and add the numerator. Write this sum over the denominator.

Examples **3 Write $\frac{8}{5}$ as a mixed number.**

$\frac{8}{5}$ means $8 \div 5$.

$$\begin{array}{r} 1 \\ 5\overline{)8} \\ \underline{-5} \\ 3 \end{array} \rightarrow 1 \text{ R3 or } 1\frac{3}{5}$$

4 Write $2\frac{3}{4}$ as an improper fraction.

$$2\frac{3}{4} = \frac{(2 \times 4) + 3}{4}$$
$$= \frac{8 + 3}{4}$$
$$= \frac{11}{4}$$

Write each fraction in simplest form.

1. $\frac{4}{8}$
2. $\frac{3}{9}$
3. $\frac{2}{12}$
4. $\frac{9}{15}$
5. $\frac{8}{20}$
6. $\frac{6}{21}$
7. $\frac{25}{30}$
8. $\frac{30}{40}$
9. $\frac{12}{16}$
10. $\frac{16}{36}$

Write each improper fraction as a mixed number.

11. $\frac{7}{6}$
12. $\frac{5}{3}$
13. $\frac{9}{2}$
14. $\frac{12}{5}$
15. $\frac{10}{4}$
16. $\frac{11}{2}$
17. $\frac{21}{10}$
18. $\frac{16}{6}$
19. $\frac{24}{14}$
20. $\frac{30}{8}$

Write each mixed number as an improper fraction.

21. $2\frac{1}{3}$
22. $6\frac{1}{2}$
23. $2\frac{4}{5}$
24. $4\frac{2}{5}$
25. $6\frac{1}{7}$
26. $3\frac{5}{8}$
27. $2\frac{7}{10}$
28. $3\frac{5}{12}$
29. $1\frac{10}{11}$
30. $5\frac{8}{9}$

Multiplying and Dividing Fractions

- To multiply fractions, multiply the numerators and multiply the denominators.
- When a numerator and denominator have a common factor, you can simplify before multiplying.

Examples **Find each product.**

1 $\frac{1}{2} \cdot \frac{3}{5}$

$\frac{1}{2} \cdot \frac{3}{5} = \frac{1 \cdot 3}{2 \cdot 5}$

$= \frac{3}{10}$

2 $\frac{4}{9} \cdot \frac{3}{5}$

$\frac{4}{9} \cdot \frac{3}{5} = \frac{4}{\cancel{9}_3} \cdot \frac{\cancel{3}^1}{5}$ *Divide by the GCF, 3.*

$= \frac{4 \cdot 1}{3 \cdot 5}$

$= \frac{4}{15}$

- To divide fractions, multiply by the reciprocal of the divisor.

Examples **Find each quotient.**

3 $\frac{5}{6} \div \frac{6}{7}$

$\frac{5}{6} \div \frac{6}{7} = \frac{5}{6} \cdot \frac{7}{6}$ *Multiply by the reciprocal.*

$= \frac{5 \cdot 7}{6 \cdot 6}$

$= \frac{35}{36}$

4 $\frac{2}{5} \div 4$

$\frac{2}{5} \div 4 = \frac{2}{5} \div \frac{4}{1}$

$= \frac{2}{5} \cdot \frac{1}{4}$ *Multiply by the reciprocal.*

$= \frac{\cancel{2}^1}{5} \cdot \frac{1}{\cancel{4}_2}$ *Divide by the GCF, 2.*

$= \frac{1 \cdot 1}{5 \cdot 2}$

$= \frac{1}{10}$

Find each product or quotient.

1. $\frac{1}{3} \cdot \frac{1}{2}$
2. $\frac{2}{5} \div \frac{1}{2}$
3. $\frac{1}{4} \div \frac{4}{5}$
4. $\frac{2}{7} \cdot \frac{1}{3}$
5. $\frac{1}{8} \div \frac{1}{5}$
6. $\frac{4}{9} \cdot \frac{2}{3}$
7. $\frac{2}{7} \div 7$
8. $\frac{6}{7} \cdot \frac{1}{5}$
9. $\frac{1}{2} \cdot \frac{2}{7}$
10. $\frac{2}{3} \cdot \frac{1}{6}$
11. $\frac{5}{9} \div \frac{2}{3}$
12. $\frac{1}{8} \div \frac{1}{2}$
13. $\frac{2}{9} \div \frac{4}{7}$
14. $\frac{3}{4} \cdot \frac{5}{6}$
15. $\frac{5}{12} \cdot \frac{3}{7}$
16. $\frac{3}{4} \div \frac{5}{6}$
17. $\frac{2}{9} \div \frac{2}{3}$
18. $\frac{2}{3} \div 4$
19. $\frac{7}{9} \cdot \frac{6}{7}$
20. $\frac{3}{10} \cdot \frac{5}{18}$
21. $\frac{5}{12} \cdot \frac{8}{25}$
22. $\frac{3}{8} \div 12$
23. $\frac{9}{20} \cdot \frac{4}{15}$
24. $\frac{8}{21} \div \frac{12}{15}$

Adding and Subtracting Fractions

- To add or subtract fractions with like denominators, add or subtract the numerators. Simplify if necessary.

Examples **Find each sum or difference.**

1 $\frac{1}{5} + \frac{3}{5}$

$\frac{1}{5} + \frac{3}{5} = \frac{1 + 3}{5}$ *Add the numerators.*

$= \frac{4}{5}$

2 $\frac{7}{12} - \frac{5}{12}$

$\frac{7}{12} - \frac{5}{12} = \frac{7 - 5}{12}$ *Subtract the numerators.*

$= \frac{2}{12}$

$= \frac{1}{6}$ *Simplify.*

- To add or subtract fractions with unlike denominators, first find the least common denominator (LCD). Rewrite each fraction with the LCD, and then add or subtract the numerators. Simplify if necessary.

Examples **Find each sum or difference.**

3 $\frac{1}{3} - \frac{2}{9}$

The LCD of 3 and 9 is 9.

$\frac{1}{3} - \frac{2}{9} = \frac{3}{9} - \frac{2}{9}$ *Rewrite $\frac{1}{3}$ as $\frac{3}{9}$.*

$= \frac{3 - 2}{9}$ *Subtract the numerators.*

$= \frac{1}{9}$

4 $\frac{3}{10} + \frac{1}{4}$

The LCD of 10 and 4 is 20.

$\frac{3}{10} + \frac{1}{4} = \frac{6}{20} + \frac{5}{20}$ *Rewrite the fractions using the LCD, 20.*

$= \frac{6 + 5}{20}$ *Add the numerators.*

$= \frac{11}{20}$

Find each sum or difference.

1. $\frac{1}{5} + \frac{2}{5}$ **2.** $\frac{7}{8} - \frac{2}{8}$ **3.** $\frac{1}{9} + \frac{4}{9}$ **4.** $\frac{6}{7} - \frac{2}{7}$

5. $\frac{1}{15} + \frac{3}{15}$ **6.** $\frac{4}{13} + \frac{1}{13}$ **7.** $\frac{5}{6} - \frac{4}{6}$ **8.** $\frac{9}{11} - \frac{3}{11}$

9. $\frac{3}{8} + \frac{1}{8}$ **10.** $\frac{5}{6} - \frac{1}{6}$ **11.** $\frac{7}{10} - \frac{3}{10}$ **12.** $\frac{1}{12} + \frac{7}{12}$

13. $\frac{1}{3} + \frac{1}{5}$ **14.** $\frac{5}{8} - \frac{1}{4}$ **15.** $\frac{4}{9} + \frac{1}{3}$ **16.** $\frac{9}{10} - \frac{3}{5}$

17. $\frac{3}{7} + \frac{1}{2}$ **18.** $\frac{3}{4} - \frac{2}{5}$ **19.** $\frac{2}{9} + \frac{2}{3}$ **20.** $\frac{9}{10} - \frac{3}{4}$

21. $\frac{2}{15} + \frac{7}{15}$ **22.** $\frac{1}{2} + \frac{1}{6}$ **23.** $\frac{5}{6} - \frac{3}{4}$ **24.** $\frac{3}{4} + \frac{1}{12}$

25. $\frac{4}{7} - \frac{1}{6}$ **26.** $\frac{2}{15} + \frac{2}{3}$ **27.** $\frac{5}{6} + \frac{1}{8}$ **28.** $\frac{7}{8} - \frac{3}{12}$

Fractions and Decimals

- To change a fraction to a decimal, divide the numerator by the denominator.
- A **terminating decimal** is a decimal like 0.75, in which the division ends, or terminates, when the remainder is zero.
- A **repeating decimal** is a decimal like 0.545454 . . . whose digits do not end. Since it is impossible to write all the digits, you can use bar notation to show that 54 repeats. We can write 0.545454 . . . as $0.\overline{54}$.

Examples **Write each fraction as a decimal.**

1 $\frac{2}{5}$

$$\begin{array}{r} 0.4 \\ 5\overline{)2.0} \\ \underline{2\,0} \\ 0 \end{array} \qquad \frac{2}{5} = 0.4$$

2 $\frac{4}{9}$

$$\begin{array}{r} 0.44 \\ 9\overline{)4.00} \\ \underline{36} \\ 40 \\ \underline{36} \\ 4 \end{array} \qquad \frac{4}{9} = 0.\overline{4}$$

The pattern is repeating.

- Every terminating decimal can be expressed as a fraction with a denominator of 10, 100, and so on.
- Every repeating decimal can be expressed as a fraction.

Examples **Write each decimal as a fraction.**

3 **0.8**

$$0.8 = \frac{8}{10} = \frac{4}{5}$$

4 $\mathbf{0.\overline{1}}$

Let $N = 0.\overline{1}$ or 0.111 . . . .

Then $10N = 1.\overline{1}$ or 1.111 . . . .

$$\begin{array}{r} 10N = 1.111\ldots \\ -\ 1N = 0.111\ldots \\ \hline 9N = 1 \\ N = \frac{1}{9} \end{array}$$

Subtract. *N = 1N*

So, $0.\overline{1} = \frac{1}{9}$.

Write each fraction as a decimal.

1. $\frac{3}{4}$ **2.** $\frac{1}{5}$ **3.** $\frac{3}{8}$ **4.** $\frac{2}{9}$

5. $\frac{4}{11}$ **6.** $\frac{7}{10}$ **7.** $\frac{2}{15}$ **8.** $\frac{1}{6}$

9. $\frac{4}{15}$ **10.** $\frac{3}{20}$ **11.** $\frac{5}{6}$ **12.** $\frac{5}{8}$

Write each decimal as a fraction in simplest form.

13. 0.9 **14.** 0.6 **15.** 0.25 **16.** $0.\overline{3}$

17. $0.\overline{5}$ **18.** 0.4 **19.** 0.16 **20.** $0.\overline{6}$

21. 0.125 **22.** 0.35 **23.** $0.\overline{8}$ **24.** $0.\overline{7}$

Decimals and Percents

- To express a decimal as a percent, first express the decimal as a fraction with a denominator of 100. Then express the fraction as a percent.

Examples **Write each decimal as a percent.**

1 **0.38**

$0.38 = \frac{38}{100}$

$= 38\%$

2 **0.4**

$0.4 = 0.40$

$= \frac{40}{100}$

$= 40\%$

Shortcut

To write a decimal as a percent, multiply by 100 and add the % symbol.
0.62 = 0.62 = 62%

3 **0.07**

$0.07 = \frac{7}{100}$

$= 7\%$

4 **2.55**

$2.55 = \frac{255}{100}$

$= 255\%$

- To express a percent as a decimal, rewrite the percent as a fraction with a denominator of 100. Then express the fraction as a decimal.

Examples **Write each percent as a decimal.**

5 **26%**

$26\% = \frac{26}{100}$

$= 0.26$

6 **9%**

$9\% = \frac{9}{100}$

$= 0.09$

Shortcut

To write a percent as a decimal, divide by 100 and remove the % symbol.
62% = 62% = 0.62

7 **87.5%**

$87.5\% = \frac{87.5}{100}$

$= 0.875$

8 **125%**

$125\% = \frac{125}{100}$

$= 1.25$

Write each decimal as a percent.

1. 0.62 **2.** 0.99 **3.** 0.14 **4.** 0.20
5. 0.15 **6.** 0.8 **7.** 0.5 **8.** 0.06
9. 0.42 **10.** 0.03 **11.** 0.1 **12.** 1.76
13. 0.08 **14.** 3.10 **15.** 1.05 **16.** 2.6

Write each percent as a decimal.

17. 12% **18.** 37% **19.** 86% **20.** 51%
21. 30% **22.** 90% **23.** 2% **24.** 55%
25. 5% **26.** 12.5% **27.** 73.6% **28.** 134%
29. 208% **30.** 120% **31.** 60.5% **32.** 200%

Fractions and Percents

- To express a percent as a fraction, write the percent as a fraction with a denominator of 100 and simplify.

Example 1 **Express 24% as a fraction in simplest form.**

$24\% = \frac{24}{100}$ *Write as a fraction with a denominator of 100.*

$= \frac{\overset{6}{\cancel{24}}}{\underset{25}{\cancel{100}}}$ *Divide by the GCF, 4.*

$= \frac{6}{25}$

- To express a fraction as a percent, write a proportion and solve.

Example 2 **Express $\frac{3}{5}$ as a percent.**

$\frac{3}{5} = \frac{n}{100}$ *Write a proportion.*

$3 \times 100 = 5 \times n$ *Find the cross products.*

$300 = 5n$

$\frac{300}{5} = \frac{5n}{5}$ *Divide each side by 5.*

$60 = n$

So, $\frac{3}{5} = \frac{60}{100}$ or 60%.

Write each percent as a fraction in simplest form.

1. 25%
2. 80%
3. 70%
4. 15%
5. 6%
6. 56%
7. 40%
8. 98%
9. 19%
10. 35%
11. 22%
12. 64%

Write each fraction as a percent.

13. $\frac{3}{10}$
14. $\frac{1}{5}$
15. $\frac{3}{4}$
16. $\frac{13}{25}$
17. $\frac{37}{50}$
18. $\frac{1}{20}$
19. $\frac{19}{20}$
20. $\frac{42}{50}$
21. $\frac{2}{25}$
22. $\frac{9}{10}$
23. $\frac{8}{32}$
24. $\frac{22}{40}$

Comparing and Ordering Rational Numbers

- To compare rational numbers, it is usually easier and faster if you write the numbers as decimals. In some cases, it may be easier if you write the numbers as fractions having the same denominator or as percents.

Examples **Replace each ● with <, >, or = to make a true sentence.**

1 $\frac{3}{5}$ ● **0.65**

$\frac{3}{5} = 0.60$

Since $0.60 < 0.65$, $\frac{3}{5} < 0.65$.

2 **0.18 ● 2%**

$2\% = 0.02$

Since $0.18 > 0.02$, $0.18 > 2\%$.

- To order rational numbers, first write the numbers as decimals. Then write the decimals in order from least to greatest and write the corresponding rational numbers in the same order.

Example **3** **Write $\frac{1}{2}$, 55%, and 0.20 in order from least to greatest.**

Write each number as a decimal.

$\frac{1}{2} = 0.50$ $\quad 55\% = 0.55$ $\quad 0.20 = 0.20$

Write the decimals in order from least to greatest. $\quad 0.20 < 0.50 < 0.55$

$\downarrow \quad \downarrow \quad \downarrow$

Write the corresponding rational numbers in the same order. $\quad 0.20 < \frac{1}{2} < 55\%$

The numbers in order from least to greatest are 0.20, $\frac{1}{2}$, and 55%.

Replace each ● with <, >, or = to make a true sentence.

1. 0.35 ● $\frac{1}{4}$

2. 65% ● 0.7

3. 80% ● $\frac{4}{5}$

4. $\frac{1}{10}$ ● 1%

5. 12.5% ● $\frac{1}{8}$

6. 36% ● 3.6

7. $\frac{2}{3}$ ● 0.7

8. 0.2 ● 20%

9. $\frac{5}{6}$ ● $\frac{5}{7}$

10. $\frac{3}{8}$ ● 0.375

11. 0.9 ● 10%

12. 30% ● 3.9%

13. 0.9 ● $0.\overline{9}$

14. 0.15 ● $\frac{2}{15}$

15. 51% ● 51

16. 78% ● $\frac{7}{8}$

17. $\frac{6}{11}$ ● 50%

18. $0.\overline{4}$ ● $\frac{4}{9}$

Write the numbers in each set in order from least to greatest.

19. 0.52, 5%, $\frac{1}{2}$

20. $\frac{1}{3}$, 40%, $\frac{1}{4}$

21. 20%, 0.1, $\frac{3}{10}$

22. 0.19, $\frac{1}{5}$, 15%

23. 63%, $\frac{2}{3}$, 0.06

24. $\frac{1}{9}$, $0.\overline{2}$, 19%

Extra Practice

Pages 692–707 are contained in Volume One.

EXTRA PRACTICE

Lesson 8–1 (*Pages 336–340*) **Write each expression using exponents.**

1. $7 \cdot 7 \cdot 7$
2. $(-3)(-3)(-3)$
3. $5 \cdot 5 \cdot 5 \cdot 6 \cdot 4 \cdot 4$
4. 6 squared
5. $m \cdot m \cdot n \cdot n \cdot n$
6. 2 cubed
7. $4 \cdot r \cdot r \cdot r \cdot s \cdot s$
8. $(-2)(k)(k)(j)(j)$

Write each power as a multiplication expression.

9. 2^3
10. $(-4)^4$
11. $8^4 2^3$
12. k^5
13. w^2z^2
14. $-6xy^2$

Evaluate each expression if $x = 4$, $y = -1$, $z = -2$, and $w = 1.5$.

15. y^3
16. $w(yz + 4)$
17. $z^3 + 2xy$
18. $2x^2 - z^5$
19. $-2(y^3 + w)$
20. wxy

Lesson 8–2 (*Pages 341–346*) **Simplify each expression.**

1. $2^2 \cdot 2^5$
2. $6^3 \cdot 6^4$
3. $x^4 \cdot x^4$
4. $y \cdot y^5$
5. $(a^3)(a^4)$
6. $(g^6h^3)(g^4h^3)$
7. $(r^2t^2)(rt^3)$
8. $(2k^2j)(-3kj)$
9. $(6m^3)(7m^2)$
10. $(2a^2b)(9ab^5)$
11. $(6a^3b)(4ac)$
12. $(8y^4)(3y^4)$
13. $(-3p^5q^2)(-pq^7)$
14. $(3xyz)(-4x^2y^3)$
15. $\frac{4^5}{4^2}$
16. $\frac{8^8}{8^7}$
17. $\frac{15m^{12}}{3m^5}$
18. $\frac{-4r^4s^5}{-rs^4}$
19. $\frac{-16m^5n^9p^{10}}{2m^2n^3p^6}$
20. $\left(\frac{2}{3}x^5y\right)(-12x^3y^2)$
21. Evaluate m^0.
22. Find the product of mn and $-3m^2n$.

Lesson 8–3 (*Pages 347–351*) **Write each expression using positive exponents. Then evaluate the expression.**

1. 8^{-3}
2. 2^{-4}
3. 10^{-3}
4. 5^{-1}
5. 9^{-3}
6. 7^{-2}
7. 3^{-5}
8. 2^{-6}

Simplify each expression.

9. $(m^6)(m^{-2})$
10. $r^0s^1t^4$
11. $6x^{-2}y^{-4}z$
12. $(a^{-3})(a^2)$
13. $\frac{1}{k^{-4}}$
14. $\frac{m}{m^{-5}}$
15. $\frac{6b^4}{3b^{-2}}$
16. $(r^{-3})(s^4)$
17. $-\frac{r^2s^5}{rs^6}$
18. $\frac{2a^2b^5}{ab^7}$
19. $\frac{-7cd^0}{14c}$
20. $-\frac{10a^7b^4}{2ab}$
21. $-\frac{4w^5z^4}{10z^7}$
22. $-\frac{2p^2q^7}{8p^4q}$
23. $\frac{56yz^4}{14y^5z^4}$
24. $\frac{x^{-3}y^4z^{-2}}{xy^2z^{-2}}$
25. Evaluate $7m^{-2}n^{-3}$ if $m = 4$ and $n = 2$.
26. Evaluate $5a^3b^{-1}$ if $a = -1$ and $b = 3$.

Lesson 8–4 (*Pages 352–356*) **Express each measure in standard form.**

1. 1.5 megaohms
2. 168 billion dollars
3. 2 megahertz
4. 76 milliamperes
5. 400 nanoseconds
6. 1.2 million dollars
7. 18 kilobytes
8. 93 micrograms

Express each number in scientific notation.

9. 178
10. 0.0098
11. 0.032
12. 106,000
13. 13.8
14. 269.3
15. 0.0000083
16. 100
17. 0.000016
18. 1.2
19. 0.3
20. 400,300
21. 17
22. 1852
23. 1900
24. 0.000000103

Evaluate each expression. Express each result in scientific notation and standard form.

25. $(6 \times 10^3)(4 \times 10^5)$
26. $(7 \times 10)(3.5 \times 10^2)$
27. $(2.1 \times 10^3)(1 \times 10^4)$
28. $(4 \times 10^{-3})(7 \times 10^4)$
29. $(1.5 \times 10^5)(6 \times 10^{-3})$
30. $(5 \times 10^{-1})(2.5 \times 10^{-4})$

Lesson 8–5 *(Pages 357–361)* Simplify.

1. $\sqrt{64}$
2. $-\sqrt{9}$
3. $\sqrt{121}$
4. $-\sqrt{225}$
5. $\sqrt{\frac{144}{81}}$
6. $-\sqrt{\frac{225}{100}}$
7. $\sqrt{\frac{256}{16}}$
8. $\sqrt{\frac{36}{64}}$
9. $-\sqrt{\frac{225}{441}}$
10. $-\sqrt{\frac{121}{289}}$
11. $-\sqrt{0.36}$
12. $\sqrt{0.81}$
13. $\sqrt{0.0025}$
14. $-\sqrt{0.0049}$
15. $\sqrt{0.0289}$
16. $-\sqrt{0.000196}$
17. Find the negative square root of 25.
18. If $x = \sqrt{1024}$, what is the value of x?

Lesson 8–6 *(Pages 362–365)* Estimate each square root to the nearest whole number.

1. $\sqrt{5}$
2. $\sqrt{10}$
3. $\sqrt{11}$
4. $\sqrt{15}$
5. $\sqrt{18}$
6. $\sqrt{24}$
7. $\sqrt{61}$
8. $\sqrt{126}$
9. $\sqrt{153}$
10. $\sqrt{412}$
11. $\sqrt{483}$
12. $\sqrt{504}$
13. $\sqrt{555}$
14. $\sqrt{621}$
15. $\sqrt{709}$
16. $\sqrt{981}$
17. $\sqrt{70.3}$
18. $\sqrt{81.4}$
19. $\sqrt{121.6}$
20. $\sqrt{153.2}$
21. $\sqrt{9.35}$
22. $\sqrt{13.6}$
23. $\sqrt{0.021}$
24. $\sqrt{0.29}$
25. Is 11 closer to $\sqrt{119}$ or $\sqrt{125}$?
26. Which is closer to $\sqrt{285}$, 16 or 17?

Lesson 8–7 *(Pages 366–371)* If c is the measure of the hypotenuse and a and b are the measures of the legs, find each missing measure. Round to the nearest tenth if necessary.

1. 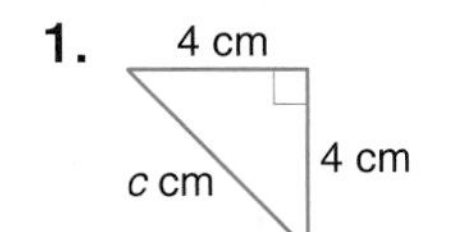

2. 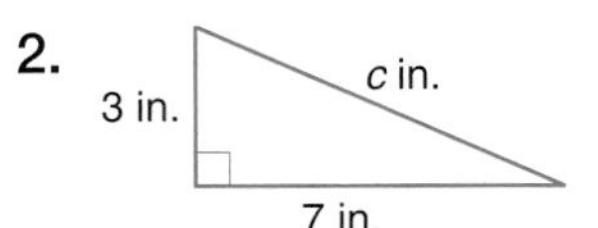

3. 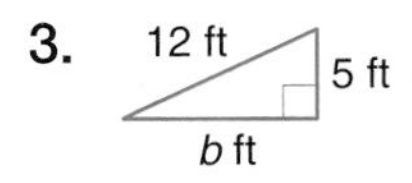

4. 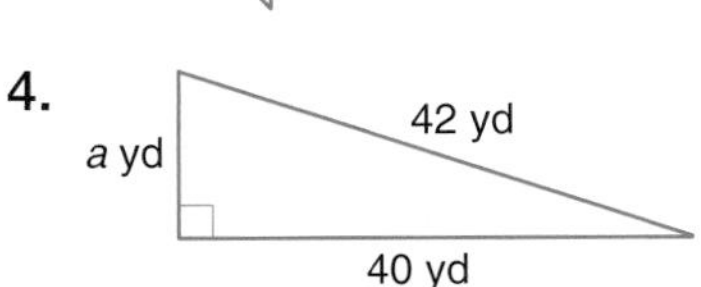

5. 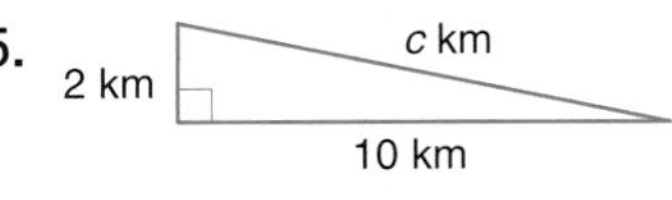
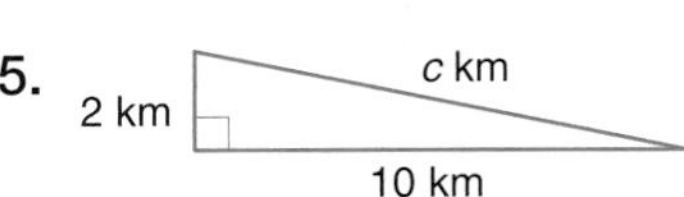

6. 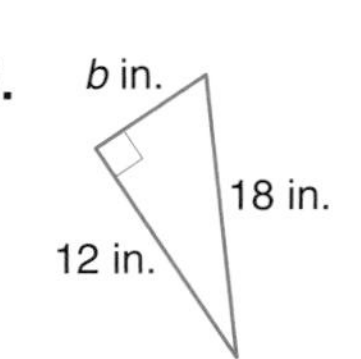

7. $a = 15, b = 17, c = ?$
8. $b = 12, c = 38, a = ?$
9. $a = 4, b = 5, c = ?$
10. $a = 10, c = 21, b = ?$
11. $a = 3, c = 8, b = ?$
12. $b = 20, c = 40, a = ?$

The lengths of three sides of a triangle are given. Determine whether each triangle is a right triangle.

13. 5 ft, 12 ft, 13 ft
14. 4 mi, 5 mi, 7 mi
15. 7 cm, 11 cm, 17 cm
16. 56 in., 70 in., 84 in.
17. 19 m, 24 m, 28 m
18. 31 mm, 37 mm, 49 mm
19. 3 ft, 6 ft, 12 ft
20. 2 in., 8 in., 10 in.
21. 6 mi, 8 mi, 10 mi
22. 10 cm, 24 cm, 26 cm
23. 20 m, 21 m, 29 m
24. 18 in., 26 in., 32 in.

EXTRA PRACTICE

Lesson 9–1 *(Pages 382–387)* Determine whether each expression is a monomial. Explain why or why not.

1. $6x$
2. $\frac{4}{k}$
3. $13b + 2a$
4. $7a^2b^{-1}$
5. m^2
6. $3x - y$
7. s^4
8. $-3x^2yz^3$

State whether each expression is a polynomial. If it is a polynomial, identify it as a *monomial, binomial,* or *trinomial.*

9. $11y$
10. $9xy^2 - y^3 + x^2$
11. $r^2 - r$
12. $\frac{1}{4}x - 2$
13. $3x - 11 + y - 2$
14. $6j^3k^2\ell - 7j^2$
15. $2.3mn^2$
16. $7r^2s^3t$

Find the degree of each polynomial.

17. y^2
18. $6x^3$
19. 4
20. $3j^2 - 2k + m$
21. $16m^2n + 14m^5n^3$
22. $11p^5q^2 - 6pq^7$
23. $x^2 + 2x + 3x^6$
24. $7a^2b^3 - 2ab$

Arrange the terms of each polynomial so that the powers of x are in descending order.

25. $4x^2 + x^3 - x$
26. $2x - 6x^5 - 3$
27. $5x^5 - 4x^2 + 3x - 2$
28. $6mx^5 - 4m^2x^6 + 4mx$
29. $3x^4y^5 + 2x^2y^3 + 6x^5y$
30. $2x - x^2 + 2$

Lesson 9–2 *(Pages 388–393)* Find each sum.

1. $\begin{array}{r} 7x - 2 \\ (+)\ x + 4 \\ \hline \end{array}$
2. $\begin{array}{r} 6x^2 - 2x - 1 \\ (+)\ 3x^2 - 4x - 7 \\ \hline \end{array}$
3. $\begin{array}{r} 2xy - 3x + 2 \\ (+)\ 4xy \qquad\ - 7 \\ \hline \end{array}$
4. $(-3y + 7) + (4y - 2)$
5. $(7ab - 2a) + (2ab + 3b)$
6. $(4x^2 - 2xy + y^2) + (xy + 3y^2)$

Find each difference.

7. $\begin{array}{r} 3x - 4 \\ (-)\ 2x + 3 \\ \hline \end{array}$
8. $\begin{array}{r} 5m^2 - \ m + 2 \\ (-)\ 2m^2 + 2m + 5 \\ \hline \end{array}$
9. $\begin{array}{r} 11a^2 - 2a - 1 \\ (-)\ 5a^2 + 6a + 2 \\ \hline \end{array}$
10. $(13x - 2) - (9x + 4)$
11. $(7m + 4n) - (2m - n)$
12. $(6a + b) - (2b + 4)$

Find each sum or difference.

13. $(2x + 3) + (4x^2 + x - 7)$
14. $(mn + 2pn - mp) - (2pn - mp)$
15. $(y^2 + 2y - 3) + (2y^3 - y - 3)$
16. $(x^2y + 5xy - 9y^2) - (3x^2 - 2y^2)$
17. $(3x^2 + 2x - 4) + (5x^2 - 9)$
18. $(x^2y + 9xy - y^2) - (2x^2y - y^2)$

Lesson 9–3 *(Pages 394–398)* Find each product.

1. $4(3x + 2)$
2. $-3(7b - 4)$
3. $a(2a - 4)$
4. $2n(n - 5)$
5. $-y(10y - 5)$
6. $-3x(5x + 8)$
7. $7p(p^2 - 3)$
8. $-5m(3m^3 - 2m)$
9. $2r^3(7 - r + r^2)$
10. $-6x^3(x^2 - 4x)$
11. $0.75k(8k^3 - k^2)$
12. $2.4z^2(2z^2 - 3)$

Solve each equation.

13. $-3(4 - x) = 18$
14. $21 = 7(y - 11)$
15. $10x - 9 = 4(x - 1) + 1$
16. $7(r + 8) - 4 = -4(-7 - r)$
17. $6(s - 7) = 3(4s + 4)$
18. $-a + 6(a + 4) = 3(a + 5) + 1$

Lesson 9–4 *(Pages 399–404)* **Find each product. Use the Distributive Property or the FOIL method.**

1. $(m + 3)(m + 2)$
2. $(x - 4)(x + 6)$
3. $(y - 5)(y - 7)$
4. $(r + 9)(r - 3)$
5. $(2s - 3)(s + 5)$
6. $(a - 2)(4a + 8)$
7. $(7p - 3)(4p + 2)$
8. $(2k + 2)(2k + 1)$
9. $(d + 7)(3d - 5)$
10. $(8x + 2y)(4x + 3y)$
11. $(3m + n)(m - 3)$
12. $(7r + 8s)(5r - 3s)$
13. $(6x - 1)(x - 2)$
14. $(7p - 2)(3p + n)$
15. $(2p - 1)(p - 3)$
16. $(b + a)(a - b)$
17. $(5c + d)(3c - 2d)$
18. $(p + 1)(p + m)$
19. $(j^2 - 2)(j + 5)$
20. $(z^2 + r)(z^2 - r)$
21. $(n^2 + 1)(2n^2 - 3)$

Lesson 9–5 *(Pages 405–409)* **Find each product.**

1. $(m + 5)^2$
2. $(n - 3)^2$
3. $(2x + 3)^2$
4. $(3y - x)^2$
5. $(4a + b)(4a - b)$
6. $(2k + 2p)^2$
7. $(4x - 2)(4x + 2)$
8. $(1 + p)^2$
9. $(a - 2b)(a + 2b)$
10. $(6 + 3m)^2$
11. $(2 - 4t)^2$
12. $(r - 2s)(3r + 2s)$
13. $(2x - 7y)^2$
14. $(m + 3n)^2$
15. $2(p - q)^2$
16. $4(r + s)^2$
17. $k(2 + k)^2$
18. $4s(s - 1)^2$
19. $3r(r - 2)^2$
20. $y(y + 1)(y - 2)$
21. $p(p + 2)(2p - 3)$
22. $2(j + 3)(j - 3)$
23. $(x + 3)(x - 1)(x + 2)$
24. $6(m + 5)(m - 1)$
25. The area of a circle is given by the formula $A = \pi r^2$, where r is the radius of the circle. Suppose a circle has a radius of $k - 4$ inches.
    a. Write an equation to find the area of the circle.
    b. Find the area to the nearest hundredth if $k = 6$.

Lesson 10–1 *(Pages 420–425)* **Find the factors of each number. Then classify each number as *prime* or *composite*.**

1. 57
2. 22
3. 65
4. 17
5. 104
6. 18
7. 81
8. 73

Factor each monomial.

9. $12x^2$
10. $28m^2n$
11. $-33j^2k^2$
12. $54p^3$
13. $81ab^2$
14. $75xy$
15. $-13p^2q^2$
16. $105r^4$

Find the GCF of each set of numbers or monomials.

17. 13, 33
18. 50, 75
19. 32, 84, 144
20. 32, 64, 96
21. $-21, 15xy$
22. $4x^2, 2x, 8x$
23. $-3r^2s^2, -17rs$
24. $14rs, 12rst, 6t$
25. $7kr, 21k^2, 2kr$
26. $-16c^2d, -4cd, -8cd^2$
27. $24m^2n, 51m, 63m^2n^2$
28. $-1x^3yz, -7x^2y^2, -2x^3yz$

Lesson 10–2 *(Pages 428–433)* **Factor each polynomial. If the polynomial cannot be factored, write *prime*.**

1. $4m + 12$
2. $13n + n^2$
3. $2k^2 + 6k$
4. $7s^2t + 3$
5. $9pq^3 - 21pq^2$
6. $x^2y + 7y$
7. $3k - 4j$
8. $14m^2n - 18m$
9. $17cd - 14mn$
10. $16a^2b^2 - 3ab$
11. $16b^2c - 2ac + 4bc$
12. $12mn - 14m^2 + 16n$
13. $6ab - 7bc + 12ac$
14. $m^2n^3p + mn - mp$
15. $x^2y + 15xy + 5x$

Find each quotient.

16. $(16x + 4y^2) \div 4$
17. $(2rs + r) \div r$
18. $(7xy - 3x) \div x$
19. $(15a + 3b^2) \div 3$
20. $(6m^2n - 9m) \div 3m$
21. $(18cd^2 - 9cd) \div 9cd$
22. $(21a^2b - 14a^2) \div 7a^2$
23. $(16xy - 12xy) \div 4xy$
24. $(12a^2b + 4b) \div 4b$
25. $(36st^2 - 9st) \div 9st$
26. $(15rs^2 + 12r^2st) \div 3rs$
27. $(5xy^2z + 10x^2z) \div 5xz$
28. $(13r^2s - 26r^2) \div 13r^2$
29. $(32np + m^2np^2) \div np$
30. $(20r^2st + 15rs) \div 5rs$

Lesson 10–3 *(Pages 434–439)* **Factor each trinomial. If the trinomial cannot be factored, write *prime*.**

1. $x^2 + 4x + 4$
2. $n^2 + 6n + 9$
3. $t^2 - 8t + 16$
4. $w^2 + 5w - 2$
5. $r^2 + r - 12$
6. $p^2 - 8p + 4$
7. $s^2 - 7s - 8$
8. $d^2 + 21d + 110$
9. $y^2 - 10y + 3$
10. $q^2 - 3q - 28$
11. $4z^2 - 16z - 18$
12. $6c^2 - 3c - 3$
13. $2m^2 - 3m - 20$
14. $8y^2 + 6y - 2$
15. $3r^2 - 12r - 15$

Lesson 10–4 *(Pages 440–444)* **Factor each trinomial. If the trinomial cannot be factored, write *prime*.**

1. $5x^2 + 13x + 6$
2. $4w^2 + 7w + 3$
3. $2a^2 - 9a + 4$
4. $12t^2 + 18t + 2$
5. $4c^2 - 8c - 3$
6. $3r^2 + 15r + 12$
7. $8y^2 - 16y + 6$
8. $20n^2 - 40n + 20$
9. $10m + 3 + 3m^2$
10. $13 + 2m^2 + 14m$
11. $3q^2 - 4 - 4q$
12. $12d + 8 - 8d^2$
13. $2m^2 - 3mn - 2n^2$
14. $2a^2 + 4ab + 2b^2$
15. $12p^2 + 10mp + 2m^2$
16. $18x^2 + 15xy + 3y^2$
17. A rectangle has dimensions of $(y + 5)$ inches and $(y - 4)$ inches.
    a. Express the area as a trinomial.
    b. If y units are removed from the length, express the new area.

Lesson 10–5 *(Pages 445–449)* **Determine whether each trinomial is a perfect square trinomial. If so, factor it.**

1. $x^2 - 14x + 49$
2. $a^2 - 4a + 4$
3. $y^2 - 16y + 3$
4. $d^2 + 10d + 25$
5. $r^2 + 24r + 144$
6. $9c^2 + 12c + 4$
7. $4k^2 - 28k + 49$
8. $16m^2 - 8m + 1$

Determine whether each trinomial is the difference of squares. If so, factor it.

9. $x^2 - 9$
10. $4 - 49z^2$
11. $36x^2 - 4$
12. $17 - 3p^2$
13. $s^2 - 16r^2$
14. $9 - m^2n^2$
15. $32k^2 - 50$
16. $12b^2 - 48$

Factor each polynomial. If the polynomial cannot be factored, write *prime*.

17. $a^2 - a - 12$
18. $2s^2 - rs - r^2$
19. $2r^2 - 3$
20. $4m^2 + 48m + 144$
21. $6m^2 - 16m - 6$
22. $2xy - 8x$
23. $3a^2b^3 - 27b$
24. $16x^2 - 8xy + y^2$

Lesson 11–1 *(Pages 458–463)* **Graph each quadratic function by making a table of values.**

1. $y = 6x^2$
2. $y = 8x^2$
3. $y = -2x^2$
4. $y = -3x^2$
5. $y = 2x^2 + 1$
6. $y = x^2 - 3$
7. $y = -3x^2 + 5$
8. $y = -4x^2 - 3$
9. $y = x^2 + 3x - 6$
10. $y = -2x^2 - x + 5$
11. $y = 3x^2 - 2x$
12. $y = 0.25x^2 - 2x - 4$

Write the equation of the axis of symmetry and the coordinates of the vertex of the graph of each quadratic function. Then graph the function.

13. $y = 2x^2$
14. $y = 3x^2$
15. $y = 8x^2$
16. $y = x^2 + 2x$
17. $y = 3x^2 - 6x$
18. $y = -2x^2 + 1$
19. $y = 4x^2 - x + 2$
20. $y = -x^2 + 6x - 4$
21. $y = \frac{1}{2}x^2 - x + 1$
22. $y = -\frac{1}{2}x^2 - 2x + 2$
23. $y = 2x^2 + 3x - 1$
24. $y = -x^2 - 2x - 1$

Match each function with its graph.

25. $y = (x + 2)^2 - 2$
26. $y = -2x^2 - x - 1$
27. $y = -2x^2 + x + 2$

A.

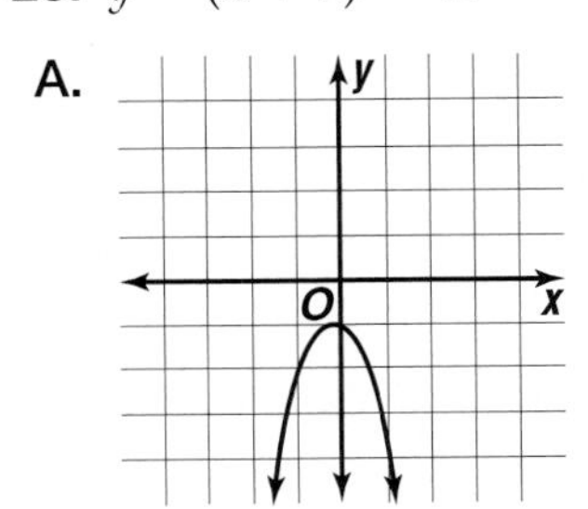

B.

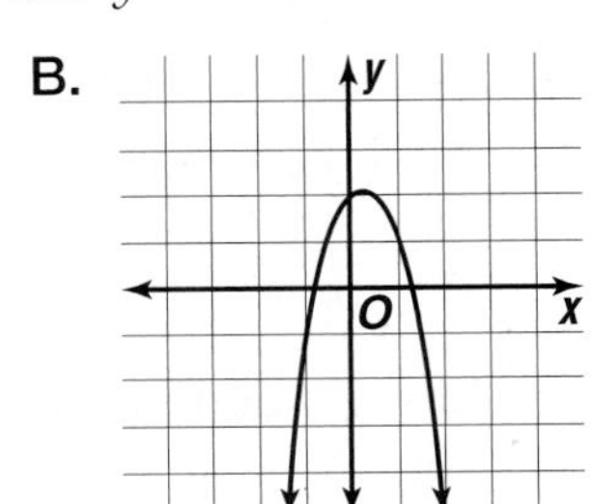

C.

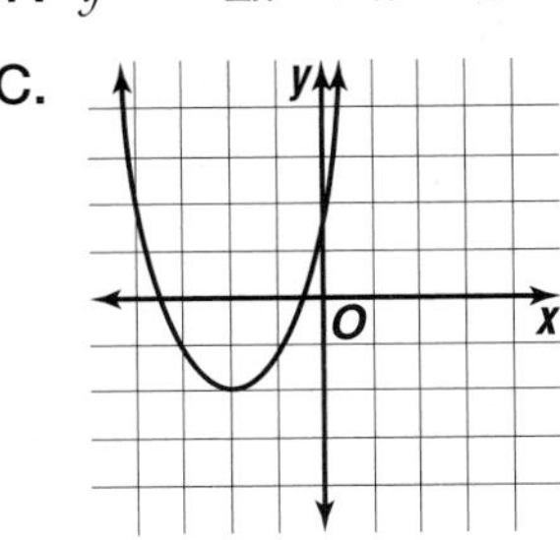

Lesson 11–2 *(Pages 464–467)* **Graph each group of equations on the same screen. Compare and contrast the graphs.**

1. $y = x^2$
   $y = 2x^2$
   $y = 4x^2$
2. $y = x^2 + 1$
   $y = x^2 - 4$
   $y = x^2 - 8$
3. $y = (2x - 1)^2$
   $y = (2x - 2)^2$
   $y = (2x - 3)^2$

Describe how each graph changes from the parent graph of $y = x^2$. Then name the vertex of each graph.

4. $y = 8x^2$
5. $y = -2x^2$
6. $y = \frac{1}{2}x^2$
7. $y = (x - 1)^2$
8. $y = (x + 3)^2$
9. $y = (3x + 2)^2$
10. $y = -(x + 3)^2$
11. $y = -2x^2 - 3$
12. $y = 0.25x^2 + 0.5$
13. $y = -3x^2 + 8$
14. $y = (4x - 1)^2 + 3$
15. $y = (x + 2)^2 - 2$

Lesson 11–3 *(Pages 468–473)* **Solve each equation by graphing the related function. If exact roots cannot be found, state the consecutive integers between which the roots are located.**

1. $-x^2 - 2x + 24 = 0$
2. $x^2 + 4x - 5 = 0$
3. $x^2 - 4x - 2 = 0$
4. $x^2 + 7x + 3 = 0$
5. $x^2 + 7x + 5 = 0$
6. $x^2 - 6x + 6 = 0$
7. $-x^2 - 3x - 2 = 0$
8. $4x^2 + 2x + 1 = 0$
9. $-x^2 - 6x - 6 = 0$
10. $x^2 - 11x + 4 = 0$
11. $-2x^2 - 2x + 10 = 0$
12. $x^2 + 12x + 20 = 0$
13. $-x^2 - 12x - 3 = 0$
14. $-3x^2 - x + 8 = 0$
15. $x^2 + 3x + 4 = 0$
16. $x^2 + 5x - 9 = 0$

Use a quadratic equation to determine the two numbers that satisfy each situation.

17. Their sum is 21 and their product is 104.
18. Their difference is 8 and their product is 20.
19. Their sum is 13 and their product is 22.
20. Their sum is 32 and their product is 135.

Lesson 11–4 *(Pages 474–477)* **Solve each equation. Check your solution.**

1. $2r(r - 4) = 0$
2. $4k(k + 5) = 0$
3. $(s + 4)(s - 3) = 0$
4. $(m - 4)(m - 5) = 0$
5. $(3x - 4)(x - 2) = 0$
6. $(2y + 2)(2y - 4) = 0$
7. $(t + 2)(6t + 1) = 0$
8. $n^2 + n - 6 = 0$
9. $k^2 + 4k + 4 = 0$
10. $p^2 - 5p + 6 = 0$
11. $q^2 - 2q - 15 = 0$
12. $x^2 + 2x - 3 = 0$
13. $j^2 + 9j + 20 = 0$
14. $r^2 - 16r = 0$
15. $z^3 - 25z = 0$

For each problem, define a variable. Then use an equation to solve the problem.

16. Find two integers whose sum is 15 and whose product is 36.
17. The length of a swimming pool is 15 feet longer than it is wide. The area in square feet is 1350. Find the dimensions of the pool.
18. Find two integers whose difference is 12 and whose product is 13.

Lesson 11–5 *(Pages 478–482)* **Find the value of *c* that makes each trinomial a perfect square.**

1. $r^2 - 16r + c$
2. $k^2 + 12k + c$
3. $p^2 - 4p + c$
4. $n^2 + 2n + c$
5. $f^2 - 8f + c$
6. $s^2 + 18s + c$
7. $x^2 + 20x + c$
8. $r^2 + 14r + c$
9. $w^2 + 30w + c$
10. $h^2 + 10h + c$
11. $z^2 - 2z + c$
12. $m^2 - 6m + c$
13. $q^2 + 26q + c$
14. $t^2 + 28t + c$
15. $y^2 + 22y + c$
16. $z^2 + 24z + c$

Solve each equation by completing the square.

17. $z^2 + 10z + 12 = 0$
18. $h^2 - 8h - 15 = 0$
19. $y^2 + 3y + 1 = 0$
20. $w^2 + 15w = 5$
21. $m^2 + 2m = 0$
22. $t^2 + 2t = 18$
23. $r^2 - 20r + 24 = 0$
24. $p^2 - 2p = 32$
25. $q^2 - 7q + 12 = 0$
26. $n^2 - 4n - 16 = 0$
27. $x^2 + 10x = 12$
28. $r^2 + 12r = 0$

Lesson 11–6 (Pages 483–488) Use the Quadratic Formula to solve each equation.

1. $j^2 + 3j - 4 = 0$
2. $w^2 + 9w + 20 = 0$
3. $m^2 - 7m + 12 = 0$
4. $n^2 + 5n - 6 = 0$
5. $k^2 + 2k - 15 = 0$
6. $z^2 - 4z + 4 = 0$
7. $d^2 - 3d - 18 = 0$
8. $s^2 + 8s - 14 = 0$
9. $x^2 - 5x - 24 = 0$
10. $t^2 + 5t + 6 = 0$
11. $r^2 + 6r + 9 = 0$
12. $y^2 - 2y - 8 = 0$
13. $y^2 - y - 6 = 0$
14. $d^2 + 4d - 21 = 0$
15. $s^2 + 3s + 2 = 0$
16. $m^2 - 11m = -18$
17. $4k^2 - 4 = 8k$
18. $-3r^2 - 15r = -18$
19. $3p^2 + 5p + 2 = 0$
20. $6x^2 + 8x = 8$
21. $2p^2 + 5p = 3$
22. $4q^2 - 2 = -2q$
23. $-2c^2 - 4c - 2 = 0$
24. $-8m^2 - 10m = 2$

Lesson 11–7 (Pages 489–493) Graph each exponential function. Then state the y-intercept.

1. $y = 2^x$
2. $y = 5^x$
3. $y = 3^x - 1$
4. $y = 4^x + 4$
5. $y = 2^x - 3$
6. $y = 2^x + 3$
7. $y = 4^x + 3$
8. $y = 2^x - 6$
9. $y = 3^x - 3$
10. $y = 2^{4x} + 1$
11. $y = 4^{0.5x}$
12. $y = 5^{3x} + 1$
13. $y = 4^{3x} - 2$
14. $y = 3^{3x} - 2$
15. $y = 2^{0.5x} - 1$
16. $y = 2^{3x} - 4$

Find the amount of money in a bank account given the following conditions.

17. initial deposit = \$6000, annual rate = 6.5%, time = 3 years
18. initial deposit = \$1000, annual rate = 12%, time = 15 years
19. initial deposit = \$2500, annual rate = 2%, time = 6 years
20. initial deposit = \$5100, annual rate = 9%, time = 4 years

Lesson 12–1 (Pages 504–508) Write an inequality to describe each number.

1. a number less than 10
2. a number that is at least −4
3. a number greater than −2
4. a number less than or equal to 5
5. a number greater than 3
6. a number more than 12
7. a minimum number of 7
8. a number less than −1
9. a maximum number of 8
10. a number greater than −6
11. a minimum number of −8
12. a number more than −11

Graph each inequality on a number line.

13. $m < 8$
14. $n \le -4$
15. $x > 2$
16. $z < -7$
17. $r \ge 15$
18. $m \le -5$
19. $s > 8.4$
20. $y > 6.2$
21. $w \le 1.3$
22. $\ell \ge -2.4$
23. $p < -3.2$
24. $j > 4.3$
25. $t < \frac{1}{3}$
26. $r \le -2\frac{1}{4}$
27. $y \le \frac{1}{2}$
28. $q \ge 3\frac{5}{8}$

Write an inequality for each graph.

29. 5 6 7 8 9

30. 0 1 2 3 4

31. 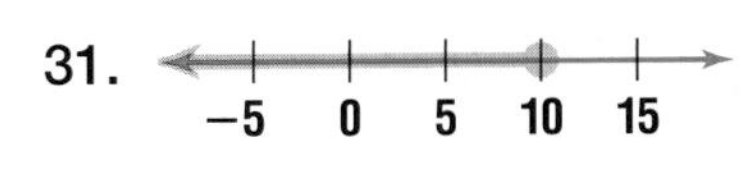

32. 6 7 8 9 10

33. −6 −5 −4 −3 −2

34. −5 −4 −3 −2 −1

35. 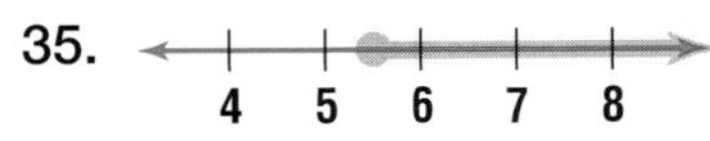

36. 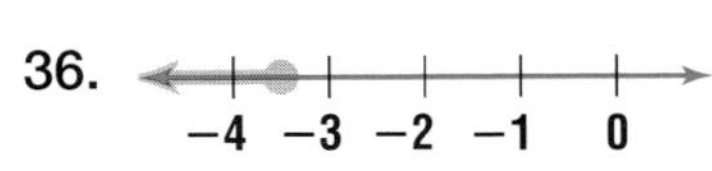

37. 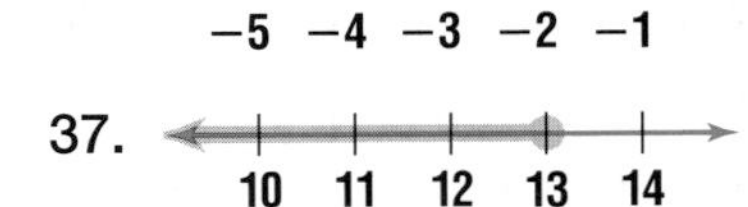

Lesson 12–2 *(Pages 509–513)* Solve each inequality. Check your solution.

1. $r + 3 < 8$
2. $m + 4 > -2$
3. $j - 3 > 5$
4. $k - 6 < -13$
5. $-12 + w < 15$
6. $p + 11 \geq 5$
7. $0.4 + p \geq 1.2$
8. $x - 6.2 < 4$
9. $4.3 < 2.1 + y$
10. $\frac{1}{4} + r \leq 3\frac{3}{4}$
11. $k + 10.6 \geq -3.4$
12. $\frac{1}{6} + s \leq \frac{1}{3}$

Solve each inequality. Graph the solution.

13. $2n > n + 3$
14. $8 + 5x > 6x$
15. $3y - 2 < 4y$
16. $5d < 8 + 4d$
17. $9 \geq 3t - 4 - 2t$
18. $-2a < -3(a - 2)$
19. $6u \leq 5(u + 1)$
20. $11p \leq 2(5p + 4)$
21. $7s < 3(2s - 3)$
22. $3(x - 3) \leq 4x$
23. $7b < 4(2b - 3)$
24. $2c + 3 \leq 3(c - 5)$

Lesson 12–3 *(Pages 514–518)* Solve each inequality. Check your solution.

1. $-3k < 15$
2. $2r > -10$
3. $-4d \geq -12$
4. $\frac{x}{5} \geq 6$
5. $-\frac{y}{4} < 8$
6. $\frac{p}{7} \leq -3$
7. $-9\ell < 27$
8. $2s > 20$
9. $-t > 11$
10. $-\frac{d}{2} > 12$
11. $\frac{m}{6} \leq -5$
12. $-\frac{b}{4} < -9$
13. $-3p < 2$
14. $-8y \geq -4$
15. $7v \leq 3$
16. $2 < \frac{2}{3}j$
17. $\frac{5}{8}t \geq -5$
18. $-\frac{1}{4}w < 6$
19. $0.01r \leq 8$
20. $2.1m > 6.3$
21. $-2.25j \geq 9$
22. $\frac{k}{7.2} < -3$
23. $-\frac{r}{2.5} > 4$
24. $\frac{y}{0.4} \geq 2$

Lesson 12–4 *(Pages 519–523)* Solve each inequality. Check your solution.

1. $2a + 7 \leq 11$
2. $5r - 3 > 27$
3. $6 - 4m > 10$
4. $1 - 3s \geq 13$
5. $3 + 5d \leq -12$
6. $-8x - 1 < 15$
7. $6 + 2b \geq -2.4$
8. $5.1x - 2.4 < -7.5$
9. $3.3 + 4k > -8.7$
10. $\frac{6-r}{8} < 7$
11. $\frac{j}{3} + 7 > 9$
12. $\frac{4 + 3m}{7} \leq -5$
13. $2x - 1 < 8x + 2$
14. $11 \leq -(t + 4)$
15. $4(3 - 6r) > 18$
16. $\frac{2}{3}(k - 6) > 4$
17. $\frac{1}{3}(y + 3) < \frac{1}{2}(y - 2)$
18. $\frac{1}{8}(m + 3) \geq \frac{1}{4}(m - 3)$

Write and solve an inequality for each situation.

19. Three fifths times the sum of a number and 5 is greater than 15.
20. Four times the difference of a number and 3 is less than 24.

Lesson 12–5 *(Pages 524–529)* Write each compound inequality without using *and*.

1. $m > 2$ and $m < 6$
2. $j > -12$ and $j < 4$
3. $r < 6$ and $r \geq -1$
4. $x < 3$ and $x \geq 2$
5. $y \leq 5$ and $y \geq -3$
6. $s \geq -7$ and $s < -4$

Graph the solution of each compound inequality.

7. $w > 6$ or $w < 2$
8. $a < -4$ or $a \geq 4$
9. $z > 12$ and $z \leq 15$
10. $h \leq 20$ and $h \geq -3$
11. $s < -7$ or $s > -5$
12. $f \leq 8$ or $f > 9$

Solve each compound inequality. Graph the solution.

13. $4 < 2n < 10$
14. $-4 \leq x + 7 < 9$
15. $12 > v - 6 > -2$
16. $9 \leq 3q \leq 21$
17. $24 > b - 5 > -2$
18. $8 > c + 4 \geq 5$
19. $t + 3 > 15$ or $t - 5 < -12$
20. $-12 \leq -3d < 9$
21. $5.4 < 0.2k < 6$
22. $u - 12 > -3$ or $u + 11 < 2$
23. $-4 \leq g - \frac{1}{3} < 1$
24. $\frac{p}{3} > 3$ or $\frac{p}{3} \leq -2$

Lesson 12–6 *(Pages 530–534)* Solve each inequality. Graph the solution.

1. $|d + 7| > 15$
2. $|5c| > 30$
3. $|a - 2| < 17$
4. $|z + 3| > 12$
5. $|j - 12| < 10$
6. $|\ell + 8| \leq -14$
7. $|t - 6| < 9$
8. $|m - 4| \leq 3$
9. $|4x| < -20$
10. $|s + 5| < 8$
11. $|n + 1| \geq 7$
12. $|-2v| > 14$
13. $|y - 3| > 16$
14. $|r - 7| \leq -11$
15. $|-6b| > 18$
16. $|w - 4| \leq 1.5$

For each graph, write an inequality involving absolute value.

17. −7 −6 −5 −4 −3 −2 −1 0 1 2 3

18. −3 −2 −1 0 1 2 3 4 5 6 7

19. −1 0 1 2 3 4 5 6 7 8 9

Write an inequality involving absolute value for each statement. Do not solve.

20. Alli's quiz score was within 5 points of her average of 85.
21. The 5-inch-wide picture frame was made with an accuracy of 0.1 inch.

Lesson 12–7 *(Pages 535–539)* Graph each inequality.

1. $x < -3$
2. $y > -6$
3. $y \geq x - 3$
4. $y < x + 2$
5. $y \geq -4x$
6. $6 \geq 2x + 4y$
7. $y > -2x + 6$
8. $x + 5y \geq 15$
9. $x + y \leq -2$
10. $-3x + 2 > y$
11. $y \leq -2(x - 1)$
12. $8 \geq y - 4x$
13. $y > -x - 8$
14. $7 + 3y \leq x$
15. $2x - y \geq -6$
16. $x - y \leq 5$
17. $3(6x + y) < 4$
18. $y \leq 4(x - 2)$
19. $-(4x - 3) \geq 2y$
20. $-2(3x + y) < 1$

For Exercises 21–24, write an inequality and graph the solution.

21. The difference of a number and three is less than or equal to eight.
22. Three times a number is greater than negative six.
23. One half the sum of a number and eight is greater than or equal to twelve.
24. A number minus three is less than another number.

Lesson 13–1 *(Pages 550–553)* Solve each system of equations by graphing.

1. $x = -2$; $y = 3$
2. $x = 5$; $y = x$
3. $x = -1$; $y = x - 2$
4. $x = -3$; $y = x + 4$
5. $y = -x - 4$; $y = x + 4$
6. $y = x + 2$; $y = 2x - 1$
7. $y = -x - 1$; $y = x - 1$
8. $x = 2$; $x + 2y = 4$
9. $y = 4x + 1$; $y = 3x$
10. $y = -3x - 2$; $2x - y = 2$
11. $y = 2$; $x - y = 3$
12. $x - y = 6$; $2x + y = 3$
13. $y = -\frac{1}{2}x$; $y = 3x + 7$
14. $y = \frac{1}{2}x - 3$; $x - y = 6$
15. $\frac{1}{4}x - 2 = y$; $x = -4y$
16. $2x + 3y = 12$; $4x + y = 4$

Lesson 13–2 *(Pages 554–559)* State whether each system is *consistent and independent, consistent and dependent,* or *inconsistent.*

1.

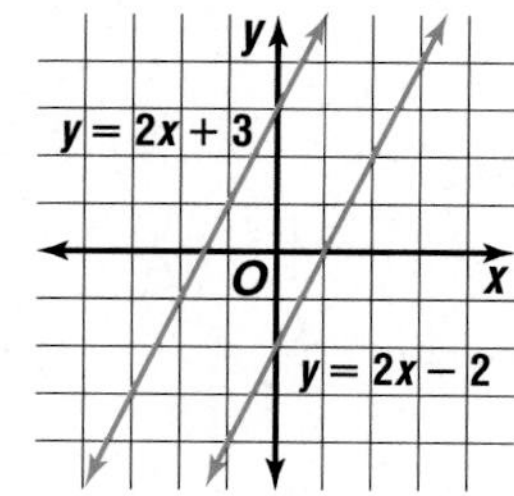

2.

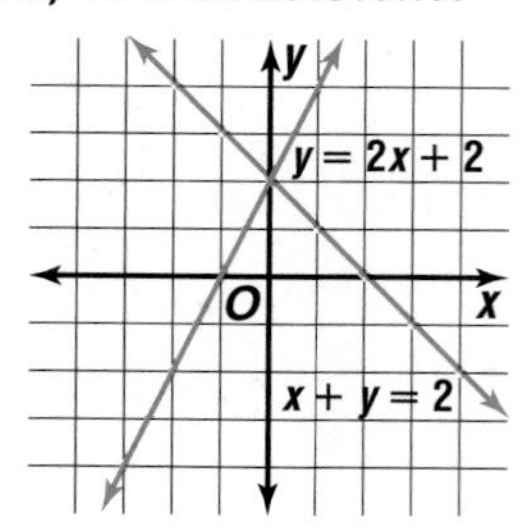

3.

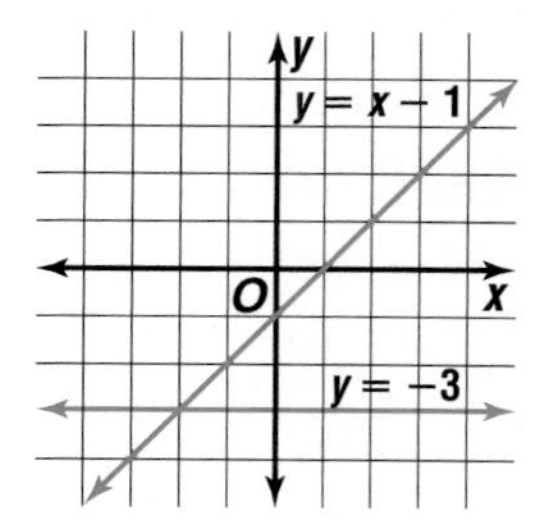

Determine whether each system of equations has *one* solution, *no* solution, or *infinitely many* solutions by graphing. If the system has one solution, name it.

4. $y = 2x + 1$; $y = -x - 2$
5. $x = 5$; $x - 4y = 1$
6. $y = 3$; $x - y = 2$
7. $y = x + 7$; $x = 7 - y$
8. $y = 2x - 6$; $y = 2x + 4$
9. $y = 10x - 16$; $y = 4x - 4$
10. $x + 2y = 5$; $x + y = 4$
11. $x + 4y = 5$; $2x + 6y = 6$
12. $y = \frac{1}{4}x - 1$; $y = -x + 4$
13. $\frac{1}{3}y = -x - \frac{1}{3}$; $y = -3x - 1$
14. $\frac{1}{3}x - 6y = 9$; $x - 18y = 27$
15. $y = \frac{3}{2}x$; $y = -\frac{2}{3}x + 13$
16. $x + y = 12$; $2x - y = 3$
17. $x - 2y = -3$; $-2x + 4y = 8$
18. $4x + 6y = 12$; $2x - 6 = -3y$
19. $3x + 3y = 6$; $4x - y = 3$

Lesson 13–3 *(Pages 560–565)* Use substitution to solve each system of equations.

1. $y = -x$; $x - y = -12$
2. $x = 5 - y$; $-3x + y = -3$
3. $y = x + 1$; $-x + 3y = -15$
4. $y = x + 2$; $2x - y = 1$
5. $y = 4x$; $x + y = 3$
6. $x = 3y$; $x + 3y = 4$
7. $y = 2x$; $2x - y = -1$
8. $y = 3x - 1$; $x + y = 4$
9. $y = 6x - 7$; $-2x - y = 3$
10. $x = 6 - y$; $x - 2y = 1$
11. $x + 4y = -8$; $3x - 6y = 0$
12. $x = 3 + y$; $3y + 2 = x$
13. $x - 7y = 0$; $2x + y = 0$
14. $x - 4y = 8$; $6y + 8 = 2x$
15. $6x - 3y = -2$; $y + 1 = x$
16. $x - \frac{1}{2}y = 14$; $x + \frac{1}{2}y = 2$

Lesson 13–4 *(Pages 566–571)* **Use elimination to solve each system of equations.**

1. $x + y = 4$
   $x - y = 16$
2. $x + 2y = 4$
   $x + 3y = 6$
3. $x = 13 - y$
   $x - y = -3$
4. $x - 6y = -1$
   $x + 3y = -10$
5. $x - 3y = 7$
   $x + 2y = 8$
6. $x - 4y = 5$
   $-5x + 10y = -5$
7. $x - y = -2$
   $2x + 2y = 16$
8. $x - y = 14$
   $3x - 6y = 15$
9. $x - 13y = -2$
   $7y - x = 5$
10. $x - 6y = 9$
    $4x - 8y = 12$
11. $x + y = 8$
    $16 + 3x = y$
12. $x - y = 9$
    $4x - 2y = -8$
13. $x = 5 - y$
    $6x - 4y = 0$
14. $x = 6 - y$
    $2x - 8y = 2$
15. $4x + 3y = 12$
    $8x - 4y = -12$
16. $6x - 3y = 18$
    $6x + 3y = 12$

Lesson 13–5 *(Pages 572–577)* **Use elimination to solve each system of equations.**

1. $x + 5y = 10$
   $x + y = -6$
2. $x + 4y = 5$
   $x - 2y = -7$
3. $x - 6y = 9$
   $x + 3y = -9$
4. $-x - y = 4$
   $x + y = -4$
5. $2x + y = -2$
   $2x - y = 4$
6. $6x + y = -9$
   $3x - y = 6$
7. $3x - y = 0$
   $4x + 2y = 10$
8. $2x + 12y = 24$
   $x + 7y = 18$
9. $4x - 8y = 0$
   $x + 3y = -10$
10. $2x - 8y = 0$
    $-x + 7y = -3$
11. $x - 8y = 5$
    $-2x + 8y = -2$
12. $15x - y = 4$
    $-6x + y = 5$
13. $4x - 16y = 24$
    $2x + 2y = 12$
14. $x - 13y = 4$
    $-2x + 10y = -4$
15. $-7x - y = 6$
    $-3x + y = 4$
16. $6x - 3y = 18$
    $10x + y = -12$

Lesson 13–6 *(Pages 580–585)* **Solve each system of equations by graphing.**

1. $x = 4$
   $y = x - 1$
2. $x = -2$
   $y = x^2$
3. $x = -3$
   $y = x^2 + 3$
4. $y = x^2$
   $y = 3x$
5. $y = x^2$
   $y = 2x$
6. $y = x^2 + 2$
   $y = x + 4$
7. $y = x^2 + 1$
   $y = -x + 1$
8. $y = x^2 + 3$
   $y = 5x - 3$
9. $y = 4x^2$
   $y = 8x$
10. $y = x^2 + 12$
    $y = -x - 8$
11. $y = x^2 + 4$
    $y = 3x + 2$
12. $y = -x^2 + 3$
    $y = 3x + 5$

Use substitution to solve each system of equations.

13. $y = 2x^2$
    $x = -2$
14. $y = -x^2$
    $y = 5x + 6$
15. $y = -2x^2$
    $y = 4x - 6$
16. $y = -3x^2$
    $x = -5$
17. $y = \frac{1}{2}x^2 - 1$
    $x = 4$
18. $y = -\frac{1}{2}$
    $y = \frac{1}{2}x^2 - \frac{5}{2}$
19. $y = -\frac{1}{2}x^2 + \frac{7}{2}$
    $x = 2x + 1$
20. $y = \frac{1}{4}x^2 - 2$
    $y = 2$
21. $y = -x + 4$
    $y = x^2 + 2$
22. $y = x^2 + 3$
    $y = -7x^2 + 5$
23. $y = x^2 + x + 4$
    $y = 6$
24. $y = 3x^2 + 6x - 9$
    $y = -12$

EXTRA PRACTICE

Lesson 13–7 *(Pages 586–591)* **Solve each system of inequalities by graphing.**

1. $y > -2$, $y < 3$
2. $x \leq -4$, $y \geq -1$
3. $y < 0$, $x \geq -1$
4. $y \geq -2$, $x > 2$
5. $y > x$, $y > x + 1$
6. $y < x - 2$, $y > -1$
7. $y < -2$, $y > -2x$
8. $y \geq x$, $y \leq -3x$
9. $x < 3$, $y < x + 1$
10. $y \geq -2$, $y \leq -2x - 1$
11. $x + y > 0$, $2x + y \leq 1$
12. $y \geq x + 1$, $y < 2x + 2$
13. $y - 2 \leq x$, $y + 4 \geq x$
14. $y \leq 3x - 1$, $2x + 4 < y$
15. $2x + 4 > y$, $y - 2 \leq x$
16. $-3y - x > 2$, $2y + x < 0$

Lesson 14–1 *(Pages 600–605)* **Name the set or sets of numbers to which each real number belongs. Let N = natural numbers, W = whole numbers, Z = integers, Q = rational numbers, and I = irrational numbers.**

1. $\frac{21}{7}$
2. $\sqrt{13}$
3. $\sqrt{121}$
4. 47.13013001 . . .
5. $-\sqrt{38}$
6. 0.631
7. $\frac{1}{4}$
8. $-\frac{8}{2}$
9. -3
10. $-\frac{1}{6}$
11. 0.949949994 . . .
12. $-\sqrt{64}$

Find an approximation, to the nearest tenth, for each square root. Then graph the square root on a number line.

13. $-\sqrt{5}$
14. $\sqrt{14}$
15. $\sqrt{18}$
16. $\sqrt{29}$
17. $\sqrt{63}$
18. $-\sqrt{71}$
19. $\sqrt{82}$
20. $-\sqrt{93}$
21. $\sqrt{102}$
22. $\sqrt{145}$
23. $-\sqrt{201}$
24. $\sqrt{305}$

Determine whether each number is *rational* or *irrational*. If it is irrational, find two consecutive integers between which its graph lies on the number line.

25. $\sqrt{4}$
26. $-\sqrt{19}$
27. $-\sqrt{62}$
28. $\sqrt{33}$
29. $-\sqrt{54}$
30. $\sqrt{49}$
31. $-\sqrt{225}$
32. $-\sqrt{15}$
33. $\sqrt{196}$
34. $-\sqrt{152}$
35. $-\sqrt{181}$
36. $\sqrt{8}$

Lesson 14–2 *(Pages 606–611)* **Find the distance between each pair of points. Round to the nearest tenth, if necessary.**

1. $Q(2, 16)$, $R(3, 15)$
2. $C(-4, -2)$, $D(-6, -5)$
3. $P(-8, 1)$, $Q(10, -7)$
4. $A(3, -9)$, $B(1, -2)$
5. $G(-6, -14)$, $H(-7, 2)$
6. $X(0, 0)$, $Y(9, 9)$
7. $E(12, -2)$, $F(-3, -4)$
8. $M(15, 3)$, $N(8, -11)$
9. $V(-4, 4)$, $W(6, -6)$

Find the value of *a* if the points are the indicated distance apart.

10. $A(a, 3)$, $B(6, 5)$; $d = 2$
11. $G(-1, 5)$, $H(-8, a)$; $d = \sqrt{85}$
12. $X(9, a)$, $Y(5, -2)$; $d = 4$
13. $P(6, 1)$, $Q(a, -7)$; $d = \sqrt{113}$
14. $C(-9, -2)$, $D(0, a)$; $d = \sqrt{90}$
15. $Q(a, -1)$, $R(4, 5)$; $d = 10$
16. $E(7, a)$, $F(-2, 4)$; $d = \sqrt{90}$
17. $M(a, 3)$, $N(-1, 5)$; $d = \sqrt{8}$
18. $V(-3, -3)$, $W(a, 4)$; $d = \sqrt{50}$

Lesson 14–3 *(Pages 614–619)* Simplify each expression. Leave in radical form.

1. $\sqrt{54}$
2. $\sqrt{80}$
3. $\sqrt{75}$
4. $\sqrt{300}$
5. $\sqrt{3} \cdot \sqrt{12}$
6. $\sqrt{4} \cdot \sqrt{8}$
7. $2\sqrt{6} \cdot \sqrt{18}$
8. $\sqrt{10} \cdot \sqrt{15}$
9. $\frac{\sqrt{18}}{\sqrt{9}}$
10. $\frac{\sqrt{12}}{\sqrt{4}}$
11. $\frac{\sqrt{20}}{\sqrt{8}}$
12. $\frac{\sqrt{75}}{\sqrt{5}}$
13. $\frac{4}{2 + \sqrt{12}}$
14. $\frac{6}{3 - \sqrt{8}}$
15. $\frac{3}{5 + \sqrt{3}}$
16. $\frac{5}{4 + \sqrt{10}}$

Simplify each expression. Use absolute value symbols if necessary.

17. $\sqrt{18a^2b}$
18. $\sqrt{21m^8}$
19. $\sqrt{120x^2y}$
20. $\sqrt{50jk}$
21. $\sqrt{16r^2s^3}$
22. $\sqrt{17n^3}$
23. $\sqrt{32c^4d^3}$
24. $\sqrt{12xy^2}$
25. $\sqrt{53m^2n^4}$
26. $\sqrt{51g^2h^2}$
27. $\sqrt{66mn^4}$
28. $\sqrt{48j^6k^2}$
29. $\sqrt{8a^2bc^2}$
30. $\sqrt{75x^6y^6}$
31. $\sqrt{30rs^2t^4}$
32. $\sqrt{196a^6}$

Lesson 14–4 *(Pages 620–623)* Simplify each expression.

1. $5\sqrt{3} + 7\sqrt{3}$
2. $4\sqrt{6} - 2\sqrt{6}$
3. $5\sqrt{7} + 2\sqrt{7}$
4. $13\sqrt{5} - 5\sqrt{5}$
5. $-11\sqrt{3} + 6\sqrt{2}$
6. $-6\sqrt{5} - 5\sqrt{5}$
7. $4\sqrt{2} + 3\sqrt{2} - 5\sqrt{2}$
8. $-3\sqrt{3} - 2\sqrt{3} - 4\sqrt{3}$
9. $2\sqrt{7} + 8\sqrt{7} - 14\sqrt{7}$
10. $4\sqrt{12} - 7\sqrt{3}$
11. $-2\sqrt{24} + 3\sqrt{6}$
12. $-6\sqrt{32} + 4\sqrt{8}$
13. $4\sqrt{2} - 3\sqrt{3} + 6\sqrt{2}$
14. $-7\sqrt{5} + 6\sqrt{3} - 2\sqrt{3}$
15. $\sqrt{105} - \sqrt{12} - \sqrt{18}$
16. $-8\sqrt{24} + 6\sqrt{12} - 3\sqrt{2}$
17. $4\sqrt{27} - 2\sqrt{48} + 3\sqrt{20}$
18. $\sqrt{52} - \sqrt{18} + \sqrt{120}$
19. If an equilateral triangle has a side of length $5\sqrt{6}$, what is the measure of the perimeter?

Lesson 14–5 *(Pages 624–629)* Solve each equation. Check your solution.

1. $\sqrt{m} = 4$
2. $\sqrt{y} = -2$
3. $\sqrt{2d} = 4$
4. $-\sqrt{5k} = 15$
5. $\sqrt{8t} = 4$
6. $\sqrt{r - 9} = 9$
7. $\sqrt{t + 5} = 7$
8. $\sqrt{n - 3} = 2$
9. $\sqrt{h + 5} = 0$
10. $\sqrt{2x + 6} = 6$
11. $\sqrt{4m + 5} - 6 = 2$
12. $\sqrt{3r - 2} + 5 = 4$
13. $r = \sqrt{r + 12}$
14. $k = \sqrt{3k + 10}$
15. $x - 10 = \sqrt{x + 2}$
16. $m = \sqrt{9m + 4} - 2$
17. $2 + \sqrt{4d + 4} = d$
18. $5 - \sqrt{5r - 6} = 9 - r$

EXTRA PRACTICE

Lesson 15–1 *(Pages 638–643)* Find the excluded value(s) for each rational expression.

1. $\frac{y}{y-6}$
2. $\frac{2m}{3m+6}$
3. $\frac{-n}{-8+4n}$
4. $\frac{j+1}{2j-6}$
5. $\frac{6}{r(r-2)}$
6. $\frac{8k}{k(k+1)}$
7. $\frac{2x}{(x-3)(x+1)}$
8. $\frac{7m}{(m+1)(m+1)}$
9. $\frac{2s}{s^2+2s}$
10. $\frac{q-2}{q^2-25}$
11. $\frac{3+p}{p^2-p-2}$
12. $\frac{x^2-2x+3}{x^2+x-6}$

Simplify each rational expression.

13. $\frac{8}{22}$
14. $\frac{6}{24}$
15. $\frac{30a}{36b}$
16. $\frac{2y^2}{4y}$
17. $\frac{2(x+1)}{x(x+1)}$
18. $\frac{(m-1)(m+2)}{(m+2)(m+3)}$
19. $\frac{2d^2-2d}{d^2-1}$
20. $\frac{-4j+12}{j^2-9}$
21. $\frac{n^2+2n-15}{n^2+3n-10}$
22. $\frac{2y^2+2y}{y^2-y-2}$
23. $\frac{x^2+2x-8}{x^2-7x+10}$
24. $\frac{r^2-r-6}{r^3-6r^2+9r}$

Lesson 15–2 *(Pages 644–649)* Find each product.

1. $\frac{xy}{x} \cdot \frac{z}{y}$
2. $\frac{2s}{3t} \cdot \frac{6}{4s^2}$
3. $\frac{7b^2}{c^2} \cdot \frac{3c}{b^2}$
4. $\frac{2p+1}{p^2} \cdot \frac{2p^2}{4p+2}$
5. $\frac{x(x+5)}{3} \cdot \frac{3}{x(2x+10)}$
6. $\frac{6r}{r+2} \cdot \frac{4r+8}{18}$
7. $\frac{3n+6}{n} \cdot \frac{n^2}{n^2+4n+4}$
8. $\frac{d^2+8d+16}{d^3} \cdot \frac{d^2}{d+4}$
9. $\frac{m}{m^2+4m+3} \cdot \frac{m+1}{m}$

Find each quotient.

10. $\frac{m}{n^2} \div \frac{2}{mn}$
11. $\frac{cd}{3a^2b} \div \frac{c^2}{3a^2}$
12. $6rs \div \frac{3r^2}{s}$
13. $\frac{2}{t-1} \div \frac{t}{2t-2}$
14. $\frac{y^2-2y+1}{2y} \div (y-1)$
15. $\frac{c^2-4}{3} \div \frac{c-2}{c+2}$
16. $\frac{x^2-9}{x^2} \div \frac{x+3}{x}$
17. $\frac{d}{d^2-16} \div \frac{6d}{d^2+2d-8}$
18. $\frac{b^2-4}{16b^2} \div \frac{2b+4}{4b}$

Lesson 15–3 *(Pages 650–655)* Find each quotient.

1. $(6x-3) \div (2x-1)$
2. $(m^2+4m) \div (m+4)$
3. $(16a^2+8a) \div (4a+2)$
4. $(k^2-5k-6) \div (k+1)$
5. $(s^2-5s+6) \div (s-3)$
6. $(2y^2-2y-24) \div (2y-8)$
7. $(m^2-6m+9) \div (m-3)$
8. $(r^2+3r-18) \div (r-3)$
9. $(s^2-3s-10) \div (s+2)$
10. $(4r^3-12r^2) \div (r-3)$
11. $(6p^3-10p^2) \div (3p-5)$
12. $(x^2+12) \div (x-3)$
13. $(t^2+8) \div (t-2)$
14. $(y^2+16) \div (y+4)$
15. $(4p^3-5p-3) \div (2p+3)$

Lesson 15-4 *(Pages 656–661)* Find each sum or difference. Write in simplest form.

1. $\frac{9}{r} + \frac{6}{r}$
2. $\frac{2a}{3} + \frac{5a}{3}$
3. $\frac{4}{j} - \frac{3}{j}$
4. $\frac{2b}{4} - \frac{4b}{4}$
5. $\frac{3}{2s} + \frac{1}{2s}$
6. $\frac{6}{3b} - \frac{9}{3b}$
7. $\frac{2a}{a} + \frac{7a}{a}$
8. $\frac{y}{16y} - \frac{y}{16y}$
9. $\frac{15}{8x} + \frac{1}{8x}$
10. $\frac{2}{9k} + \frac{5}{9k}$
11. $\frac{4}{6p} + \frac{10}{6p}$
12. $\frac{14}{25r} - \frac{19}{25r}$
13. $\frac{11}{10s} - \frac{12}{10s}$
14. $\frac{9}{2y} - \frac{3}{2y}$
15. $\frac{6}{6 + m} + \frac{m}{6 + m}$
16. $\frac{6s}{2 - s} + \frac{3}{2 - s}$
17. $\frac{2p}{3 + j} - \frac{p}{3 + j}$
18. $\frac{16}{m - 5} - \frac{10}{m - 5}$
19. $\frac{3r}{r + 3} + \frac{9}{r + 3}$
20. $\frac{-2x - 4}{x + 7} + \frac{2x + 11}{x + 7}$
21. $\frac{8}{6m - 3} - \frac{4}{6m - 3}$
22. $\frac{4x + 5}{2x + 6} + \frac{2 - 4x}{2x + 6}$
23. $\frac{8 - 6r}{5r - 2} + \frac{11r - 10}{5r - 2}$
24. $\frac{12t + 1}{t + 1} + \frac{3 - 8t}{t + 1}$

Lesson 15-5 *(Pages 662–667)* Find the LCM for each pair of expressions.

1. $4mn, 6m^2$
2. $8x^2y, 20xy$
3. $4cd, 18d$
4. $d^2 - 9, d + 1$
5. $a^2 + a - 2, a^2 + 5a - 6$
6. $x^2 + x - 6, x^2 - x - 12$

Write each pair of rational expressions with the same LCD.

7. $\frac{2}{6mn}, \frac{1}{30}$
8. $\frac{5}{4p}, \frac{3}{2p}$
9. $-\frac{1}{6rs}, \frac{3}{4st}$
10. $\frac{5a}{3a^2b}, -\frac{4}{2a}$
11. $\frac{1}{y - 2}, -\frac{3}{y^2 - 4}$
12. $\frac{m^2}{2m + 4}, \frac{6m}{4m + 8}$

Find each sum or difference in simplest form.

13. $\frac{r}{8} - \frac{r}{3}$
14. $\frac{2p}{6} - \frac{p}{5}$
15. $-\frac{r}{4} + \frac{3r}{7}$
16. $\frac{6t}{9} + \frac{t}{6}$
17. $\frac{2m}{5p^2} + \frac{2}{6p}$
18. $\frac{5x}{x - 1} + \frac{7}{x}$
19. $\frac{6m}{j^2} + \frac{7m}{3j}$
20. $\frac{2t}{3 + r} - \frac{9}{r^2}$
21. $\frac{4}{b^3} + \frac{3}{b}$
22. $\frac{6y + 1}{2y + 1} - \frac{1}{4y + 2}$
23. $\frac{1}{x^2 + x - 2} + \frac{3}{x + 2}$
24. $\frac{4}{b^2 - 16} + \frac{8}{b + 4}$

Lesson 15-6 *(Pages 668–673)* Solve each equation. Check your solution.

1. $2 = \frac{2x}{5} + \frac{x}{10}$
2. $\frac{y}{5} + \frac{3y}{5} = \frac{4}{5}$
3. $\frac{m}{8} - \frac{3m}{4} = \frac{1}{2}$
4. $\frac{d}{5} = \frac{6d}{5} - \frac{3}{10}$
5. $\frac{8}{7v} + \frac{3}{4v} = -2$
6. $\frac{s + 1}{2} + \frac{s}{4} = -2$
7. $\frac{2b - 1}{2} = \frac{b}{3} - \frac{1}{4}$
8. $\frac{m - 1}{3} + \frac{m + 4}{3} = -5$
9. $\frac{4}{c + 1} + \frac{2}{c + 1} = -3$
10. $\frac{7}{3d} - \frac{5}{9d} = -3$
11. $\frac{1}{2m} + \frac{3}{m} = \frac{1}{4}$
12. $\frac{1}{r} + \frac{2}{r - 5} = \frac{3}{r}$
13. $\frac{1}{3a + 5} - \frac{4a}{6a + 10} = \frac{1}{2}$
14. $\frac{1}{k - 3} + \frac{k}{2} = \frac{k - 2}{2}$
15. $\frac{r + 2}{r^2 - 4} + \frac{1}{r - 2} = \frac{1}{r}$

Graphing Calculator Tutorial

General Information

- Any yellow commands written above the calculator keys are accessed with the [2nd] key, which is also yellow. Similarly, any green characters or commands above the keys are accessed with the [ALPHA] key, which is also green. In this text, commands that are accessed by the [2nd] and [ALPHA] keys are shown in brackets. For example, [2nd] [QUIT] means to press the [2nd] key followed by the key below the yellow QUIT command.
- [2nd] [ENTRY] copies the previous calculation so it can be edited or reused.
- [2nd] [ANS] copies the previous answer so it can be used in another calculation.
- [2nd] [QUIT] will return you to the home (or text) screen.
- [2nd] [A-LOCK] allows you to use the green characters above the keys without pressing [ALPHA] before typing each letter. (This is handy for programming.)
- Negative numbers are entered using the [(–)] key, not the minus sign, [–].
- The variable x can be entered using the [X,T,θ,n] key, rather than using [ALPHA] [X].
- [2nd] [OFF] turns the calculator off.

Basic Keystrokes

Some commonly used mathematical functions are shown in the table below. As with any scientific calculator, graphing calculators observe the order of operations.

Mathematical Operation	Example	Keys	Display
evaluate expressions	Evaluate 2 + 5.	2 [+] 5 [ENTER]	2 + 5 7
multiplication	Evaluate 3(9.1 + 0.8).	3 [(] 9.1 [+] .8 [)] [ENTER]	3(9.1 + .8) 29.7
division	Evaluate $\frac{8-5}{4}$.	[(] 8 [−] 5 [)] [÷] 4 [ENTER]	(8 − 5)/4 .75
exponents	Find 3^5.	3 [^] 5 [ENTER]	3^5 243
roots	Find $\sqrt{14}$.	[2nd] [√] 14 [ENTER]	√(14 3.741657387
opposites	Enter −3.	[(−)] 3	−3
variable expressions	Enter $x^2 + 4x - 3$.	[X,T,θ,n] [x^2] [+] 4 [X,T,θ,n] [−] 3	X² + 4X − 3

Key Skills

Each Graphing Calculator Exploration in the Student Edition requires the use of certain key skills. Use this section as a reference for further instruction.

A: Entering and Graphing Equations

Press [Y=]. Use the [X,T,θ,n] key to enter *any* variable for your equation. To see a graph of the equation, press [GRAPH].

B: Setting Your Viewing Window

Press [WINDOW]. Use the arrow or [ENTER] keys to move the cursor and edit the window settings. Xmin and Xmax represent the minimum and maximum values along the x-axis. Similarly, Ymin and Ymax represent the minimum and maximum values along the y-axis. Xscl and Yscl refer to the spacing between tick marks placed on the x- and y-axes. Suppose Xscl = 1. Then the numbers along the x-axis progress by 1 unit. Set Xres to 1.

C: The Standard Viewing Window

A good window to start with to graph an equation is the **standard viewing window.** It appears in the [WINDOW] screen as follows.

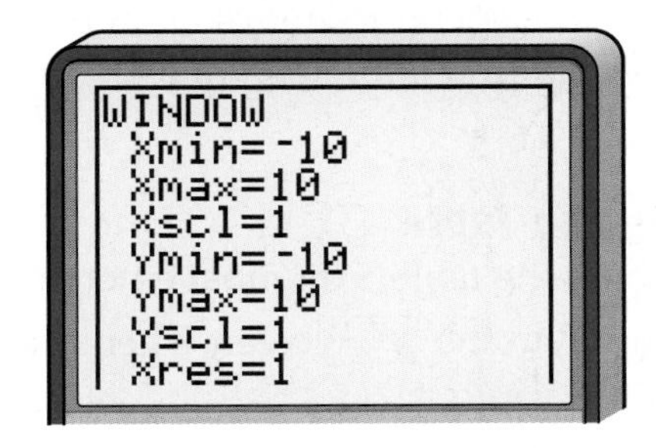

To easily set the values for the standard viewing window, press [ZOOM] 6.

D: Zoom Features

To easily access a viewing window that shows only integer coordinates, press [ZOOM] 8 [ENTER].

To easily access a viewing window for statistical graphs of data you have entered, press [ZOOM] 9.

E: Using the Trace Feature

To trace a graph, press TRACE. A flashing cursor appears on a point of your graph. At the bottom of the screen, x- and y-coordinates for the point are shown. At the top left of the screen, the equation of the graph is shown. Use the left and right arrow keys to move the cursor along the graph. Notice how the coordinates change as the cursor moves from one point to the next. If more than one equation is graphed, use the up and down arrow keys to move from one graph to another.

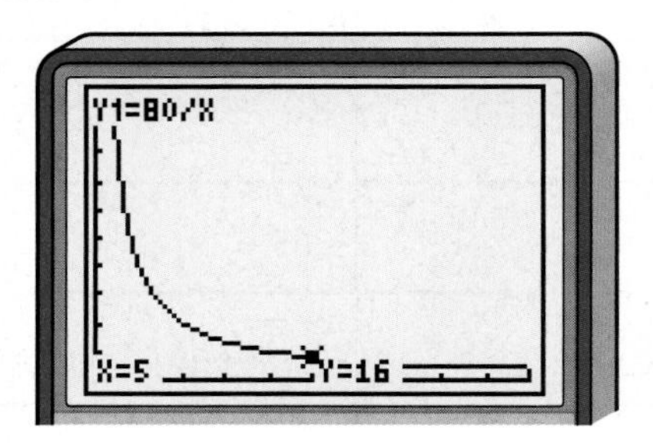

F: Setting or Making a Table

Press 2nd [TBLSET]. Use the arrow or ENTER keys to move the cursor and edit the table settings. Indpnt represents the x-variable in your equation. Set Indpnt to *Ask* so that you may enter any value for x into your table. Depend represents the y-variable in your equation. Set Depend to *Auto* so that the calculator will find y for any value of x.

G: Using the Table

Before using the table, you must enter at least one equation in the Y= screen. Then press 2nd [TABLE]. Enter any value for x as shown at the bottom of the screen. The function entered as Y_1 will be evaluated at this value for x. In the two columns labeled X and Y_1, you will see the values for x that you entered and the resulting y-values.

H: Entering Inequalities

Press 2nd [TEST]. From this menu, you can enter the $=$, $\neq$, $>$, $\geq$, $<$, and $\leq$ symbols.

Chapter	Page(s)	Key Skills
2	61	B
3	106	I
5	214	K
6	272	A, B, E
7	317	A, D, E
8	307	A, B, E
10	422	K
11	471, 491	A, D, E, F, G, I, J
12	521	A, D, E, H
13	551	A, D
14	625	A, B
15	638–639	A, B, E

I: Entering and Deleting Lists

Press STAT ENTER. Under L_1, enter your list of numerical data. To delete the data in the list, use your arrow keys to highlight L_1. Press CLEAR ENTER. Remember to clear all lists before entering a new set of data.

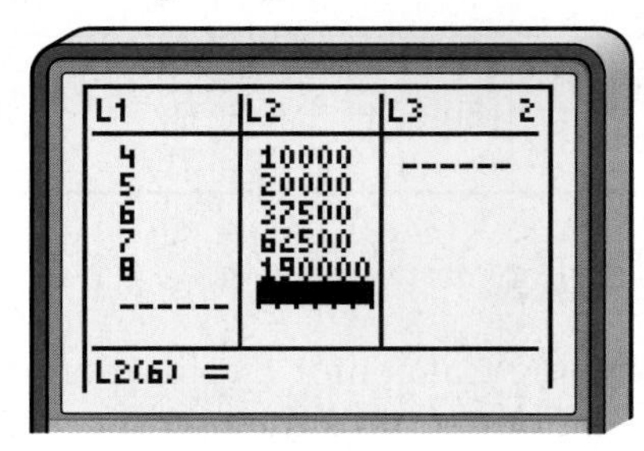

J: Plotting Statistical Data in Lists

Press Y=. If appropriate, clear equations. Use the arrow keys until Plot1 is highlighted. Plot1 represents a Stat Plot, which enables you to graph the numerical data in the lists. Press ENTER to turn the Stat Plot on and off. You may need to display different types of statistical graphs. To set the details of a Stat Plot, press 2nd [STAT PLOT] ENTER. A screen like the one below appears.

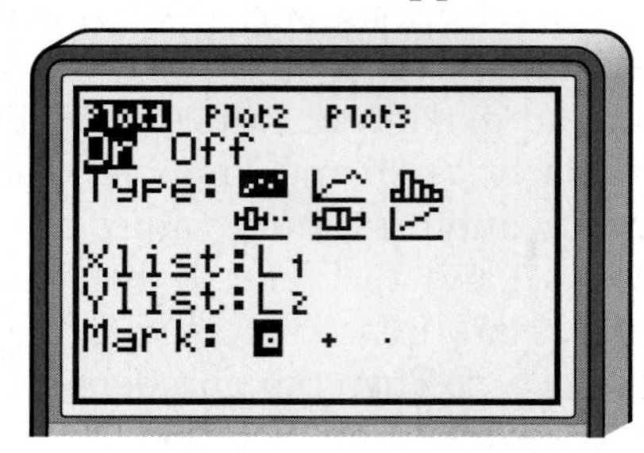

At the top of the screen, you can choose from one of three plots to store settings. The second line allows you to turn a Stat Plot on and off. Then you may select the type of plot: scatter plot, line plot, histogram, two types of box-and-whisker plots, or a normal probability plot. For this text, you will mainly use the scatter plot, line plot, and histogram. Next, choose which lists of data you would like to display along the x- and y-axes. Finally, choose the symbol that will represent each data point. To see a graph of the statistical data, press ZOOM 9.

K: Programming on the TI–83 Plus

The TI–83 Plus has programming features that allow you to write and execute a series of commands to perform tasks that may be too complex to perform otherwise. Each program is given a name. Commands begin with a colon (:), followed by an expression or an instruction. Most calculator features are accessible from the program mode.

When you press PRGM, you see three menus: EXEC, EDIT, and NEW. EXEC allows you to execute a stored program by selecting the name of the program from the menu. EDIT allows you to edit or change an existing program. NEW allows you to create a new program.

The following example illustrates how to create and execute a new program that stores an expression as Y and evaluates the expression for the designated value of X.

1. Press PRGM ▶ ▶ ENTER to create a new program.
2. Type EVAL ENTER to name the program. (Be sure that the A-LOCK is on.) You are now in the program editor, which allows you to enter commands. The colon (:) in the first column of the line indicates that it is the beginning of a command line.
3. The first command line will ask the user to choose a value for x. Press PRGM ▶ 3 2nd [A-LOCK] "ENTER X" ENTER. (To enter a space between words, press the 0 key when the A-LOCK is on.)
4. The second command line will allow the user to enter any value for x into the calculator. Press PRGM ▶ 1 X,T,θ,*n* ENTER.
5. The expression to be evaluated for the value of x is $x - 7$. To store the expression as Y, press X,T,θ,*n* − 7 STO▸ ALPHA [Y] ENTER.
6. Finally, we want to display the value for the expression. Press PRGM ▶ 3 ALPHA [Y] ENTER. At this point, you have completed writing the program. It should appear on your calculator like the screen shown below.

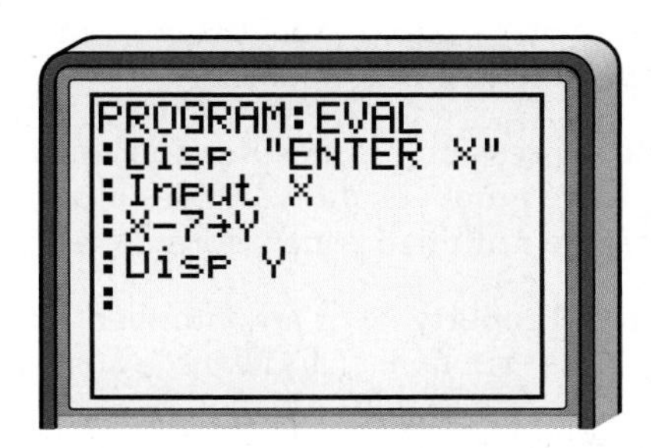

7. Now press 2nd [QUIT] to return to the home screen.
8. To execute the program, press PRGM. Then press the down arrow to locate the program name and press ENTER twice. The program asks for a value for x. Input any value for which the expression is defined and press ENTER. To immediately re-execute the program, simply press ENTER when the word *Done* appears on the screen. To break during program execution, press ON.
9. To delete a program, press 2nd [MEM] 2 7. Use the arrow and ENTER keys to select a program you wish to delete. Then press DEL 2.

While a graphing calculator cannot do everything, it can make some tasks easier. To prepare for whatever lies ahead, you should try to learn as much as you can. The future will definitely involve technology. Using a graphing calculator is a good start toward becoming familiar with technology.

Glossary

This glossary contains terms from Volumes One and Two.
Pages 1–379 are contained in Volume One. Pages 344–681 are contained in Volume Two.

absolute value The absolute value of a number is its distance from zero on a number line. *(p. 55)*

Addition Property of Equality For any numbers a, b, and c, if $a = b$, then $a + c = b + c$. *(p. 122)*

additive inverses Two numbers are additive inverses if their sum is 0. *(p. 65)*

algebraic expression An expression consisting of one or more numbers and variables along with one or more arithmetic operations. *(p. 4)*

Associative Property For any numbers a, b, and c, $(a + b) + c = a + (b + c)$ and $(ab)c = a(bc)$. *(p. 14)*

axis of symmetry The vertical line containing the vertex of a parabola. *(p. 459)*

base **1.** The number that is divided into the percentage in the percent proportion. *(p. 199)* **2.** In an expression of the form x^n, the base is x. *(p. 336)*

binomial A polynomial with two terms. *(p. 383)*

boundary A line that separates the coordinate plane into half-planes. *(p. 535)*

circle graph A graph that shows the relationship between parts of the data and the whole. *(p. 200)*

Closure Property A set of numbers is closed under an operation if the result of that operation on two numbers is included in the same number system. *(p. 16)*

coefficient The numerical part of a term. *(p. 20)*

Commutative Property For any numbers a and b, $a + b = b + a$ and $ab = ba$. *(pp. 14–15)*

complements Two events are complements if the sum of their probabilities is 1. *(p. 223)*

completing the square To add a constant term to a binomial of the form $x^2 + bx$ so that the resulting trinomial is a perfect square. *(p. 478)*

composite numbers A whole number that has more than two factors. *(p. 420)*

compound event Two or more simple events that are connected by the words *and* or *or*. *(p. 224)*

compound inequality Two or more inequalities that are connected by the words *and* or *or*. *(p. 524)*

conjugates Two binomials of the form $a\sqrt{b} + c\sqrt{d}$ and $a\sqrt{b} - c\sqrt{d}$. *(p. 616)*

consecutive integers Integers in counting order. *(p. 167)*

consistent A system of equations is said to be consistent when it has at least one ordered pair that satisfies both equations. *(p. 554)*

constant of variation The number k in equations of the form $y = kx$ and $xy = k$. *(p. 264)*

coordinate The number that corresponds to a point on a number line. *(p. 53)*

coordinate plane The plane containing the x- and y-axes. *(p. 58)*

coordinate system The grid formed by the intersection of two perpendicular number lines that meet at their zero points. *(p. 58)*

counterexample An example showing that a statement is not true. *(p. 16)*

cross products When two fractions are compared, the cross products are the products of the terms on the diagonals. *(p. 95)*

cumulative frequency histogram A histogram organized using a cumulative frequency table. *(p. 39)*

cumulative frequency table A table in which the frequencies are accumulated for each item. *(p. 33)*

D

data Numerical information. *(p. 32)*

degree **1.** The degree of a monomial is the sum of the exponents of its variables. **2.** The degree of a polynomial is the greatest of the degrees of its terms. *(p. 384)*

dependent A system of equations is said to be dependent when it has an infinite number of solutions. *(p. 554)*

dependent variable The variable in a relation whose value depends on the value of the independent variable. *(p. 264)*

difference of squares Two perfect squares separated by a subtraction sign. *(p. 447)*

$$a^2 - b^2 = (a + b)(a - b)$$

digit problems Problems that explore the relationships between digits. *(p. 569)*

dimensional analysis The process of carrying units throughout a computation. *(p. 190)*

direct variation An equation of the form $y = kx$, where $k \neq 0$. *(p. 264)*

discount The amount by which the regular price of an item is reduced. *(p. 213)*

Distance Formula The distance d between any two points with coordinates (x_1, y_1) and (x_2, y_2) is given by the formula $d = \sqrt{(x_2 - x_1)^2 + (y_2 - y_1)^2}$. *(p. 607)*

Distributive Property For any numbers a, b, and c, $a(b + c) = ab + ac$ and $a(b - c) = ab - ac$. *(p. 19)*

Division Property for Inequalities For all numbers a, b, and c, the following are true. *(p. 514)*

1. If c is positive and $a < b$, then $\frac{a}{c} < \frac{b}{c}$, and if c is positive and $a > b$, then $\frac{a}{c} > \frac{b}{c}$.
2. If c is negative and $a < b$, then $\frac{a}{c} > \frac{b}{c}$, and if c is negative and $a > b$, then $\frac{a}{c} < \frac{b}{c}$.

Division Property of Equality For any numbers a, b, and c, where $c \neq 0$, if $a = b$, then $\frac{a}{c} = \frac{b}{c}$. *(p. 160)*

domain The set of all first coordinates from the ordered pairs in a relation. *(p. 238)*

E

elimination The elimination method of solving a system of equations uses addition or subtraction to eliminate one of the variables and solve for the other variable. *(p. 566)*

empirical probability The most accurate probability based upon repeated trials in an experiment. *(p. 220)*

empty set A set with no members. *(p. 130)*

equation A mathematical sentence that contains an equals sign, =. *(p. 5)*

equation in two variables An equation that contains two unknown values. *(p. 244)*

equivalent equations Equations that have the same solution. *(p. 122)*

equivalent expressions Expressions whose values are the same. *(p. 20)*

evaluating To find the value of an expression when replacing the variables with known values. *(p. 10)*

event A subset of the possible outcomes in a counting problem. *(p. 147)*

excluded value A value is excluded from the domain if it is substituted for a variable and the result has a denominator of 0 or the square root of a negative number. *(p. 638)*

experimental probability What actually occurs when conducting a probability experiment. *(p. 220)*

exponent In an expression of the form x^n, the exponent is n. It tells how many times x is used as a factor. *(p. 336)*

exponential function A function that can be described by an equation of the form $y = a^x$, where $a > 0$ and $a \neq 1$. *(p. 489)*

F

factoring To express a polynomial as the product of monomials and polynomials. *(p. 428)*

factors In a multiplication expression, the quantities being multiplied are called factors. *(p. 4)*

family of graphs Graphs and equations of graphs that have at least one characteristic in common. *(p. 316)*

FOIL method To multiply two binomials, find the sum of the products of
F the First terms,
O the Outside terms,
I the Inside terms, and
L the Last terms. *(p. 401)*

formula An equation that states a rule for the relationship between quantities. *(p. 24)*

frequency table A table of tally marks used to record and display how often events occur. *(p. 33)*

function A relation in which each element of the domain is paired with exactly one element of the range. *(p. 256)*

functional notation In functional notation, the equation $y = x + 5$ is written as $f(x) = x + 5$. *(p. 258)*

functional value The element in the range that corresponds to a specific element in the domain. *(p. 258)*

Fundamental Counting Principle If event M can occur in m ways and is followed by event N that can occur in n ways, then the event M followed by event N can occur in $m \times n$ ways. *(p. 147)*

G

graph To draw, or plot, the points named by certain numbers or ordered pairs on a number line or coordinate plane. *(p. 53)*

greatest common factor (GCF) The greatest common factor of two or more integers is the product of the prime factors common to the integers. *(p. 422)*

H

half-plane The region of the graph of an inequality on one side of a boundary. *(p. 535)*

histogram A graph that displays data from a frequency table over equal intervals. *(p. 39)*

hypotenuse The side of a right triangle opposite the right angle. *(p. 366)*

I

identity An equation that is true for every value of the variable. *(p. 172)*

inclusive Two events that can occur at the same time are inclusive. *(p. 227)*

inconsistent A system of equations is said to be inconsistent when no ordered pair satisfies both equations. *(p. 555)*

independent A system of equations is said to be independent if it has exactly one solution. *(p. 554)*

independent events The outcome of one event does not affect the outcome of the other event. *(p. 224)*

independent variable The variable in a function whose value is subject to choice. *(p. 264)*

inequality A statement used to compare two nonequal measures. *(p. 95)*

integers The set of numbers $\{\ldots, -3, -2, -1, 0, 1, 2, 3, \ldots\}$. *(p. 52)*

intersection The intersection of two inequalities is the set of elements common to both inequalities. *(p. 524)*

inverse variation An equation of the form $xy = k$, where $k \neq 0$. *(p. 270)*

irrational number A number whose decimal value does not terminate or repeat. *(p. 362)*

L

least common denominator (LCD) The least common multiple of the denominators of two or more fractions. *(p. 663)*

least common multiple (LCM) The least number that is a common multiple of two or more numbers. *(p. 662)*

legs The sides of a right triangle that form the right angle. *(p. 366)*

like terms Terms that contain the same variables raised to the same exponents. *(p. 20)*

linear equation An equation whose graph is a straight line. *(p. 250)*

line graph Numerical data displayed to show trends or changes over time. *(p. 38)*

M

maximum The highest point on the graph of a curve. *(p. 459)*

mean The mean of a set of data is the sum of the data divided by the number of items of data. *(p. 104)*

measures of central tendency Numbers known as measures of central tendency are often used to describe sets of data because they represent a centralized, or middle, value. *(p. 104)*

measures of variation Measures of variation are used to describe the distribution of the data. *(p. 106)*

median The middle number when data are arranged in numerical order. *(p. 104)*

minimum The lowest point on the graph of a curve. *(p. 459)*

mixture problems Problems in which two or more parts are combined into a whole. *(p. 206)*

mode The item of data that occurs most often in the set. *(p. 104)*

monomial A number, a variable, or a product of numbers and variables that have only positive exponents and no variable exponents. *(p. 382)*

Multiplication Property for Inequalities For all numbers a, b, and c, the following are true. *(p. 515)*

1. If c is positive and $a < b$, then $ac < bc$, and if c is positive and $a > b$, then $ac > bc$.
2. If c is negative and $a < b$, then $ac > bc$, and if c is negative and $a > b$, then $ac < bc$.

Multiplicative Inverse Property For every nonzero number $\frac{a}{b}$, where $a, b \neq 0$, there is exactly one number $\frac{b}{a}$ such that $\frac{a}{b} \cdot \frac{b}{a} = 1$. *(p. 154)*

multiplicative inverses Two numbers are multiplicative inverses if their product is 1. *(p. 154)*

Multiplicative Property of −1 The product of −1 and any number is the number's additive inverse. *(p. 143)*

Multiplicative Property of Equality For any numbers a, b, and c, if $a = b$, then $a \cdot c = b \cdot c$. *(p. 161)*

Multiplicative Property of Zero For any number a, $a \cdot 0 = 0 \cdot a = 0$. *(p. 10)*

mutually exclusive events Two events that cannot occur at the same time. *(p. 226)*

negative exponent For any nonzero number a and any integer n, $a^{-n} = \frac{1}{a^n}$. *(p. 342)*

negative number Any number that is less than zero. *(p. 52)*

number line A line with equal distances marked off to represent numbers. *(p. 52)*

numerical expression An expression containing only numbers and mathematical operations. *(p. 4)*

odds The ratio of the number of ways an event can occur (successes) to the number of ways the event cannot occur (failures). *(p. 221)*

open sentences Mathematical statements with one or more variables. *(p. 112)*

opposites The opposite of a number is its additive inverse. *(p. 65)*

order of operations

1. Find the values of the expressions inside grouping symbols, such as parentheses, brackets, and braces, and as indicated by fraction bars. Start with the innermost grouping symbols.
2. Evaluate all powers in order from left to right.
3. Do all multiplications and/or divisions from left to right.
4. Do all additions and/or subtractions from left to right. *(pp. 8, 338)*

ordered pair A pair of numbers used to locate any point on a coordinate plane. *(p. 58)*

origin The point of intersection of the two axes in the coordinate plane. *(p. 58)*

outcomes All possible combinations of a counting problem or the results of an experiment. *(p. 146)*

parabola The general shape of the graph of a quadratic function. *(p. 458)*

parallel lines Lines in the plane that never intersect and have the same slope. *(p. 322)*

parent graph The simplest of the graphs in a family of graphs. *(p. 318)*

percent A ratio that compares a number to 100. *(p. 198)*

percentage The number that is divided by the base in a percent proportion. *(p. 199)*

percent equation Percentage = Base · Rate *(p. 204)*

percent of decrease The ratio of an amount of decrease to the previous amount, expressed as a percent. *(p. 212)*

percent of increase The ratio of an amount of increase to the previous amount, expressed as a percent. *(p. 212)*

percent proportion $\frac{\text{percentage}}{\text{base}} = \frac{r}{100}$ *(p. 199)*

perfect square The product of a number and itself. *(p. 336)*

perfect square trinomial A trinomial which, when factored, has the form $(a + b)^2 = (a + b)(a + b)$ or $(a - b)^2 = (a - b)(a - b)$. *(p. 445)*

perpendicular lines Lines that meet to form right angles. *(p. 324)*

point-slope form An equation of the form $y - y_1 = m(x - x_1)$, where m is the slope and (x_1, y_1) is any point on a nonvertical line. *(p. 290)*

polynomial A monomial or sum of monomials. *(p. 383)*

population A large group of data usually represented by a sample. *(p. 32)*

power An expression of the form x^n. *(p. 336)*

prime factorization A whole number expressed as a product of factors that are all prime numbers. *(pp. 358, 421)*

prime number A whole number whose only factors are 1 and itself. *(p. 420)*

prime polynomial A polynomial that cannot be written as a product of two polynomials with integral coefficients. *(p. 420)*

probability The ratio that compares the number of favorable outcomes to the number of possible outcomes. *(p. 219)*

product In a multiplication expression, the result is called the product. *(p. 4)*

proportion An equation of the form $\frac{a}{b} = \frac{c}{d}$ stating that two ratios are equivalent. *(p. 188)*

Pythagorean Theorem If a and b are the measures of the legs of a right triangle and c is the measure of the hypotenuse, then $c^2 = a^2 + b^2$. *(p. 366)*

Q

quadrant One of the four regions into which the x- and y-axes separate the coordinate plane. *(p. 60)*

quadratic equation An equation of the form $ax^2 + bx + c = 0$, where $a \neq 0$. A quadratic equation is one in which the value of the related quadratic function is 0. *(p. 468)*

Quadratic Formula The roots of a quadratic equation in the form $ax^2 + bx + c = 0$, where $a \neq 0$, are given by the formula $x = \frac{-b \pm \sqrt{b^2 - 4ac}}{2a}$. *(p. 484)*

quadratic function A function that can be described by an equation of the form $y = ax^2 + bx + c$, where $a \neq 0$. *(p. 458)*

quadratic-linear system of equations A system of equations involving a linear and a quadratic function. *(p. 580)*

R

radical equations Equations that contain radicals with variables in the radicand. *(p. 624)*

radical expression An expression that contains a square root. *(p. 358)*

radical sign The symbol $\sqrt{\ }$, used to indicate the positive square root. *(p. 357)*

random When all outcomes have an equally likely chance of happening. *(p. 220)*

range **1.** The set of all second coordinates from the ordered pairs in a relation. *(p. 238)* **2.** The difference between the greatest and the least values of a set of data. *(p. 106)*

rate **1.** The ratio of two measurements having different units of measure. *(p. 190)* **2.** In the percent proportion, the rate is the decimal form of the percent. *(p. 204)*

rate problems Problems involving distance, rate, and time. The formula $d = rt$ is used to solve rate problems. *(p. 266)*

ratio A comparison of two numbers by division. *(p. 188)*

rational equation An equation that contains at least one rational expression. *(p. 668)*

rational expression An algebraic fraction whose numerator and denominator are polynomials. *(p. 638)*

rational function A function that contains rational expressions. *(p. 638)*

rationalizing the denominator A method used to remove or eliminate radicals from the denominator of a fraction. *(p. 615)*

rational numbers A number that can be expressed in the form of a fraction $\frac{a}{b}$, where a and b are integers and $b \neq 0$. *(p. 94)*

real numbers The set of rational numbers and the set of irrational numbers together form the set of real numbers. *(p. 600)*

reciprocal The multiplicative inverse of a number. *(p. 154)*

relation A set of ordered pairs. *(p. 238)*

replacement set A set of numbers from which replacements for a variable may be chosen. *(p. 112)*

roots The solutions of a quadratic equation. *(p. 468)*

S

sales tax A tax added to the cost of an item. *(p. 213)*

sample A group that is used to represent a much larger population. *(p. 32)*

sample space The list of all possible outcomes of a counting problem. *(p. 146)*

sampling A method used to gather data in which a small group, or sample, of a population is polled so that predictions can be made about the population. *(p. 32)*

scale drawing A drawing that represents an object too large or too small to be drawn at actual size. *(p. 194)*

scale model A model that represents an object too large or too small to be built at actual size. *(p. 194)*

scatter plot Two sets of data plotted as ordered pairs in the coordinate plane. *(p. 302)*

scientific notation A number of the form $a \times 10^n$, where $1 \leq a < 10$ and n is an integer. *(p. 353)*

set-builder notation A notation used to describe the members of a set. For example, $\{y \mid y < 17\}$ represents the set of all numbers y such that y is less than 17. *(p. 510)*

simple interest The amount paid or earned for the use of money. The formula $I = prt$ is used to solve simple interest problems. *(p. 205)*

simplest form An expression is in simplest form when there are no longer any like terms or parentheses. *(p. 20)*

simplify In an expression, eliminate all parentheses and then add, subtract, multiply, or divide. *(p. 15)*

slope The ratio of the change in the y-coordinates to the corresponding change in the x-coordinates as you move from one point to another along a line. *(p. 284)*

slope-intercept form An equation of the form $y = mx + b$, where m is the slope and b is the y-intercept of a given line. *(p. 296)*

solution A replacement for the variable in an open sentence that results in a true sentence. *(p. 112)*

solving an open sentence Finding a replacement for the variable that results in a true sentence. *(p. 112)*

square root One of two equal factors of a number. *(p. 357)*

statement Any sentence that is either true or false, but not both. *(p. 112)*

stem-and-leaf plot In a stem-and-leaf plot, each piece of data is separated into two numbers that are used to form a stem and a leaf. The data are organized into two columns. The column on the left contains the stems and the column on the right contains the leaves. *(p. 40)*

substitution The substitution method of solving a system of equations uses substitution of one equation into the other equation to solve for the other variable. *(p. 560)*

Subtraction Property of Equality For any numbers a, b, and c, if $a = b$, then $a - c = b - c$. *(p. 124)*

system of equations A set of equations with the same variables. *(p. 550)*

system of inequalities A set of two or more inequalities with the same variables. *(p. 586)*

tally marks A type of mark used to display data in a frequency table. *(p. 33)*

term A number, a variable, a product, or a quotient of numbers and variables. *(p. 20)*

theoretical probability The probability that should occur in an experiment. *(p. 220)*

tree diagram A diagram used to show the total number of possible outcomes. *(p. 146)*

trinomial A polynomial with three terms. *(p. 383)*

uniform motion problems When an object moves at a constant speed, or rate, it is said to be in uniform motion. *(p. 670)*

union The union of two inequalities is the set of elements in each inequality. *(p. 525)*

unit cost The cost of one unit of something used to compare the costs of similar items. *(p. 97)*

unit rate A simplified rate with a denominator of 1. *(p. 190)*

variable Symbols used to represent unknown numbers. *(p. 4)*

Venn diagrams Diagrams that use circles or ovals inside a rectangle to show relationships. *(p. 53)*

vertex The maximum or minimum point of a parabola. *(p. 459)*

vertical line test If any vertical line passes through no more than one point of the graph of a relation, then the relation is a function. *(p. 257)*

whole numbers The set of numbers {0, 1, 2, 3, . . . }. *(p. 16)*

***x*-axis** The horizontal number line on a coordinate plane. *(p. 58)*

***x*-coordinate** The first number in an ordered pair. *(pp. 59, 238)*

***x*-intercept** The coordinate at which a graph intersects the x-axis. *(p. 296)*

***y*-axis** The vertical number line on a coordinate plane. *(p. 58)*

***y*-coordinate** The second number in an ordered pair. *(pp. 59, 238)*

***y*-intercept** The coordinate at which a graph intersects the y-axis. *(p. 296)*

zero pair The result of a positive algebra tile paired with a negative algebra tile. *(p. 65)*

Zero Product Property For all numbers a and b, if $ab = 0$, then $a = 0$, $b = 0$, or both a and b equal zero. *(p. 474)*

zeros The roots, or x-intercepts, of a function. *(p. 468)*

Selected Answers

Chapter B

Pages B3–B5 Pretest
1. C **3.** A **5.** A **7.** B **9.** C **11.** B **13.** B **15.** A **17.** D **19.** B **21.** A **23.** A **25.** B

Page B6 Review Lesson 1-1
1. $n + 14$ **3.** $12 + n$ **5.** $n + 17$ **7.** $3 + 13n$
9. Sample answer: one more than z **11.** Sample answer: fifty-seven decreased by the product of 3 and q **13.** $20 \div y = 10$ **15a.** $20x$ **15b.** $25x$ **15c.** $25x - 20x$

Page B7 Review Lesson 1-2
1. 12 **3.** 39 **5.** 28 **7.** 46 **9a.** $15(d + j) + 9(d + j)$ **9b.** \$810

Page B8 Review Lesson 1-3
1. Commutative Property (+) **3.** Associative Property (×) **5.** $6b + 17$ **7.** $3q + 8$ **9.** $6st + 16$ **11.** $24s$ **13.** $24v$ **15.** $32q$ **17.** No, the order of these steps cannot be reversed.

Page B9 Review Lesson 1-4
1. 12, 7, 3, 4; $12r$, $7r$, $3r$ **3.** $35w + 21$ **5.** $7c$
7. $4p + 3$ **9.** $17t + 7y$ **11.** $1 - 3x$ **13.** $20z - 18$
15. in simplest form **17a** $12(23 + 9.95)$
17b. \$395.40

Page B10 Review Lesson 1-5
1a. $P = 4s$ **1b.** 20 in.

Page B11 Review Lesson 1-6
1. No, a sample of eight is small. Also, the first people to leave the movie may not have liked it enough to stay until the end. **3.** Yes, the sample is random and would accurately represent the larger population.

Page B12 Review Lesson 1-7
1a. 800 feet **1b.** 500 feet **1c.** 600 feet

Page B13 Chapter 1 Test
1. $16 - x = 2$ **3.** 11 **5.** $4c + 4g$ **7.** $10b$
9. $16h + 12$ **11.** Associative Property (+)
13. Commutative Property (×) **15.** 5
17. $98 - 70 = 28$ **19.** No, the students in a British literature course may read more often than the average student. The sample does not represent the larger population well.

Page B14 Review Lesson 2-1
1. 2 **3.** 8 **5.** −11

7.
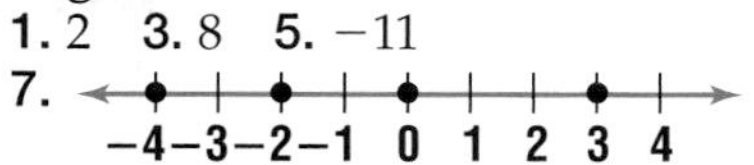

9.
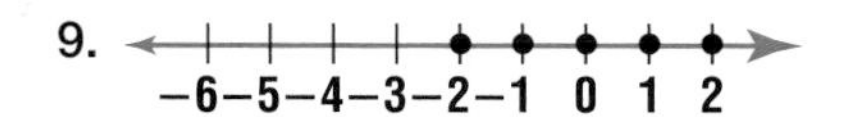

11. −4 −3 −2 −1 0 1 2 3 4

13. < **15.** < **17.** >

Page B15 Review Lesson 2-2
1–6.
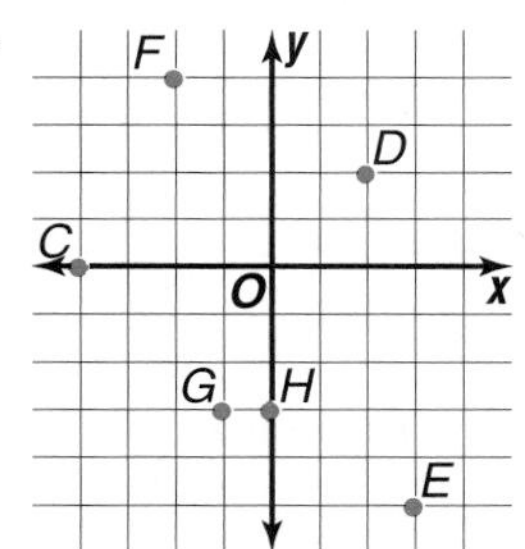

7. (1, 2); I **9.** (−2, 3); II **11.** (0, 1); none

Page B16 Review Lesson 2-3
1. −20 **3.** 101 **5.** −14 **7.** $-5a$ **9.** $9x$ **11.** −22 **13.** 14

Page B17 Review Lesson 2-4
1. 9 **3.** −10 **5.** 20 **7.** 4 **9.** −3 **11.** 15 **13.** 8 **15.** 4 **17.** 1 **19.** y **21.** $-4v$ **23.** $-17r$ **25.** \$9

Page B18 Review Lesson 2-5
1. 24 **3.** −32 **5.** 9 **7.** −15 **9.** −30 **11.** 10 **13.** 12

Page B19 Review Lesson 2-6
1. −7 **3.** 9 **5.** −5 **7.** 4 **9.** 7 **11.** −5 **13a.** 3 **13b.** 5

Page B20 Chapter 2 Test
1. −3 −2 −1 0 1 2 3 4 5

3.

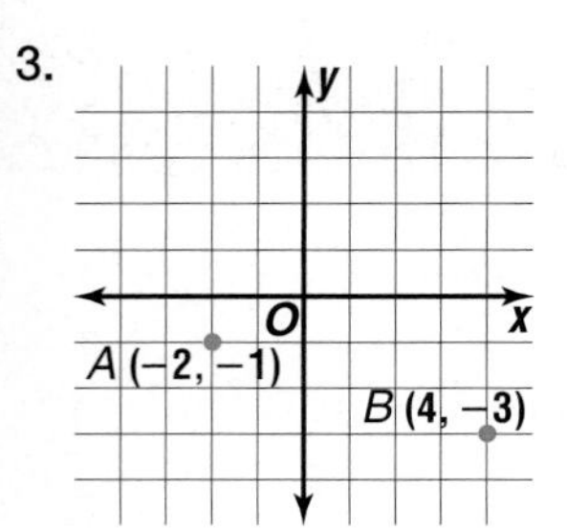

5. IV **7.** 2 **9.** −9 **11.** −3, −1, 0, 1 **13.** −18 **15.** −17 **17.** −4 **19.** 8

Page B21 Review Lesson 3-1

1. < **3.** > **5.** $-\frac{9}{7}, -1, \frac{1}{8}$

7. 20-ounce box

Page B22 Review Lesson 3-2

1. $-\frac{11}{13}$ **3.** 0.05 **5.** 2.42 **7.** 0.31 **9.** −2.75

11. 5.8 **13.** $\frac{3}{8}$

Page B23 Review Lesson 3-3

1. 7.6; 8; 9; 4 **3.** 14; 14; 10 and 18; 8 **5.** 8; 7; 5 and 7; 8 **7.** 67; 63; 62 and 77; 26 **9a.** 5.1; 5; 5; 4
9b. Sample answer: No, all three measures of central tendency represent the data well since they are all either 5 or very close to 5.

Page B24 Review Lesson 3-4

1. 8 **3.** 2 **5.** 1 **7.** 34 **9.** 3

Page B25 Review Lesson 3-5

1. 4 **3.** 8 **5.** 1 **7.** 3 **9.** 3

Page B26 Review Lesson 3-6

1. 4 **3.** −25 **5.** 16 **7.** −27 **9.** −6.2

11. $\frac{5}{6}$ **13a.** $l = \frac{1}{8} + (d + t)$ **13b.** $\frac{3}{16}$ inch

Page B27 Review Lesson 3-7

1. {±13} **3.** {0} **5.** ∅ **7.** {±2} **9.** ∅ **11.** {−1, 7}
13. $x = 13$ or $x = -17$

Page B28 Chapter 3 Test

1. > **3.** < **5.** 2.2 **7.** 4.04 **9.** $\frac{13}{12}$ or $1\frac{1}{12}$ **11.** 2

13. $\frac{1}{2}$ **15.** −4 **17.** 10.8 **19.** −6, 6 **21.** −14, −4

23. $70 **25.** Mean; only two prices are below the mean, whereas seven are above it.

Page B29 Review Lesson 4-1

1. 7.5 **3.** −4 **5.** $-\frac{5}{3}$ **7.** $-39.6m$ **9.** $9g$

Page B30 Review Lesson 4-2

1.

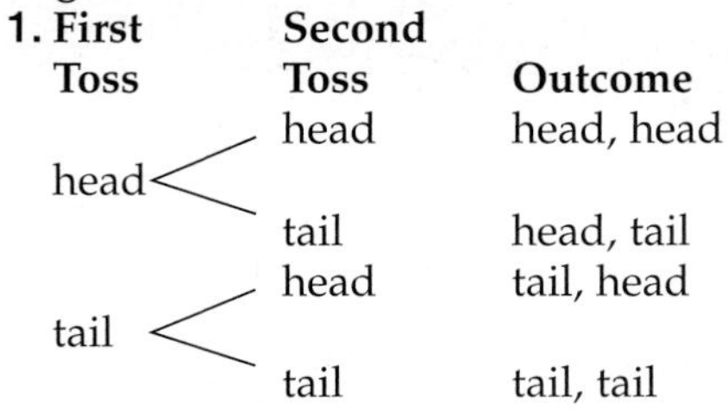

First Toss	Second Toss	Outcome
head	head	head, head
	tail	head, tail
tail	head	tail, head
	tail	tail, tail

There are 4 outcomes.

3.

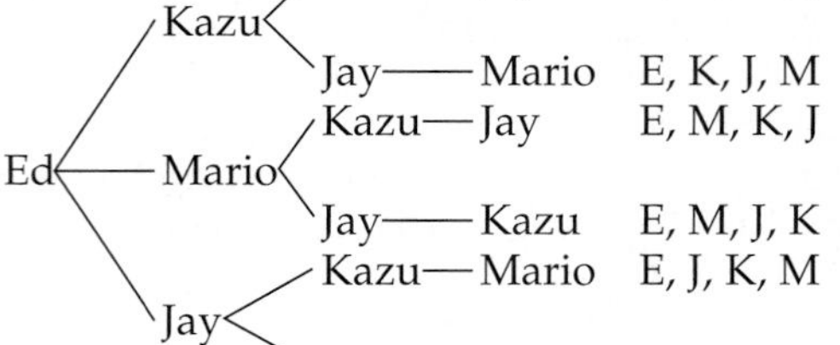

Leg 1	Leg 2	Leg 3	Leg 4	Outcome
Ed	Kazu	Mario	Jay	E, K, M, J
		Jay	Mario	E, K, J, M
	Mario	Kazu	Jay	E, M, K, J
		Jay	Kazu	E, M, J, K
	Jay	Kazu	Mario	E, J, K, M
		Mario	Kazu	E, J, M, K
Kazu	Ed	Mario	Jay	K, E, M, J
		Jay	Mario	K, E, J, M
	Mario	Ed	Jay	K, M, E, J
		Jay	Ed	K, M, J, E
	Jay	Ed	Mario	K, J, E, M
		Mario	Ed	K, J, M, E
Mario	Ed	Kazu	Jay	M, E, K, J
		Jay	Kazu	M, E, J, K
	Kazu	Ed	Jay	M, K, E, J
		Jay	Ed	M, K, J, E
	Jay	Ed	Kazu	M, J, E, K
		Kazu	Ed	M, J, K, E
Jay	Ed	Kazu	Mario	J, E, K, M
		Mario	Kazu	J, E, M, K
	Kazu	Ed	Mario	J, K, E, M
		Mario	Ed	J, K, M, E
	Mario	Kazu	Ed	J, M, K, E
		Ed	Kazu	J, M, E, K

There are 24 outcomes.

Page B31 Review Lesson 4-3

1. 0.6 **3.** 0.2 **5.** $-\frac{14}{3}$ or $-4\frac{2}{3}$ **7.** $\frac{10}{9}$ or $1\frac{1}{9}$ **9.** −1

11. 32 **13a.** $1\frac{1}{8}$ cups **13b.** $1\frac{3}{4}$ cups

Page B32 Review Lesson 4-4

1. 12 **3.** −7 **5.** $\frac{1}{14}$ **7.** $7x = 63; 9$

9. $-3y = -33; 11$ **11.** 14 **13.** 10

Page B33 Review Lesson 4-5
1. 13 **3.** 1 **5.** 11 **7.** 25 **9.** 8 **11.** $n + (n + 2) = 76$; 37, 39

Page B34 Review Lesson 4-6
1. no solution **3.** 5 **5.** 12 **7.** -66 **9.** -3
11. identity **13.** 10 years

Page B35 Review Lesson 4-7
1. 8 **3.** -4 **5.** 1 **7.** 4 **9.** -38 **11.** -2
13a. x, $2x$ **13b.** 5 ft, 10 ft

Page B36 Chapter 4 Test
1. 14.1 **3.** -0.5 **5.** 36 **7.** $-4.8x$ **9.** $-\frac{3s}{5}$ **11.** -6
13. -40 **15.** 4 **17.** 1 **19.** -8

Page B37 Review Lesson 5-1
1. 33 **3.** 8 **5.** 5 **7.** $\frac{3}{5}$ **9.** 1900 miles

Page B38 Review Lesson 5-2
1. 80 miles **3.** 360 miles **5.** 150 miles **7.** 24

Page B39 Review Lesson 5-3
1. 75% **3.** 36 **5.** 125 **7.** 60% **9a.** about 338
9b. 41 **9c.** about 342

Page B40 Review Lesson 5-4
1. 60 **3.** 1.8 **5.** 100 **7.** 250 **9.** 135 **11.** \$12.75
13. \$10.36

Page B41 Review Lesson 5-5
1. \$9.60 **3.** \$243.60 **5.** \$103.58 **7.** about 7.1%

Page B42 Review Lesson 5-6
1. $\frac{1}{3}$ **3.** $\frac{3}{10}$ **5.** 1:3

Page B43 Review Lesson 5-7
1. $\frac{3}{4}$ **3.** $\frac{9}{64}$

Page B44 Chapter 5 Test
1. 12 **3.** 4 **5.** 15 **7.** 20 **9.** 15% **11.** 432
13. \$38.22 **15.** \$5.21 **17.** 8:7

Page B45 Review Lesson 6-1
1. D = {−3, −2, 4, 5}; R = {0, 4, 5}
3. $D = \left\{-\frac{7}{9}, \frac{1}{3}, 2\frac{1}{5}\right\}$; $R = \left\{-\frac{1}{4}, \frac{3}{8}, 4\right\}$
5. {(−2, 0), (−1, 3), (2, 4), (3, 0); D = {−2, −1, 2, 3}; R = {0, 3, 4}

Page B46 Review Lesson 6-2
1. b, c, d **3.** {(−2, −5), (−1, −4), (0, −3), (1, −2), (2, −1)} **5.** {(−2, 11), (−1, 9), (0, 7), (1, 5), (2, 3)}
7. {(−2, −13), (−1, −11), (0, −9), (1, −7), (2, −5)}

9a. $y = x + 1$
9b.

Page B47 Review Lesson 6-3
1. yes; $A = 5$, $B = -3$, $C = 6$
3. yes; $A = 0$, $B = 9$, $C = 3$
5.

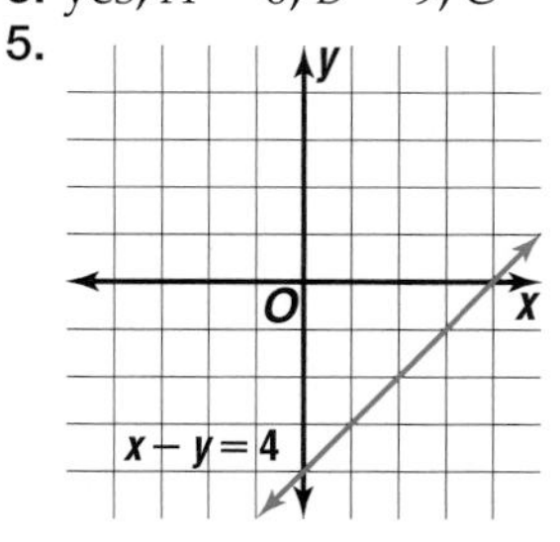

7.

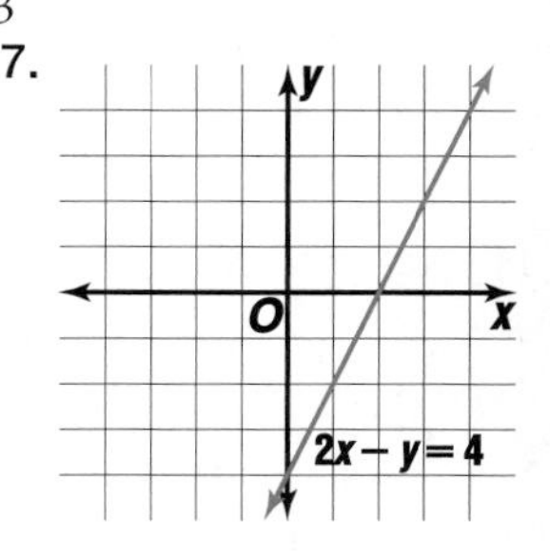

9.

Page B48 Review Lesson 6-4
1. no **3.** no **5.** no **7.** yes **9.** yes **11.** -14
13. 3 **15.** -3

Page B49 Review Lesson 6-5
1. yes

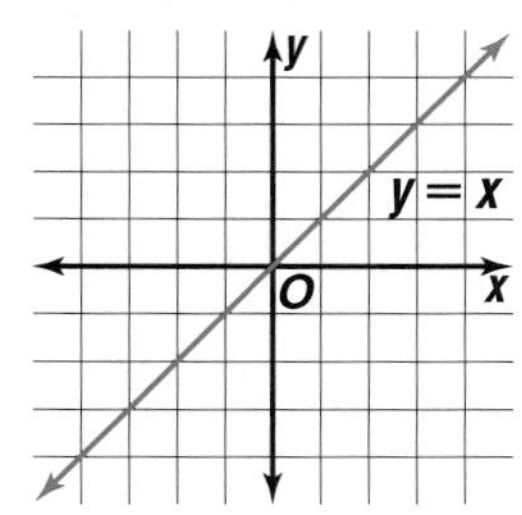

3. no

5. no

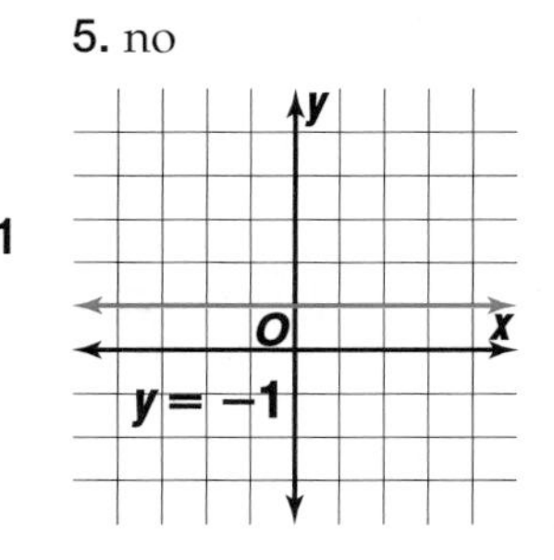

7. 40 **9.** 0.375 **11.** $83\frac{3}{5}$ pounds

Page B50 Review Lesson 6-6

1. 36; $y = \frac{36}{x}$ **3.** 2; $y = \frac{2}{x}$ **5.** 21 **7.** $\frac{1}{4}$

Page B51 Chapter 6 Test

1. {(3, 12), (0, 5), (−4, −8)} **3.** c **5.** {(−2, 4), (0, 10), (2, 16), (4, 22)}

7.

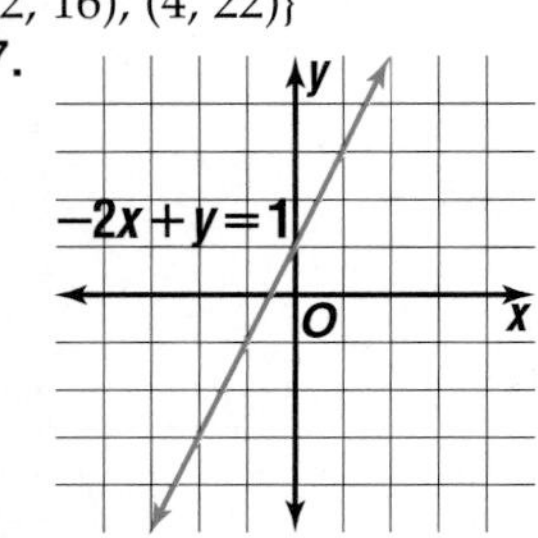

9.

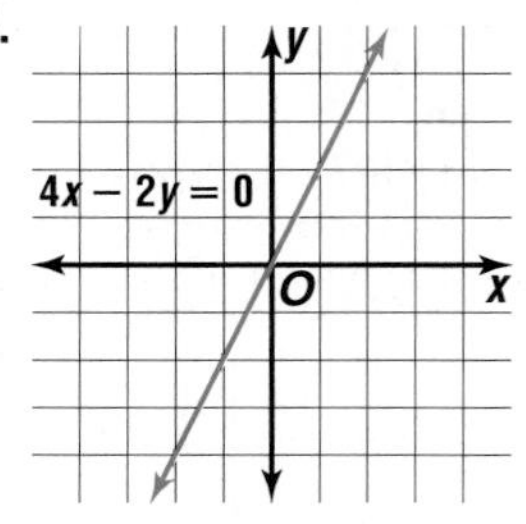

11. yes **13.** yes **15.** −9 **17.** 33 **19.** 1.8

Page B52 Review Lesson 7-1

1. $-\frac{1}{3}$ **3.** 1 **5.** $-\frac{5}{4}$ **7.** $-\frac{7}{4}$ **9.** $-\frac{3}{13}$

Page B53 Review Lesson 7-2

1. $y - 6 = \frac{7}{10}(x - 5)$ **3.** $y + 8 = -3(x - 7)$

5. $y + 5 = \frac{1}{6}(x - 6)$ **7.** $y - 5 = \frac{3}{10}(x - 7)$

9. $y - 4 = 6(x - 4)$ **11.** $y - 6 = -1(x - 7)$

13a. $y - 2 = \frac{5}{7}(x - 3)$ **13b.** No; the slope of the line through points (3, 2) and (6, 5) is 1 so it does not lie on the line whose slope is $\frac{5}{7}$.

Page B54 Review Lesson 7-3

1. $y = 5$ **3.** $y = -9x + 8$ **5.** $y = \frac{1}{3}x - 7$

7. $y = -\frac{5}{3}x + 10$ **9.** $y = -2x + 4.5$

11. $y = -\frac{1}{2}x + 3\frac{1}{2}$

Page B55 Review Lesson 7-4

1a.

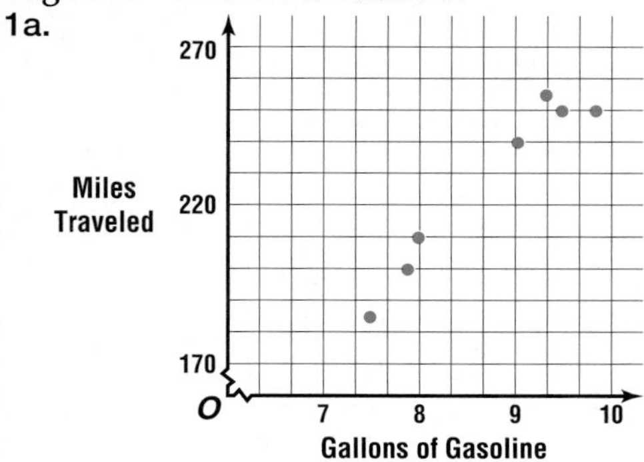

1b. gallons of gasoline and the miles traveled
1c. Yes, there is a strong positive correlation.

3a.

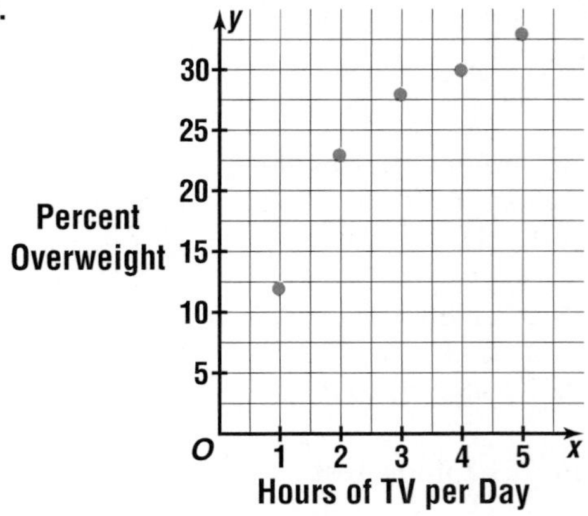

3b. positively **3c.** The more television a child watches, the greater the likelihood that he or she will be overweight.

Page B56 Review Lesson 7-5

1. 5, −3

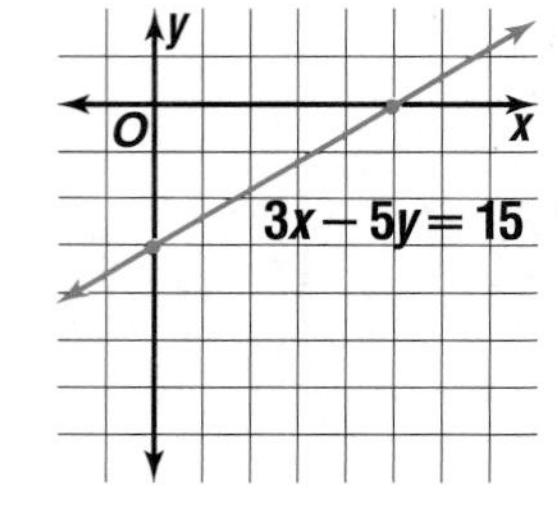

3. $m = \frac{1}{5}$, $b = 2$

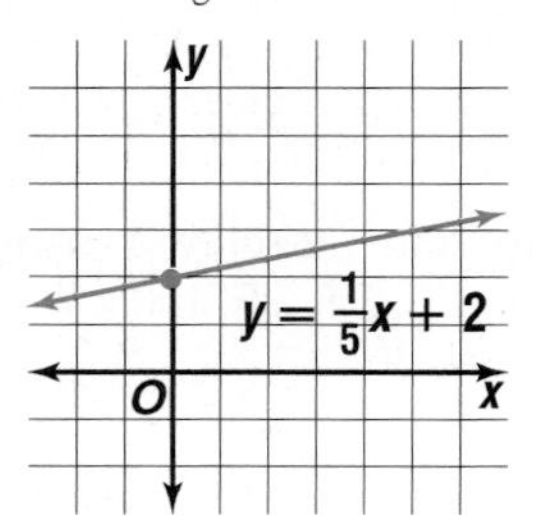

5a. $T = \frac{1}{4}c + 40$

5b. 73°

Page B57 Review Lesson 7-6

1.

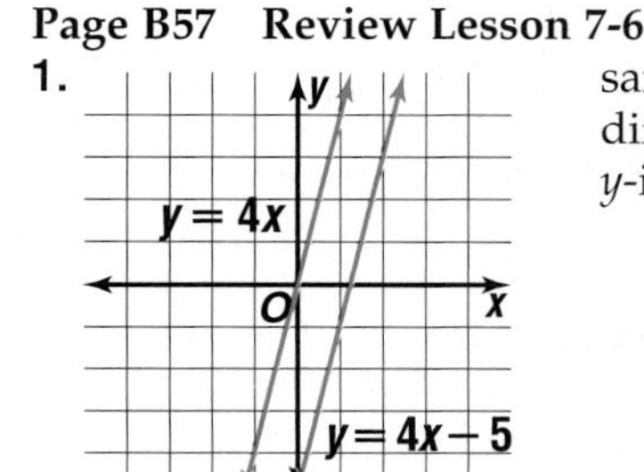

same slope different y-intercepts

3. 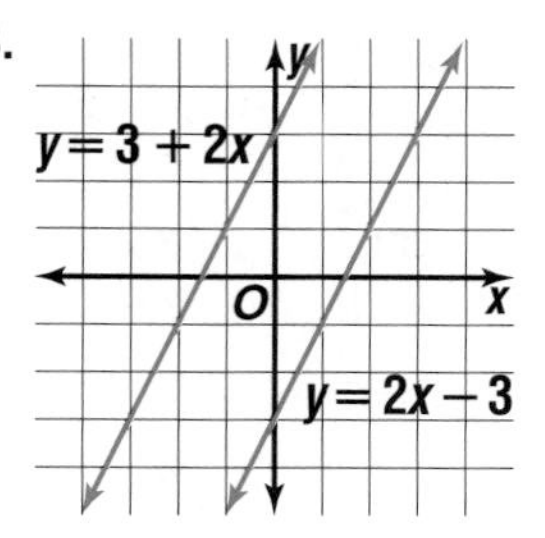

same slope
different
y-intercepts

5. 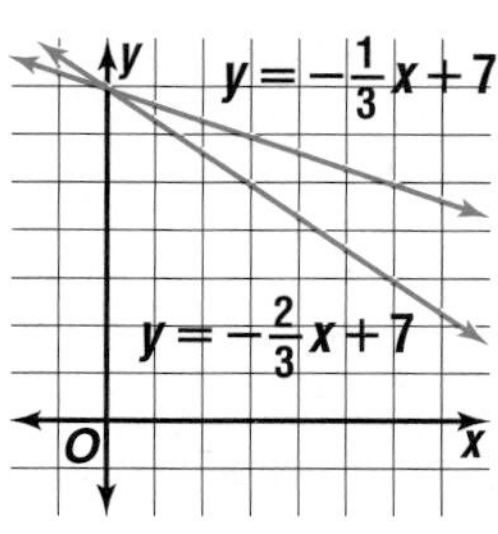

same y-intercept
different slopes

7a. Yes; they have the same y-intercept. **7b.** cost per plastic sheet **7c.** vendor 2

Page B58 Review Lesson 7-7
1. neither **3.** perpendicular
5. $y = 5x + 32$; $y = -\frac{1}{5}x + \frac{4}{5}$
7. $y = -x + 1$; $y = x - 5$

Page B59 Chapter 7 Test
1. $-\frac{1}{5}$ **3.** 2 **5.** $y - 3 = \frac{5}{8}(x - 1)$
7. $y - 7 = \frac{5}{7}(x - 4)$ or $y - 2 = \frac{5}{7}(x + 3)$
9. positive **11.** $m = -4, b = 2$

13. 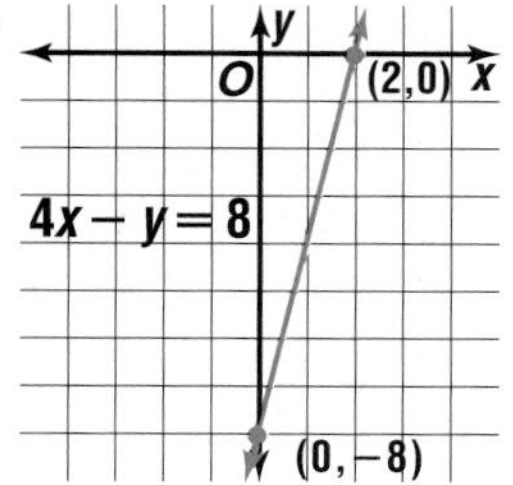

15. $y = -\frac{3}{4}x - 11$

Pages B60–B62 Posttest
1. A **3.** C **5.** B **7.** C **9.** A **11.** C **13.** B **15.** A **17.** C **19.** B **21.** D **23.** D **25.** B

Chapter 8 Powers and Roots

Pages 339–340 Lesson 8–1
1. A perfect square is the product of a number and itself. **3.** Becky; $(6n)(6n)(6n) = 6 \cdot 6 \cdot 6 \cdot n \cdot n \cdot n$, not $6n^3$. **5.** a^5 **7.** $12 \cdot 12 \cdot 12 \cdot 12$ **9.** $m \cdot m \cdot m \cdot m \cdot n \cdot n \cdot n$ **11.** 162 **13.** $2^3 3^2 5$ **15.** $(-2)^4$ **17.** 7^3 **19.** $2^2 3^3 5$ **21.** $3x^2y^2$ **23.** $3 \cdot 3 \cdot 3 \cdot 3 \cdot 3$ **25.** $2 \cdot 2 \cdot 2 \cdot 2 \cdot 3 \cdot 3$ **27.** $y \cdot y \cdot y$ **29.** $6 \cdot a \cdot b \cdot b \cdot b \cdot b$ **31.** −32 **33.** −128 **35.** 24 **37.** 0.5 **39.** 4 **41.** 486 **43a.** 200.96 ft^2 **43b.** 21 bags **45.** $y = -2x + 6$
47. 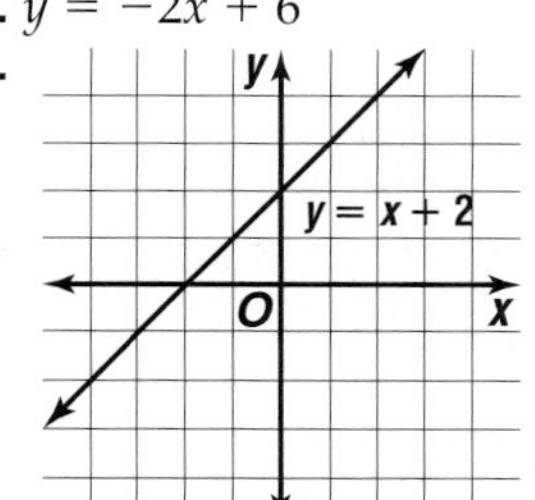

49. 31.2 **51.** 50% **53.** B

Pages 344–345 Lesson 8–2
1. In the first case, the bases are the same, but in the second case, the bases are different. **3.** y^7 **5.** t^5

7. $12a^5$ **9.** n^3 **11.** xy^3 **13.** b^5 **15.** $10^3 = 1000$ **17.** 5^4 **19.** d^6 **21.** a^3b^5 **23.** r^7t^8 **25.** $15a^2$ **27.** $8x^3y^5$ **29.** m^7b^2 **31.** $m^3n^3a^2$ **33.** 9 **35.** k^5 **37.** $-12ab^4$ **39.** 4 **41.** $15a^3c$ **43.** $2x^3$ **45.** $-16x^2$ **47.** 10^6 or 1,000,000 times **49a.** 10^{12} **49b.** 4^{15} **49c.** x^8 **49d.** Multiply the exponents; $(x^a)^b = x^{ab}$. **51.** 3 **53.** 15 **55.** 9.5 **57.** 3 **59.** D

Pages 349–351 Lesson 8–3

1. $\frac{1}{5^2} = \frac{1}{25}$ **3a.** Booker **3b.** Antonio **5.** $\frac{1}{3^3} = \frac{1}{27}$ **7.** $\frac{1}{n^3}$ **9.** $\frac{r^2}{pq^2}$ **11.** a^7 **13.** $\frac{c^3}{9a}$ **15.** $\frac{1}{2^5} = \frac{1}{32}$ **17.** $\frac{1}{4}$ **19.** $\frac{1}{r^{10}}$ **21.** a^2 **23.** $\frac{t^4}{s^2}$ **25.** $\frac{15r}{s^2}$ **27.** m^6 **29.** $\frac{1}{k^8}$ **31.** $\frac{a}{n^2}$ **33.** $\frac{1}{y^9}$ **35.** $3b^9$ **37.** $\frac{x^3}{4}$ **39.** $\frac{x^2}{7}$ **41.** $\frac{5}{8b^5c}$ **43.** 18 **45.** 2^{-4} **47.** radar microwaves **49.** $-10a^7$ **51.** n^6 **53.** z^4 **55.** 9^1 **57.** B

Page 351 Quiz 1

1. 6^3 **3.** 10^1 **5.** $2^5 3$ **7.** x^5 **9.** $-\frac{1}{3b^3}$

Pages 355–356 Lesson 8–4

1. no, 2.35×10^4 **5.** 68,000 **7.** 0.64 **9.** 0.0003 **11.** 0.0065 **13.** 5.69×10 **15.** 2.3×10^{-5} **17.** $2.6 \times 10^3 = 2600$ **19.** 5,800,000,000 **21.** 0.0000000039 **23.** 0.009 **25.** 5.28×10^3 **27.** 2.683×10^2 **29.** 3.2×10^{-4} **31.** 4.296×10^{-3} **33.** 1.2×10^{-2} **35.** 2.2×10^{-8} **37.** $6 \times 10^6 =$ 6,000,000 **39.** $7.5 \times 10^2 = 750$ **41.** $2 \times 10^3 = 2000$ **43.** $5.5 \times 10^{-5} = 0.000055$ **45.** 5,000,000 **47.** Earth, Venus, Mars **49.** between 0.3 and 2 mm **51.** y^3 **53.** $-6c^6$ **55.** x^6 **57.** a^3b^3 **59.** $-6x^2y^6$ **61.** C

Pages 360–361 Lesson 8–5

1. 1, 4, 9, 16, 25, 36, 49, 64, 81, 100 **3.** $3^2 = 9$ and $\sqrt{9} = 3$ **5.** 11^2 **7.** $2^2 \cdot 7^2$ **9.** 2 **11.** −11 **13.** $\frac{1}{2}$ **15.** 52.5 mi **17.** 10 **19.** −14 **21.** 23 **23.** −26 **25.** −22 **27.** 17 **29.** $-\frac{3}{10}$ **31.** $-\frac{1}{4}$ **33.** $\frac{6}{7}$ **35.** $\frac{3}{4}$ **37.** −0.05 **39.** 0.03 **41.** 6 **43.** 52 m **45.** false; $-6(-6) \neq -36$ **47.** 6.3×10^4 **49.** 7.6×10^{-4} **51.** $\frac{1}{x^4}$ **53.** $\frac{1}{x^2y^3}$ **55.** Sample answer: $2x + 6 = x + 1$

Page 361 Quiz 2

1. 7×10^5 **3.** 21 **5.** square house, 30 ft × 30 ft

Pages 364–365 Lesson 8–6

1. Its decimal value does not terminate or repeat. **3.** Sample answer: The area of a square garden is 200 square feet. Estimate the length of one side of the garden. **5.** 81, 100 **7.** 484, 529 **9.** 8 **11.** 16 **13.** 2 **15.** 4 **17.** 6 **19.** 11 **21.** 20 **23.** 24 **25.** 8 **27.** 11 **29.** 12 **31.** 0 **33.** $\sqrt{34}$ **35.** 17 h **37.** 31 mi^2 **39.** 10, 11, 12, 13, 14, 15 **41.** 16 **43.** $\frac{13}{11}$ **45.** 0.000000004 **47.** $y = 5x - 17$ **49.** $y = -5x + 29$ **51.** C

Pages 369–371 Lesson 8–7

1. Sample answer:

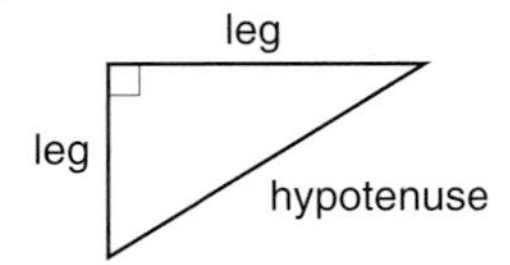

3. false **5.** false **7.** 14.1 **9.** 8.1 **11.** yes **13.** 13 **15.** 8.2 **17.** 4.2 **19.** 6.7 **21.** 21 **23.** 11.4 **25.** no **27.** no **29.** yes **31.** yes **33.** 9.4 m **35.** 6.9 ft **37.** 8 **39.** 14 **41.** 15 **43.** 10 **45a.** \$100 **45b.** \$5 **47.** C

Pages 374–376 Chapter 8 Study Guide and Assessment

1. prime factorization **3.** irrational numbers **5.** exponent **7.** hypotenuse **9.** perfect squares **11.** 9^5 **13.** $2^2 3^3$ **15.** $12 \cdot 12 \cdot 12$ **17.** $y \cdot y$ **19.** $5 \cdot x \cdot x \cdot y \cdot y \cdot y \cdot z \cdot z$ **21.** b^7 **23.** a^4b^3 **25.** y^4 **27.** $3x^3$ **29.** $\frac{1}{2^3} = \frac{1}{8}$ **31.** $\frac{1}{x^4}$ **33.** y^7 **35.** $-5b^3$ **37.** 0.0000000015 **39.** 2.4×10^5 **41.** 4.88×10^9 **43.** 6×10^{11} **45.** 11 **47.** $\frac{2}{9}$ **49.** 4 **51.** 14 **53.** 7 **55.** 11.2 **57.** 10.4 **59.** 9 m

Page 379 Preparing for Standardized Tests

1. C **3.** B **5.** C **7.** D **9.** 25

Chapter 9 Polynomials

Pages 385–387 Lesson 9–1

1. Sample answer: $\frac{2}{x}$; it includes division. **3.** $2y^2 + 5xy + 3x^2$ **5.** yes; product of numbers and variables **7.** yes; binomial **9.** no **11.** 1 **13.** 2 **15a.** 20 **15b.** 9 **17.** no; includes addition **19.** yes; product of numbers and variables **21.** yes; product of numbers and variables **23.** yes; trinomial **25.** no **27.** yes; binomial **29.** yes; binomial **31.** yes; monomial **33.** yes; monomial **35.** 2 **37.** 2 **39.** 1 **41.** 3 **43.** 6 **45.** 5 **47.** $-x^5 + x^2 - x + 25$ **49.** $5x^7 - 10x^6 + 3wx^2 + 6w^3x$ **51.** $7 + 2x - x^2 + 5x^3$ **53.** $y^6 + 5xy^4 - x^2y^3 + 3x^3y$ **55.** false **57a.** $-0.006t^4 + 0.14t^3 - 0.53t^2 + 1.79t$ **57b.** 4 **59.** yes **61.** yes **63.** 7 **65.** 11 **67.** $\frac{x}{y^3}$ **69.** B

Pages 392–393 Lesson 9–2

1. Arrange like terms in column form. **3.** $-2a - 9b$ **5.** $-x^2 - 8x - 5$ **7.** $-4xy^2 - 6x^2y + y^3$ **9.** $8y + 2$ **11.** $3x^2 - x + 2$ **13.** $2x + 2$ **15.** $x^2 + 3x - 3$ **17.** $x^2 - x + 5$ **19.** $4x^2 + 3x + 3$ **21.** $14x + y$ **23.** $3n^2 + 13n + 11$ **25.** $7n^2 - 8n + 5$ **27.** $5x + 8$ **29.** $-x^2 + 3x - 3$ **31.** $-3x + y$ **33.** $a^2 + 3a - 1$ **35.** $5x^2 - xy - 2y^2$ **37.** $2pq + pr$ **39.** $-2x^2y - 7x^2y^2$ **41.** $-2x^2 + 6x - 7$ **43a.** $2w + 4$ in. **43b.** 34 in. **45.** 2 **47.** 3 **49.** x^5y^3 **51.** $10xy^2$

Pages 396–398 Lesson 9–3

1. Distributive Property **3.** Consuelo; Shawn forgot to multiply $2x$ and 4. **5.** $x^2 + 2x$ **7.** $-4x + 8$ **9.** $4z^3 - 8z^2$ **11.** $24y^2 + 9y - 15$ **13.** 4 **15.** 2 **17.** $3x^2$ **19.** $-3y - 9$ **21.** $x^2 - 5x$ **23.** $3z^2 - 2z$ **25.** $8x^2 - 12x$ **27.** $-2a^2 + 4a$ **29.** $-30y + 10y^2$ **31.** $5d^3 + 15d$ **33.** $-10a^3 + 14a^2 - 4a$ **35.** $40n^4 + 35n^3 - 15n^2$ **37.** $-14a^2 + 35a - 77$ **39.** $1.2c^2 - 12$ **41.** $x^2 - 2x$ **43.** $4y^2 - 6y + 2$ **45.** 5 **47.** -2 **49.** -1 **51.** 2 **53.** -3 **55.** 17 **57.** $4x^3 - 8x^2 + 4x$ **59.** $-4a^3 - 10a^2 - 10a + 66$ **61.** $21n^3 - 6n^2 - 46n + 28$ **63.** $8t^2 + t$ **65a.** $7y^2 + 35y$ units^2 **65b.** 1050 in^2 **67.** $7x - 2$ **69.** $3x^2 - 2x - 3$ **71.** 5 **73.** -11 **75.** $\frac{3}{10}$ **77.** 1:1

Page 398 Quiz 1

1. 4 **3.** $-2x^2 + 8x - 2$ **5a.** x ft, $2x + 40$ ft **5b.** $2x^2 + 40x$ ft^2

Pages 402–404 Lesson 9–4

1. $(x - 2)(x + 3) = x^2 + x - 6$

	x	3
	x^2	$3x$
	$-2x$	-6

3. The Distributive Property is used to multiply both two binomials and a binomial and a monomial. But when you multiply two binomials, there are four multiplications to perform. With a binomial and a monomial there are only two multiplications to perform. **5.** $6x$ **7.** $13m$ **9.** $14y$ **11.** $w^2 - 2w - 35$ **13.** $2y^2 + y - 3$ **15.** $6y^2 + 11y - 35$ **17.** $2a^2 - ab - 10b^2$ **19.** $m^4 + 3m^3 - 2m^2 - 6m$ **21.** $x^2 + 12x + 32$ **23.** $a^2 + 4a - 21$ **25.** $n^2 - 16n + 55$ **27.** $z^2 + 2z - 24$ **29.** $3x^2 - 10x - 8$ **31.** $3z^2 - 17z + 20$ **33.** $5n^2 - 13n - 6$ **35.** $9a^2 + 6a + 1$ **37.** $12h^2 - h - 6$ **39.** $2x^2 + x - 21$ **41.** $6y^2 - 24yz + 24z^2$ **43.** $6x^2 + 13xy - 5y^2$ **45.** $x^3 - 2x^2 + x - 2$ **47.** $3x^4 + 8x^2 - 3$ **49.** $x^3 - x^2 - 6x$ **51.** $x^2 - 9$ **53a.** $3x + 2$ **53b.** $3x^2 + 4x - 1$ **55a.** $4x^3 + 2x^2 - 20x$ in. **55b.** 450 in^3 **57a.** $10 - 2t$ **57b.** 6 in. **59.** 2^3x^3 **61.** $3^2(-2)^3$

63.

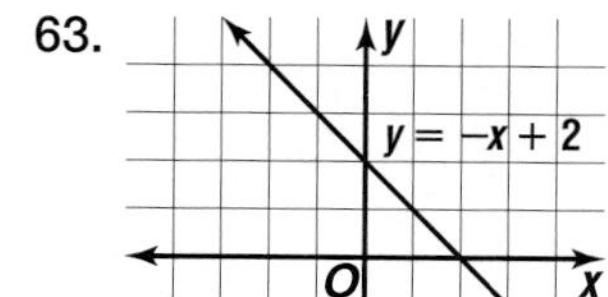

65. 84 students

Pages 408–409 Lesson 9–5

1a. ii **1b.** iii **1c.** i **3.** $y^2 + 4y + 4$ **5.** $m^2 - 18m + 81$ **7.** $x^2 - 49$ **9.** 90.25% **11.** $x^2 + 10x + 25$ **13.** $w^2 - 16w + 64$ **15.** $m^2 - 6mn + 9n^2$ **17.** $x^2 - 16y^2$ **19.** $25 + 10k + k^2$ **21.** $y^2 - 9z^2$ **23.** $9x^2 - 25$ **25.** $16 + 16x + 4x^2$ **27.** $a^2 - 6ab + 9b^2$ **29.** $x^3 + 2x^2 + x$ **31.** $x^2 - 4y^2$ **33.** $(40 - 1)(40 + 1) = 1600 - 1$ or 1599 **35a.** $\frac{1}{2}x^2 + \left(-\frac{9}{2}\right)$ **35b.** 8 square units **35c.** about 9.5 units **37.** $3x^2 - 4x - 3$ **39.** $y = 3x + 4$ **41.** $y = \frac{2}{5}x - 3$ **43.** B

Page 409 Quiz 2

1. $c^2 + 10c + 16$ **3.** $c^2 - 2c + 1$ **5.** $12x$ cm^2

Pages 412–414 Chapter 9 Study Guide and Assessment

1. i **3.** j **5.** c **7.** a **9.** e **11.** yes; monomial **13.** yes; trinomial **15.** 0 **17.** 4 **19.** $7cd - d^2 + 3d^3$ **21.** $9x^2 - 3x + 6$ **23.** $7s^2 - 6s - 2$ **25.** $4g^2 + 8g$ **27.** $-10s^2t^2 + 2st^2$ **29.** $3x - 15$ **31.** $2m^3 + 8m^2$ **33.** $10x^2 + 15x - 10$ **35.** -3 **37.** $y^2 + 4y - 12$ **39.** $x^2 - 4x + 3$ **41.** $2m^2 - 9m + 4$ **43.** $12y^2 + 10y + 2$ **45.** $10x^2 - 7x - 12$ **47.** $4x^2 + 12x + 9$ **49.** $x^2 - 4x + 4$ **51.** $25m^2 - 30mn + 9n^2$ **53.** $4a^2 - 9$ **55.** $25a^2 + 10a + 1$ **57.** $10\frac{1}{2}x^2 + 22x - 16$ **59a.** $64x^2 - 32x + 4$ **59b.** $(48x - 12)(48x - 12)$

Page 417 Preparing for Standardized Tests

1. C **3.** D **5.** B **7.** A **9.** 265

Chapter 10 Factoring

Pages 424–425 Lesson 10–1

1. 23, 29, 31, 37, 41, 43, 47 **3.** Write $8x^2$ and $16x$ as the product of prime numbers and variables where no variable has an exponent. Then multiply the common prime factors and variables. The GCF is $8x$. **5.** $3 \cdot 7$ **7.** $3 \cdot 17$ **9.** $2^2 \cdot 3^3$ **11.** 1, 2, 3, 6, 7, 14, 21, 42; composite **13.** $2 \cdot 2 \cdot 2 \cdot 3 \cdot x \cdot x \cdot y$ **15.** 5 **17.** 1 **19.** $7y^2$ **21.** 72 cm, 1 cm **23.** 1, 2, 4, 5, 10, 20; composite **25.** 1, 3, 5, 9, 15, 45; composite **27.** 1, 7, 13, 91; composite **29.** $-1 \cdot 3 \cdot 5 \cdot a \cdot a \cdot b$ **31.** $2 \cdot 5 \cdot 5 \cdot m \cdot m \cdot n \cdot n$ **33.** $2 \cdot 3 \cdot 3 \cdot 5 \cdot y \cdot z \cdot z$ **35.** 4 **37.** 18 **39.** 9 **41.** $3y$ **43.** 9 **45.** 1 **47.** 1 **49.** 5 **51.** 5 **53.** 5 **55.** 12 in. by 12 in. **57.** Every even number has a factor of 2. So, any even number greater than 2 has at least 3 factors—the number itself, 1, and 2—and is therefore composite. **59.** $4y^2 + 4y + 1$ **61.** $z^2 + 7z + 12$ **63.** $2n^2 + 9n + 4$ **65.** $6a - 2a^3$ **67.** 8.17×10^8

Pages 431–433 Lesson 10–2

1. Sample answer:

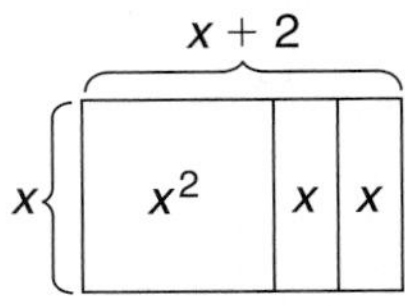

3. The Distributive Property is used to factor out a common factor from the terms of an expression. Examples: $5xy + 10x^2y^2 = 5xy(1 + 2xy)$, $2a^2 + 8a + 10 = 2(a^2 + 4a + 5)$ **5.** x **7.** $6y$ **9.** $5m^2n$ **11.** $2x(x + 2)$ **13.** prime **15.** $2ab(a^2b + 4 + 8ab^2)$ **17.** $c + 3$ **19.** $3(3x + 5)$ **21.** $2x(4 + xy)$ **23.** $3c^2d(1 - 2d)$ **25.** $mn(36 - 11n)$ **27.** prime **29.** $y^3(12x + y)$ **31.** $3y(8x + 6xy - 1)$ **33.** $x(1 + xy^3 + x^2y^2)$ **35.** $2a(6xy - 7y + 10x)$ **37.** $9x^2 - 7y^2$ **39.** $a + 2$ **41.** $2xy^2 + 3z$ **43.** $2x^2 + 3$ **45.** $(4x + y)$ ft **47a.** $8(a + b + 8)$ **47b.** $3(4a + b + 12)$ **49.** prime **51.** composite **53.** composite **55.** $3x^2 - 2x - 12$ **57.** Sample answer: $x^2 + 3x - 4$

Page 433 Quiz 1

1. $2^3 \cdot 3$ **3.** x^3; $x^3(a + 7b + 11c)$ **5.** $8(x + 1)$

Pages 438–439 Lesson 10–3

1.

	x	6
x	x^2	$6x$
1	$1x$	6

$(x + 1)(x + 6)$

3. $-10, -3$ **5.** $-4, -3$ **7.** $-6, 1$ **9.** $(x + 3)(x + 2)$ **11.** $(a - 1)(a - 4)$ **13.** $(x + 5)(x - 2)$ **15.** prime **17.** $2(c - 7)(c + 1)$ **19.** $(b + 4)(b + 1)$ **21.** $(a + 3)(a + 4)$ **23.** $(y + 9)(y + 3)$ **25.** $(x - 5)(x - 3)$ **27.** $(c - 9)(c - 4)$ **29.** prime **31.** $(c + 3)(c - 1)$ **33.** $(r - 6)(r + 3)$ **35.** $(n + 15)(m - 2)$ **37.** $(x - 9)(x - 8)$ **39.** $(r + 24)(r - 2)$ **41.** $3(y - 4)(y - 3)$ **43.** $m(m + 1)(m + 2)$ **45.** $2a(a + 8)(a - 1)$ **47.** Sample answer: $x^2 + 7x - 9$ **49a.** $x^2 - 5x - 24$ **49b.** $(x - 8)(x + 3)$

51a.

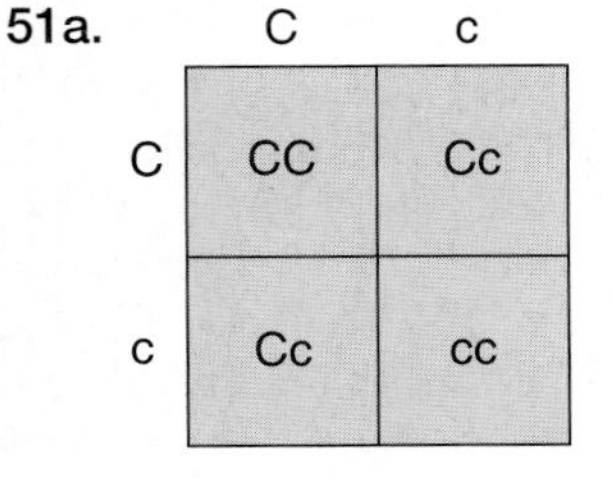

51b.

	C	c
c	Cc	cc
c	Cc	cc

53. $2x^2 + 5y^2$ **55.** $3a^2 + 4ab - 3b^2$ **57.** 4 **59.** 1 **61a.** $y = 0.2x - 398$ **61b.** the average increase each year **61c.** \$3 billion

Pages 443–444 Lesson 10–4

1a. $2x^2 - 5x - 12 = (x - 4)(2x + 3)$ **1b.** $6x^2 - 11x + 3 = (2x - 3)(3x - 1)$ **1c.** $5x^2 + 28x + 15 = (5x + 3)(x + 5)$ **3.** $(2a + 3)(a + 1)$ **5.** $(5x + 3)(x + 2)$ **7.** $(2x + 7)(x - 3)$ **9.** $(5a - 2)(2a - 1)$ **11.** $2(3x + 5)(x + 1)$ **13.** $(2y + 1)(y + 3)$ **15.** $(2a + 1)(2a + 3)$ **17.** $(2q + 3)(q - 6)$ **19.** $(7a + 1)(a + 3)$ **21.** $(3x + 2)(x + 4)$ **23.** $(3x + 5)(x + 3)$ **25.** prime **27.** $(7x - 1)(2x + 5)$ **29.** $(2m - 1)(4m - 3)$ **31.** $3(2x + 5)(x - 2)$ **33.** $x(2x - 3)(x + 4)$ **35.** $(6y - 1)(y + 2)$ **37.** $(2a - b)(a + 3b)$ **39.** $(3k + 5m)(3k + 5m)$ **41.** $3x, 2x - 1, x + 3$ **43a.** $2(y^2 - 7y - 2)$ **43b.** $2(y - 9)(y - 1)$ **43c.** 56 in^2, 18 in^3 **45.** $(x + 16)(x - 2)$ **47.** $3a(a^2 - 5a + 2)$ **49.** 25 ft **51.** 10

Page 444 Quiz 2

1. $(x + 5)(x - 2)$ **3.** $(2x + 7)(x + 1)$ **5.** $(x + 3)(x - 2)$

Pages 448–449 Lesson 10–5

1. No; 7 is not a perfect square. **3.** $(y + 7)^2$ **5.** $(a - 5)^2$ **7.** $2(2x - 5y)(2x + 5y)$ **9.** $3(x^2 + 5)$ **11.** $3(y - 1)(y + 8)$ **13.** $(r + 4)^2$ **15.** $(a + 1)^2$ **17.** $(2z - 5)^2$ **19.** $(3a + 4)^2$ **21.** $(7 + z)^2$ **23.** $(a - 6)(a + 6)$ **25.** $(1 - 3m)(1 + 3m)$ **27.** no **29.** $2(z - 7)(z + 7)$ **31.** Sample answer: $25x^2 - 4$; $(5x - 4)(5x + 4)$ **33.** $5(x^2 + 5)$ **35.** $(y - 3)(y - 2)$ **37.** $2(x - 6)(x + 6)$ **39.** $xy(8y - 13x)$ **41.** prime **43.** $2(5n + 1)(2n + 3)$ **45.** $(4w - 3)(2w + 5)$ **47.** $5(x + 1)(x + 2)$ **49.** $2x(x - 4)(x + 4)$ **51.** 2, 4 **53.** 4 **55.** $(4x - 1)(x + 3)$ **57.** $(m - 7)(m + 2)$ **59.** a^2 **61.** $\frac{d}{3c^2}$ **63.** C

Pages 450–452 Chapter 10 Study Guide and Assessment

1. false, $2^2 \cdot 3$ **3.** true **5.** true **7.** false, $x^2 - 9$ **9.** true **11.** 5 **13.** 1 **15.** $9x$ **17.** $5(x + 6y)$ **19.** $6a(2b - 3a)$ **21.** $3xy(1 + 4xy)$ **23.** $5ab - 1$ **25.** $(x - 5)(x - 3)$ **27.** $(x - 4)(x + 2)$ **29.** $(x + 7)(x - 5)$ **31.** $2(n - 6)(n + 2)$ **33.** $(3x + 5)(x + 1)$ **35.** $(3a - 2)(2a + 1)$ **37.** $(2y + 3)(y - 6)$ **39.** $5(3a - 1)(a - 1)$ **41.** $(a - 6)^2$ **43.** $(5x + 2)^2$ **45.** $(2x + 3)(2x - 3)$ **47.** $12(c + 1)(c - 1)$

49.

	H	h
H	HH	Hh
H	HH	Hh

Page 455 Preparing for Standardized Tests
1. C **3.** D **5.** D **7.** B **9.** 0.5

Chapter 11 Quadratic and Exponential Functions

Pages 461–463 Lesson 11–1
1. Sample answer: Quadratic functions have a degree of 2, but linear functions have a degree of 1. The graphs of quadratic functions are not straight lines as with linear functions. **3.** the vertex
5. 1, 0, 4 **7.** 3, 4, 0

9.

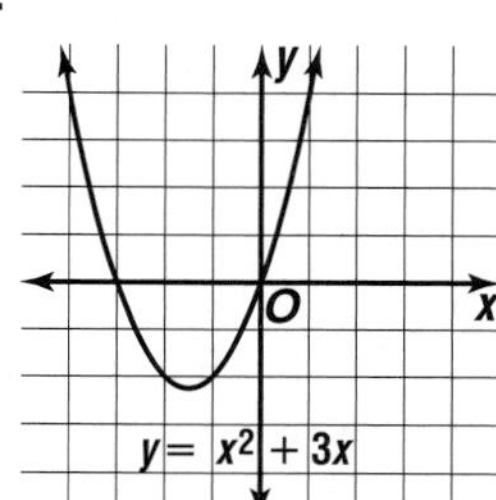

11. $x = 0$; (0, 2)

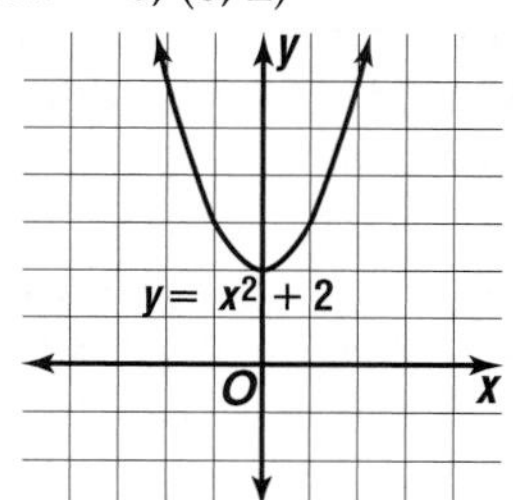

13. $x = 2$; (2, 7)

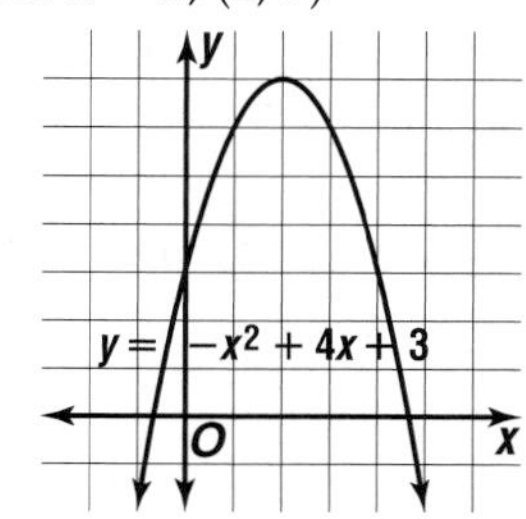

15a.

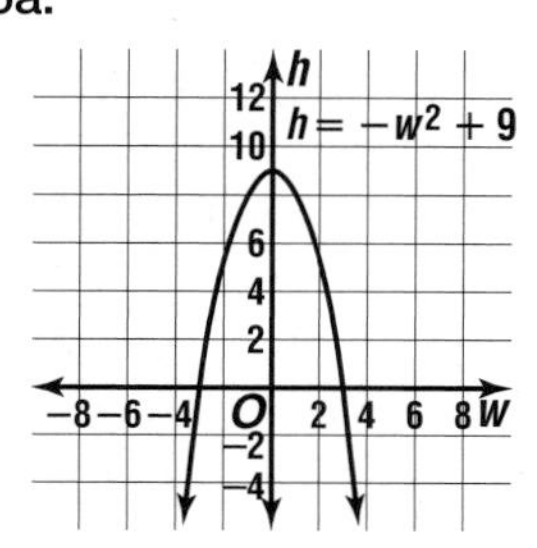

15b. 9 ft **15c.** 6 ft

17.

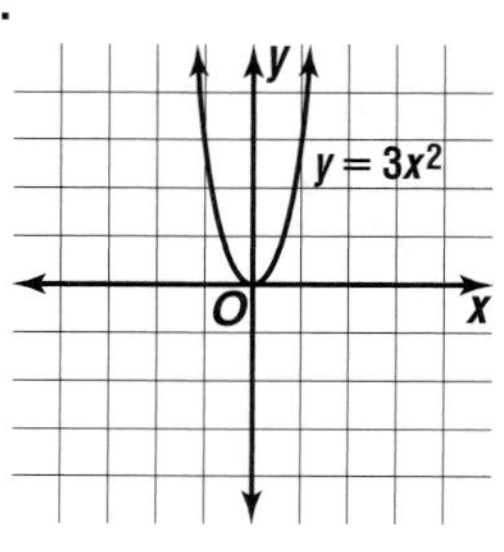

19.

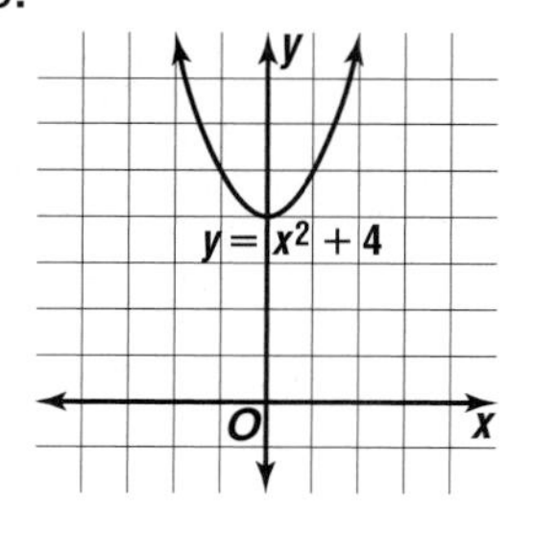

21.

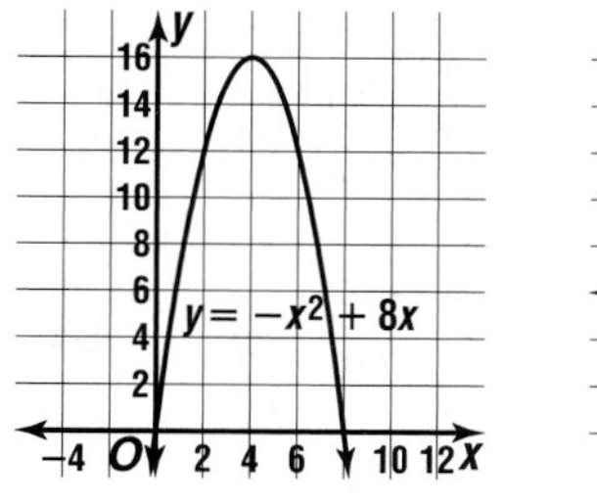

23.

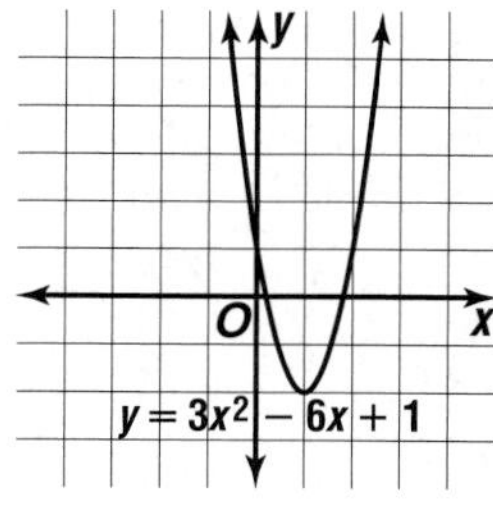

25.

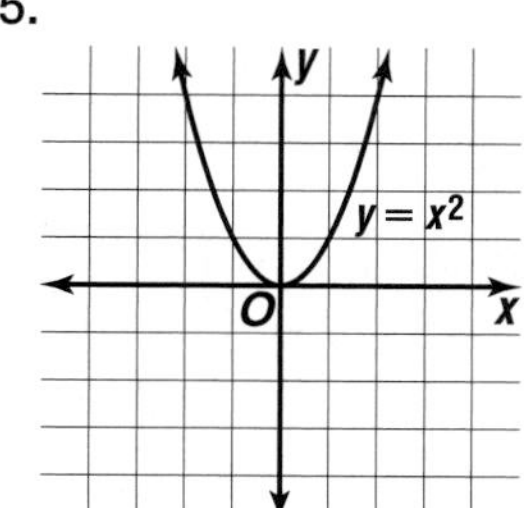

27.

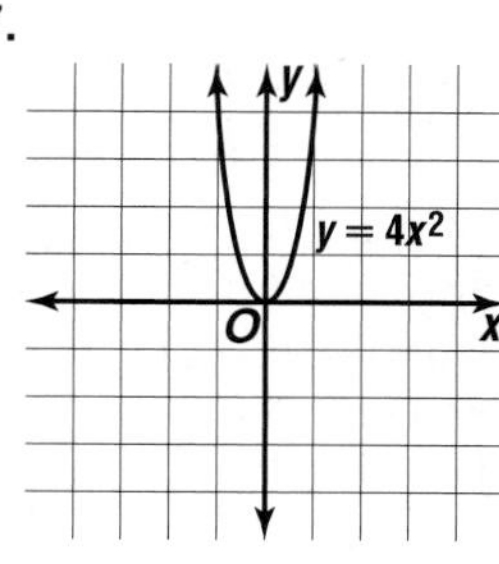

29.

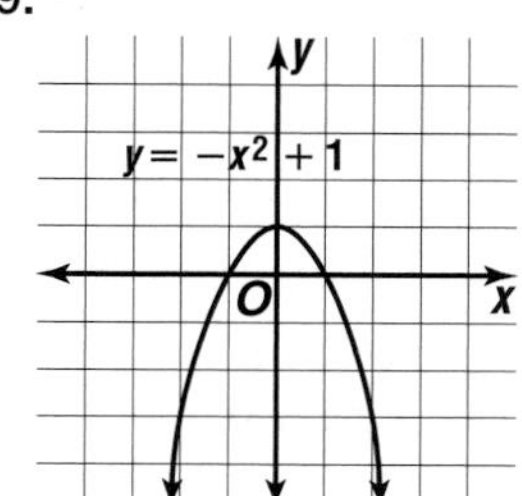

31.

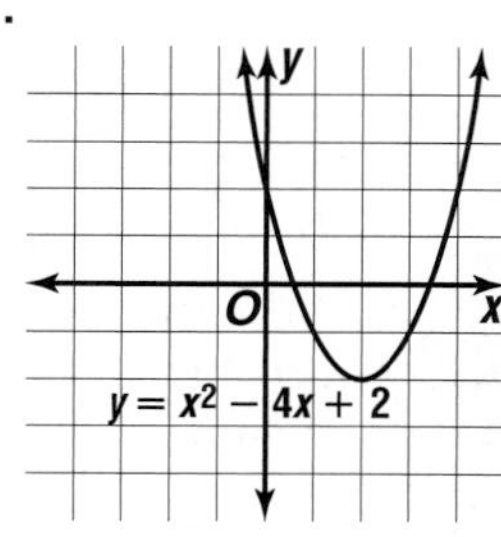

33.

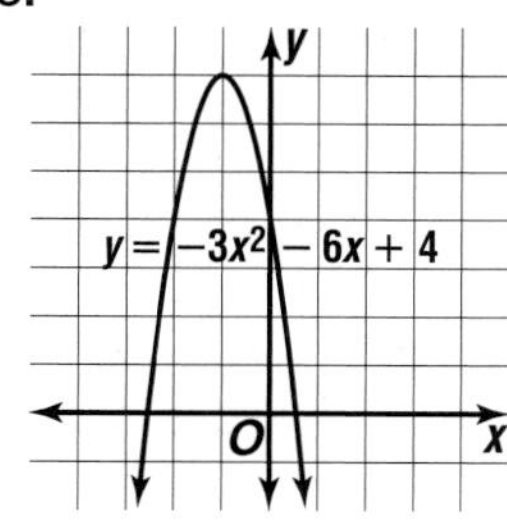

35.

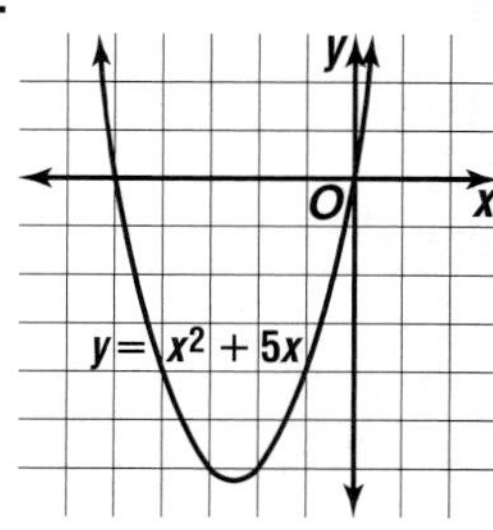

37. C **39.** B **41.** $x = -1$ **43a.** 150 **43b.** \$22,500 **45.** $(x - 7)(x + 7)$ **47.** $2(g - 4)^2$ **49.** $-x^3 - 4x^2 - 8x + 5$; 3 **51.** 1.4 mi

Pages 466–467 Lesson 11–2
1. extremely narrow with vertex at (0, 0), opening up **3.** B, A, C **5.** widens; (0, 0) **7.** left 4 units; (−4, 0) **9.** $y = 0.5(x - 4)^2 + 3$ **11.** shifts left, opens up **12.** same axis of symmetry, shifts down, opens down **13.** narrows; (0, 0) **15.** right 7 units; (7, 0) **17.** opens down, narrows; (0, 0) **19.** narrows, up 1 unit; (0, 1) **21.** widens, down 8 units; (0, −8) **23.** left 1 unit, down 5 units; (−1, −5) **25.** $y = (x + 8)^2$ **27.** $h(t) = -4.9(t - 34)^2 + 80$ **29.** $y = (x - 2)^2 - 4$ **31.** (−1.5, −1.5) **33.** $a^2 + 3a - 28$ **35.** B

Pages 471–473 Lesson 11–3

1. Sample answer: The x-intercepts are the values for x where $f(x)$ equals 0. Before solving a quadratic function, you always set it equal to zero.
3. Sample answer:

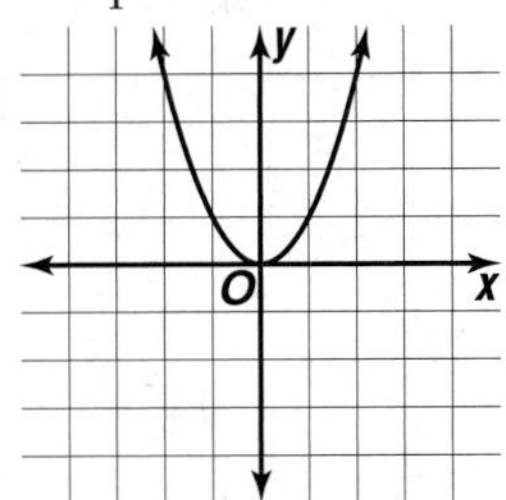

5. $-1, 4$ **7.** 2, 3 **9.** between -2 and -1; between 2 and 3 **11.** -1 **13.** $-2, 7$ **15.** 5 **17.** between -5 and -4; between 0 and 1 **19.** between -1 and 0; between 2 and 3 **21.** $-4, -8$ or 4, 8
23.

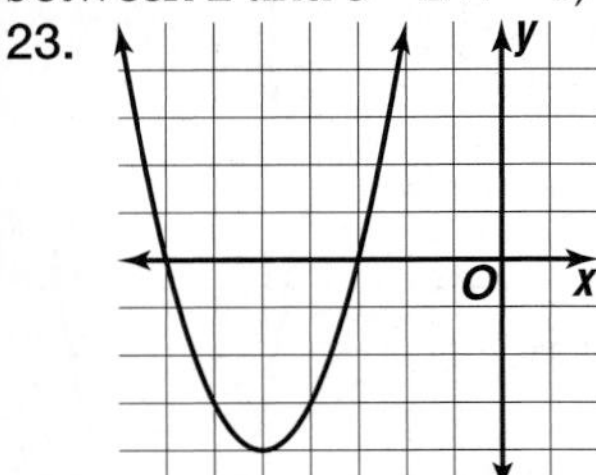

25. between 46 and 47 s **27.** The value of the function changes from negative when $x = 1$ to positive when $x = 2$. To do this, the value of the function would have to be 0 for some value of x between 1 and 2. Thus, the value of x represents the x-intercept of the function. This is the root of the related equation. **29.** up 4 units; (0, 4) **31.** 4

Page 473 Quiz 1

1. $x = -3$; $(-3, -14)$ **3.** $x = -1$; $(-1, 7)$

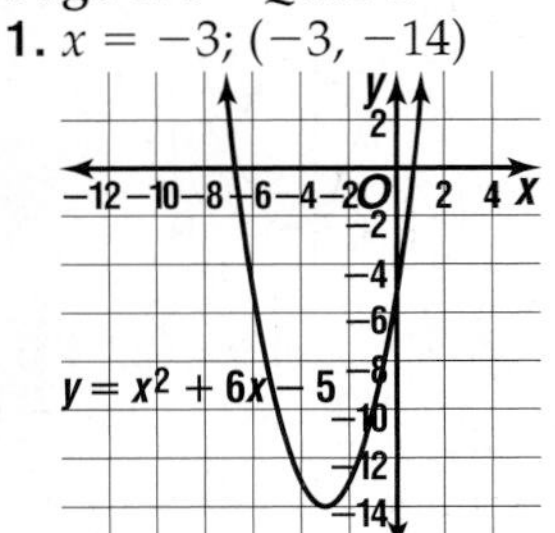

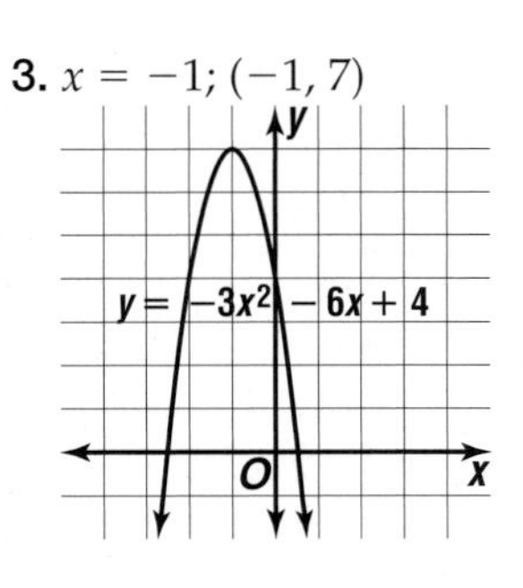

5. right 6 units; (6, 0) **7.** 2, 3 **9.** between -1 and 0; between 2 and 3

Pages 476–477 Lesson 11–4

1. Any number times 0 is 0. Therefore, if two or more numbers are multiplied and the result is 0, at least one of those numbers has to equal 0.
3. They are both correct, but factoring can only be used when the equation is not prime. **5.** 0, 5
7. 3 **9.** $-2, 4$ **11.** $0, -2$ **13.** $-7, 6$ **15.** $-1, 4$
17. $-8, -2$ **19.** 3, 8 **21.** $-3, 4$ **23.** 0, 3 **25.** 2
27. $-1, 5$ **29.** 25 ft by 35 ft **31.** 8, 11; $-11, -8$
33. 3 ft by 1 ft by 1 ft **35a.** $-5, 2$ **35b.** Sample answer: $y = (x + 5)(x - 2)$ **37.** no real solution
39. The graph of $y = 0.25x^2$ is wider than the graph of $y = x^2$. The graph of $y = 4x^2$ is narrower than the graph of $y = x^2$.

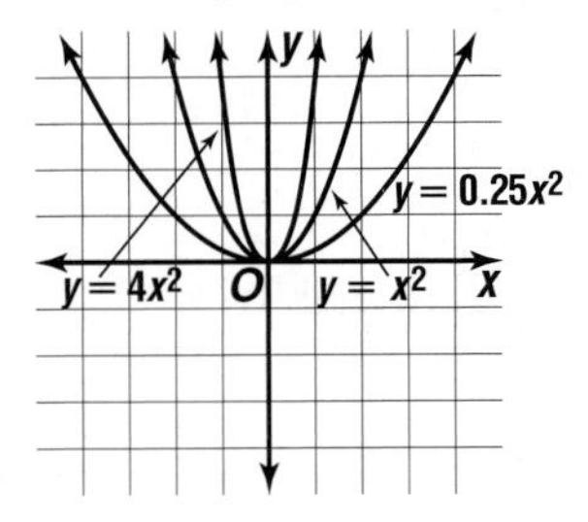

41. $(b + 7)(b - 6)$

Pages 481–482 Lesson 11–5

1. The equation is not factorable with integers and the coefficient of x^2 is 1. **3.** yes **5.** no **7.** 1
9. $\frac{9}{4}$ **11.** $-3, -4$ **13.** 9 ft by 11 ft **15.** 36 **17.** 1
19. 144 **21.** $\frac{25}{4}$ **23.** $-3, 1$ **25.** $-5, -1$ **27.** 0, 14
29. $1 \pm \sqrt{31}$ **31.** $3 \pm \sqrt{19}$ **33.** $-5, 2$ **35.** 3 s or 5 s **37.** -18 or 18 **39.** 2, 5 **41.** -3
43a. $2b(a + 4b)$ **43b.** $2a + 12b$ **43c.** 9 in. by 2 in., 18 in^2, 22 in.

Pages 486–487 Lesson 11–6

1. $b^2 - 4ac < 0$ **3.** 48 **5.** 192 **7.** ± 5 **9.** $-3, -1$
11. $-6, -1$ **13.** 2, 3 **15.** no real solutions **17.** 0, 8
19. $\frac{-1 \pm \sqrt{29}}{2}$ **21.** $\frac{-2 \pm \sqrt{7}}{2}$ **23.** $\frac{-7 \pm \sqrt{69}}{2}$
25. $\frac{5 \pm \sqrt{249}}{16}$ **27a.** 6.25 s **27b.** about 1.02 s
29. 4 **31.** 81 **33.** B

Page 487 Quiz 2

1. $-6, 2$ **3.** 3, 7 **5.** $3 \pm \sqrt{2}$ **7.** no real solutions
9. $\frac{-5 \pm \sqrt{13}}{2}$

Pages 491–493 Lesson 11–7

1. Sample answer: The graph begins almost flat, but then for increasing x-values, it becomes more and more steep. **3.** 256 **5.** 1.26
7. -5

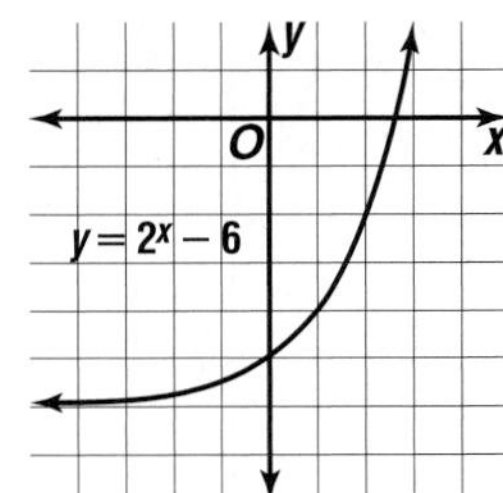

9a.

Year	Balance	Pattern
0	500	$500(1.02)^0$
1	$500(1.02)1.02 = 510$	$500(1.02)^1$
2	$(500 \cdot 1.02)1.02 = 520.20$	$500(1.02)^2$
3	$(500 \cdot 1.02 \cdot 102)1.02 = 530.604$	$500(1.02)^3$

9b. $T(x) = 500(1.02)^x$ **9c.** \$714.12 **9d.** \$1428.25
11. 1 **13.** 2

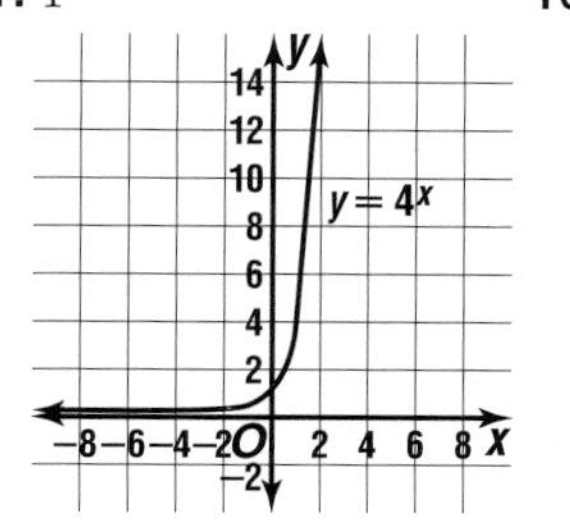

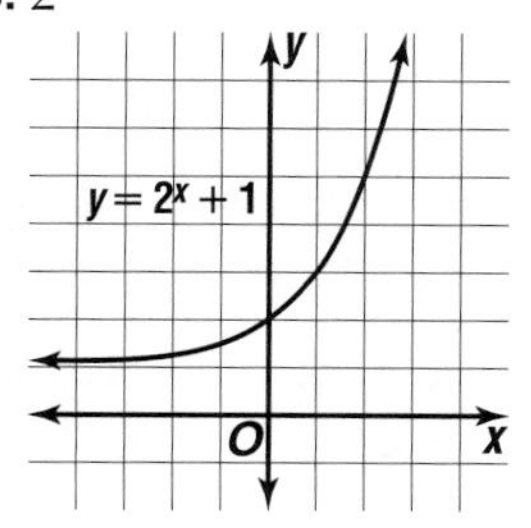

15. −2 **17.** 0

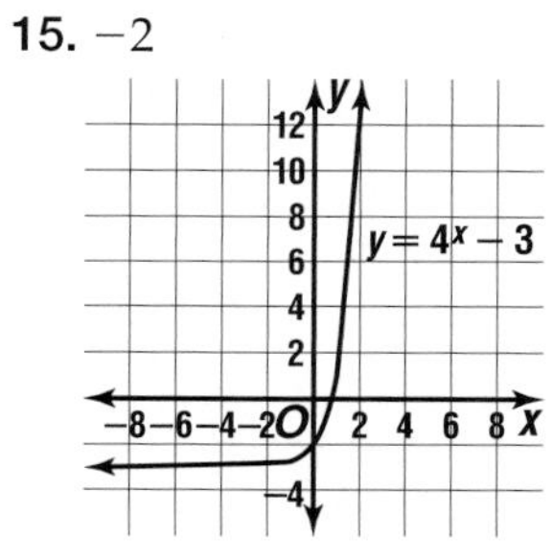

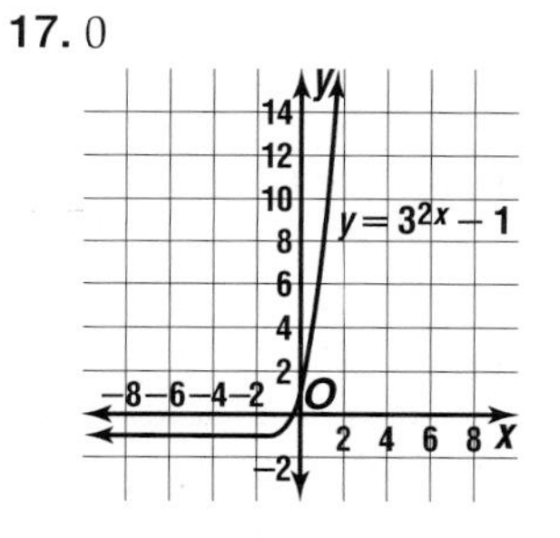

19. \$5304.50 **21.** \$3920.88 **23.** \$7986.70

25a.

Day	Cents	Pattern
1	1	2^0
2	$2 \cdot 1$	2^1
3	$2 \cdot 2 \cdot 1$	2^2
4	$2 \cdot 2 \cdot 2 \cdot 1$	2^3
5	$2 \cdot 2 \cdot 2 \cdot 2 \cdot 1$	2^4

25b. $y = 2^{x-1}$ **25c.** January 17 **27.** −2, 2
29a. $(2x + 4)x = 30$ **29b.** 3 yd by 10 yd
31. $y - 6 = 0$

Pages 496–498 Chapter 11 Study Guide and Assessment

1. f **3.** c **5.** j **7.** a **9.** d **11.** $x = 0$; $(0, -3)$
13. $x = -1$; $(-1, 20)$
15.

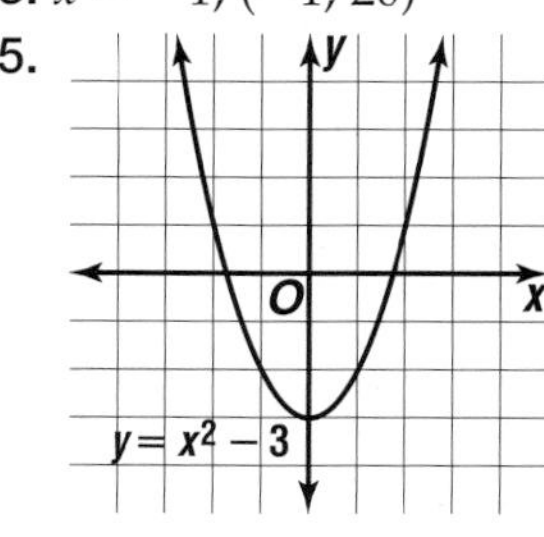

17.

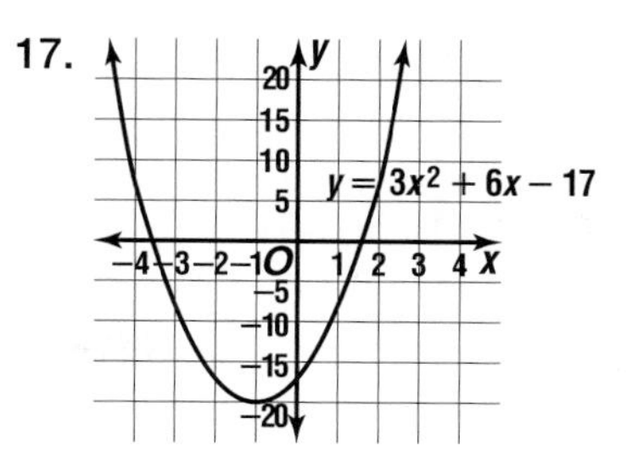

19. shifts right; (8, 0) **21.** widens; (0, 0)
23. shifts right and up; (1, 7) **25.** −3, 4
27. no real solutions **29.** $-1 < x < 0$, 3
31. −4, 5 **33.** −8 **35.** −1, 1 **37.** 4 **39.** −3, 7
41. −1, −9 **43.** $-\frac{1}{3}$, 1 **45.** $\frac{-5 \pm \sqrt{13}}{2}$
47. 1 **49.** 4

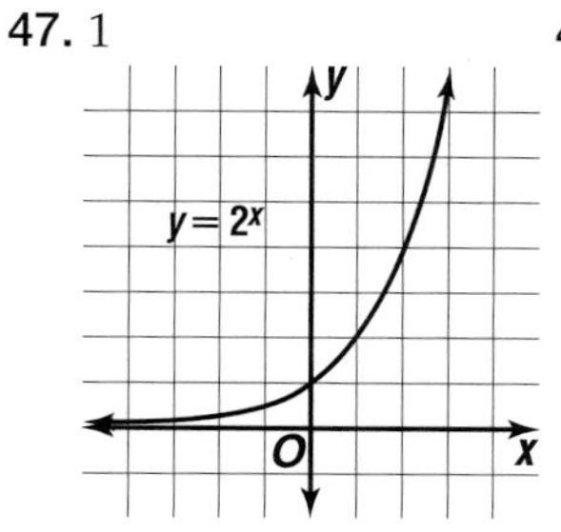

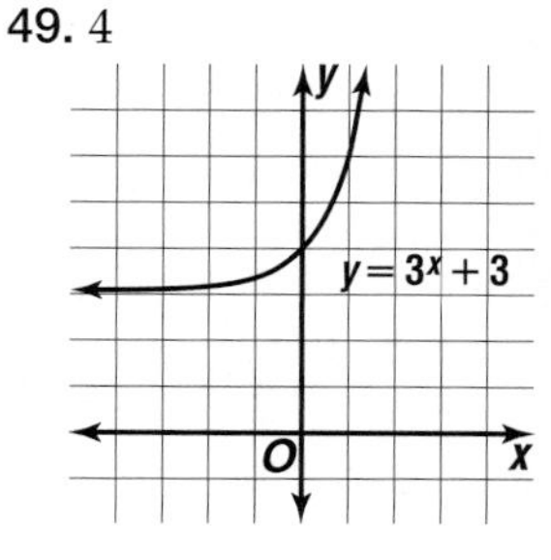

51. \$5.50

Page 501 Preparing for Standardized Tests
1. D **3.** A **5.** C **7.** C **9.** 2

Chapter 12 Inequalities

Pages 506–508 Lesson 12–1

1. i. b; ii. a; iii. d; iv. c **3.** The graph of $x \leq 4$ has a bullet at 4. The graph of $x < 4$ has a circle at 4.
5. $x < 14$
7. 4 5 6 7 8
9. 3 4 5 6 7
11. $x < -1$ **13.** $x > 4$ **15.** $x < 9$ **17.** $x \geq -8$
19. 5 6 7 8 9
21. −2 −1 0 1 2
23. −6 −5 −4 −3 −2
25. 2 3 4
27. −4 −3.5 −3
29. 1 2 3
31. $x > -6$ **33.** $x \leq 3$ **35.** $x \leq 5$ **37.** $x \geq 2.5$

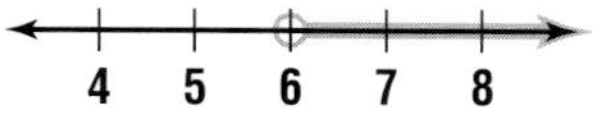

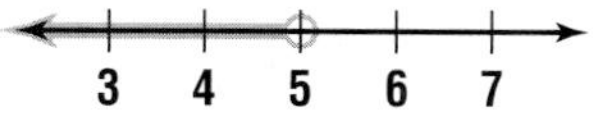

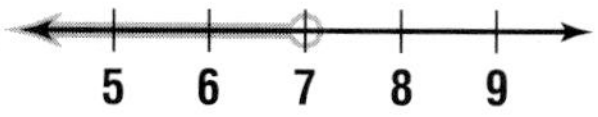

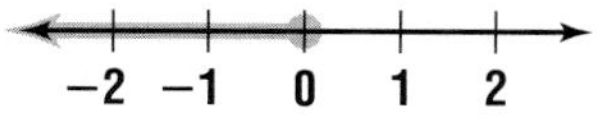

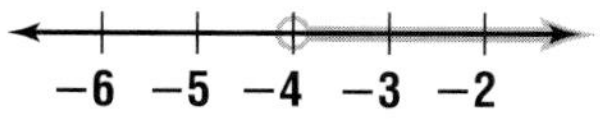

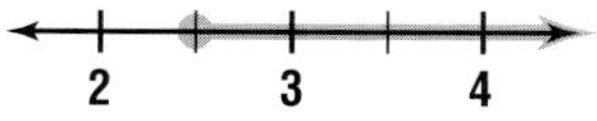

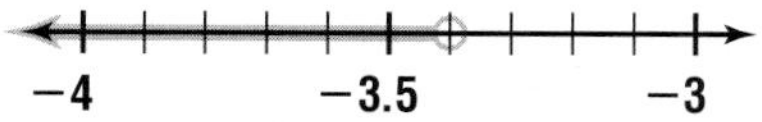

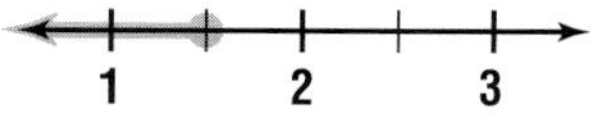

39. $f < 3$;

1 2 3 4 5

41. $w \leq 3600$;

3400 3600 3800

43. 7 yr 45. no real solutions 47. $2^3 \cdot 3$; two
49. $-6x^2y$

Pages 511–513 Lesson 12–2

1. They are the same. 5. $8 + n \leq 12$ 7. $\{v \mid v \geq 17\}$
9. $\{h \mid h \leq 21\}$ 11. $\left\{y \mid y < 1\frac{5}{6}\right\}$
13. $\{x \mid x < -1\}$;

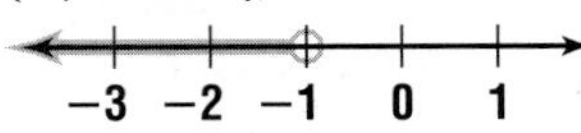

15. $\{n \mid n > 3\}$ 17. $\{b \mid b < -5\}$ 19. $\{r \mid r < 19\}$
21. $\{w \mid w > 22\}$ 23. $\{t \mid t \leq 0\}$ 25. $\{c \mid c > -4\}$
27. $\{d \mid d < 5.4\}$ 29. $\{z \mid z \leq -0.5\}$ 31. $\left\{v \mid v \leq 2\frac{1}{2}\right\}$
33. $\{g \mid g < 13\}$;

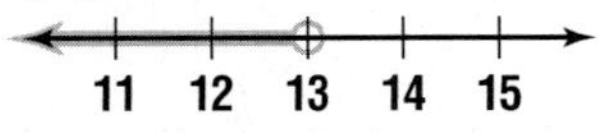

35. $\{x \mid x \geq -16\}$;

−18 −17 −16 −15 −14

37. $\{t \mid t \geq 9\}$;

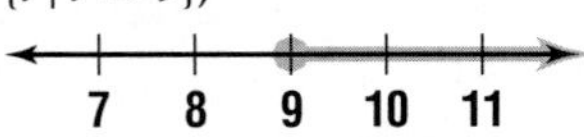

39. $x - 8 \leq 12$; $x \leq 20$ 41. $t \geq 99$ 43. $p \geq 99$
45. Yes; examples are $x > x$ and $2x + 1 \leq 2x$.
47. $x < -8$ 49. $x \geq 5$ 51. $(3x + 2)(x - 5)$ 53. 32

Pages 516–518 Lesson 12–3

1. For both, reverse the sign when dividing or multiplying by a negative number.
3.

2 3 4 5 6

5. $\{x \mid x \geq -9\}$ 7. $\{x \mid x \geq 20\}$ 9. $\{a \mid a \leq -30\}$
11. $\{h \mid h \geq -2\}$ 13. $\{d \mid d \leq -3\}$ 15. $\{g \mid g > -36\}$
17. $\{x \mid x \geq 7\}$ 19. $\{z \mid z > -3\}$ 21. $\{c \mid c > -54\}$
23. $\left\{k \mid k > \frac{5}{3}\right\}$ 25. $\left\{w \mid w > \frac{1}{2}\right\}$ 27. $\{y \mid y \geq 6\}$
29. $\{v \mid v \geq 2\}$ 31. $\{s \mid s \leq 280\}$ 33. $\{n \mid n < 105\}$
35. $x < 45$ 37. at least 17 lawns 39. $\{z \mid z \leq 4\}$
41. $p > 600$ 43. $4v^2 - 1$ 45. $\frac{1}{400}$

Page 518 Quiz 1

1. $x < -1$
3. $\{x \mid x > 5\}$;

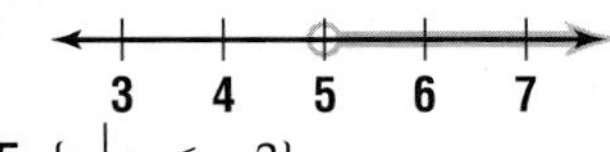

5. $\{z \mid z < -3\}$;

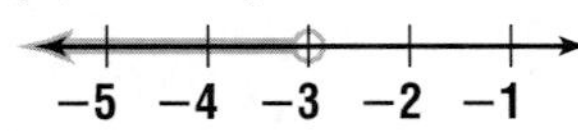

7. $\{x \mid x > -5\}$ 9. $\left\{v \mid v \leq \frac{15}{2}\right\}$

Pages 522–523 Lesson 12–4

1. subtraction, then division 3. −7 5. −3.1
7. $\{y \mid y > 4\}$ 9. $\{x \mid x < -1\}$ 11. $\{w \mid w > -2\}$
13. $s \geq 92$ 15. $\{b \mid b < 12\}$ 17. $\{x \mid x \geq 4\}$
19. $\{g \mid g > 2\}$ 21. $\{v \mid v \leq 0.5\}$ 23. $\{m \mid m \leq -3\}$
25. $\{d \mid d < 1\}$ 27. $\{x \mid x \leq 2\}$ 29. $\{y \mid y < 1\}$
31. $\{b \mid b < 11\}$ 33. $\{x \mid x < 1\}$ 35. at least \$260
37. $x > \$1484$ 39. $\{p \mid p > 7\}$ 41. $\{x \mid x \leq -40\}$
43. $m \leq \$19.50$ 45. A

Pages 526–529 Lesson 12–5

1. an inequality made from two inequalities
3. intersection 5. union 7. $5 < \ell < 10$
9.

11. $\{c \mid 3 < c < 8\}$;

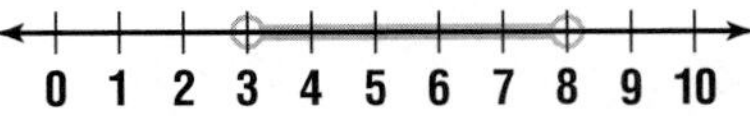

13. $\{v \mid 0 \leq v \leq 4\}$;

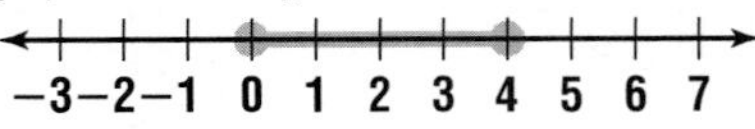

15. $\{j \mid j \leq -1 \text{ or } j > 0\}$;

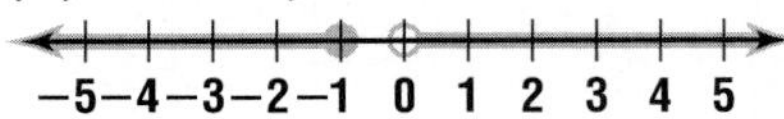

17. $9 \leq t \leq 11.5$;

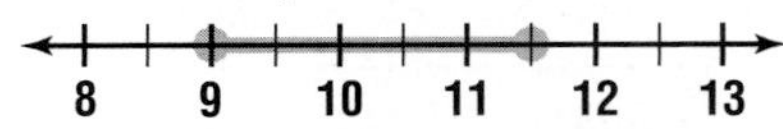

19. $-8 < h < 8$ 21. $2 \leq g \leq 5$ 23. $6 < x \leq 8$
25.

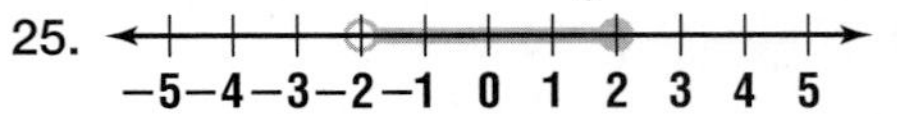

27.

13 14 15 16 17 18 19 20 21 22 23

29.

2 3 4 5 6 7 8 9 10 11 12

31. $\{x \mid 4 \leq x \leq 8\}$;

0 1 2 3 4 5 6 7 8 9 10

33. $\{w \mid 0 < w \leq 3\}$;

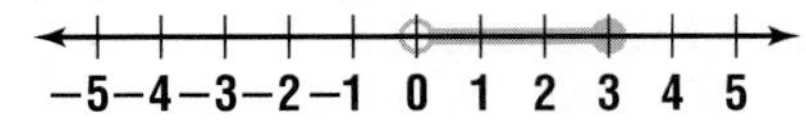

35. $\{h \mid -3 \leq h \leq 3\}$;

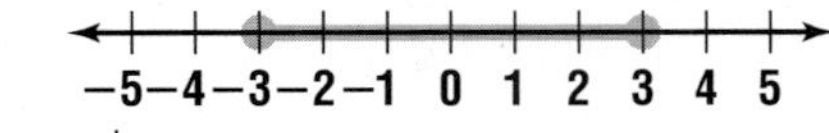

37. $\{c \mid -5 < c < -4\}$;

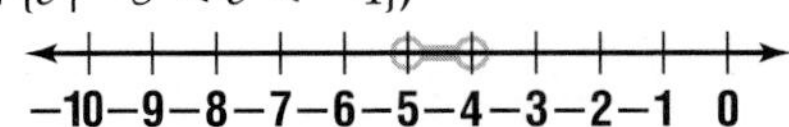

39. $\{z \mid z \leq -7 \text{ or } z > 4\}$;

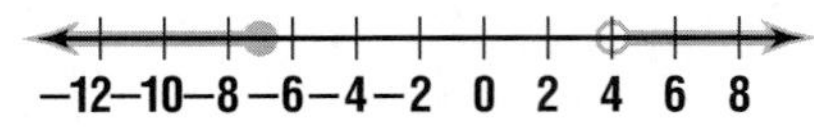

41. $\{r \mid r \leq -1 \text{ or } r > 0\}$;

−5 −4 −3 −2 −1 0 1 2 3 4 5

43. $\{p \mid p \geq -1\}$;

−5 −4 −3 −2 −1 0 1 2 3 4 5

45. $\{d \mid d > 1.5\}$;

0 1 2 3 4 5

47. $\{w \mid w \leq -6 \text{ or } w > -2\}$;

−7 −6 −5 −4 −3 −2 −1 0 1 2 3

49. $x < -3$ or $x > 3$ **51.** $-3 \leq x < 5$ **53.** $\{x \mid x \leq 3$ or $x > 9\}$ **55.** $\{k \mid 11 \leq k \leq 13\}$ **57a.** Sample answer: $11 \leq h \leq 14$

57b.

8 9 10 11 12 13 14 15 16 17 18

59. $6.6 < p < 9.4$ **61.** $\{y \mid y < 0\}$ **63.** $\{x \mid x \geq 5\}$
65. $\{b \mid b > 6\}$

67a.

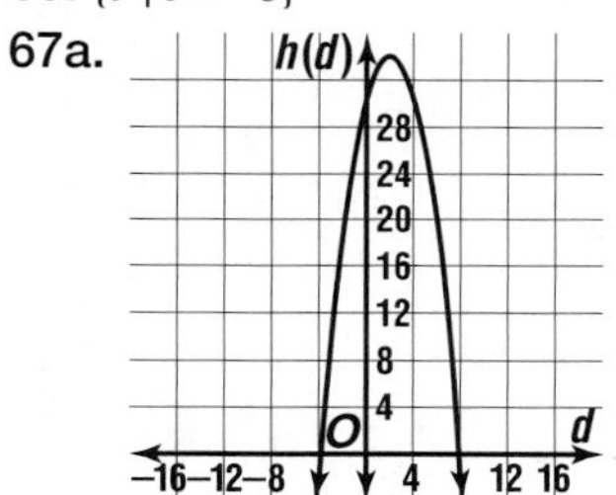

67b. 34 in. **69.** $3b - 18$ **71.** C

Pages 532–534 Lesson 12–6

1. $|x| < 7$ includes all numbers between −7 and 7. $|x| > 7$ includes all numbers to the left of −7 and to the right of 7. **3.** Mia; $|x| \leq 0$ includes only 0. $|x| \geq 0$ includes all real numbers. **5.** $x \leq 3$ and $x \geq -3$ **7.** $x > 8$ or $x < -8$

9. $\{x \mid -4 < x < 8\}$;

−8 −6 −4 −2 0 2 4 6 8 10 12

11. $\{t \mid t \geq 8 \text{ or } t \leq 2\}$;

0 1 2 3 4 5 6 7 8 9 10

13. $\{s \mid s \geq -1 \text{ or } s \leq -7\}$;

−9 −8 −7 −6 −5 −4 −3 −2 −1 0 1

15. $\{m \mid -6 < m < 4\}$;

−6 −5 −4 −3 −2 −1 0 1 2 3 4

17. $\{z \mid -9 \leq z \leq -5\}$;

−12 −11 −10 −9 −8 −7 −6 −5 −4 −3 −2

19. $\{p \mid -1 < p < 3\}$;

−5 −4 −3 −2 −1 0 1 2 3 4 5

21. $\{t \mid -2 \leq t \leq 2\}$;

−5 −4 −3 −2 −1 0 1 2 3 4 5

23. $\{k \mid -5.5 < k < 1.5\}$;

−6 −5 −4 −3 −2 −1 0 1 2 3 4

25. $\{y \mid y \leq -9 \text{ or } y \geq 3\}$;

−14 −12 −10 −8 −6 −4 −2 0 2 4 6

27. $\{x \mid x < -2 \text{ or } x > 2\}$;

−5 −4 −3 −2 −1 0 1 2 3 4 5

29. $\{r \mid r \leq 3 \text{ or } r \geq 5\}$;

−2 −1 0 1 2 3 4 5 6 7 8

31. $\{h \mid h < -6 \text{ or } h > 12\}$;

−6 −4 −2 0 2 4 6 8 10 12 14

33. $|s - 90| < 4$ **35.** $|s - 65| \leq 3$
37. $|x - 1| \geq 1$

39. $\{x \mid 0 < x < 8\}$;

−1 0 1 2 3 4 5 6 7 8 9

41. $\{x \mid x < -10 \text{ or } x > 8\}$;

−10 −8 −6 −4 −2 0 2 4 6 8 10

43. $|m - 3.25| \leq 0.05$; $3.20 \leq m \leq 3.30$

45.

−8 −7 −6 −5 −4 −3 −2 −1 0 1 2

47.

−7 −6 −5 −4 −3 −2 −1 0 1 2 3

49. $\{y \mid y \leq -1\}$ **51.** −2, 6

Page 534 Quiz 2

1. $\{x \mid x < 1\}$ **3.** $\{d \mid d \geq 1.2\}$

5. $\{n \mid -5 \leq n < 1\}$;

−7 −6 −5 −4 −3 −2 −1 0 1 2 3

7. $\{y \mid 5 < y < 9\}$;

2 3 4 5 6 7 8 9 10 11 12

9. $\{x \mid -4 < x < 4\}$;

−5 −4 −3 −2 −1 0 1 2 3 4 5

Pages 538–539 Lesson 12–7

1. Graph a solid line for ≤ and ≥. Graph a dashed line for < and >.

3.

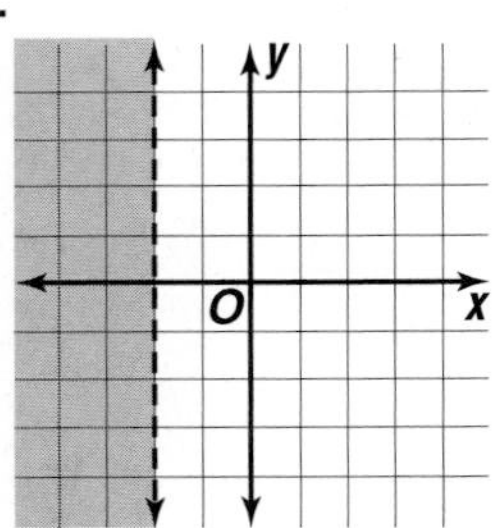

5.

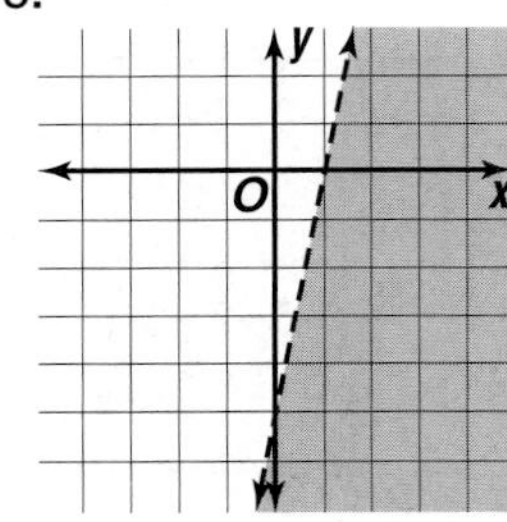

7.

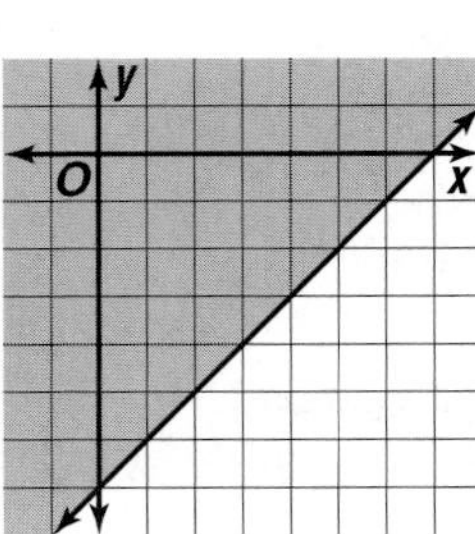

9.

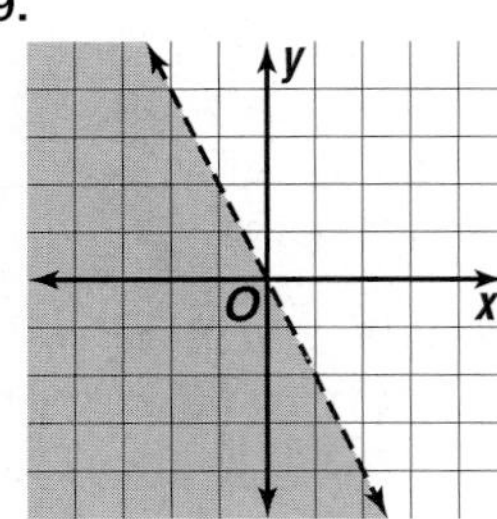

11.

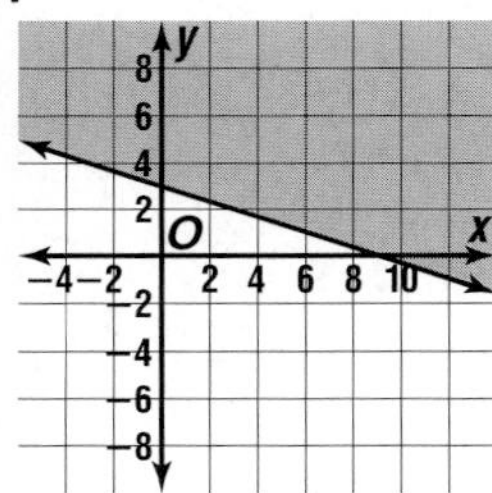

13.

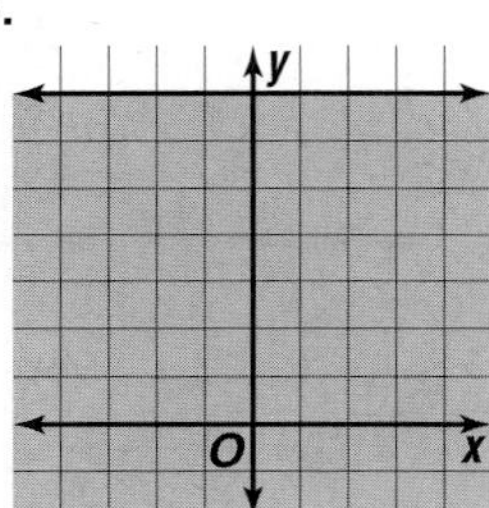

15.

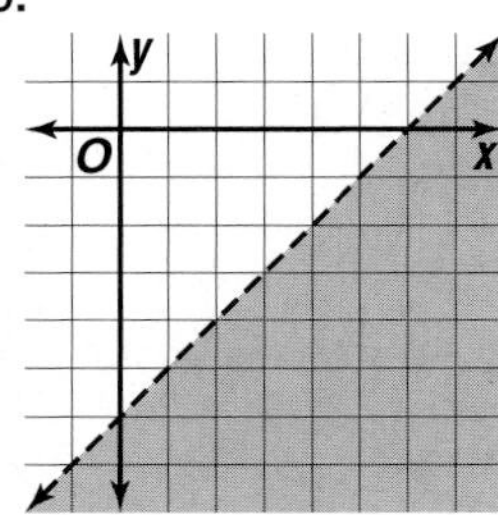

17.

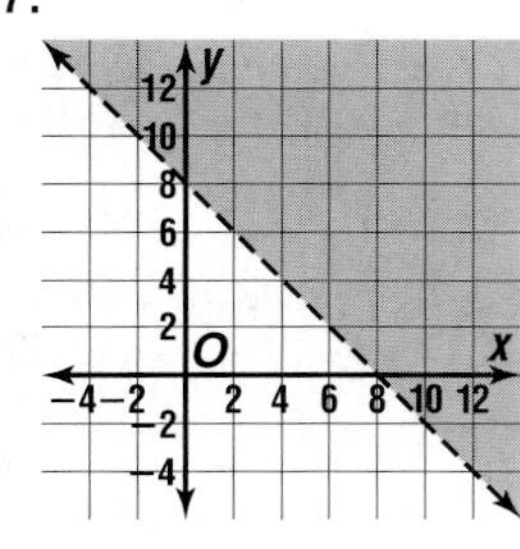

19.

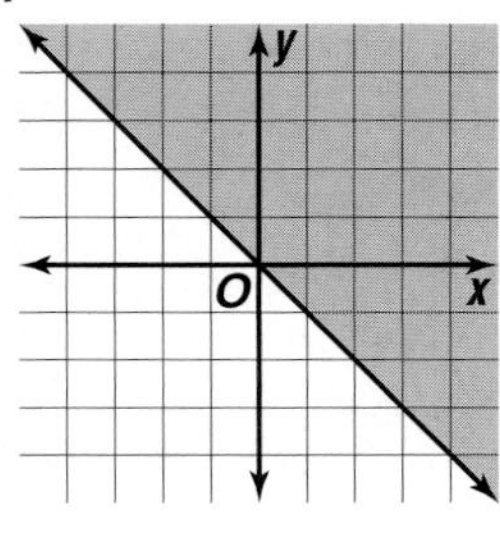

21.

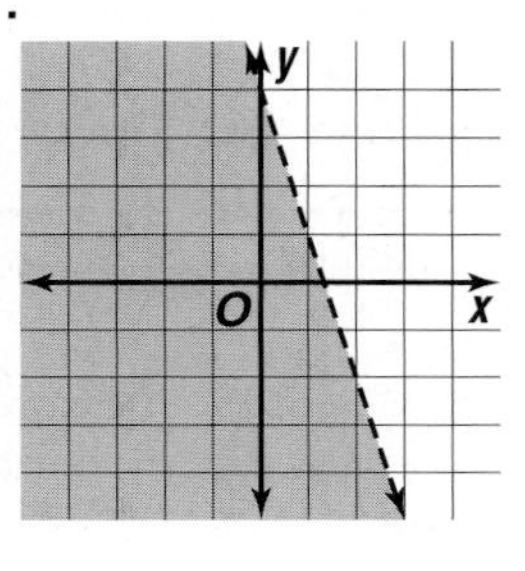

23.

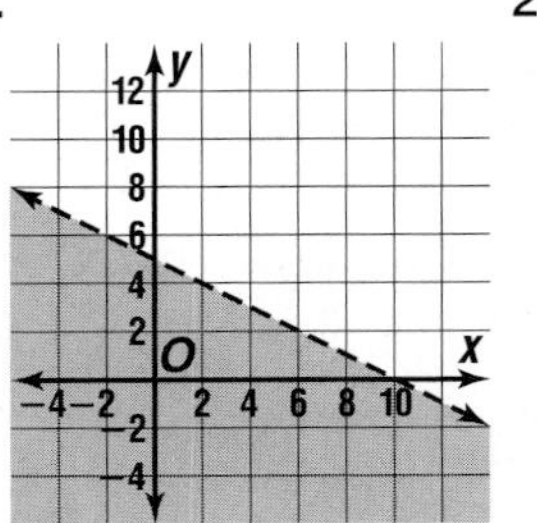

25.

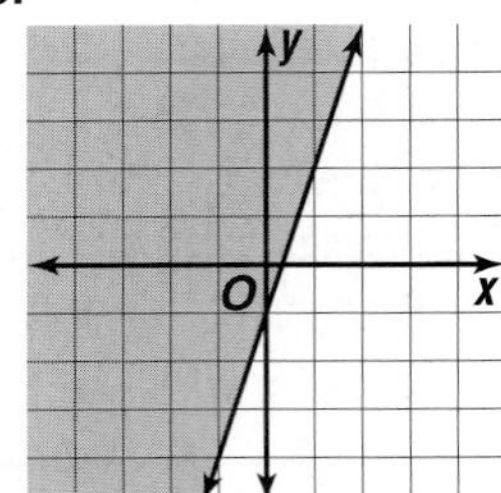

27.

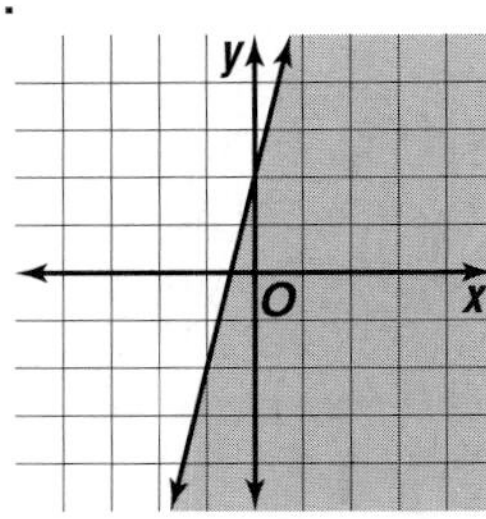

29.

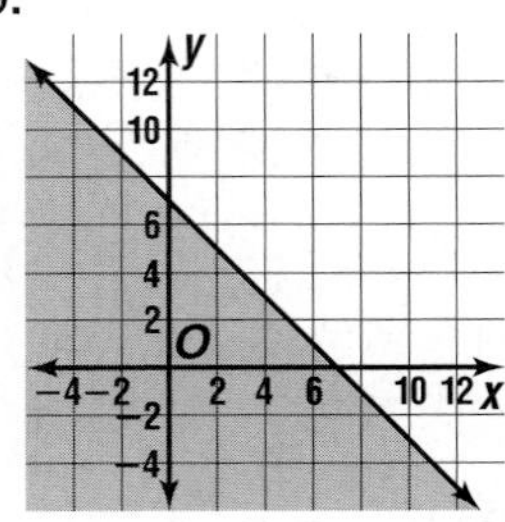

31.

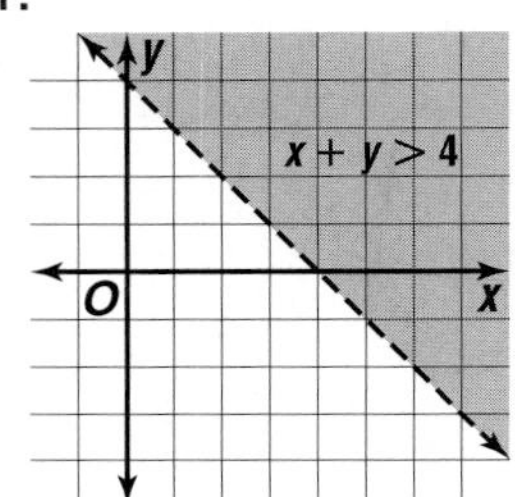

33a.

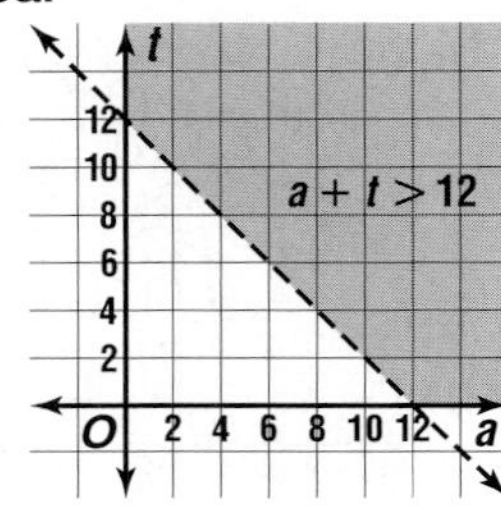

33b. Sample answer: {(4, 9), (2, 11), (3, 16)}
33c. positive, whole

35.

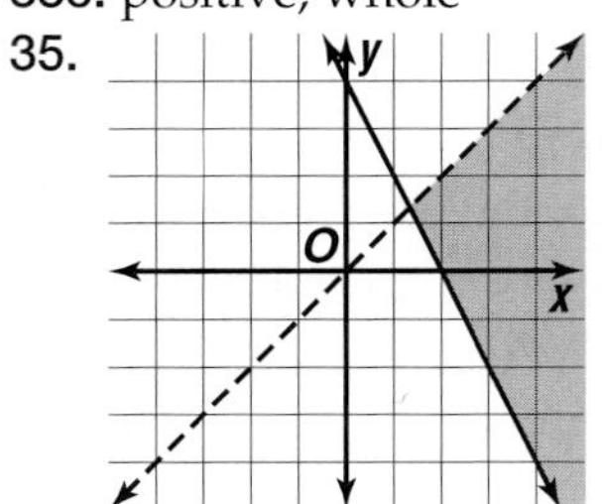

37. $\{h \mid 5 < h < 9\}$;

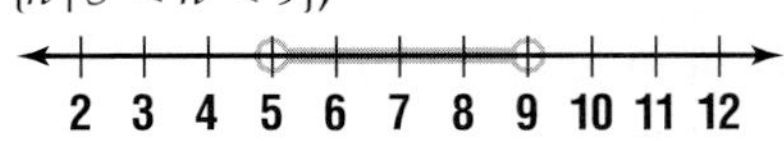

39. $\{b \mid b < -2 \text{ or } b > 1\}$;

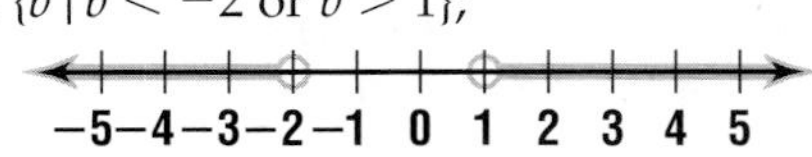

41. 8.483×10^2

Pages 542–544 Chapter 12 Study Guide and Assessment

1. union **3.** half-plane **5.** boundary
7. half-plane **9.** set-builder notation

11. −5 −4 −3 −2 −1

13. 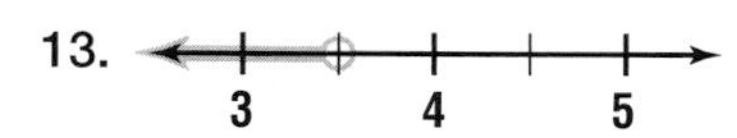

15. $x \geq -5$ 17. $\{x \mid x > 4\}$ 19. $\left\{y \mid y \geq 1\frac{1}{6}\right\}$
21. $\{x \mid x \leq 3\}$ 23. $\{n \mid n < -5\}$ 25. $\{x \mid x > 15\}$
27. $\{w \mid w \leq 8\}$ 29. $\{x \mid x \geq 4\}$ 31. $\{t \mid t \leq 9\}$
33. $4x - 3 > 25$; $\{x \mid x > 7\}$

35. $\{y \mid -7 \leq y < -1\}$;

−8 −7 −6 −5 −4 −3 −2 −1 0 1 2

37. $\{a \mid a \leq 1 \text{ or } a \geq 2\}$;

−5 −4 −3 −2 −1 0 1 2 3 4 5

39. $\{t \mid -4 \leq t \leq 6\}$;

−4 −3 −2 −1 0 1 2 3 4 5 6

41. $\{x \mid -4 < x < -2\}$;

−8 −7 −6 −5 −4 −3 −2 −1 0 1 2

43. $\{y \mid y \neq 2\}$;

−5 −4 −3 −2 −1 0 1 2 3 4 5

45. $|s - 80| > 5$

47. 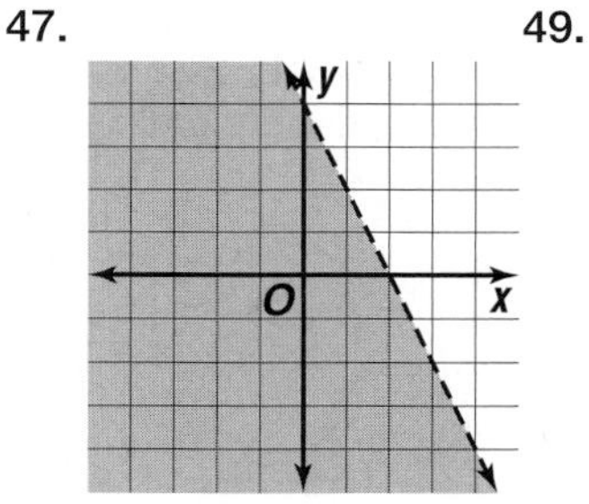

49. 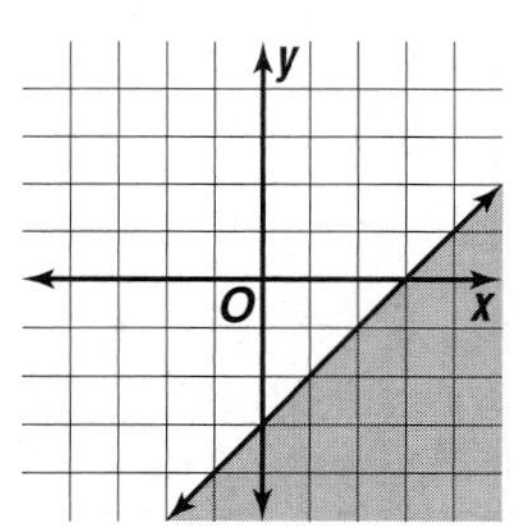

51. $\{x \mid 30 < x < 60\}$

Page 547 Preparing for Standardized Tests
1. A 3. C 5. A 7. B 9. 270

Chapter 13 Systems of Equations and Inequalities

Pages 552–553 Lesson 13–1
1. the ordered pair for the point at which the graphs of the equations intersect 3a. $(-4, 2)$
3b. $(2, -2)$ 3c. $(2, 4)$

5. $(-4, 1)$

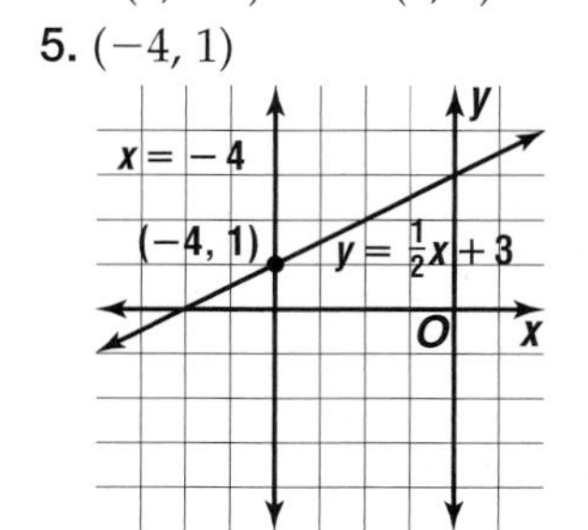

7. $(1, -3)$

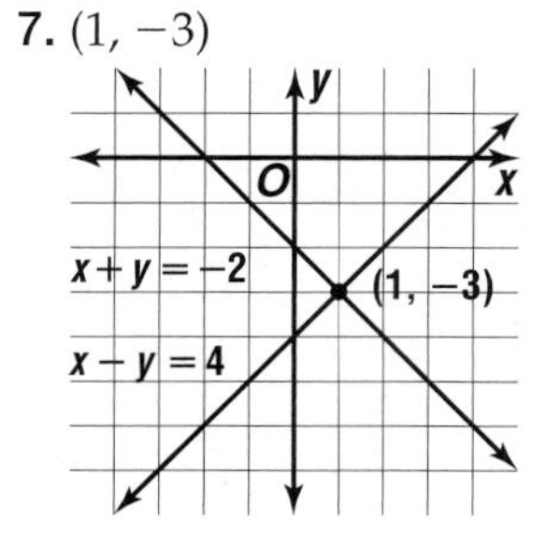

9. $(5, -4)$

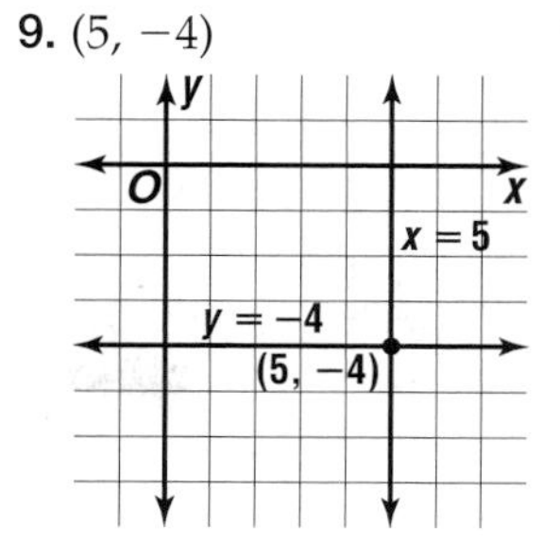

11. $(0, 2)$

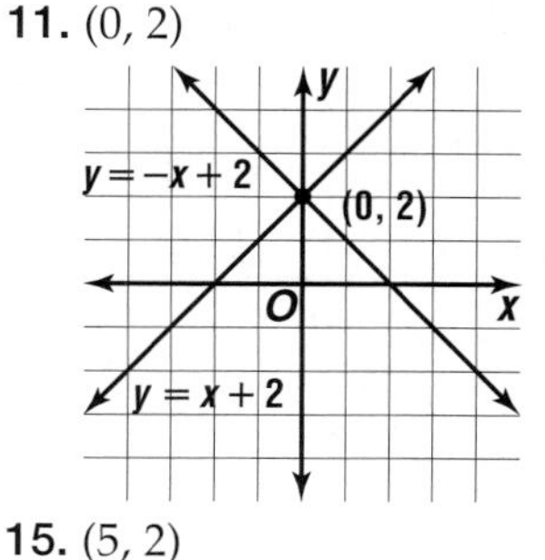

13. $(3, 6)$

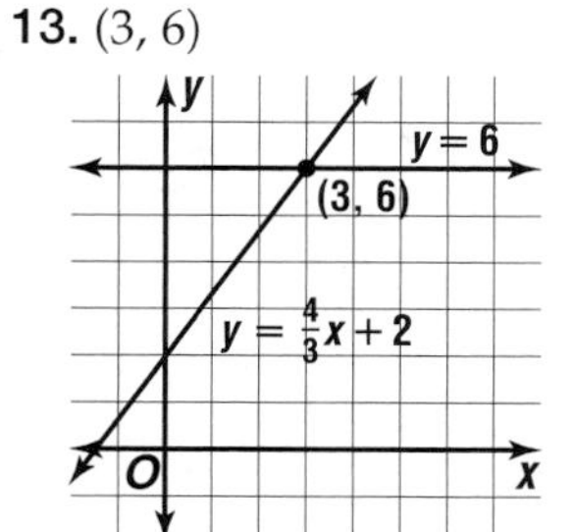

15. $(5, 2)$

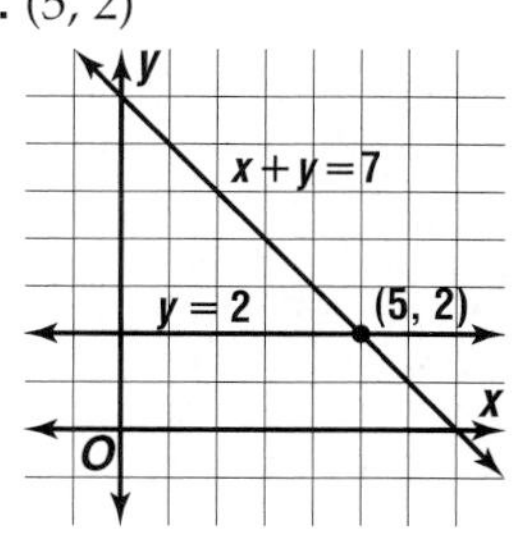

17. $(-2, -3)$

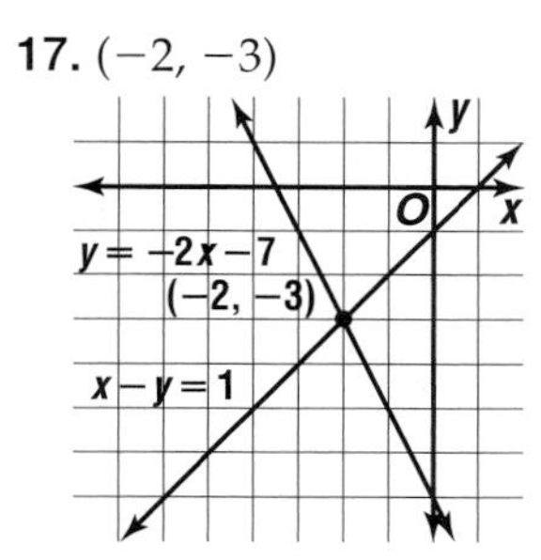

19. $(-6, 3)$

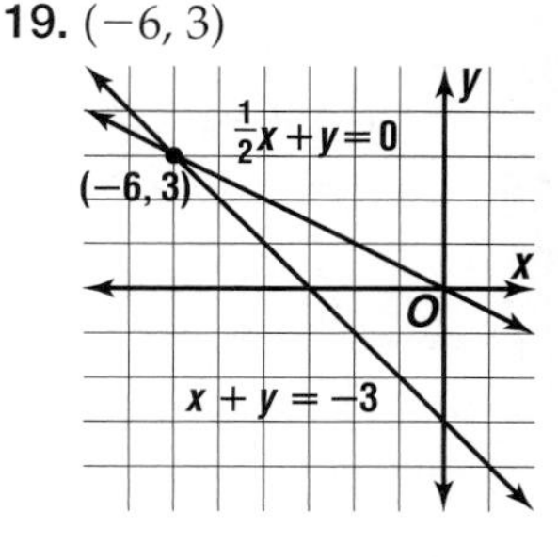

21. $(2, 6)$ 23. $(10, -6)$
25. Sample answer: $y = x + 4$ and $y = -x + 4$
27. $\{x \mid 1 < x < 5\}$

−5 −4 −3 −2 −1 0 1 2 3 4 5

29. $\{m \mid m > 2 \text{ or } m < -4\}$

−5 −4 −3 −2 −1 0 1 2 3 4 5

31. 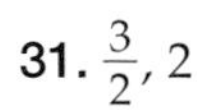 $\frac{3}{2}, 2$

Pages 557–559 Lesson 13–2
1. two parallel lines, two intersecting lines, the same line 5. consistent and dependent
7. $(1, -1)$

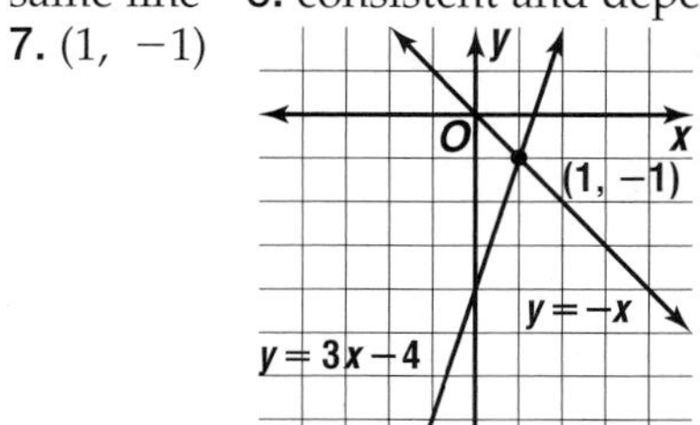

9a. Consistent and independent; there is one solution, (3, 150). 9b. After 3 seconds, both the cat and dog will be 150 feet from the dog's original starting position. 11. inconsistent 13. consistent and dependent 15. consistent and independent

17. infinitely many

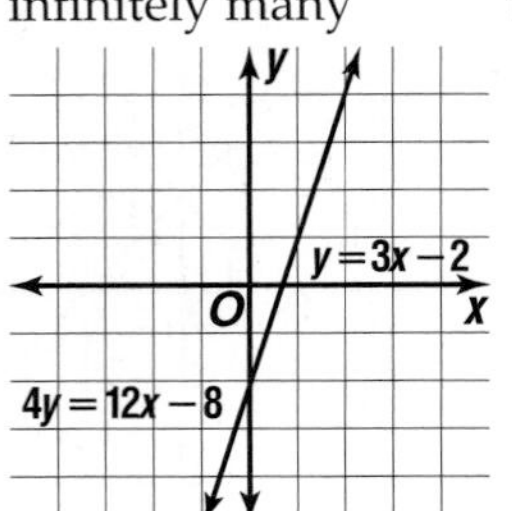

19. (5, −2)

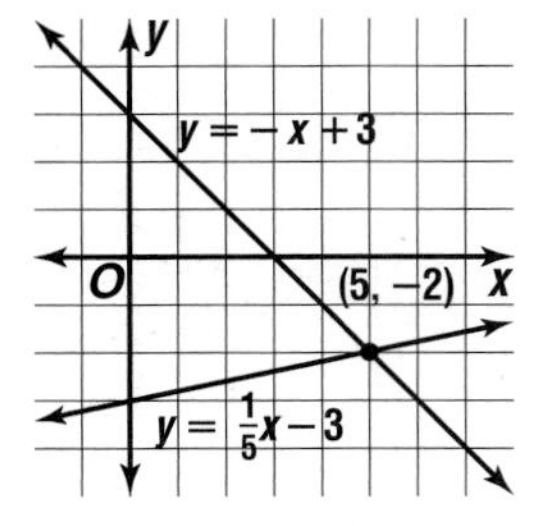

21. no solution

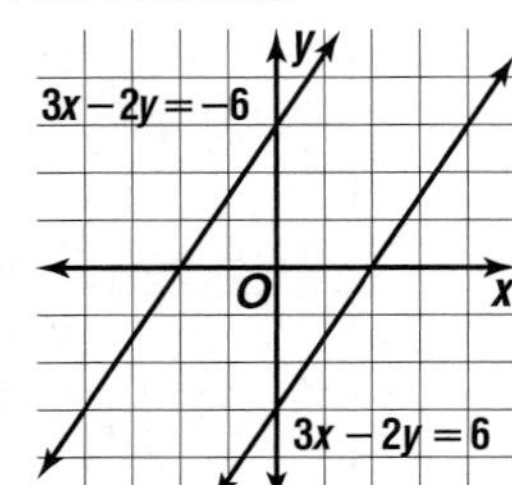

23. (−1, 3)

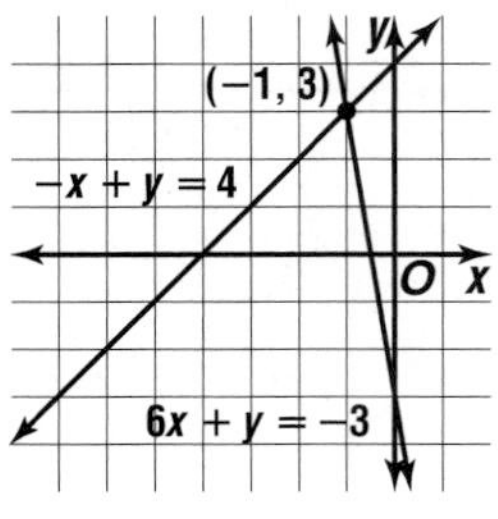

25. one; (5, 1) **27.** There is no solution because the graphs of the equations are parallel lines. The second dog never catches up with the first.

29. Sample answer: $y = -\frac{1}{3}x$

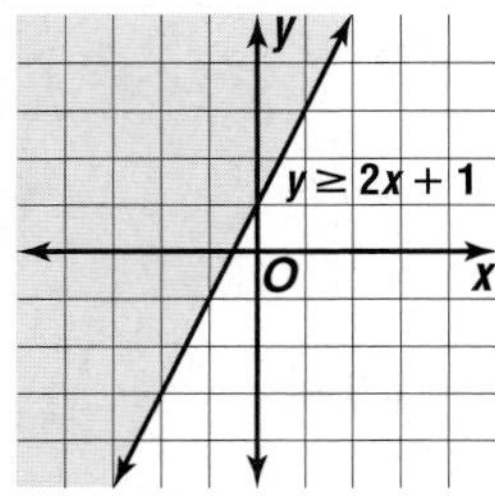

33. $4(x - 2)$ **35.** $3a(ab^2 + 2b - 3)$

Pages 564–565 Lesson 13–3

1. when the exact coordinates are not easily determined from the graph **3.** Both are correct because you can substitute for either variable. **5.** (4, 12) **7.** (4, 0) **9.** no solution **11.** (1, 1) **13.** (−3, −9) **15.** no solution **17.** (−4, 1) **19.** (3, 4) **21.** (2, 2) **23.** (2, −10) **25.** $\left(4, \frac{1}{3}\right)$ **27.** infinitely many **29.** (13, 30) **31a.** 5.5 hr **31b.** 357.5 mi **33.** No; the graphs of equations *A*, *B*, and *C* could form a triangle. Any two of the lines would be intersecting, but there is no point where all three lines intersect.

35. (−1, 4)

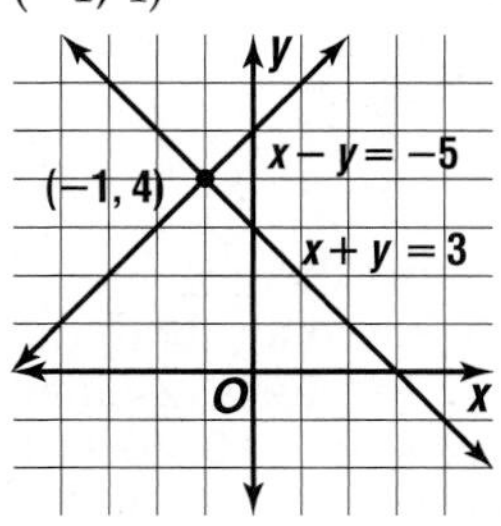

37. $-3 < x \leq 2$ **39.** 460

Page 565 Quiz 1

1. (4, −1)

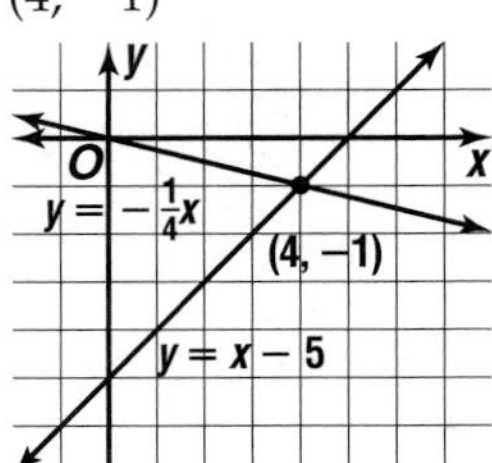

3. no solution

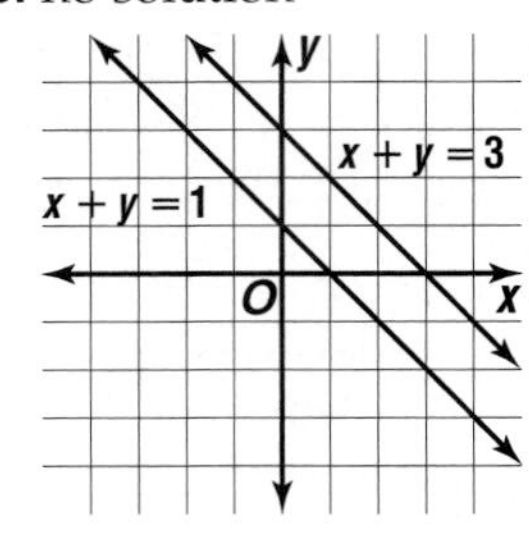

5. 12.5 lb of the \$3.90, 37.5 lb of the \$4.30

Pages 569–571 Lesson 13–4

1. when the coefficients of terms with the same variable are the same or additive inverses **3.** addition **5.** subtraction **7.** (8, −1) **9.** (2, 2) **11.** $\left(\frac{16}{3}, 3\right)$ **13a.** $x + y = 42$ and $x = 2y + 3$ **13b.** 29, 13 **15.** (14, 5) **17.** (−5, 15) **19.** (0, 6) **21.** (8, −1) **23.** (−15, −22) **25.** $\left(5, \frac{5}{2}\right)$ **27.** infinitely many **29.** (−11, −4) **31.** (2, 8) **33.** $\left(2, \frac{1}{2}\right)$ **35a.** $13x + 2y = 137.50$ and $9x + 2y = 103.50$ **35b.** \$8.50 child, \$13.50 adult **37a.** $x + 2y = 29$ and $y = 3x - 10$ **37b.** 7 and 11 **39.** (11, −38) **41.** consistent and dependent **43.** $y < 4$ **45.** $x < -2$ **47.** $a^2 - 6a + 9$ **49.** $y - 1 = -5(x - 6)$ or $y + 4 = -5(x - 7)$

Pages 575–577 Lesson 13–5

1. when neither of the variables in a system of equations can be eliminated by simply adding or subtracting the equations **3a.** substitution **3b.** elimination (×) **3c.** elimination (+) **3d.** elimination (−) **5.** Multiply second equation by 2. Then add. **7.** (2, 2) **9.** (−3, 2) **11.** no solution **13a.** 2 mph **13b.** 14 mph **15.** (2, 0) **17.** (−1, 3) **19.** $\left(\frac{1}{2}, \frac{3}{4}\right)$ **21.** $\left(\frac{5}{2}, 1\right)$ **23.** (−6, −3) **25.** infinitely many **27.** (−2, −1) **29.** $\left(\frac{7}{9}, 0\right)$ **31.** Sample answer: substitution or multiplication; (−2, 4) **33.** Sample answer: graphing; no solution **35a.** $6a + 10b = 108$ and $4a + 12b = 104$ **35b.** \$8, \$6 **37.** (7, −1) **39.** $\left(\frac{3}{2}, 1\right)$ **41.** $m < -4$ **43.** $a \geq -4$ **45.** 5

Pages 583–585 Lesson 13–6

1. (−1, −7), (2, 0) **3.** Sample answer: The following methods can be used to solve linear systems of equations: graphing, substitution, elimination using addition or subtraction, and elimination using multiplication. Graphing and substitution can be used to solve quadratic-linear systems of equations. Graphing is useful when you want to estimate the solution. The other methods are useful when you want to find the exact solution.

5. (0, 0), (2, 4)

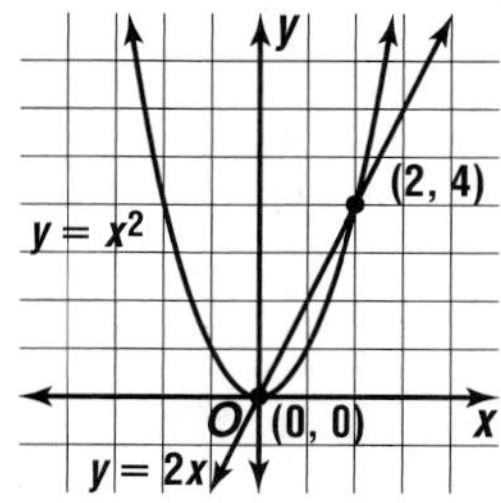

7. no solution

9. (2, 0), (0, −4)

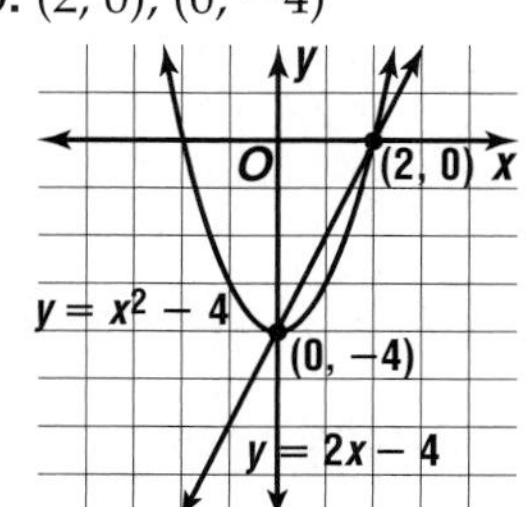

11. no solution

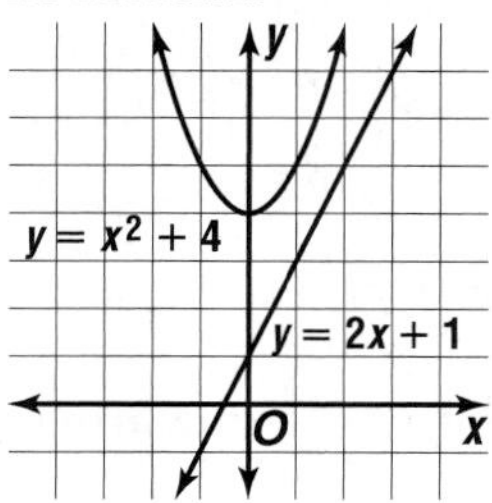

13. (−1, 4), (2, 1)

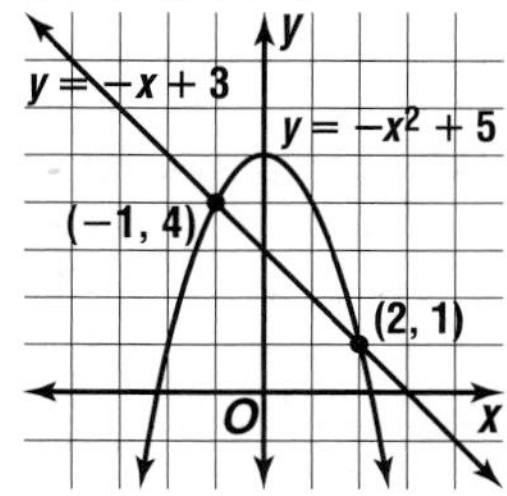

15. (−4, 7) **17.** (8, 32), (−2, 2) **19.** no solution **21.** (3, 0), (0, −9) **23a.** $A = 16$ and $A = 4x^2 - 28x + 40$ **23b.** $\ell = 8$ ft, $w = 2$ ft, $h = 1$ ft **23c.** 16 ft^3 **25.** Side length 4 units; solve the quadratic-linear system of equations $p = s^2$ and $p = 4s$, where p represents perimeter and s represents side length. The solutions are (0, 0) and (4, 16). **27.** (2, −1) **29.** {−7, 7} **31.** {4, −8}

Page 585 Quiz 2

1. (4, 2) **3.** (6, 2) **5a.** $x - y = 38$, $x = 3y - 2$ **5b.** 58, 20

Pages 589–590 Lesson 13–7

1. A system of linear equations may have at most one solution if the equations are distinct. A system of inequalities may have an infinite number of solutions. **3.** Kyle; all of the points in region B satisfy both inequalities. **5.** no

7.

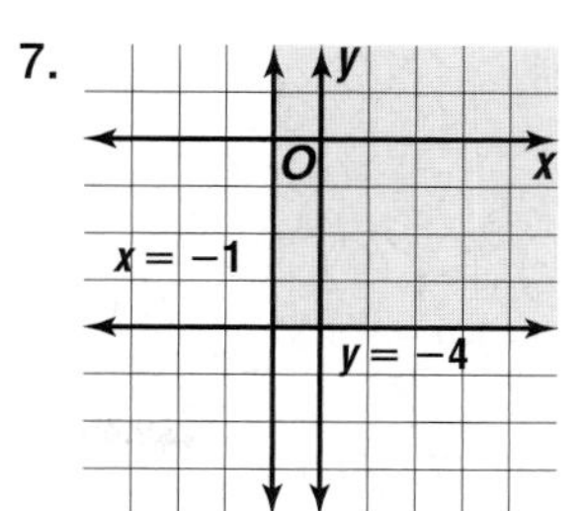

9.

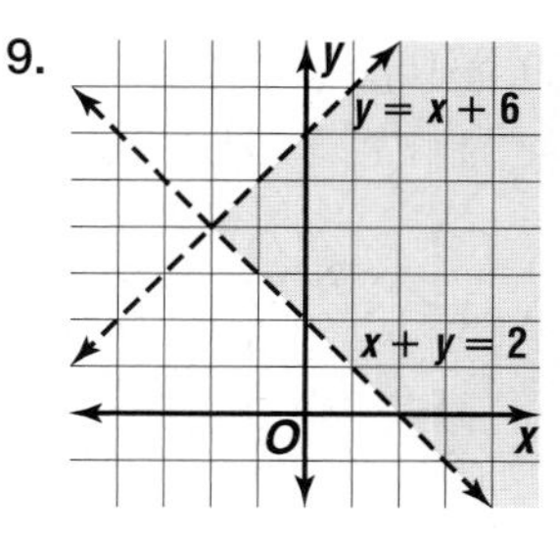

11. Sample answer: 2 dozen sugar, 4 dozen choc. chip; 4 dozen sugar, 3 dozen choc. chip; 5 dozen sugar, 2 dozen choc. chip

13.

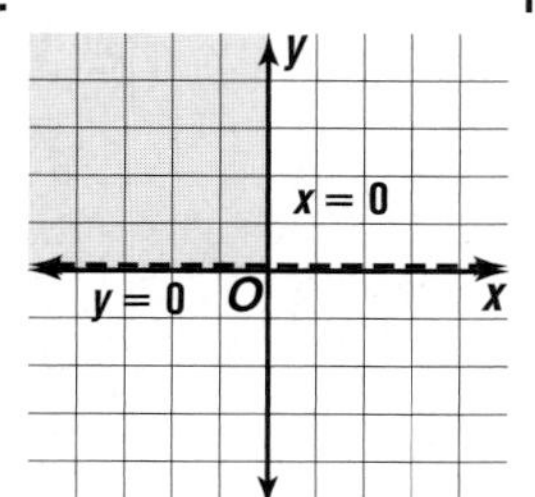

15.

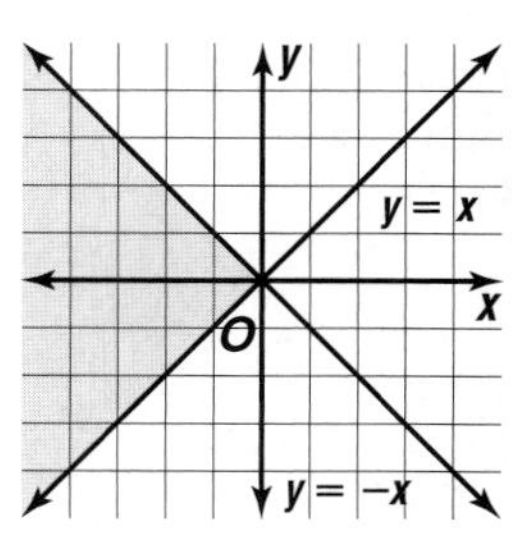

17.

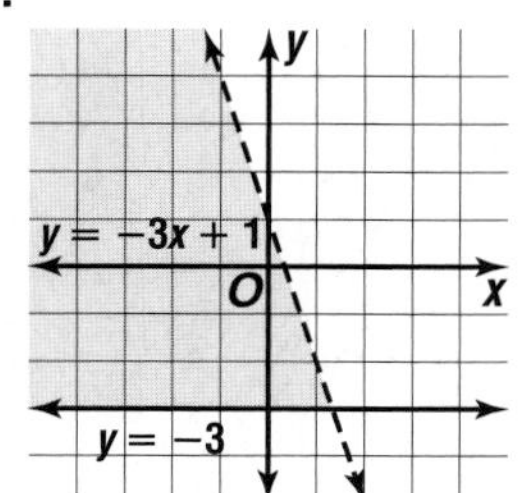

19. no solution

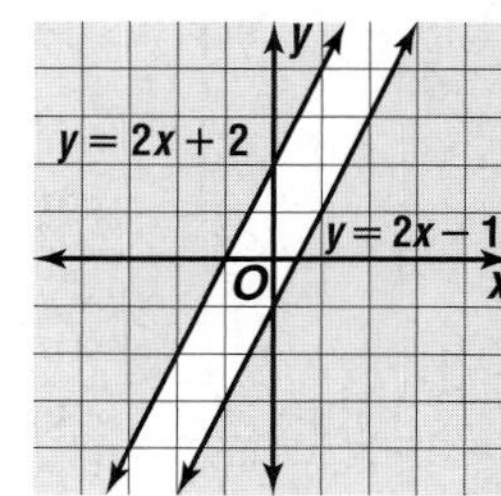

21.

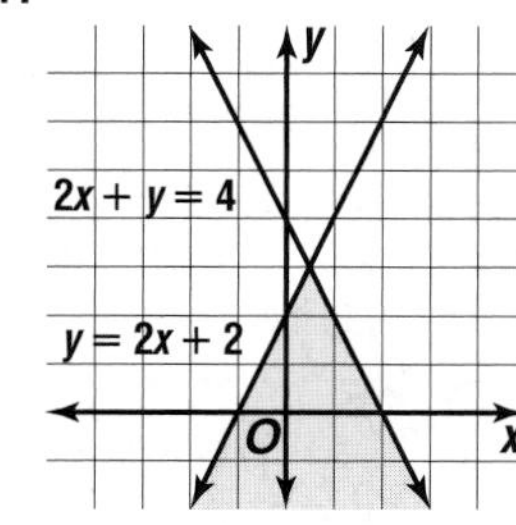

23.

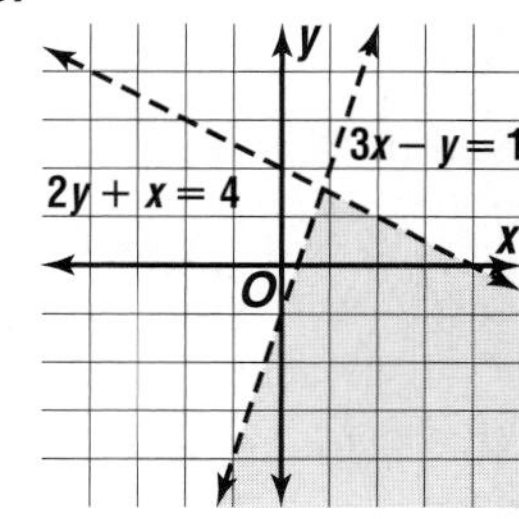

25.

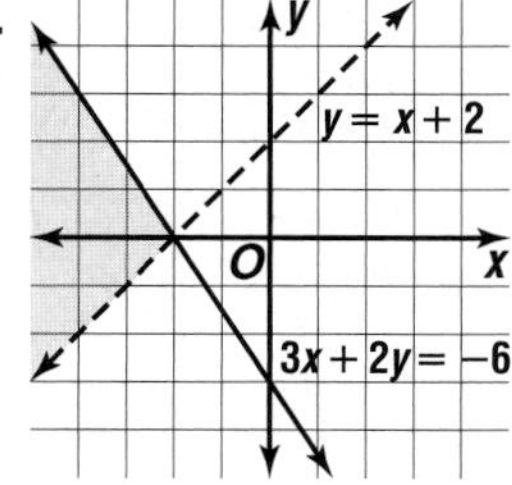

27. People who use the phone less than 75 minutes a month should select the $0.15 per minute plan. People who use the phone more than 75 minutes a month should select the $0.05 per minute plan. The middle plan is never the cheapest.

29. (2, 4), (−2, 4)

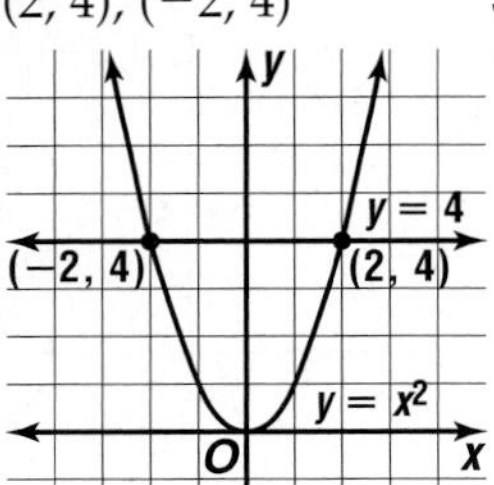

31. no solution

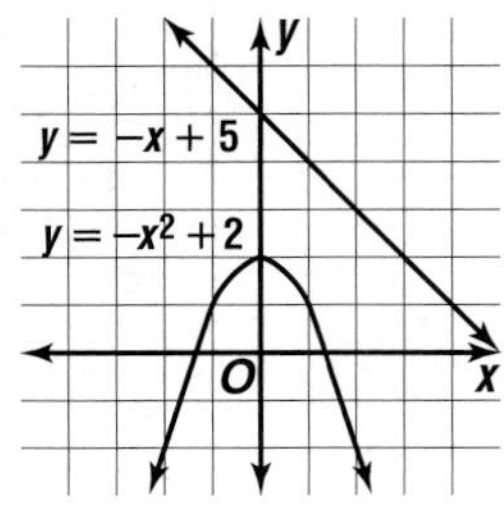

33. 4 lb of the $5.25, 1 lb of the $6.50 35. 5 37. A

Pages 592–594 Chapter 13 Study Guide and Assessment

1. true 3. false; all 5. true 7. true 9. false; elimination

11. (1, 4)

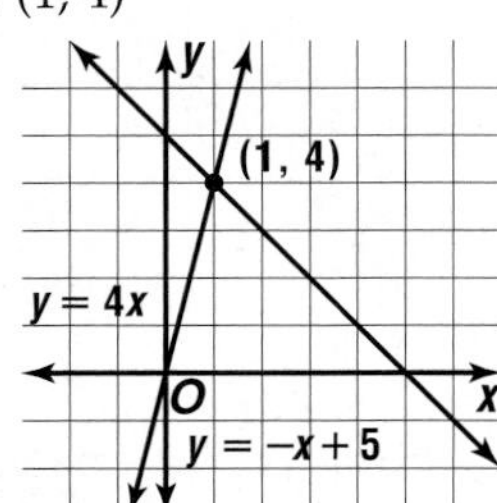

13. (−2, 2)

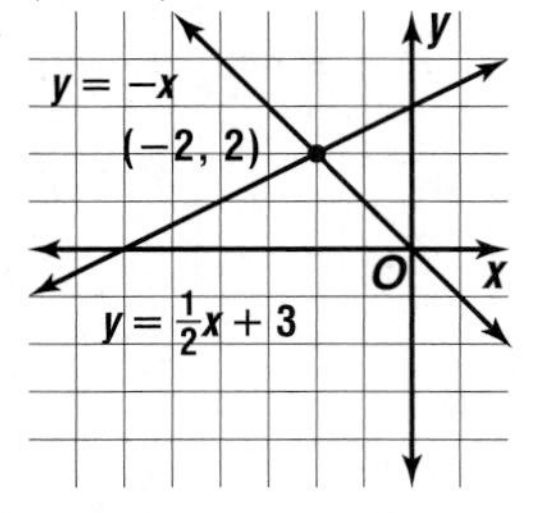

15. (−6, 0)

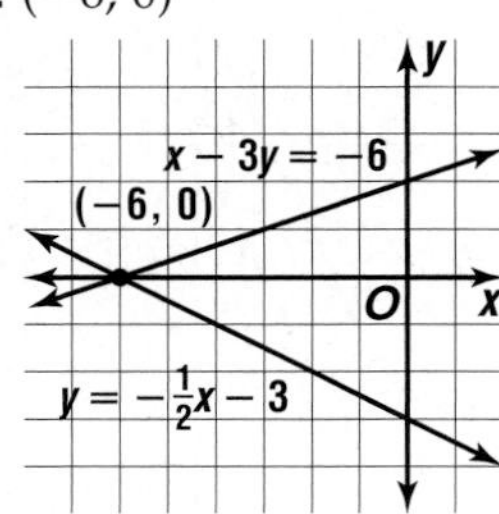

17. (3, −2)

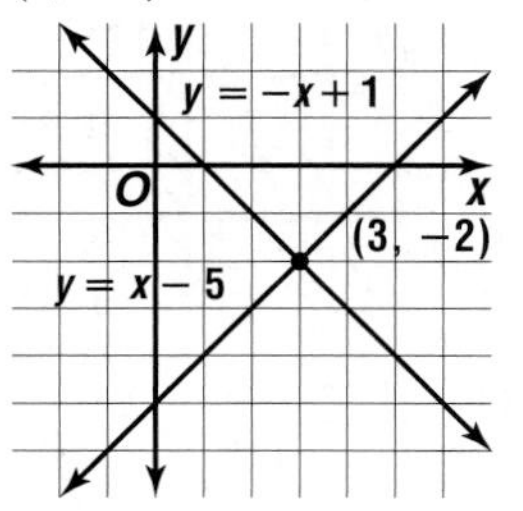

19. infinitely many

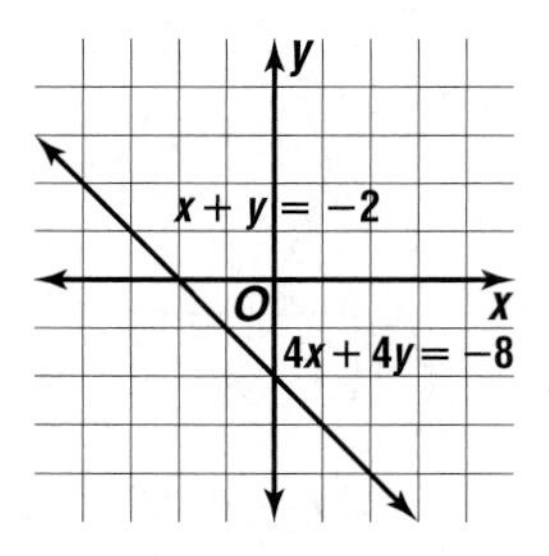

21. $\left(\frac{1}{2}, 1\right)$ 23. no solution 25. infinitely many

27. (4, −6) 29. no solution 31. infinitely many

33. (2, 0) 35. $\left(\frac{6}{11}, \frac{1}{11}\right)$ 37. (22, 31)

39. (−2, −4), (1, −1)

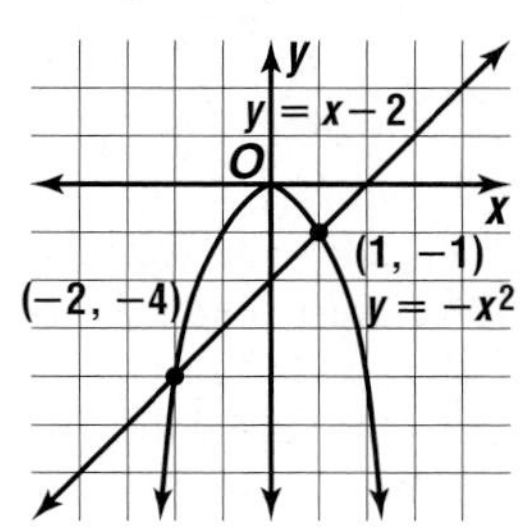

41. no solution

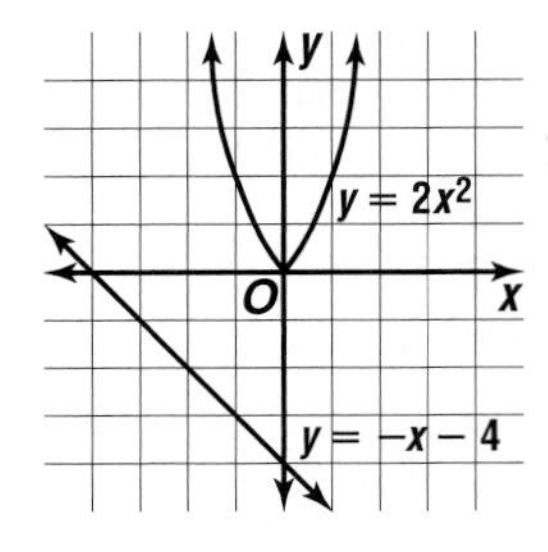

43. (4, 18), (−1, 3) 45. (3, 3)

47.

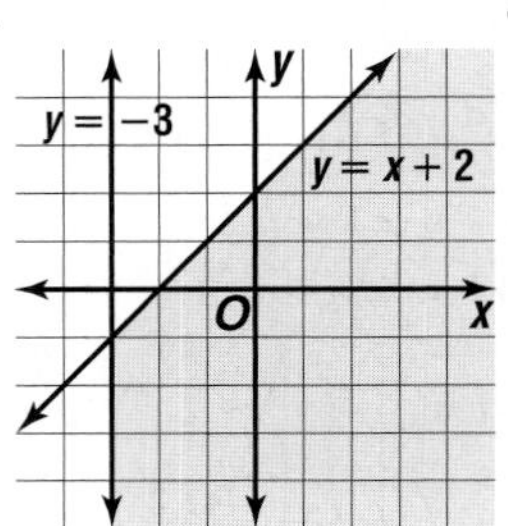

49.

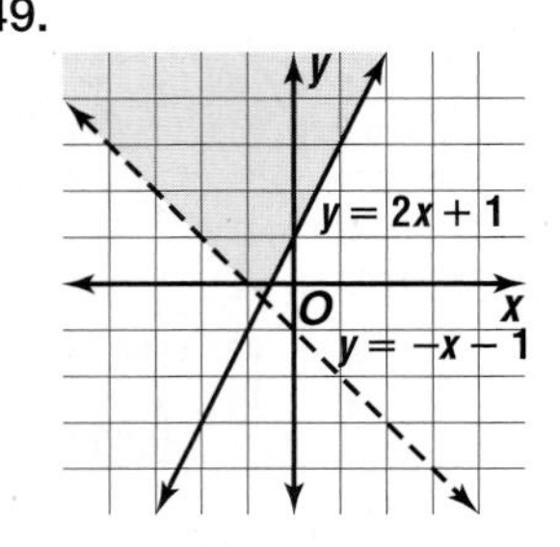

51. 8 lb of almonds, 12 lb of cashews 53. 16 ft by 9 ft

Page 597 Preparing for Standardized Tests

1. D 3. A 5. B 7. D 9. 934

Chapter 14 Radical Expressions

Pages 603–605 Lesson 14–1

1. Sample answer: −4 3. Sample answer: $\sqrt{13}$

5. I 7. Q

9. −4.5

$-\sqrt{20}$

−5 −4 −3 −2 −1 0 1 2 3 4 5

11. 7.3

13. rational 15. rational 17. Q 19. W, Z, Q

21. Z, Q 23. N, W, Z, Q 25. Z, Q 27. Q

29. 1.7

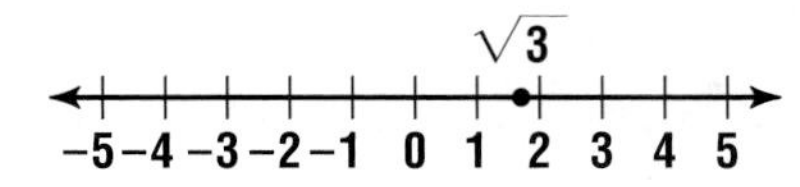

31. −3.2

$-\sqrt{10}$

−5 −4 −3 −2 −1 0 1 2 3 4 5

33. −6.3

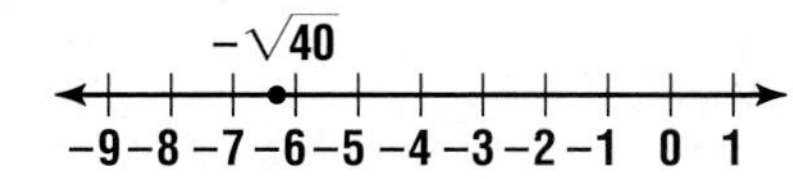

35. 7.2

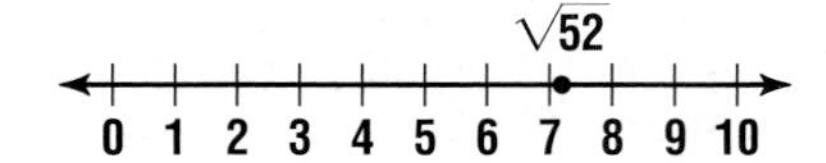

37. 10.4

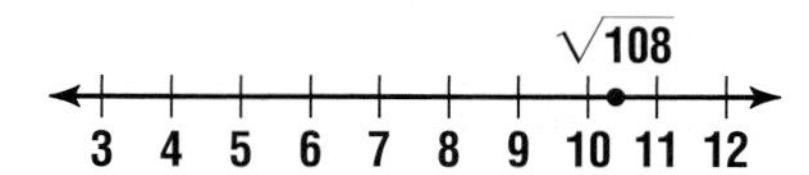

39. 15.8

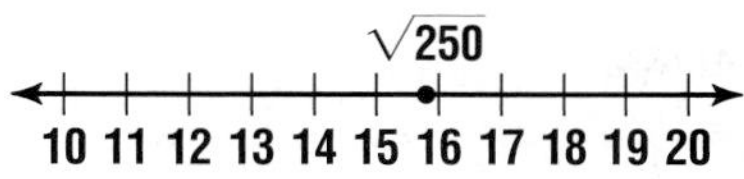

41. rational **43.** irrational; 6 and 7 **45.** rational **47.** irrational; -2 and -3 **49.** irrational; 11 and 12 **51.** rational

53.

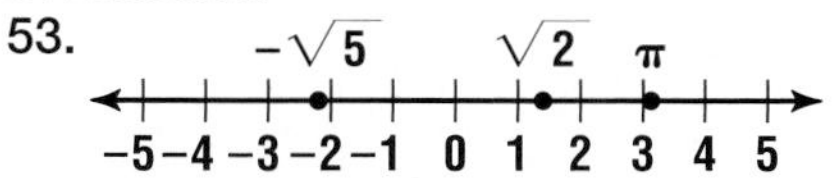

55. about 8.5 s **57.** 5.2 ohms and 1200 watts or 4.6 ohms and 1500 watts

59a. $\sqrt{17}$, $\sqrt{18}$, $\sqrt{19}$, $\sqrt{20}$, $\sqrt{21}$, $\sqrt{22}$, $\sqrt{23}$, $\sqrt{24}$ **59b.** $\sqrt{19}$ and $\sqrt{20}$ **61.** (3, 11), (1, 3)

63. $\{x \mid x < -7\}$ **65.** $\{m \mid m \leq -1.8\}$

67. 0

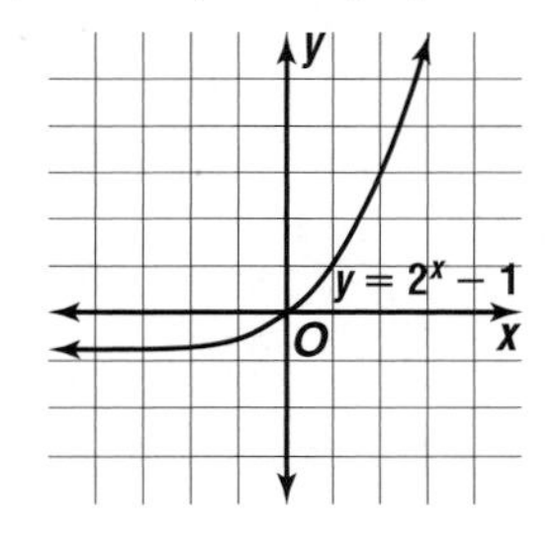

69. $4(a + 3)(a - 3)$ **71.** 4 **73.** A

Pages 608–611 Lesson 14–2

1. Pythagorean Theorem **3.** Both are correct since it does not matter which ordered pair is first when using the Distance Formula as long as the coordinates are used in the same order. **5.** 5

7. 9 or -3 **9.** 9.5 mi **11.** 12 **13.** $\sqrt{58}$ or 7.6 **15.** $\sqrt{185}$ or 13.6 **17.** $\sqrt{18}$ or 4.2 **19.** 17 or -13 **21.** -12 or 6 **23.** 2 or -14 **25.** $\sqrt{194}$ or 13.9 **27.** 8 or 32 **29.** No; no two sides have the same measure. **31.** 16.3 mi

33. 3.5

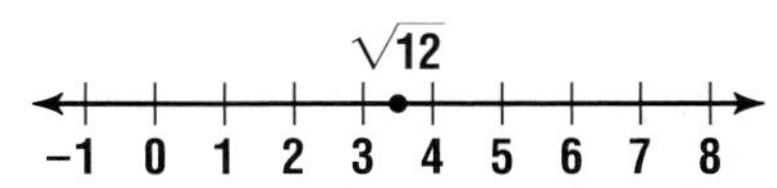

35. 11.6

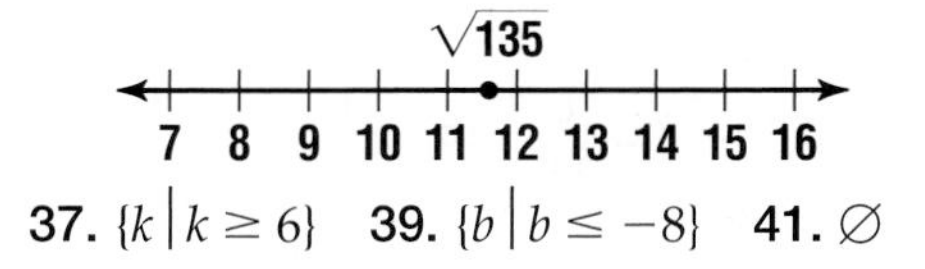

37. $\{k \mid k \geq 6\}$ **39.** $\{b \mid b \leq -8\}$ **41.** $\varnothing$

Page 611 Quiz 1

1. W, Z, Q **3.** I **5.** 1.22, $1.\overline{2}$, $\sqrt{2}$ **7.** $\sqrt{98}$ or 9.9 **9.** 0 or 6

Pages 617–619 Lesson 14–3

1. to ensure nonnegative results **3.** Greg is correct. In its simplest form, the expression is written as $\frac{3 - \sqrt{5}}{2}$ since the numbers 6, 2, and 4 in $\frac{6 - 2\sqrt{5}}{4}$ are all divisible by 2. **5.** $2 - \sqrt{8}; -4$ **7.** $5\sqrt{3}$

9. $3\sqrt{10}$ **11.** $\frac{\sqrt{21}}{7}$ **13.** $6|x|\sqrt{y}$ **15.** $4\sqrt{2}$ units **17.** $7\sqrt{2}$ **19.** $10\sqrt{5}$ **21.** 4 **23.** 15 **25.** 4 **27.** $2\sqrt{10}$ **29.** $\frac{2\sqrt{5}}{5}$ **31.** $\frac{8 + 2\sqrt{3}}{13}$ **33.** $\frac{24 + 4\sqrt{7}}{29}$ **35.** $2|m|\sqrt{10}$ **37.** $3|a|b\sqrt{6b}$ **39.** $6|x|y^2z^2\sqrt{xz}$ **41a.** $2\sqrt{3}$ mph **41b.** 10 in. **43.** 2 or -4 **45.** Z, Q **47.** no solution **49.** $\sqrt{79}$

Pages 621–623 Lesson 14–4

1. Sample answer: Add like terms.
3. The Distributive Property allows you to add like terms. Radicals with like radicands can be added or subtracted. **5.** none **7.** $8\sqrt{3}, -18\sqrt{3}$ **9.** $2\sqrt{5}$ **11.** $-5\sqrt{7}$ **13.** $-6\sqrt{2}$ **15.** $6\sqrt{3}$ **17.** $14\sqrt{5}$ **19.** $4\sqrt{7}$ **21.** $-2\sqrt{5}$ **23.** $-9\sqrt{2} + 8\sqrt{5}$ **25.** $7\sqrt{6}$ **27.** $14\sqrt{3}$ **29.** $-4\sqrt{2}$ **31.** $8\sqrt{2} + 6\sqrt{3}$ **33.** $-2\sqrt{5} - 6\sqrt{6}$ **35.** $5\sqrt{7}$ ft **37a.** $4\sqrt{3}$ in. **37b.** $6\sqrt{3} + 6$ in. **39.** $3n|p|\sqrt{5mn}$ **41.** $\sqrt{101}$ or 10.0 **43.** $-1 \cdot 2 \cdot 13 \cdot a \cdot a \cdot a \cdot b \cdot b$ **44.** $2 \cdot 2 \cdot 3 \cdot 3 \cdot x \cdot y \cdot y \cdot y \cdot y \cdot z$ **45.** $a^2 - 4c^2$ **46.** A

Pages 627–629 Lesson 14–5

1. Isolate the radical. **5.** $a + 5 = 4$ **7.** 121 **9.** 4 **11.** 35.5 ft **13.** no solution **15.** -12 **17.** 2 **19.** 75 **21.** 63 **23.** no solution **25.** 2 and 3 **27.** 6 **29.** 10 **31.** 140 **33a.** 1.6 m/s^2 **33b.** 9.8 m/s^2 **35.** $-5\sqrt{5}$ **37.** $-\sqrt{6} - 3\sqrt{2}$ **39.** $\sqrt{5}$

41.

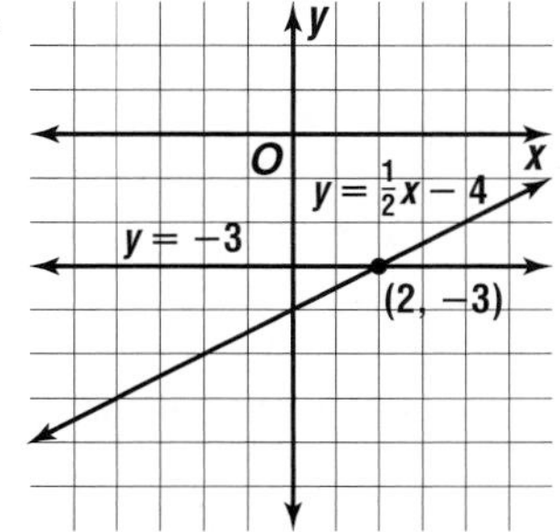

43. $x > 1$ **45.** 22.5

Page 629 Quiz 2

1. $3\sqrt{10}$ **3.** $\frac{7 - \sqrt{5}}{44}$ **5.** $12\sqrt{5}$ **7.** $7\sqrt{2}$ in. **9.** no solution

SELECTED ANSWERS

Pages 630–632 Study Guide and Assessment

1. $\sqrt{(x_2 - x_1)^2 + (y_2 - y_1)^2}$ **3.** radicals **5.** $\sqrt{7ab}$ **7.** whole **9.** no solution **11.** Q **13.** Z, Q **15.** 2.2

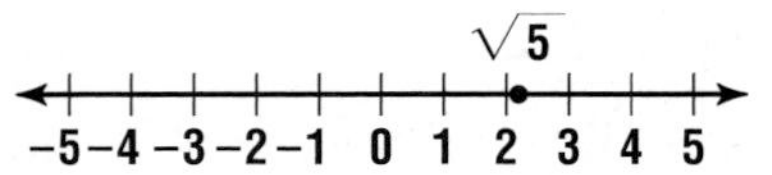

17. 10.5

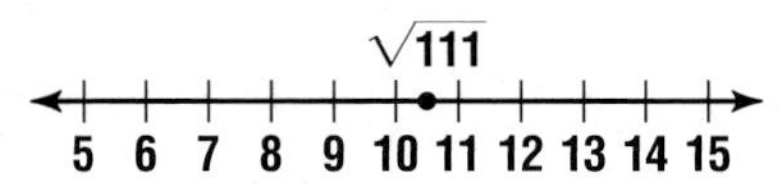

19. 5 **21.** $\sqrt{72}$ or 8.5 **23.** −1 or −5 **25.** $5\sqrt{2}$ **27.** $3\sqrt{3}$ **29.** $\frac{4+\sqrt{5}}{11}$ **31.** $10|a|c^2\sqrt{3b}$ **33.** $7\sqrt{5}$ **35.** 0 **37.** $22\sqrt{2}$ **39.** $-14\sqrt{5}$ **41.** 25 **43.** 9 **45.** no solution **47.** 6 **49.** 150 **51.** integers, rational numbers **53.** $10\sqrt{2}$ lb **55a.** 3844 m **55b.** 2601 m

Page 635 Preparing for Standardized Tests

1. A **3.** A **5.** A **7.** C **9.** 50

Chapter 15 Rational Expressions and Equations

Pages 641–643 Lesson 15–1

1. The denominator is 0 when $x = 2$. **3.** Darnell; x is not a factor of $x + 3$ and $x + 4$. Therefore, it cannot be divided out. **5.** $4x(4x - 5)$ **7.** $(a - 5)(a + 4)$ **9.** $(x - 6)(x - 4)$ **11.** 0, −6 **13.** $\frac{2}{5}$ **15.** $\frac{2y}{3z^3}$ **17.** $\frac{3}{4}$ **19.** $\frac{x-2}{x+1}$ **21.** $\frac{-(x+5)}{x+6}$ **23.** −5 **25.** 10 **27.** −2, 3 **29.** 0, −5 **31.** 4, −4 **33.** $\frac{1}{3}$ **35.** $\frac{2c}{3d}$ **37.** $-4bc$ **39.** $\frac{1}{5}$ **41.** $\frac{x+3}{x+1}$ **43.** $\frac{x}{2}$ **45.** $\frac{r+2}{3}$ **47.** $\frac{y+3}{y-4}$ **49.** $\frac{y+2}{y-1}$ **51.** $\frac{z-5}{z+3}$ **53.** $\frac{8m}{m-2}$ **55.** $\frac{c-5}{c(c+6)}$ **57.** $\frac{-x}{4+x}$ **59.** $\frac{4}{3+a}$ **61.** 9 yr **63.** 8 lumens/m^2 **65.** 16 **67.** 19 **69.** $7\sqrt{3}$ **71.** $6\sqrt{3}$ **73.** 5 **75.** D

Pages 647–649 Lesson 15–2

1. The reciprocal of $\frac{a^2-25}{3a}$ was used instead of the reciprocal of $\frac{a+5}{15a^2}$. **3.** $\frac{4}{x^2}$ **5.** $\frac{x+5}{x-4}$ **7.** $\frac{a^2}{bd}$ **9.** $\frac{2}{x+4}$ **11.** $4a$ **13.** $\frac{3x^2}{2}$ **15.** 12 flags **17.** $\frac{xy}{4}$ **19.** $3a$ **21.** $\frac{2s}{3}$ **23.** $\frac{x}{(x+4)^2}$ **25.** $\frac{2(x-3)}{x+3}$ **27.** $\frac{n}{n+2}$ **29.** $6a^2b$ **31.** $2a$ **33.** $\frac{y+4}{y^2}$ **35.** $\frac{(m+1)(m-1)}{2}$ **37.** $\frac{-(x+2)}{x-3}$ **39.** $\frac{-3a^2(a+3)}{2}$ **41.** $\frac{1}{x-y}$ **43a.** $\frac{x^4}{625}$ **43b.** $\frac{81}{625}$ or about 13% **45.** $\frac{7b^3}{a}$ **47.** $\frac{-4}{x-2}$ **49.** (1, 3) **51.** (−3, −4) **53a.** (5, −75), (8, 0), (10, 50)

53b.

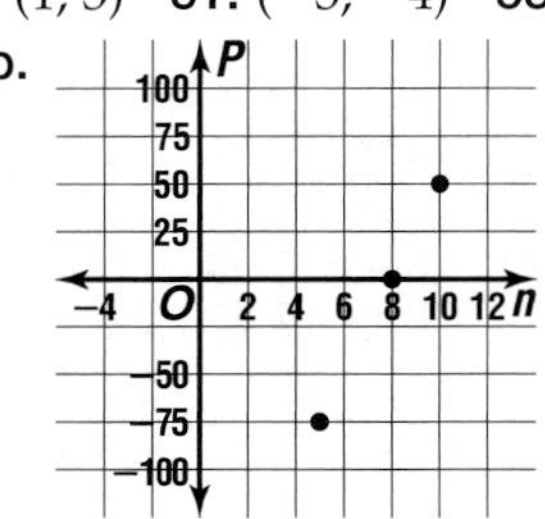

Pages 653–655 Lesson 15–3

1. dividend: $3k^2 - 7k - 5$; divisor: $3k + 2$; quotient: $k - 3$; remainder: $\frac{1}{3k+2}$ **3.** The quotient is a factor of the dividend if the remainder is 0. **5.** $3a$ **7.** x **9.** $x + 2 - \frac{1}{2x-1}$ **11.** $2x + 3$ *in.* **13.** x **15.** $3r^2$ **17.** $x + 4$ **19.** $a + 5$ **21.** $t - 2 - \frac{32}{3t-4}$ **23.** $b + 2 - \frac{4}{2b-1}$ **25.** $x^2 - 3x + 9$ **27.** $4x^3 + 4x^2 + 2x + 3 + \frac{4}{x-1}$ **29.** $x - 3$ **31a.** 936 h **31b.** 140 h **31c.** 51 h **31d.** 5 h **33.** 27 **35.** 1 **37.** one solution **39.** infinitely many **41.** $y \geq -6$

Page 655 Quiz 1

1. 2; m **3.** $\frac{1}{2}$ **5.** $\frac{-2(a+1)}{a-1}$ **7.** y **9.** $x^2 + x + 1$

Pages 659–661 Lesson 15–4

1. They are additive inverses.

3.

a	b	$a+b$	$a-b$	$a \cdot b$	$a \div b$
$\frac{2}{t}$	$\frac{1}{t}$	$\frac{3}{t}$	$\frac{1}{t}$	$\frac{2}{t^2}$	2
$\frac{12}{y}$	$\frac{4}{y}$	$\frac{16}{y}$	$\frac{8}{y}$	$\frac{48}{y^2}$	3
$\frac{x}{x-1}$	$\frac{1}{x-1}$	$\frac{x+1}{x-1}$	1	$\frac{x}{x^2-2x+1}$	x

5. $\frac{3}{4}$ **7.** $\frac{7}{x}$ **9.** −1 **11.** 2 **13.** $\frac{10y}{7x-2y}$ in. **15.** $\frac{7x}{8}$ **17.** $\frac{t}{2}$ **19.** $\frac{7}{6r}$ **21.** n **23.** $\frac{x}{12}$ **25.** $-\frac{1}{z}$ **27.** 0 **29.** 2 **31.** 3 **33.** $\frac{2y}{y-2}$ **35a.** Sample answer: $\frac{15{,}000}{5} - \frac{0.3(15{,}000)}{5}$ **35b.** \$2100 **37.** $a + 4$ **39.** $y - 5$ **41.** $\frac{3axy}{10}$ **43.** $9x^2 + 6xy + y^2$

Pages 665–667 Lesson 15–5

1. Sample answer: $\frac{1}{2x}, \frac{1}{4x}$ **3.** Malik; he found the LCD and then added the terms. Ashley incorrectly added the numerators and the denominators. **5.** 72 **7.** $30xy^2$ **9.** $\frac{4}{a^2}, \frac{5a}{a^2}$ **11.** $\frac{5t}{12}$ **13.** $\frac{2+3b}{ab^2}$ **15.** $\frac{-3x+8}{x^2-4}$ **17.** 2042 **19.** $12a^2b^2$ **21.** $(y+2)(y-2)$ **23.** $(x+3)(x-3)(3x+1)$ **25.** $\frac{35}{10t}, \frac{8}{10t}$ **27.** $\frac{5z}{xyz}, \frac{6x}{xyz}$ **29.** $\frac{30k}{3(3k+1)}, \frac{k}{3(3k+1)}$ **31.** $\frac{3d}{10}$ **33.** $\frac{15}{4x}$ **35.** $\frac{m+3}{3t}$ **37.** $\frac{14x-1}{6x^2}$ **39.** $\frac{2y+x^2+3x}{xy}$ **41.** $\frac{5x+2y^2}{20x^2y}$ **43.** $\frac{2a+3}{a+3}$ **45.** $\frac{1}{(y-1)^2}$ **47.** $\frac{22x+5}{4(2x-1)}$ **49.** $\frac{x^2+6x-11}{(x+1)(x-3)}$ **51.** $\frac{m^2-5m+5n}{m(m-n)}$ **53.** 17.5 min **55.** $\frac{16}{a}$ **57.** 2 **59.** $9\sqrt{19}$ **61.** in simplest form **63.** (2, 5)

Page 667 Quiz 2

1. $x+8$ **3.** $\frac{8a}{9}$ **5.** 168

Pages 671–673 Lesson 15–6

1. Sample answer: In a linear equation, the variable cannot appear in the denominator and it cannot be raised to a power higher than 1. **3.** $-\frac{7}{3}$ **5.** $-\frac{10}{3}$ **7.** 3 **9.** 30 mph **11.** $-\frac{2}{3}$ **13.** 7 **15.** -3 **17.** 9 **19.** $\frac{4}{3}$ **21.** 7 **23.** $\frac{1}{12}$ **25.** 3 **27.** $\frac{1}{6}$ **29.** 6 **31.** -13 **33.** $-\frac{3}{2}$ **35.** $\frac{7}{5}$ **37.** 5 **39.** 2 **41.** 7 **43.** $\frac{9}{2n^2}$ **45.** Z, Q **47a.** $c+s \leq 16$, $10c+15s \leq 200$

47b.

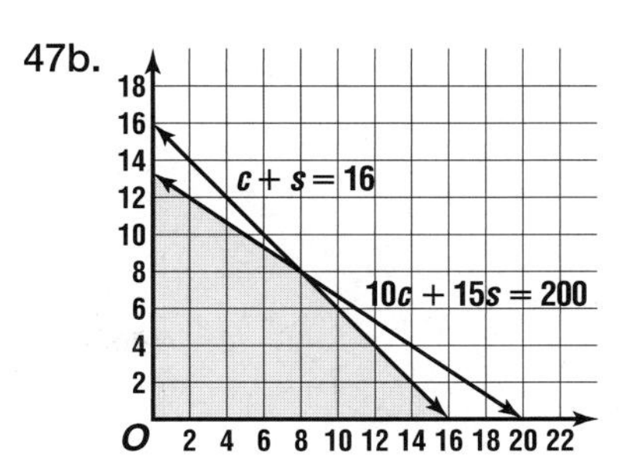

47c. Sample answer: (7, 8), he could plant corn for 7 days and soybeans for 8 days; (2, 12), he could plant corn for 2 days and soybeans for 12 days. **49.** (4, −1);

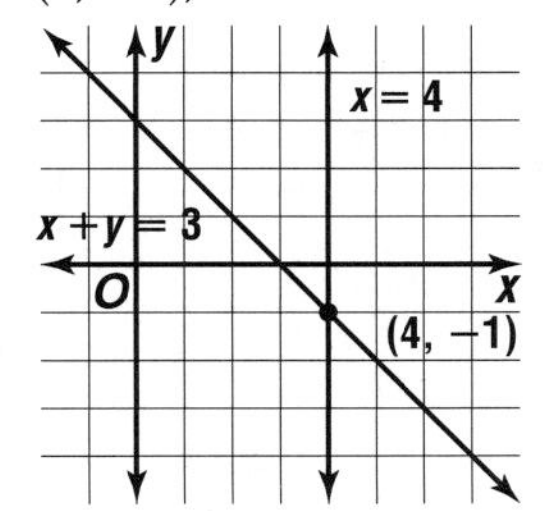

Pages 676–678 Study Guide and Assessment

1. denominator **3.** $\frac{x}{3x+6y}$ **5.** greatest common factor **7.** $\frac{1}{a+b}$ **9.** $2x^2$ **11.** $\frac{xz}{4y^2}$ **13.** $\frac{n-4}{n-2}$ **15.** $\frac{2x^2}{3y^2}$ **17.** $\frac{x}{(x+3)(x-3)}$ **19.** $\frac{3(y-2)}{y+2}$ **21.** a **23.** $2y+1+\frac{2}{2y+3}$ **25.** $\frac{3x}{4}$ **27.** $\frac{5x}{x+4}$ **29.** $\frac{6x+5}{8x^2}$ **31.** $\frac{7a-8}{a(a-2)}$ **33.** $-\frac{1}{5}$ **35.** -4 **37.** 80 pieces **39a.** $x+1$ **39b.** 4 in., 3 in., 9 in., 108 in^3

Page 681 Preparing for Standardized Tests

1. A **3.** B **5.** D **7.** C **9.** 35

Photo Credits

Photo credits for pages 2–333 are contained in Volume One.

Cover (globe and background) Imtek Imagineering/Masterfile, (roller coaster) Lester Lefkowitz/The Stock Market; **v** file photo; **ix** Elizabeth Simpson/FPG; **x** Chip Simons/FPG; **xi** Grafton Marshall Smith/ The Stock Market; **xii** KS Studio; **xiii** Alan Carey; **xiv** Rich Brommer; **xv** Larry Tackett/ Masterfile; **xvi** Rob Atkins/Image Bank; **xvii** SuperStock; **xviii** Lynn M. Stone; **xix** Garry Black/Masterfile; **xx** Jeri Gleiter/ FPG; **xxi** K.L. Gay/©Arizona State Parks; **xxii** Mark Reinstein/FPG; **xxiii** JPL; **B1** (l) Dominic Oldershaw, (r) Mark Burnett; **334–335** Corbis Images; **337** (l)SSEC/UW-Madison, (r)file photo; **340** SuperStock; **345** Rob Atkins/Image Bank; **346** SuperStock; **349** Alfred Pasieka/Peter Arnold, Inc.; **351** SuperStock; **354** Albert Normandin/ Masterfile; **356** Hal Horwitz/CORBIS; **357** ©Tribune Media Services, Inc. All Rights Reserved. Reprinted with permission; **359** Bill Brooks/Masterfile; **364** NASA/The Stock Market; **368** SuperStock; **371** Mark Tomalty/Masterfile; **373** Icon Images; **380–381** Aaron Haupt; **385** Syd Greenberg/ Photo Researchers; **387** Doug Martin/Photo Researchers; **393** Courtesy Carol McAdoo; **410** J.A. Kraulis/Masterfile; **411** SuperStock; **418–419** Toyohiro Yamada/FPG; **427** Mark Burnett; **431** Aaron Haupt; **434** Lynn M. Stone; **456–457** Gisela Damm/Leo de Wys; **467** Garry Black/Masterfile; **468** G. Randall/ FPG; **480** Aaron Haupt; **485** Aaron Haupt; **488** (t)Aidan O'Rourke, (b)Mark Burnett; **493** E. Alan McGee/FPG; **494–495** Mark Burnett; **502–503** Aaron Haupt; **509** Telegraph Colour Library/FPG; **517** Jeri Gleiter/FPG; **523** Ron Holt/Aristock; **529** Peter Christopher/Masterfile; **532, 534** Mark Burnett; **535** Richard Pasley/Stock Boston; **539** Mark Burnett; **541** Aaron Haupt; **548–549** Christie's Images/ ©2000 Artists Rights Society (ARS), New York/ADAGP, Paris; **557** Josph DiChello; **563** Photo 20-20; **567** K.L. Gay/©Arizona State Parks; **571** Mak-1; **575** David R. Frazier Photolibrary/Photo Researchers; **579** Sarah Jones (Debut Art)/FPG; **585** VCG/FPG; **591** Mug Shots/The Stock Market; **598–599** Bachmann/PhotoEdit; **604** David Parker/Science Photo Library/Photo Researchers; **608** Mark Reinstein/FPG; **610** Gordon R. Gainer/The Stock Market; **611** Miles Ertman/Masterfile; **613** Mark Burnett; **614** FOX TROT ©1998 Bill Amend. Reprinted with permission of UNIVERSAL PRESS SYNDICATE. All rights reserved; **623** CLOSE TO HOME ©1994 John McPherson. Reprinted with permission of UNIVERSAL PRESS SYNDICATE. All rights reserved; **628** Peter Brock/Liaison; **636–637** Tim Davis/Stone; **640** Aaron Haupt; **646** John Zimmerman/FPG; **666** JPL; **668** Aaron Haupt; **671** Roberto Soncin Gerometta/Photo 20-20; **673** Science Photo Library/Photo Researchers; **675** David Lissey/FPG; **683** (t)(br)Aaron Haupt, (bl)Patrix Ravaux/Masterfile, (bc)Gary Randall/FPG; **724** Aaron Haupt.

Index

This index contains terms from Volumes One and Two. Pages 1–379 are contained in Volume One. Pages 344–681 are contained in Volume Two.

A

B

C

D

E

F

G

N

O

P

INDEX

Q

R

S

T

U

Z

Properties

Substitution (=)	If $a = b$, then a may be replaced by b.
Reflexive (=)	$a = a$
Symmetric (=)	If $a = b$, then $b = a$.
Transitive (=)	If $a = b$ and $b = c$, then $a = c$.
Additive Identity	For any number a, $a + 0 = 0 + a = a$.
Multiplicative Identity	For any number a, $a \cdot 1 = 1 \cdot a = a$.
Multiplicative (0)	For any number a, $a \cdot 0 = 0 \cdot a = 0$.
Multiplicative (−1)	For any number a, $-1 \cdot a = -a$.
Additive Inverse	For any number a, there is exactly one number $-a$ such that $a + (-a) = 0$.
Multiplicative Inverse	For any number $\frac{a}{b}$, where $a, b \neq 0$, there is exactly one number $\frac{b}{a}$ such that $\frac{a}{b} \cdot \frac{b}{a} = 1$.
Commutative (+)	For any numbers a and b, $a + b = b + a$.
Commutative (×)	For any numbers a and b, $a \cdot b = b \cdot a$.
Associative (+)	For any numbers a, b, and c, $(a + b) + c = a + (b + c)$.
Associative (×)	For any numbers a, b, and c, $(a \cdot b) \cdot c = a \cdot (b \cdot c)$.
Distributive	For any numbers a, b, and c, $a(b + c) = ab + ac$ and $a(b - c) = ab - ac$.
Comparison	For any numbers a and b, exactly one of the following sentences is true: $a < b$, $a > b$, or $a = b$.
Addition (=)	For any numbers a, b, and c, if $a = b$, then $a + c = b + c$.
Subtraction (=)	For any numbers a, b, and c, if $a = b$, then $a - c = b - c$.
Division and Multiplication (=)	For any numbers a, b, and c, with $c \neq 0$, if $a = b$, then $ac = bc$ and $\frac{a}{c} = \frac{b}{c}$.
Product Property of Square Roots	For any numbers a and b, with $a, b \geq 0$, $\sqrt{ab} = \sqrt{a} \cdot \sqrt{b}$.
Quotient Property of Square Roots	For any numbers a and b, with $a \geq 0$ and $b > 0$, $\sqrt{\frac{a}{b}} = \frac{\sqrt{a}}{\sqrt{b}}$.
Zero Product	For any numbers a and b, if $ab = 0$, then $a = 0$, $b = 0$, or both a and b equal 0.
Addition (>)*	For any numbers a, b, and c, if $a > b$, then $a + c > b + c$.
Subtraction (>)*	For any numbers a, b, and c, if $a > b$, then $a - c > b - c$.
Division and Multiplication (>)*	For any numbers a, b, and c, **1.** if $a > b$ and $c > 0$, then $ac > bc$ and $\frac{a}{c} > \frac{b}{c}$. **2.** if $a > b$ and $c < 0$, then $ac < bc$ and $\frac{a}{c} < \frac{b}{c}$.

* *These properties are also true for $<$, $\geq$, and $\leq$.*